DRUG DELIVERY APPLICATIONS OF NONINVASIVE IMAGING

DRUG DELIVERY APPLICATIONS OF NONINVASIVE IMAGING

Validation from Biodistribution to Sites of Action

Edited by

CHUN LI
Department of Cancer Systems Imaging
The University of Texas MD Anderson Cancer Center
Houston, Texas, USA

MEI TIAN
Department of Nuclear Medicine and Medical PET Center
The Second Hospital of Zhejiang University School of Medicine
Hangzhou, Zhejiang, China

WILEY

Published by John Wiley & Sons, Inc., Hoboken, New Jersey
Published simultaneously in Canada

Library of Congress Cataloging-in-Publication Data:

Drug delivery applications of noninvasive imaging : validation from biodistribution to sites of action / edited by Chun Li, Mei Tian.
p. ; cm.
Includes bibliographical references and index.
ISBN 978-0-470-63347-2 (cloth)
I. Li, Chun, 1963– II. Tian, Mei.
[DNLM: 1. Drug Delivery Systems. 2. Diagnostic Imaging. QV 785]
RS431.R34
615.1′9–dc23

2013014002

Printed in the United States of America

ISBN: 9780470633472

10 9 8 7 6 5 4 3 2 1

CONTENTS

PREFACE

Modern imaging technologies offer high-sensitivity and high-resolution visualization of imaging probes. In the last decade, in vivo imaging has been gradually growing in importance as an aid for the development of new drug delivery systems and approaches to controlled dosage release. These two fields are powerful in their own right, but together they can generate considerable new knowledge in a cost-effective manner with regard to the pharmacokinetics, biodistribution, bioavailability, local concentration, and clearance of drug substances for the treatment of a variety of diseases including, in particular, the treatment of cancer. In the pharmaceutical industry, controlled release formulations are widely used, and become increasingly popular for reasons of improving patient compliance, reducing side effects, and extending patent protection. Noninvasive imaging techniques allow rapid, repetitive (and thus potentially high throughput) assessment of the drug deposition in various tissues in the body, which can vastly facilitate the eventual translation of novel dosage forms into the clinic. Giving the popularity and importance of both imaging and drug delivery fields, we are extremely pleased to present to you this book entitled *Drug Delivery Applications of Noninvasive Imaging: Validation from Biodistribution to Sites of Action*.

The book starts with an introduction to molecular imaging. Chapters 2–4 focus on *in vivo* imaging techniques including PET, SPECT, and optical imaging. Given the importance of quantitative analysis on the basis of imaging data, a whole chapter (Chapter 5) is dedicated to quantitative analysis. Chapters 6–13 highlight imaging drugs and drug carriers at the sites of action. These therapeutic agents include small-molecular weight radiopharmaceuticals, peptide and proteins, siRNA, cells, and nanoparticles. Many different types of nanoparticles with new and improved diagnostic and therapeutic effects have been developed over the last 20 years. A more broad-based nanotechnology application will come from a better understanding of how nanoparticles interact with biological systems and how multiple functions, including imaging and therapeutic functions, can be integrated into a single nanoplatform. This volume has attempted to represent several major classes of nanoparticles that are being developed for image-guided therapy (theranostics), such as liposomes (Chapter 11), polymeric micelles (Chapter 12), perfluorocarbon nanoparticles (Chapter 13), and gold nanoparticles (Chapter 14). The next two chapters (Chapters 14, 15) highlight the applications of imaging techniques in routes of administration other than intravenous injection, including pulmonary and oral delivery. Finally, no comprehensive volume of imaging and drug delivery would be complete without translational research towards clinical applications. Chapter 16 offers valuable information on imaging drug delivery in large animal models. Chapters 18–20 provide examples on clinical applications of imaging techniques in guiding drug development and drug delivery.

Thanks to excellent contributors and the Wiley publication team, this book has turned out to be the only one of its kind bringing synergy between imaging and drug delivery. With 20 authoritative chapters highlighting the contemporary use of imaging in preclinical and clinical evaluation of drug and drug delivery systems, we have no doubt that this book will be extremely helpful for a broad spectrum of readers ranging from students, teachers, practitioners, and followers of pharmaceutics and drug delivery.

Chun Li
Mei Tian

CONTRIBUTORS

Adil Al-nahhas, Department of Nuclear Medicine, Imperial College Healthcare Trust, Hammersmith Hospital, London, UK

Mansoor Amiji, Department of Pharmaceutical Sciences, School of Pharmacy, Northeastern University, Boston, MA, USA

Ande Bao, Nanoparticle Imaging/Therapeutic Systems, Department of Radiology, University of Texas HSC - San Antonio, San Antonio, TX, USA

Jeff W. M. Bulte, Russell H. Morgan Department of Radiology and Radiological Science, Division of MR Research, Cellular Imaging Section and Vascular Biology Program, Institute for Cell Engineering, The Johns Hopkins University School of Medicine, Baltimore, MD, USA

S. D. Caruthers, Department of Medicine, Washington University Medical School, St. Louis, MO, USA

Tatyana Chernenko, Center for Biophotonics Science and Technology, University of California

Bo Youn Cho, Department of Internal Medicine, Seoul National University College of Medicine, Korea

June-key Chung, Department of Nuclear Medicine, Cancer Research Institute, Tumor Immunity Medical Research Center, Research Center of Radiation Medicine, Seoul National University College of Medicine, Seoul, Korea

Max Diem, Department of Chemistry and Chemical Biology, Northeastern University, Boston, MA, USA

Lars Farde, Translational Science, AstraZeneca R&D, Södertälje, Sweden

Department of Clinical Neuroscience, Karolinska Institutet, Stockholm, Sweden

Frederik L. Giesel, Department of Nuclear Medicine, University of Heidelberg, Heidelberg, Germany

Beth Goins, Nanoparticle Imaging/Therapeutic Systems, Department of Radiology, University of Texas HSC – San Antonio, San Antonio, TX, USA

Edward F. Jackson, Department of Imaging Physics, The University of Texas MD Anderson Cancer Center, Houston, TX, USA

Keon Wook Kang, Department of Nuclear Medicine, Cancer Research Institute, Research Center of Radiation Medicine, Seoul National University College of Medicine, Seoul, Korea

Clemens Kratochwil, Department of Nuclear Medicine, University of Heidelberg, Heidelberg, Germany

Vikas Kundra, Department of Diagnostic Radiology and Cancer Systems Imaging, The University of Texas MD Anderson Cancer Center, Houston, TX, USA

G. M. Lanza, Department of Medicine, Washington University Medical School, St. Louis, MO, USA

Chi-ming Lee, Translational Science, AstraZeneca R&D, Södertälje, Sweden

Ho Young Lee, Department of Nuclear Medicine, Research Center of Radiation Medicine, Seoul National University College of Medicine Seoul, Korea

Jong Jin Lee, Department of Nuclear Medicine, Seoul National University College of Medicine, Seoul, Korea

Myung Chul Lee, Department of Nuclear Medicine, Seoul National University College of Medicine, Seoul, Korea

Chun Li, Department of Cancer Systems Imaging–Unit 59, The University of Texas MD Anderson Cancer Center, Houston, TX, USA

Shihong Li, Nanoparticle Imaging/Therapeutic Systems, Department of Radiology, University of Texas HSC – San Antonio, San Antonio, TX, USA

Yajie Liang, Russell H. Morgan Department of Radiology and Radiological Science, Division of MR Research, Cellular Imaging Section and Vascular Biology Program, Institute for Cell Engineering, The Johns Hopkins University School of Medicine, Baltimore, MD, USA

Wei Lu, Department of Biomedical and Pharmaceutical Sciences, College of Pharmacy, The University of Rhode Island, Kingston, RI, USA

Christian Matthäus, Institute of Photonic Technology, Jena, Germany

Fiona J. McInnes, Drug Delivery International, Glasgow, Scotland, UK

Zdravka Medarova, Molecular Imaging Laboratory, MGH/MIT/HMS Athinoula A. Martinos Center for Biomedical Imaging, Department of Radiology, Massachusetts General Hospital/Harvard Medical School, Boston, MA, USA

Lara Milane, The Nano Doctor, North Easton, MA, USA

Stephen M. Moerlein, Radiological Chemistry and Imaging Laboratory, Mallinckrodt Institute of Radiology, Washington University in St. Louis, St. Louis, MO, USA

Anna Moore, Molecular Imaging Laboratory, MGH/MIT/HMS Athinoula A. Martinos Center for Biomedical Imaging, Department of Radiology, Massachusetts General Hospital/Harvard Medical School, Boston, MA, USA

Meeran Naji, Department of Nuclear Medicine, Imperial College Healthcare Trust, Hammersmith Hospital, London, UK

Stephen P. Newman, Scientific Consultant, Norfolk, UK

So Won Oh, Department of Nuclear Medicine, Seoul National University College of Medicine, Korea

Do Joon Park, Department of Internal Medicine, National University College of Medicine, Seoul, Korea

William T. Phillips, Nanoparticle Imaging/Therapeutic Systems, Department of Radiology, University of Texas HSC – San Antonio, San Antonio, TX, USA

A. H. Schmieder, Department of Medicine, Washington University Medical School, St. Louis, MO, USA

Kooresh I. Shoghi, Radiological Chemistry and Imaging Laboratory, Mallinckrodt Institute of Radiology, Washington University in St. Louis, St. Louis, MO, USA

Liang Song, Optical Imaging Laboratory, Department of Biomedical Engineering, Washington University in St. Louis, St. Louis, MO, USA

Institute of Biomedical and Health Engineering, Shenzhen Institutes of Advanced Technology, Chinese Academy of Sciences, Shenzhen, China

Keitaro Sou, Center for Advanced Biomedical Sciences, TWIns, Waseda University, Tokyo, Japan

Howard N. E. Stevens, Bio-Images Group, Glasgow, Scotland, UK

Yuan-chuan Tai, Radiological Chemistry and Imaging Laboratory, Mallinckrodt Institute of Radiology, Washington University in St. Louis, St. Louis, MO, USA

Mei Tian, Department of Nuclear Medicine and Medical PET Center, The Second Affiliated Hospital of Zhejiang University School of Medicine, Hangzhou, Zhejiang, China

Zhude Tu, Radiological Chemistry and Imaging Laboratory, Mallinckrodt Institute of Radiology, Washington University in St. Louis, St. Louis, MO, USA

Rajesh K. Uthamanthil, Comparative Medicine, Fred Hutchinson Cancer Research Center, Seattle, WA, USA

Katarina Varnäs, Department of Clinical Neuroscience, Karolinska Institutet, Stockholm, Sweden

Lihong V. Wang, Optical Imaging Laboratory, Department of Biomedical Engineering, Washington University in St. Louis, St. Louis, MO, USA

S. A. Wickline, Department of Medicine, Washington University Medical School, St. Louis, MO, USA

P. M. Winter, Department of Radiology, Cincinnati Children's Hospital, Cincinnati, OH, USA

Imene Zerizer, Department of Nuclear Medicine, Imperial College Healthcare Trust, Hammersmith Hospital, London, UK

Jun Zhao, Department of Cancer Systems Imaging–Unit 59, The University of Texas MD Anderson Cancer Center, Houston, TX, USA

CHAPTER 1

INTRODUCTION TO MOLECULAR IMAGING

VIKAS KUNDRA

Molecular imaging may be defined as the imaging of molecules either delivered to the body or already present in the body. Generally, this refers to *in vivo* imaging, that is, imaging within a living multicellular organism. The process requires an imaging instrument, a subject, and, commonly, an imaging agent. These tools enable longitudinal assessment of delivered materials, specific targets, mechanisms of action, and biological processes. Molecular imaging has already found utility in the clinic.

An example of a clinically useful molecular imaging system is positron emission tomography (PET) imaging using ^{18}F-fluorodeoxyglucose (^{18}F-FDG). The agent is a glucose analogue labeled with a radioactive substance, ^{18}F, with known decay characteristics. Glucose is the primary source of energy in animals. Most foods that we ingest are broken down to or converted to glucose. This sugar enters the bloodstream and then cells via one of several specific transporters, known as GLUT transporters. Once inside the cell, glucose is phosphorylated by hexokinase into glucose-6-phosphate. This is then isomerized by phosphoglucose isomerase to fructose-6-phosphate and continues along the path for energy generation. Fluorodeoxyglucose (FDG) mimics glucose and also enters the cell via GLUT transporters and is phosphorylated by hexokinase; but once inside the cell, it is a poor substrate for phosphoglucose isomerase and for glucose-6-phosphatase and remains phosphorylated. This adds a negative charge to ^{18}F-FDG, which prevents the molecule from crossing the cell membrane and entraps it inside the cell.

Cells with greater metabolic demand require more glucose and therefore entrap more ^{18}F-FDG. Cells can use glucose to generate more ATP, usable energy within the cell, using oxidative phosphorylation rather than using anaerobic respiration. Because cancer cells are constantly reproducing, they have high energy demand; however, they are also inefficient at energy production and behave as though they are in an oxygen-poor environment. They tend to use the less efficient anaerobic pathway, and thus, require more glucose. In turn, in the presence of the glucose analogue, cancer cells tend to accumulate more ^{18}F-FDG than normal cells.

Drug Delivery Applications of Noninvasive Imaging: Validation from Biodistribution to Sites of Action, First Edition. Edited by Chun Li and Mei Tian.

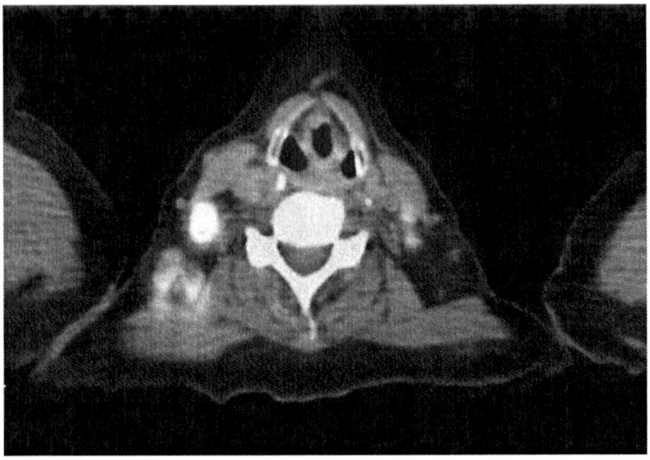

FIGURE 1.1 ^{18}F-FDG PET/CT. Axial view of the neck demonstrates increased uptake (orange) of ^{18}F-FDG in lymph nodes signifying involvement by lymphoma. For color details, please see color plate section.

In this imaging agent, the FDG provides specificity by mimicking glucose rather than another sugar such as sucrose. ^{18}F enables imaging. The known decay of ^{18}F can be imaged using a PET camera sensitive to the positron decay that is employed by ^{18}F to come to a more stable atomic state. ^{18}F-FDG enables one to study a basic biological process, cellular metabolism, that is used to identify and characterize disease. Clinically, ^{18}F-FDG imaging (Fig. 1.1) is used to identify, localize, and stage many types of cancer, such as breast cancer and lymphoma. It is also used to assess response to therapy. More recently, it has been used to predict response to certain therapies and to assess durability of response at the end of therapy [1, 2]. It has also had a significant impact in cardiology, primarily for assessing ischemia and infarction, and in neurology for locating a seizure focus.

1.1 ASSESSING THE TARGET DIRECTLY VERSUS DOWNSTREAM EFFECTS

Systems may be designed to image the target itself versus downstream effects. For example, one may image a receptor or the downstream effect of signaling pathways elicited by the receptor. Practically, one may first want to know if the target is present and how it is distributed within a site of disease and within normal tissues. If the target is present, the drug created against it may be effective. On the other hand, if the target is not present, the targeted therapy is not likely to be effective by the expected mechanism of action. Clinically, although target presence may be evaluated by biopsy, it is not without risk and patient compliance issues may arise if biopsies need to be performed in multiple locations and/or longitudinally. Biopsy samples a small portion of the tissue of interest. By imaging, heterogeneity of target expression may be assessed, for example, within a tumor. Without bystander effect, areas of tumor without target expression may not respond to the targeted drug. Likewise, heterogeneity of target expression in different metastases may be assessed. If expression is present in some metastases, but not others, there may be a mixed response to the therapy. Longitudinal imaging may be used to assess change in target expression, which may change secondary to the targeted or other therapy that is given to the subject. Reduction in target expression may make the therapeutic less effective.

With targeted imaging used in conjunction with targeted therapy, careful interpretation is necessary. The timing of delivering the imaging agent versus the therapeutic is important since the two may compete. This may be advantageous because one can use it to assess if the therapeutic can displace binding of the imaging agent to the target. On the other hand, the imaging agent may bind to a site separate to that bound by the therapeutic agent; therefore, the imaging agent may not reflect the binding site of the therapeutic agent. Moreover, the imaging agent may bind, but not inhibit function, such as that of a tyrosine kinase domain, whereas the therapeutic agent needs to bind and inhibit function in order to be effective. A caveat for delivery is that to be effective, a therapeutic agent may not need as favorable a biodistribution in terms of signal to background noise as is needed for imaging. Another potential outcome is that the imaging identifies the target and the drug inhibits target function, yet the tumor grows, that is, the drug is not efficacious. This may be because the primary growth/maintenance signaling pathway for the tumor was not targeted or was not adequately suppressed, or secondary/redundant pathways supported growth.

One example of targeted imaging is using ^{111}In-octreotide to image somatostatin receptors. Identifying the presence of this receptor predicts response to octreotide therapy for suppressing carcinoid syndrome, which is associated with neuroendocrine tumors [3]. Another example is ^{18}F-fluoroestradiol (^{18}F-FES). PET imaging with this agent correlates with estrogen receptor (ER) expression and predicts response to tamoxifen [4]. With targeted imaging, the imaging system commonly requires development of a particular contrast agent/radiopharmaceutical for each target; this provides specificity but may limit its utility to a few relevant applications/diseases.

Signaling induced by receptors may activate a variety of downstream pathways that regulate a few critical cellular functions such as growth and metabolism. Processes at the tissue level also represent central processes that may play a key role in normal physiology and diseases. These may be less specific, but because many interesting targets affect them, these downstream effects represent an opportunity for understanding a variety of normal and pathological processes. For example, one may image alterations in glucose metabolism by ^{18}F-FDG PET or changes in the function of the vasculature by dynamic contrast-enhanced magnetic resonance (MR) imaging [5–9] by a number of drugs. Because the process being imaged may be affected by various pathways, a variety of targeted therapeutic agents may be tested using a similar readout. Care must be taken since the downstream readout may not be due to the predicted mechanism of action of the targeted therapeutic. Even so, such a readout may prove practically useful and hypothesis generating. Imaging of downstream effects has the potential for wider applicability than specifically targeted agents, which would be an advantage for clinical translation.

1.2 IMAGING METHODS

Physical properties such as radioactive decay; absorbance or reflectance of light, sound, or X-rays; and behavior in a magnetic field are used to generate images. Machines sensitive to such physical changes are used to create images (Table 1.1). These require appropriate tissue and spatial resolution. If the imaging is performed from one angle, a two-dimensional (2D) image is generated. If the imaging is performed at multiple angles, with appropriate mathematical algorithms and the speed of modern computers, three-dimensional (3D) images can be created.

In vivo imaging may consist of superficial imaging such as that of skin and lumens like the epithelium of the bowel. In such situations, light-based imaging may be applied since one is not as limited by depth of penetration. Due to scatter, light-based imaging is currently limited to a few millimeters or centimeters, thus is not applicable to percutaneous imaging of deep structures. However, for small animals, this is less of an issue since their entire width may fall within this range. Light may be generated from within the animal itself, for example, using an enzyme such as luciferase that converts a substrate such as luciferin into light. Fluorescence may also be exploited in which a certain wavelength of light is input and output of a second wavelength is separated from scatter using a filter and then measured by a camera. Currently, most cameras are cooled charge-coupled device (CCD) chips of the type found in digital cameras.

Due to high sensitivity, light-based imaging is commonly used in small animal imaging, particularly of mice. Luciferase-based imaging in particular has enjoyed a large amount of success since it can be placed into vectors and delivered to cells, enabling a range of molecular biology techniques such as studying promoter function. Fluorescent proteins have the advantage that they can be seen in tissues and cells without a substrate, thus can be followed *in vitro*, *in vivo*, and often *ex vivo*.

TABLE 1.1 Imaging Modalities

Light-based imaging
Primarily CCD-based cameras
White light
With filters: fluorescence, near infrared, infrared
Raman spectroscopy
Penetration depth limited: external cameras, internal cameras such as endoscopes
Nuclear medicine
Gamma camera-based imaging
PET
Magnetization
MR
Sound
Ultrasound
Photoacoustic imaging (light input, sound output)
Thermoacoustic imaging (radiofrequency input, light output)
X-rays
Radiography
CT

A disadvantage is that fluorescence in the visible range has a higher incidence of scatter and background signal in tissues. Some of this may be obviated using near-infrared imaging, since at these wavelengths background signal from tissues is diminished. Another form of light-based imaging is Raman spectroscopy, which visualizes molecular vibrations based on inelastic scattering of monochromatic light. It may be used to interrogate intrinsic characteristics of superficial tissues without the need for contrast agents.

Among percutaneous imaging methods that may also be applicable in humans, nuclear medicine offers the highest sensitivity, in the nanomolar range. For such imaging, a radiopharmaceutical is delivered to the patient. It includes a radionuclide whose decay in the body is imaged. Radionuclides are unstable atoms. Atoms are made up of neutrons, proton, and electrons. In their ground state, nucleons (protons and neutrons) are stable, but if the ratio of neutrons to protons is not optimal or the nucleons are not in their ground state, they may release energy/particles, including gamma rays with characteristic energy. This released energy signature is used to distinguish radioactive decay arising directly from the radionuclide from background/scatter reactions that result in different energies from the source of interest. The de-excitation may be immediate or delayed. The latter is referred to as a metastable state. The decay of ^{99m}Tc (m for metastable) is commonly imaged. The released characteristic energy is imaged using a gamma camera, which may be used to perform 2D planar imaging or single photon emission computed tomography (SPECT) 3D imaging. Characteristic energy detection and collimation to avoid scatter adds specificity in imaging the radiopharmaceutical.

Nucleons may decay to more stable states by releasing particles. Positron emission reduces the number of protons in the nucleus by transforming a proton to a neutron and ejecting both a positron and a neutrino from the nucleus. A positron has the same mass but opposite charge as an electron. It loses kinetic energy after traveling a short distance (usually millimeters) and collides with an electron in an annihilation reaction that transforms their combined mass into energy, releasing not one but two gamma ray photons (each of 511 keV) traveling in opposite trajectories. This enables coincidence detection, which is capitalized upon in PET imaging to localize the annihilation event along a line called the line of response (LOR) that connects the two detectors that detected the two 511-keV gamma rays. In PET imaging, a ring of radiation detectors encircles the patient and detects the gamma rays on opposite sides within a specified period of time. Unlike SPECT, PET imaging does not require a collimator to help identify the source of activity.

Magnetic resonance imaging enables a wide range of contrast mechanisms. Most commonly hydrogen is imaged due to its abundance, although several other nuclei may be used. For MR, when a collection of nuclei with nonzero spin (odd number of protons, neutrons, or both) are placed in a strong magnetic field, a very small net greater amount align with the field creating a net magnetization which eventually decays. Additional time-varying magnetic fields are applied to convert the net magnetization to other forms such as radiofrequency (B_1 field). Magnetic field gradients are also applied in orthogonal directions to encode positional information. Different pulse sequences are used to obtain different types of MR contrast such as T1 weighting, T2 weighting, and T2* weighting. The variety of pulses allow for native tissue characterization, which is especially helpful for soft tissue imaging. Spectroscopy imaging may also be performed for evaluation of molecules such as choline and has enabled fields such as metabolomics. Contrast agents may be used to enhance tissue contrast. Targeted agents may also be employed. New developments include hyperpolarization of input molecules that increases sensitivity of MR for such molecules by tens of thousands times with low background, enabling imaging of not only their input molecules but also of their metabolic products. A disadvantage of hyperpolarization is only a few molecules hyperpolarize long enough (up to a few minutes) for practical imaging. Hyperpolarized gases have been used, including in patients, for lung imaging. The time resolution of MR and appropriate pulses also allows imaging of motion; moreover, with appropriate modeling, function can be assessed, for example, that of vessels using dynamic contrast-enhanced imaging.

Ultrasound images the reflectance of sound waves in tissues. It most commonly uses a piezoelectric transducer to create sound of different frequencies and shape. The reflected sound returns to the transducer that converts the vibrations into electrical pulses to create the image. Sound of different frequencies is used depending on the application and the species studied. Higher frequencies are used for smaller animals than those used in humans. Ultrasound is most commonly used to image intrinsic tissue reflectances. Its rapid temporal resolution enables evaluation of motion in real time, for example, for imaging cardiac motion. Doppler may be used to image blood flow. Newer applications include injectable ultrasound contrast agents that encapsulate gases such as perfluorocarbons for enhancing reflectance. They can be decorated with targeted moieties. Photoacoustic imaging delivers nonionizing laser pulses. Some of these are absorbed and converted to heat leading to thermoelastic expansion and ultrasonic emission that is detected by an ultrasound transducer. Photoacoustic imaging may be used on native tissue or with contrast agents. When radiofrequency, instead of light, is used to heat tissue, it is referred to as thermoacoustic imaging.

X-rays are commonly used in imaging. Electrons in a lower orbital shell have greater binding energy than those in a higher shell. When an electron in a lower orbital shell of an atom is lost, due to heating, for example, one from a higher orbital shell replaces it with resultant release of energy in the form of an X-ray. 2D radiography and 3D computed tomography (CT) use the absorption of X-rays by the subject to generate an image. Although excellent for anatomic imaging, these techniques have relatively poor sensitivity for molecular imaging. They are more commonly used to fuse images with other methods such as nuclear medicine to help localize signal. The temporal resolution of CT may be used for functional imaging such as that of the vasculature using dynamic contrast-enhanced CT.

1.3 IMAGING AGENTS

Molecular imaging may be performed using the native tissue contrast discerned by the instrument itself. For example, MR spectroscopy enables evaluation of molecular species in a defined volume. Clinically, metabolites such as choline and N-acetylaspartate (NAA) have been used in the brain to help distinguish malignant and live versus necrotic regions. Choline and citrate levels have been used to distinguish prostate cancer from benign prostate tissue in the peripheral zone of the prostate gland.

1.4 CONTRAST

More commonly, molecular imaging uses a contrast agent. These generally have at least two domains—one for producing contrast and one for specificity to the target. The two domains may be part of the same molecule or attached to each other using a linker or a shell. Nuclear medicine provides examples of each. Most PET agents employing ^{18}F use the ^{18}F to replace another atom or group in the molecule. For example, ^{18}F-FDG may also be written as 2-deoxy-2-[^{18}F]fluoro-D-glucose, where ^{18}F replaces the hydroxyl (OH) group in the second position of glucose. Most gamma camera agents use a chelator to join two groups. For example, in ^{111}In-octreotide, which is used to image tumors such as carcinoid that overexpress somatostatin receptors, the imaging agent, ^{111}In, is connected to a peptide via a bifunctional linker called DTPA. DTPA is a chelator that entraps the ^{111}In and has a reactive group that is covalently linked to the peptide. This is needed because many radiometals are not readily reactive for direct binding to molecules or such binding would interfere with their biological function. Chelators are also used for MR imaging. For example, DOTA is used for entrapping gadolinium, which in its free form can be toxic. The same principles are used for light-based imaging agents. Example light-based imaging agents include small molecules such as rhodamine and quantum dots. Another method for linking an imaging agent to a targeting domain is to use polymers or liposomes. These have the advantage of potentially linking several targeting agents and/or imaging agents to potentially amplify signal. An example of a polymer with multiple imaging agents connected is PG-gadolinium [10]. An example of a liposome with multiple imaging agents is dual-Gd, which has ~10,000× the relaxivity per particle compared to classic clinically used Gd chelates [11]. Liposomes may encapsulate material. This has been exploited to capture gases such as perfluorocarbon in order to enable contrast-enhanced ultrasound imaging. These "bubbles" can be decorated with targeting agents so that the externally exposed targeting moiety is tethered to the lipid bilayer and the gas is entrapped within the liposome. The increased echogenicity enables imaging by ultrasound.

The types of contrast induced depend upon the imaging system employed. For light-based agents, most commonly a fluorophore is employed, for example, fluorescein or rhodamine. To decrease tissue background, an emitter of non-visible light such as near infrared may be used, for example, Cy5.5. For nuclear medicine, both gamma- and positron-emitting substances are exploited; examples include ^{99m}Tc, ^{111}In, and ^{131}I for the former and ^{18}F, ^{68}Ga, and ^{64}Cu for the latter. For MR, both T1- and T2-shortening agents are most commonly used. T1-shortening agents are the workhorse, and among these, gadolinium is exploited most frequently in conjunction with a T1-based sequence. T2-shortening agents have found more popularity in small animal imaging than clinically, and among these, those based on iron are exploited most frequently. For ultrasound, encapsulated echogenic gases are employed. For CT, agents that alter X-ray penetration, such as chelated iodine, are employed.

1.5 TARGETING

The targeting moiety may be one of a variety of molecules. Antibody-based agents are popular because they may provide a very high degree of selectivity. The specificity of the antibody resides in the variable domain. The species of antibody used is important to avoid an immune response to the antibody. Whole antibodies tend to stay in circulation for a long time, on the order of days. Imaging with such has been performed successfully, commonly one day or more [12] after delivering the agent. For imaging, washout from normal structures is essential for visualizing the targeted material. In order to speed washout, smaller versions of antibodies/fragments have been created, including diabodies and single-chain fragments of the variable domain (scFv's). These preserve the variable region for specificity. The smaller size permits earlier imaging, including within the same day. Peptides have also found utility. These may be designed, but often are discovered, like antibodies, using a library containing many variants (10^9–10^{12}) that is screened for the one most ideal binder. Peptides may be stabilized to prevent digestion within the body. An example of a clinically important peptide-based agent is ^{111}In-octreotide. The octreotide portion mimics the hormone somatostatin. Small molecules are also commonly used and again libraries of such may be screened. These may also be designed based on the characteristics of the target to which they will bind. An example of a clinically important small molecule is ^{99m}Tc-labeled methyl diphosphonate (MDP). It mimics calcium phosphate and is incorporated into the mineralized matrix of newly formed bone. More exotic targeting moieties include aptamers, which consist of stabilized oligonucleotides or peptides of different shapes that bind specific molecules. The targeting moiety may be designed but more commonly is selected from a random library of a large number of variants and is then "panned" against the target to select those few that bind the target and wash away those that do not. This process is repeated until the best binders are selected.

Once the targeting moiety is selected, it may be labeled for imaging as mentioned earlier. This process is most applicable for evaluating delivery since these agents, like drugs, are given to the patient. Delivery may be via various routes such as intravenous, intraarterial, intratumoral, intralymphatic, intraperitoneal, intrapleural, intravesical, inhalation, or oral. The route of delivery influences bioavailability and whether the drug itself or its metabolite causes functional change. Most commonly, imaging agents are not delivered via an oral route for several reasons including to avoid portal flow and metabolism in the liver. If a metabolite is the active compound, an alternative imaging strategy may be to label it, instead of the parent drug.

1.6 GENE EXPRESSION IMAGING

Molecular imaging has made inroads into the realms of molecular biology. The central dogma teaches that DNA encodes genes that are transcribed into messenger RNA (mRNA), and this is translocated out of the nucleus into the cytoplasm where it is translated into protein by the sequential addition of appropriate amino acids based on the mRNA code. One may design agents for imaging the building blocks of these

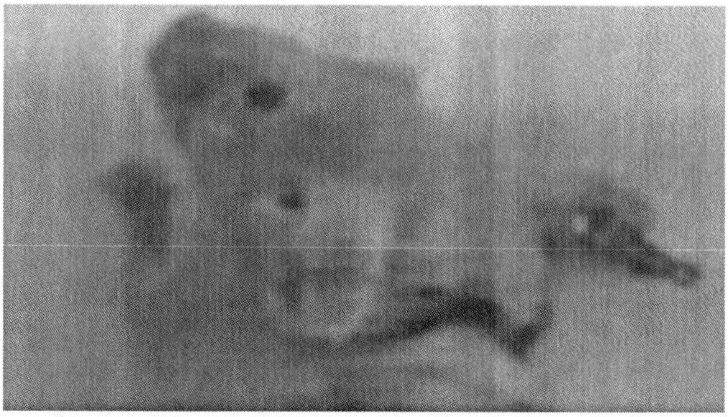

FIGURE 1.2 SPECT imaging of SSTR2-based gene expression. Coronal view of the lungs demonstrates imaging of gene expression after *in vivo* transfer of a somatostatin receptor-based reporter. The reporter was made to express in a human lung tumor on the left side and bound its ligand, ^{111}In-octreotide, as demonstrated by the increased uptake (pink). For color details, please see color plate section.

processes, such as labeled nucleotides that are components of DNA or RNA or such as labeled amino acids, like methionine, that are components of proteins. One may also want to image the end product, i.e., the protein that is built, since this will inform regarding the robustness of the entire process. Imaging of gene expression can be performed using appropriate reporter technology. A reporter is a gene product that can be imaged due to its intrinsic nature, or more commonly, because it binds or enzymatically acts on an imaging agent. Light-based reporters include green fluorescent protein and luciferase. The former fluoresces. The latter acts on a systemically delivered substrate, luciferin, to produce light. For imaging larger animals and humans percutaneously, reporters that have the greatest potential currently include those that can be imaged using nuclear medicine-based techniques such as those based upon the somatostatin receptor type 2 (SSTR2). These have the advantage of a human origin to limit immunogenicity, can be tagged to distinguish endogenous versus exogenously delivered material [13], and can be made signaling deficient to prevent disrupting the cellular milieu [14]. They can also be imaged using a labeled somatostatin analogue such as ^{111}In-octreotide and can be quantified *in vivo* using imaging [15–17], thus, have significant potential for clinical translation (Fig. 1.2).

The reporter is commonly delivered in a vector that contains elements for expressing a gene, such as a promoter for initiating and maintaining transcription, sometimes an enhancer to improve promoter activity, the gene with a stop codon for appropriate termination, and a poly A site to stabilize the transcript. The reporter concept allows evaluation of this entire process, or parts of it, for example, variation of the promoter for optimizing promoter function. The reporter gene may also be linked to other genes such as a therapeutic gene to follow its linked expression. Linkage may be performed using tools such as an IRES [18, 19] or bifunctional promoter [20]. For delivery, multiple different kinds of vectors may be used. The vector may be labeled to image delivery [20]. However, there are a large variety of vectors that would involve substantial work for labeling each; more simply and widely, delivery and expression may be evaluated using reporter technology that uses the same reporter-imaging agent pair. Thus, one would be able to decrease workload and evaluate the system from delivery to the ultimate goal of expression of the gene product in one imaging session.

Example applications of reporter imaging include evaluating delivery vectors for localization, as well as degree and duration of expression; promoter function for selectivity of expression in a particular normal tissue or pathology, degree of expression, and control of expression; monitoring expression of a linked therapeutic gene; studying mechanisms of action of therapeutics; understanding therapeutic efficacy such as onset of response and why response may have waned, for example, loss of expression; understanding toxicity such as due to inappropriately located expression. These are just some of the applications of imaging of gene expression. This technology has significant potential to positively impact research and clinical needs.

1.7 SUMMARY

Molecular imaging has already impacted research and clinical medicine. Established and new technologies are enabling understanding of molecular events and physiological and pathological process in living systems. These technologies are based on physical characteristics such as detecting light, sound, magnetization, and X-ray absorption but are often made even more powerful by adding contrast agents. These techniques enable the study of basic biological processes and their alterations in disease. Not only do they enable approaching research questions such as the mechanism of action but also clinical questions such as disease localization and prognosis. This book will explore molecular imaging techniques and their applications.

REFERENCES

1. Jerusalem G, Hustinx R, Beguin Y, et al. Evaluation of therapy for lymphoma. Semin Nucl Med 2005;35:186–196.
2. Zijlstra JM, Hoekstra OS, Raijmakers PG, et al. ^{18}FDG positron emission tomography versus ^{67}Ga scintigraphy as prognostic test during chemotherapy for non-Hodgkin's lymphoma. Br J Haematol 2003;123:454–462.
3. Lamberts SW, Hofland LJ, Nobels FR. Neuroendocrine tumor markers. Front Neuroendocrinol 2001;22:309–339.
4. Dehdashti F, Flanagan FL, Mortimer JE, et al. Positron emission tomographic assessment of "metabolic flare" to predict response of metastatic breast cancer to antiestrogen therapy. Eur J Nucl Med 1999;26:51–56.
5. Morgan B, Thomas AL, Drevs J, et al. Dynamic contrast-enhanced magnetic resonance imaging as a biomarker for the pharmacological response of PTK787/ZK 222584, an inhibitor of the vascular endothelial growth factor receptor tyrosine kinases, in patients with advanced colorectal cancer and liver metastases: results from two phase I studies. J Clin Oncol 2003;21:3955–3964.
6. Mross K, Drevs J, Muller M, et al. Phase I clinical and pharmacokinetic study of PTK/ZK, a multiple VEGF receptor inhibitor, in patients with liver metastases from solid tumors. Eur J Cancer 2005;41:1291–1299.

7. Liu G, Rugo HS, Wilding G, et al. Dynamic contrast-enhanced magnetic resonance imaging as a pharmacodynamic measure of response after acute dosing of AG-013736, an oral angiogenesis inhibitor, in patients with advanced solid tumors: results from a phase I study. J Clin Oncol 2005;23:5464–5473.
8. Anderson HL, Yap JT, Miller MP, et al. Assessment of pharmacodynamic vascular response in a phase I trial of combretastatin A4 phosphate. J Clin Oncol 2003;21:2823–2830.
9. Mayr NA, Yuh WT, Zheng J, et al. Prediction of tumor control in patients with cervical cancer: analysis of combined volume and dynamic enhancement pattern by MR imaging. AJR Am J Roentgenol 1998;170:177–182.
10. Tian M, Wen X, Jackson EF, et al. Pharmacokinetics and magnetic resonance imaging of biodegradable macromolecular blood-pool contrast agent PG-Gd in non-human primates: a pilot study. Contrast Media Mol Imaging 2011;6:289–297.
11. Ghaghada KB, Ravoori M, Sabapathy D, et al. New dual mode gadolinium nanoparticle contrast agent for magnetic resonance imaging. PLoS One 2009;4 (10):e7628.
12. Reynolds PR, Larkman DJ, Haskard DO, et al. Detection of vascular expression of E-selectin in vivo with MR imaging. Radiology 2006;241 (2):469–476.
13. Kundra V, Mannting F, Jones AG, et al. Noninvasive monitoring of somatostatin receptor type 2 chimeric gene transfer. J Nucl Med 2002;43 (3):406–412.
14. Han L, Yang D, Kundra V. Signaling can be uncoupled from imaging of the somatostatin receptor type 2. Mol Imaging 2007;6 (6):427–437.
15. Yang D, Han L, Kundra V. Exogenous gene expression in tumors: noninvasive quantification with functional and anatomic imaging in a mouse model. Radiology 2005;235 (3):950–958.
16. Singh SP, Yang D, Ravoori M, et al. In vivo functional and anatomic imaging for assessment of in vivo gene transfer. Radiology 2009;252 (3):763–771.
17. Singh SP, Han L, Murali R, et al. SSTR2-based reporters for assessing gene transfer into non-small cell lung cancer: evaluation using an intrathoracic mouse model. Hum Gene Ther 2011;22 (1):55–64.
18. Tjuvajev JG, Joshi A, Callegari J, et al. A general approach to the non-invasive imaging of transgenes using cis-linked herpes simplex virus thymidine kinase. Neoplasia 1999;1:315–320.
19. Liang Q, Gotts J, Satyamurthy N, et al. Noninvasive, repetitive, quantitative measurement of gene expression from a bicistronic message by positron emission tomography, following gene transfer with adenovirus. Mol Ther 2002;6:73–82.
20. Sun X, Annala AJ, Yaghoubi SS, et al. Quantitative imaging of gene induction in living animals. Gene Ther 2001;8:1572–1579.

CHAPTER 2

PET/SPECT: INSTRUMENTATION AND IMAGING TECHNIQUES

YUAN-CHUAN TAI

Positron emission tomography (PET) and single photon emission computed tomography (SPECT) are both nuclear imaging technologies that employ radiolabeled biomolecules to probe biological processes in a subject. The biomolecules of interest may be as simple as O-15-labeled water molecules [1] for studying perfusion of blood flow into tissues or as complex as radiolabeled cells for studying autoimmune diseases or stem cell therapy [2–4]. Since both technologies are very sensitive, a tracer amount of molecules can be measured without perturbing the biological system being studied. When the biomolecule of interest is a new drug, nuclear imaging provides a convenient and powerful way to study pharmacokinetics and pharmacodynamics *in vivo*. This is especially valuable when the drug is translated from preclinical studies in laboratory animals to clinical trials in human because (i) there is a wide variety of radionuclides that have been approved for use in human and thus readily available for labeling the pharmaceutical of interest for clinical trials; (ii) nuclear imaging technology has very high sensitivity that will allow one to use trace amount of molecules for *in vivo* imaging. This minimizes the potential risk associated with toxicity from new drugs under initial evaluation, and (iii) PET and SPECT scanners are routinely used for clinical diagnosis and are becoming widely adapted in the preclinical research using laboratory animals. Therefore, the research protocol(s) based on PET or SPECT techniques can be easily translated to human imaging research with minimal difficulty. In contrast, some imaging techniques (e.g., bioluminescence) may be extremely useful in preclinical research but significantly more difficult to translate to human applications.

This chapter introduces the basic physics, instrumentation, and technical aspects of these two imaging techniques by analyzing the source of signal, physics and detectors, system design and performances, correction techniques for quantitative imaging, and finally some practical notes for using these technologies. Additional details can be found in more complete references at the end of the chapter. Mathematical

Drug Delivery Applications of Noninvasive Imaging: Validation from Biodistribution to Sites of Action, First Edition. Edited by Chun Li and Mei Tian.

modeling of biological systems based on information extracted from PET and SPECT images is described in the next chapter.

2.1 SOURCE OF SIGNAL: RADIONUCLIDES

Nuclear imaging techniques such as PET and SPECT rely on the signal produced by the radionuclides that are used to label the biomolecule of interest. Radionuclides are those nuclides that have either too many or too few protons in their nucleus and thus have the tendency to rearrange their constituents in order to reach a more stable lower energy state [5]. The transformation of the unstable (parent) nuclides into lower-energy-state (daughter) nuclides is called radioactive decay. This process is spontaneous and cannot be accelerated or stopped. The fact that the signal from radionuclides cannot be turned on or off is a rather unique property when it comes to imaging applications. In contrast, most other imaging modalities have ways to control the signal generation and, in some cases, signal amplification. This unique limitation requires special considerations when designing nuclear imaging systems in order to catch as much signal as possible before all radionucleus in the object/sample decay to their daughter nuclide and no longer emit the signal of our interest.

Accompanying the radioactive decay is the emission of various forms of ionization radiation that releases the difference in energy and mass between the parent and daughter nuclides. The ionization radiation emitted may take the form of particulate radiation (such as alpha particles, neutrons, and beta particles) or electromagnetic radiation (such as gamma rays or characteristic X-rays). The particulate radiation interacts with tissues and loses its kinetic energy quickly after the emission. As a result, they tend to be absorbed locally and cannot be detected by external detectors for *in vivo* imaging applications. In contrast, electromagnetic radiation can penetrate tissues and be detected by external detectors for imaging purposes. One exception among particulate radiations is the use of positron-emitting radionuclides for *in vivo* imaging. Positron is the antimatter of electron. It can annihilate with an electron and produce two 511 keV gamma rays which can be detected by external gamma ray detectors. In order to employ nuclear imaging techniques to study the biomolecules of interest noninvasively, radionuclides that emit gamma rays or positrons are commonly used to label the molecules of interest.

Table 2.1 lists the most commonly used gamma-emitting radionuclides for nuclear imaging applications [6]. The energy of the emitted gamma rays ranges from a few tens to a few hundred keV, depending on the type of radionuclides used. There are several routes (or decay modes) that mother nuclides may take to reach their daughter nuclides with accompanying gamma ray emission. Electron capture (EC) is a process where an inner shell electron is captured by the nucleus during the radioactive decay. The daughter nuclide may be at a high-energy state which subsequently decays to the ground state and releases a gamma ray. Alternatively, characteristic X-rays may be emitted as the outer shell electron(s) fill in the lower-energy-state inner shell(s). Either the gamma ray or the characteristic X-rays may be used for imaging purpose. Isomeric transition (IT) is the process where a parent nuclide decays to a daughter nuclide's "long-lived" metastable state and then decays to its isomer ground state by emitting a gamma ray to release the energy. This is represented by Tc-99 m, the mostly widely used radionuclide for clinical nuclear imaging applications because of

TABLE 2.1 Radionuclides Commonly Used for SPECT Imaging

Radionuclide	Decay mode	Energy (yield)	Half-life
^{67}Ga	EC	93.31 keV (38.3%)	78.26 h
		184.6 keV (20.9%)	
		300.2 keV (16.8%)	
^{99m}Tc	IT	140.5 keV (88.9%)	6.02 h
^{111}In	EC	171.3 keV (90.5%)	2.83 day
		245.4 keV (94.0%)	
^{123}I	EC	159 keV (82.8%)	13.2 h
^{131}I	(β^-,γ)	β^-_{ave} : 191.5 keV (89.4%)	8.04 day
		γ: 364.5 keV (81.2%)	
^{201}Tl	EC	x: 70.8 keV (46.2%)	3.04 day
		x: 68.9 keV (27.2%)	
		x: 80.3 keV (10.5%)	

its favorable physical properties (described in the following text) for human imaging. Beta–gamma transition is the process where a neutron-rich nuclide decays to its daughter nuclide by emitting a beta$^-$ particle to convert a neutron to a proton in the nucleus. The daughter nuclide may be at a high energy state and subsequently decays to its ground state by emitting a gamma ray.

The choice of radionuclide for a particular application depends on the biological process (e.g., the pharmacokinetics of a new drug) that is under investigation. The half-life of a radionuclide is the time needed for the nuclide to decay to half of its original activity. If the half-life of a radionuclide is much shorter than the biological process under investigation, signal will diminish completely before the measurement can be done. If the halflife of a radionuclide is much longer than the biological process, large quantity of the radiotracer needs to be injected in order to produce sufficient signal within a reasonable amount of time because the radionuclide decays slowly. This can lead to prolonged and unnecessary radiation exposure to the subject. As a result, careful match of the radionuclide half-life with the time frame of the biological process is the first step toward a successful imaging experiment. Additional consideration may involve the energy of the gamma ray. For small animal applications, lower-energy gammas are acceptable, while higher-energy gammas may compromise image resolution because they tend to scatter in detectors (see additional information in Section 2.3). For human imaging applications, however, very low-energy gamma rays may have little probability to escape from a human body. As a result, they will only contribute to radiation dose to the subject instead of signal for imaging. Therefore, very low-energy gamma emitter is not used for human imaging applications.

Table 2.2 lists the most commonly used positron-emitting radionuclides for PET imaging applications [6]. Among them, C-11 can be incorporated into many organic molecules without altering their chemical property. Therefore, it is extremely useful for evaluating pharmacokinetics of new drugs. With its 20-min half-life, however, one would need an on-site cyclotron to produce C-11 locally. It will also require experienced radiochemist who can radiolabel molecules quickly with high yield in order to take advantage of this powerful technology. F-18 has a half-life of ~110 min, allowing radiolabeled pharmaceuticals to be manufactured in centralized facilities

TABLE 2.2 Radionuclides Commonly Used for PET Imaging

Radionuclide	Mean energy of beta (yield)	Half-life
^{11}C	385.5 keV (99.8%)	20.38 min
^{13}N	491.8 keV (99.8%)	9.965 min
^{15}O	735.3 keV (99.9%)	122.24 s
^{18}F	249.8 keV (96.7%)	109.77 min
^{82}Rb	1.157 MeV (11.6%)	1.3 min
	1.523 MeV (83.3%)	
^{64}Cu	β^{-ve}: 190.2 keV (37.2%)	12.70 h
	β^{+ve}: 278.1 keV (17.9%)	

and distributed regionally. Therefore, most clinical PET applications are based on F-18 labeled pharmaceuticals. Rb-82 can be produced by a generator instead of a cyclotron. It is gaining acceptance in clinical cardiac imaging applications. Non-conventional PET radionuclides such as Cu-64 offer longer half-life that can be used to probe slower biological processes. C-11 is commonly used to label small organic molecules without altering their chemical properties, a critical advantage for evaluating new drugs.

2.2 DETECTION PHYSICS

2.2.1 PET

Radionuclides that decay through the emission of positron can be used for PET imaging applications. Positron is the antimatter of electron. It has the same mass and carries the same amount of charge of an electron, except that the charge is positive instead of negative. When a positron is emitted from a nucleus, it carries initial kinetic energy that ranges from several hundred to several thousand keV, depending on the type of radionuclides it originates. As a positron travels through an object, it continuously interacts with tissues and rapidly loses its kinetic energy. As it slows down, the probability that a positron combines with an electron and annihilates each other increases. Upon annihilation, the mass of the positron and the electron is converted into energy and released as two 511 keV gamma rays. These two gamma rays travel in the opposite directions in order to conserve the momentum and energy. Gamma rays of 511 keV in energy are highly penetrating and may escape from the object and become detected externally. Figure 2.1 illustrates that a pair of detectors can be placed around an object to detect the annihilation gamma rays.

From the simple cartoon in Figure 2.1, it becomes clear that a pair of PET detectors (or a PET scanner) measures the location of the annihilation gamma rays, not the location of the radiolabeled biomolecules of interest. This inherent uncertainty due to the range of positron is one of the fundamental limits of the image resolution achievable by PET. To minimize the blurring of images due to positron range effect, it is preferable to use radionuclides that emit positrons with low kinetic energy, which translates to small positron range effect. Among the PET radionuclides listed in Table 2.2, F-18 has the lowest mean kinetic energy and therefore the least blurring due to positron range effect.

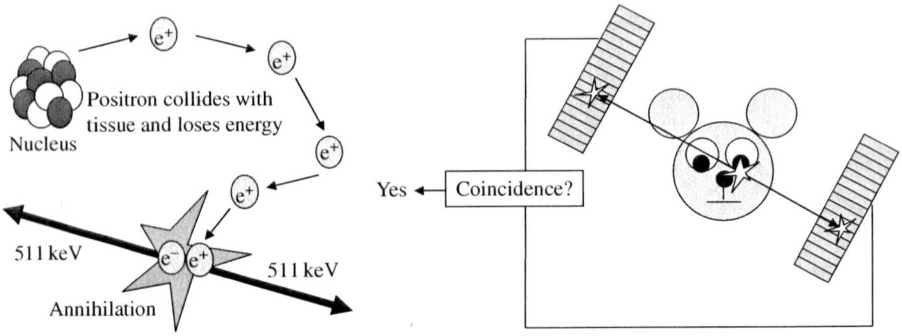

FIGURE 2.1 Detection of positron annihilation gamma rays.

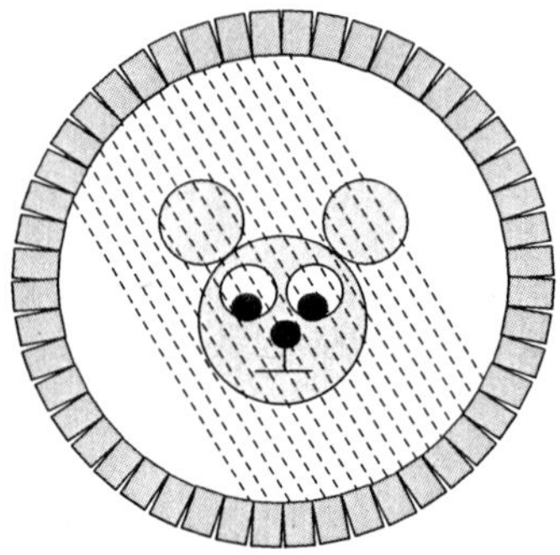

FIGURE 2.2 Most PET scanners employ ring geometry with which coincidence events can be detected from all angles simultaneously, eliminating the need to rotate detectors around an object.

Since the two gamma rays are created simultaneously, they should reach the two detectors almost at the same time (off only by the time difference for traveling from the origin of the gamma rays to the two detectors). It is the coincidence detection of the two annihilation gamma rays that forms the basis of a PET imager. When the coincidence detection circuit identifies two gamma rays that are detected simultaneously, the line joining the two gamma detectors in coincidence (known as coincidence line of response (LOR)) also defines the direction along which the two gamma rays come from. As a result, the directional information is provided by the coincidence detection circuit which electronically collimates the gamma rays to form images. In order to reconstruct tomographic images, the object needs to be sampled from multiple angles. This can be achieved by rotating the pair of detectors in Figure 2.1 around the object. Alternatively, most PET scanners adapt the ring geometry (Fig. 2.2) with which detectors are arranged into one or more rings around the object to collect coincidence events simultaneously from all angles for tomographic imaging. This avoids the need to rotate detectors that enables dynamic imaging capability (see more in Section 2.6) for most PET scanners.

Coincidence detection is commonly implemented by digitizing the arrival time of individual gamma rays at detectors. These events are fed into a centralized coincidence processor that compares the time stamps of individual gamma ray events. If

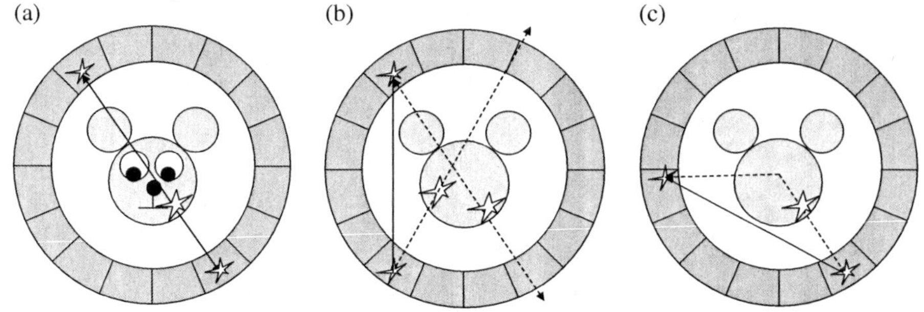

FIGURE 2.3 (a) True coincidence allows one to position an event correctly during image reconstruction; (b) random coincidence causes a random event (noise) be placed at an arbitrary location in the image; (c) scatter coincidence causes mispositioning of an event that leads to loss of resolution and contrast.

two events occur within a preselected timing window (typically within 4.5–10 ns), the two gamma rays are considered "in coincidence." The coincidence event is then recorded by a computer for further sorting and image reconstruction. This commonly employed coincidence detection scheme not only registers the "true" coincidence events from individual positron annihilations but also detects random and scattered coincidence events. Figure 2.3 illustrates the three types of coincidence events (i) true, (ii) random, and (iii) scatter coincidences. A true coincidence refers to the detection of the two annihilation gamma rays from one positron annihilation. The rate of the "true coincidence events" between any two detectors is related to the total amount of radioactivity along the LOR connecting the two detectors. Therefore, the true coincidence rate measured by two detectors relates to the "line integral of radioactivity along the LOR." Mathematically, this is equivalent to the Radon transform of the 2D radioactivity distribution function on the plane where the LOR lies [7]. Through Inverse Radon transform, one can restore the original function, that is, the radioactivity (or radiolabeled biomolecule) distribution within the object. As a result, the "true coincidence" is considered the "signal" that a PET system measures. A recently published book by Zeng [8] provides an excellent introduction to image reconstruction techniques used in medical imaging (including PET and SPECT).

A random coincidence refers to two uncorrelated gamma rays that happen to occur and be detected at the same time. These two gamma rays may be the result of the annihilation of two positrons, or annihilation gamma rays in coincidence with gamma rays from other sources nearby. Since it occurs purely by chance, it is considered noise to the desired coincidence measurement. The random coincidence rate is known to be

$$R = 2\tau S_1 S_2 \tag{2.1}$$

where τ is the preselected coincidence timing window of the system and S_1 and S_2 are singles event rate of detectors 1 and 2, respectively [9]. Therefore, the larger the coincidence timing window is, the higher the random coincidence rate will be. If the amount of activity in or near a PET scanner doubles, the singles event rate S_1 and S_2 will also double, but the random coincidence rate will quadruple. This nonlinear

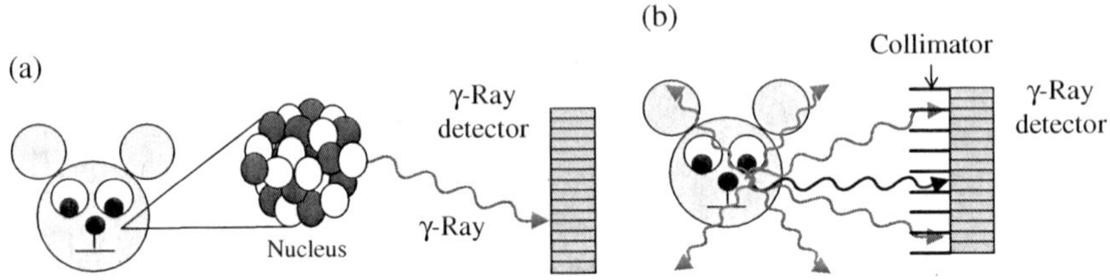

FIGURE 2.4 (a) Detection of gamma rays from gamma-emitting source. (b) A collimator restricts the directions along which gamma rays may travel and be accepted by the detector.

increase of random coincidences prohibits the use of excessive amount of radioactivity for PET imaging and will be described in more details in Section 2.5 of this chapter.

A scattered coincidence refers to the detection of two annihilation gamma rays from the same positron annihilation, with at least one (and maybe both) gamma rays being scattered by the object (tissues) before they reach the detectors. When a gamma ray undergoes Compton scatter, part of its energy is transferred to the Compton electron while the remaining energy is carried away by a secondary gamma ray that travels along a new direction that deviates from the original path. The LOR defined by the scattered coincidence will deviate from the correct LOR, depending on the amount of energy lost. The net effect is reduced resolution and contrast of the images. Therefore, it is desirable to reject scatter events by only accepting events with the correct energy (511 keV).

2.2.2 SPECT

Compared to the indirect measurement of positron (or radionuclide) locations in PET, a SPECT system directly measures the origins of the gamma-emitting radionuclides (Fig. 2.4a). Therefore, the image resolution of SPECT is only limited by the resolution of the imaging system rather than the underlying physics such as positron range and photon acolinearity effects in PET. Emission of gamma rays from a nucleus is a random process with no directional preference. Hence, gamma rays are emitted isotropically from a cluster of radiolabeled molecules in an object and can reach the entire surface of a gamma detector as illustrated in Figure 2.4b. In order to image the radioactivity distribution within an object, the imaging system needs to not only detect the gamma rays but also estimate the directions along which gamma rays come from. Without the directional information, an event registered by a detector may come from anywhere within the object, thus does not contain any positioning information. In a SPECT system, the directional information is provided by the use of a collimator [10]. A collimator is typically made of lead or tungsten. It limits the direction along which each detector element can "see." The directional constraint(s) imposed by a collimator selectively block gamma rays from reaching the gamma detectors. For example, the collimator in Figure 2.4b (represented by thick black lines) only allows gamma rays that are coming straight at the detector surface to pass through. All other gamma rays that come from oblique angles are filtered out by the collimator. The more stringent the constraints, the more directional information each detected gamma ray carries. However, this information-rich content comes at the cost of low system sensitivity because most events are filtered out by the constraints (i.e., gamma rays are blocked out by the collimator). In contrast to the physical collimator used in SPECT, coincidence detection circuit provides electronic collimation for PET. Therefore, there is no need to employ physical collimator in PET. As a result, the overall system sensitivity (the probability that a radioactive decay in the imaging field of view is detected by the scanner) of a PET system is usually much higher than that of a SPECT system (by one to two orders of magnitude), despite its inherent limitations in image resolution.

In order to reconstruct tomographic images, the projection measurement illustrated in Figure 2.4b needs to be collected from multiple angles. This can be achieved by rotating the camera head around the object or using multiple camera heads each responsible for collecting data from a smaller number of angles.

2.3 DETECTOR TECHNOLOGY

2.3.1 Types of Radiation Detector

Regardless of whether it is a gamma-emitting radionuclide for SPECT or a positron-emitting radionuclide for PET, the signal that needs to be detected for *in vivo* imaging is gamma rays of different energies. There are three major types of radiation detectors that can be used to detect gamma rays (i) gas detector, (ii) semiconductor detector, and (iii) scintillation detector [11]. Important characteristics of gamma ray detectors used for imaging applications include spatial, energy, and timing resolutions.

For imaging applications, spatial resolution of a detector is clearly an important property. Intrinsic spatial resolution of a gamma ray detector relates to how well a detector can localize the gamma ray interaction within the detector. This is often a complex property that depends on the material, dimension and geometry of the detector, as well as energy of gamma rays. The intrinsic spatial resolution of a detector can be measured by recording detected events as a narrow beam of gamma rays scans across the detector surface. The count profile measured by individual detector element corresponds to the detector response function. Intrinsic spatial resolution of a detector is expressed as the full width at half maximum (FWHM) of the detector response function.

The energy resolution of a detector is also critical for PET and SPECT applications. When a gamma ray interacts with matter, it may undergo photoelectric interaction with which all the energy carried by the gamma ray is transferred to the photoelectron and subsequently converted to detector signal. Alternatively, a gamma ray may undergo a Compton scatter interaction with which only part of the energy carried by the initial gamma ray is transferred to the Compton electron. The remaining energy is carried away by a secondary gamma ray. For nuclear imaging systems, the output signal of a detector is typically proportional to the amount of energy deposited by the gamma rays through photoelectron or Compton electrons. If one irradiates a detector with a particular gamma ray source and sorts individual events into a histogram based on signal amplitude, the result is called an energy spectrum and will look similar to the curve in Figure 2.5. The events under the peak of the energy spectrum correspond to gamma rays that undergo photoelectric interaction in the detector and deposit all of their energy. Therefore, this peak is called photopeak.

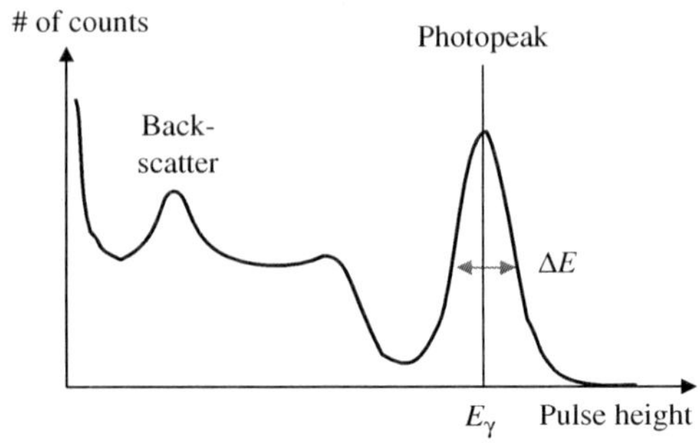

FIGURE 2.5 A typical energy spectrum from a scintillation detector exposed to a gamma ray source with energy E_γ. ΔE is the width of the photopeak at the half-maximum of its peak value level.

The energy resolution of a detector is expressed as $R_E = (\Delta E/E_\gamma)100\%$, where E_γ is the energy of the gamma ray source and the ΔE is the FWHM of the photopeak in the energy spectrum. The sharper the photopeak in an energy spectrum, the smaller the FWHM and the better the detector's ability to identify the energy of a gamma ray event. Energy resolution of a detector is critical in applications where the energy of a gamma ray must be accurately identified. For example, a SPECT system will identify an event as signal from the radiotracer labeled with Tc-99m if its energy is within the photopeak that corresponds to 140 keV gamma ray. Events that fall outside of the photopeak undergo Compton scatter either in the detector or in the object before they reach the detector. In the first case, the energy of the event is unknown, but the positioning information associated with this event may be correct. In the second case, the secondary gamma ray is deviated from its original path and is not useful for imaging purpose. Since a detector has no way of knowing which case the event belongs to, events that fall outside of the photopeak are typically discarded for nuclear imaging applications. Energy resolution of a detector is important for SPECT and PET as it helps to screen out the scattered gamma rays that carry false positioning information.

Timing resolution of a detector relates to how accurate can the detector identify the time when an interaction occurs. For SPECT, such information is not critical. For PET, this has great impact on the coincidence detection. In general, the coincidence timing window (τ in Eq. 2.1) is at least two to three times of the timing resolution of the detector. The timing resolution of typical PET detectors ranges from a few hundred picoseconds to a few nanoseconds. For systems that have sub-nanosecond timing resolution, the time that the two annihilation gamma rays arrive at two detectors can be precisely measured, and the time difference will indicate the source location along the LOR defined by the detector pair. As a result, the positioning of individual event can be more accurate, and hence the signal-to-noise ratio in the reconstructed image is improved. This so-called time-of-flight capability is most useful for clinical PET scanners where a large patient body is imaged. For animal PET applications, the currently achievable timing resolution is insufficient to make a significant difference and is often not used.

Among the three types of gamma ray detectors, gas detector has the lowest detection efficiency due to the low density of detector materials (gases). That is, it only detects a small fraction of gamma rays that pass through the detector. Gas detector is commonly used in dose calibrators that measure the total amount of radioactivity in a sample. To improve the detection efficiency of gas detector for imaging applications, one may enclose metal structures in a gas ion chamber. The metal structure will have a higher probability of interaction with the incoming gamma rays. Photoelectrons (or Compton electrons) produced from these interactions can lead to ionization of the gas molecules in the detector chamber and produce signal [12]. This type of detector can produce high spatial resolution but often has limited energy and timing resolutions. As a result, it has not been widely applied to PET and SPECT.

Semiconductor detector can produce electron–hole pairs when excited by ionization radiation. Some semiconductor detectors require very low amount of energy (a few eV) to produce the "signal" (electron–hole pair in this case) when compared to the gas and scintillation detectors which typically require a few tens of eV for ionizing a molecule or producing light signal. As a result, semiconductor detectors offer the highest energy resolution among all radiation detectors [11]. In addition, many technologies developed for consumer electronics are applicable to the fabrication of semiconductor radiation detectors. These advanced fabrication technologies make high-resolution (potentially submillimeter) semiconductor detectors available for imaging applications [13, 14]. The drawback of semiconductor detectors is its relatively low detection efficiency and slower detector response when compared to the scintillation detectors (described in the following text). For SPECT imaging applications, gamma ray energy rarely exceeds 365 keV and can be effectively detected by semiconductor detectors such as CdTe or CdZnTe. As a result, CdZnTe has been applied in both small animal and special purpose human SPECT systems [15–18]. For PET, the 511 keV gamma rays are highly penetrating and harder to stop. The aforementioned semiconductors are less effective in detecting PET signal. In addition, the slower detector response makes the system more susceptible to random coincidences. As a result, there are limited applications of semiconductor detectors for PET presently despite its potentially very high-resolution capability [14, 19].

A scintillator is a material whose atoms or molecules can be excited to a higher energy state and subsequently emit light photons as the atoms or molecules return to their ground state. When a gamma ray interacts with a scintillator, either through photoelectric interaction or Compton scatter, the energy carried by the photoelectron or Compton electron is transferred to the scintillator and subsequently converted to light photons that can be detected by a secondary light detector. Among all scintillation detectors, inorganic scintillator crystals read out by light detectors (such as photomultiplier tubes, PMT) is the most widely used technology for PET and SPECT applications. Inorganic scintillation crystals that have high density, high Z (atomic number), and high light yield are most suitable for nuclear imaging applications because (i) a detector with high density can stop and interact with high-energy gamma rays effectively; (ii) a detector with high effective Z will have a higher probability to interact with gamma rays through photoelectric interaction, which allows one to determine the energy of a gamma ray accurately; and (iii) a detector with high light yield for a given amount of energy deposition offers better signal-to-noise ratio at the output stage of the light sensor. In addition to these important characteristics

TABLE 2.3 Physical Properties of Commonly Used Scintillators for PET and SPECT

Scintillation material	Density (g/cm^3)	Effective atomic number (Z)	Light yield (% relative to NaI)	Primary decay constant (ns)	Peak emission wavelength (nm)	Attenuation coefficient at 511 keV (cm^{-1})
NaI(Tl)	3.67	51	100	230	410	0.35
BGO	7.13	75	15	300	480	0.95
LSO	7.40	65	75	40	420	0.86

that are common for both PET and SPECT applications, detectors with fast decay time can emit light photons in a short burst, allowing one to determine the exact timing of the gamma ray interaction accurately. This is critical for PET because it will allow a system to employ a small coincidence timing window to minimize the number of random coincidences. Table 2.3 lists the physical properties of the most commonly used scintillators for PET (bismuth germanate (BGO) and lutetium oxyorthosilicate (LSO)) and SPECT (NaI(Tl)).

Photomultiplier tube is the most commonly used light detector in SPECT and PET scanners. When the photocathode of a PMT interacts with a light photon, a photoelectron is emitted and accelerated toward a series of dynodes that provide multiple stages of amplification of the incoming electron flux. At the output stage, the original charge signal from the photocathode is amplified by a factor of 10^6 or higher. The high gain and fast response time make PMT ideal for reading out scintillation light output for nuclear imaging applications.

In recent years, semiconductor light detectors such as avalanche photodiode (APD) and silicon photomultiplier (SiPM) are becoming popular for PET and SPECT applications [20–26]. They are relatively immune to strong magnetic field and can be used inside MRI scanners. PET detectors based on APD or SiPM detectors have been used to develop MR-compatible PET scanners which are becoming commercially available for both human and animal imaging applications.

2.3.2 Block Detector for PET

The most common implementation of PET detector is the so-called "block detector" that consists of an array of LSO or BGO crystals read out by a 2-by-2 array of PMTs [27], as illustrated in Figure 2.6a (only the two front PMTs are visible in the drawing). A light diffuser may be inserted between the scintillator array and the light sensors to allow the scintillation light photons to reach all four PMTs. Depending on the location of the "light source" (scintillation crystal in which the gamma ray interacts), the four PMTs may detect different portion of the light photons emitted from a crystal. Based on the ratio of the total light collected by the two left PMTs versus the two right PMTs (A + C versus B + D) or the two front PMTs versus the two back PMTs (A + B versus C + D), the location of the light source along the left–right (X) or front–back (Y) direction can be identified. This simple ratiometric positioning method is implemented in most clinical PET scanners. In general, the intrinsic spatial resolution of such detector is determined by the width of the crystal elements in the crystal array.

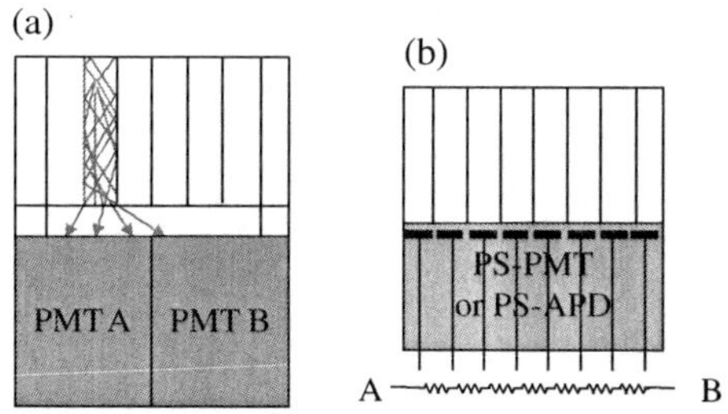

FIGURE 2.6 (a) A typical PET "block detector" is made of a scintillation crystal array and four PMTs (only the two front ones shown in drawing). (b) A typical highresolution PET detector is made of a scintillation crystal array and a PS light detector. Charge division resistive network may be used to multiplex the output signals to reduce the total number of readout channels.

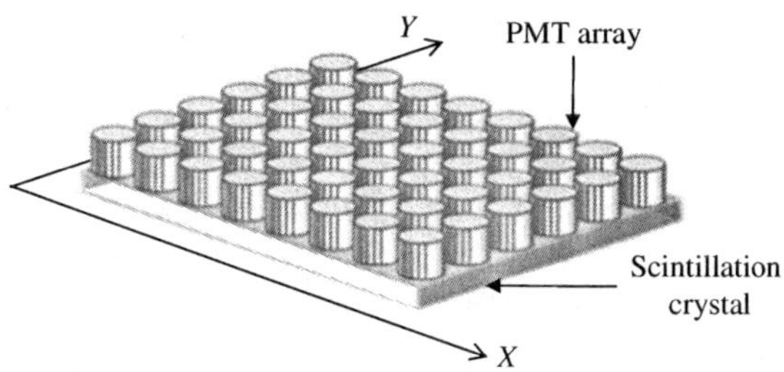

FIGURE 2.7 A scintillation camera is typically made of a monolithic scintillation crystal with its light read out by an array of PMT.

For higher-resolution PET scanners, such as those used for animal imaging applications, the 2-by-2 PMT array is replaced by a position-sensitive (PS) light detector such as a PS-PMT or a PS-APD (a semiconductor light sensor made of silicon), as illustrated in Figure 2.6b. Individual outputs from the anodes may be read out separately or multiplexed to form four position-encoded signals by a charge division resistive network to reduce the number of output signals to four. The former approach allows one to identify inter-crystal scatter events and maximize the spatial resolution of the detector, at the expense of a large number of readout channels [28]. The latter approach mimics the block detector output format and then uses the same ratiometric positioning algorithm to identify the crystal of interaction [29]. This cost-effective approach is adapted by most commercial animal PET scanners.

2.3.3 Scintillation Camera for SPECT

The most common form of detectors in a SPECT system is the so-called Anger camera or scintillation camera [30]. A scintillation camera is typically made of a large scintillation crystal (e.g., a monolithic NaI(Tl) crystal) coupled to an array of PMT, as shown in Figure 2.7. When a gamma ray interacts with the scintillator, light photons are emitted and detected by the array of PMT. The position of a gamma ray interaction within the crystal is calculated from the PMT signals weighted by individual PMT's coordinates using Equation 2.2.

$$\hat{x} = \frac{\sum_{i=1}^{N} x_i M_i}{\sum_{i=1}^{N} M_i}, \quad \hat{y} = \frac{\sum_{i=1}^{N} y_i M_i}{\sum_{i=1}^{N} M_i}, \tag{2.2}$$

where x_i and y_i are the x- and y-coordinates of the i-th PMT, while M_i is the signal amplitude of the i-th PMT. The energy of the detected gamma ray is proportional to the total signal collected by all PMT. That is,

$$\hat{E} = \sum_{i=1}^{N} M_i \tag{2.3}$$

Using such a device, the position and energy of individual gamma rays that interact with the camera can be recorded. This allows one to detect the spatial and temporal distribution of gamma-emitting radionuclides within the object of interest.

Despite the improvement in electronics and PMT performance, the basic design of a gamma camera has not changed since its invention in 1958. Although there are many newer types of scintillation materials available now, NaI(Tl) crystal remains the most popular choice because of its low cost and good performance. Gamma camera can be made cost-effectively to provide gamma ray detection over a large area with reasonable performance in both energy and spatial resolution. Room temperature semiconductor detectors such as CdTe or CdZnTe have gained some acceptance in recently years as a potential replacement for the scintillator-based gamma camera. These detectors provide higher-energy resolution (as low as 1% FWHM for 662 keV gamma rays in some cases) than that of NaI(Tl)-based detectors. They can also be made into pixelated detectors to provide higher spatial resolution. However, these detectors are still quite expensive presently, and they typically require higher-performance readout electronics. As a result, their cost is significantly higher than scintillator-based detectors, while their benefits are limited as the image resolution of a SPECT system depends not only on the detectors intrinsic spatial resolution but also on the collimator resolution and design (more details in the next section). As a result, scintillator-based detector is still the dominant technology in gamma camera and SPECT imaging systems.

2.4 SYSTEM DESIGN AND PERFORMANCE

2.4.1 PET

As mentioned earlier, most PET systems employ ring geometry by populating block detectors into multiple rings. Coincidence events measured from detector pairs form projection (or Radon transform) of the 3D radioactivity distribution function within the object. The image resolution of a PET system is affected by several physical factors and can be expressed by the following equation [31, 32]:

$$R_{img} \approx 1.25 \cdot \sqrt{R_{src}^2 + R_{180°}^2 + R_{det}^2 + BE^2} \tag{2.4}$$

where R_{src} is the effective source dimension, $R_{180°}$ is the photon acolinearity effect, R_{det} is the intrinsic spatial resolution of a PET detector, and BE is the "block effect" that is commonly observed when block detectors are employed in a PET system that results in resolution degradation due to multiplexing of signals. The first term, effective source dimension, includes the positron range effect previously described. When small range radionuclides (such as F-18 and C-11) are employed, resolution loss due to this term is in the range of 200–300 μm FWHM [33].

The second term relates to photon acolinearity effect. When a positron annihilates with an electron, it is often not completely at rest and thus carries residual kinetic energy and momentum. To preserve the energy and the momentum, the directions of the two annihilation gamma rays are not exactly 180° apart, but off by a smaller angular deviation that is estimated to have a distribution of 0.5° FWHM [34]. This angular deviation translates to an uncertainty between the annihilation location and the coincidence LOR connecting the two detectors. The acolinearity effect increases linearly as the separation of the detector pair increases. For a PET scanner that follows the ring geometry with a diameter of D, it has been shown that the acolinearity effect causes a blurring of ~ 0.0022 D. This translates to ~2 mm or ~0.4 mm of blurring for a clinical whole-body PET scanner or an animal PET scanner, respectively.

The third term relates to the intrinsic spatial resolution of detectors employed in a scanner. With the common "block detector" design [27], the detector's intrinsic spatial resolution is typically one half of the width of the scintillation crystals in the array. This is true only when the gamma rays are coming at the detectors "head-on," i.e., along the norm direction of the crystal surface. When gamma rays enter a detector along an oblique angle, the "effective crystal width" and the crystal length viewed from the incident angle need to be considered. In addition, gamma rays may interact with detector materials through Compton scatter initially and subsequently become completely absorbed by a secondary interaction with the detector. This intra-block and inter-crystal scattering is not differentiable from a non-scattered event in a block detector. Therefore, there is a fair amount of resolution loss by the use of a block detector that is attributed as "block effect" of a detector [31, 35] (the fourth term in Eq. 2.4).

In addition to image resolution, the overall system sensitivity is the other critical parameter of a PET system. Sensitivity of a system relates to how effective the system is in terms of detecting a radioactive decay within its image field of view (FOV). This important characteristic is limited by (i) the solid angle coverage of the detectors in the system and (ii) the intrinsic detection efficiency of 511 keV gamma rays by the employed PET detectors. To maximize the solid angle coverage of a system, a PET scanner is often built with multiple flat panel detectors (similar to the two detector panels in Fig. 2.1) or multiple rings of detectors (similar to the geometry in Fig. 2.2) to surround the object being imaged. To improve detector's intrinsic detection

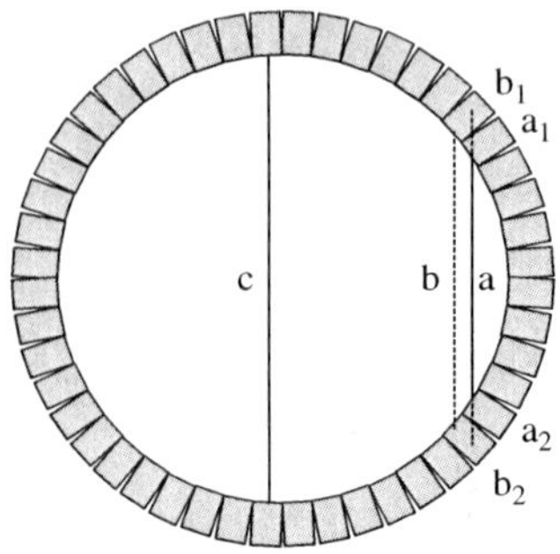

FIGURE 2.8 Parallax error: coincidence gamma rays originated along LOR "a" may be mislocated to LOR "b" if the gamma rays penetrate the initial intercepted crystals (a_1 and a_2) and become detected by crystals b_1 and b_2.

efficiency of 511 keV gamma rays, one would use high-density high-Z detectors of sufficient length in order to stop and detect gamma rays. In general, the longer the scintillation crystals used in the array, the higher the detection efficiency. However, the longer the crystals in an array are, the more susceptible the system is to the so-called parallax error [36]. Figure 2.8 illustrates the cause of the parallax error. When high-energy gamma rays enter a pair of detectors from their norm directions (e.g., line c in Fig. 2.8), the coincidence LOR is always correctly identified regardless of the depth of the gamma ray interaction in the crystals. However, when gamma rays enter detectors from an oblique angle (e.g., line a in Fig. 2.8), they may interact with the crystals of initial interception (a_1 and a_2 in Fig. 2.8). In this case, the event is identified as being originated from the coincidence LOR "a" and is correctly located. However, each of the two gamma rays may penetrate the crystals of initial interception and interact with crystals behind them (b_1 and b_2 in Fig. 2.8). When this happens, the event is identified as being originated from the coincidence LOR "b" and is mispositioned. This leads to degradation of radial resolution when the radioactive source is away from the center of the FOV in a typical PET scanner. Parallax error is most significant when (i) the scintillation crystals length is large compared to their width, (ii) the ring diameter of the scanner is small, or (iii) the radial offset of the source location is large. As a result, increasing system sensitivity through the use of longer crystals leads to compromised image resolution unless the ring diameter is large or the imaging FOV is kept to the very center of the ring. In the former case, the solid angle coverage of a given detector is reduced when the system diameter is increased; thus the overall sensitivity of the system is reduced. In the latter case, one limits the useful imaging FOV for a given ring diameter of a system. In another word, it is difficult to improve the image resolution, system sensitivity, and useful image FOV of a PET scanner all at once. This dilemma can be solved if the detector is capable of resolving the depth of interaction (DOI) of a gamma ray in the detectors, thereby minimizing the uncertainty caused by parallax error. The DOI-capable detectors often cost more than conventional block detectors and thus increase the overall cost of the system. As a result, the design of a PET scanner and the detector technology employed require a balance among image resolution, system sensitivity, useful imaging FOV, and the overall cost [37].

For clinical whole-body PET scanners, the photon acolinearity effect limits the image resolution of the system to be ≥2 mm FWHM. As a result, most scanners employ scintillation crystals of 4 mm width because crystals of smaller dimensions provide limited improvement to the image resolution while significantly increasing the cost of the scanner. For whole-body PET imaging protocol, since the scan duration for each section of the body is only 2–3 min, the image quality is often limited by the counting statistics instead of the system resolution. As a result, scintillation crystals of 2 cm or longer in length are commonly used to maximize the overall system sensitivity. Higher-resolution human PET imaging applications are currently limited to special purpose PET scanners such as positron emission mammography systems [38–40], brain imager [41], or prostate imager [42, 43]. A new technology known as "PET insert" devices that can be integrated into existing PET scanners to locally enhance the image resolution of PET is currently being developed [26, 32, 44–49]. This novel approach may offer a cost-effective solution for bringing high-resolution PET imaging capability to human PET imaging applications.

For small animal PET imaging applications, especially those that utilize F-18 or C-11 for radiolabeling, it is usually the intrinsic spatial resolution of PET detectors that limits the achievable PET image resolution. With the growing demand in high-resolution and high-sensitivity molecular imaging techniques and their applications in biomedical and pharmaceutical research, tremendous effort has been devoted to the development of higher-resolution PET detectors and systems [37, 50, 51]. Presently, most commercial animal PET scanners employ scintillation crystals of 1–2-mm width and 10–15-mm length in order to achieve high resolution and high sensitivity. To minimize the parallax error, some scanners employ detectors that can decode the DOI. This allows one to use longer crystals without losing image resolution near the edge of the imaging FOV, therefore achieving both high resolution and high sensitivity at the same time. The drawback is, of course, a more complex system and electronics that leads to higher cost. There are research prototype PET systems that employ submillimeter crystal detectors. At this level of crystal dimension, other physical factors (positron range and photon acolinearity effect) become significant and cannot be neglected. As a result, the improvement in image resolution becomes incremental when the crystal width is reduced beyond 1 mm or less. When decoding submillimeter crystals, reading out individual anodes of the light sensor is advantageous because it eliminates the block effects. However, this approach also increases the cost of the scanner and is not commonly employed by commercial systems. Overall, one should expect some level of compromise among system resolution, system sensitivity, useful FOV, and cost when choosing a system. Commercially available animal PET systems currently have an image resolution ranging from 1 to 2 mm FWHM, system sensitivity ranging from 2 to ~10%, and usable imaging FOV ranging from 5 to 20 cm transaxially and 4 to 13 cm axially. Modern clinical PET scanners typically have an image resolution of ~4 mm FWHM, system sensitivity of 5–6%, and usable imaging FOV of ~60 cm transaxially and 16–22 cm axially.

2.4.2 SPECT

SPECT systems typically consist of one or more scintillation camera heads; each includes a collimator and a detector panel. The most common types of collimators are parallel hole and pinhole collimators [10]. Shown in Figure 2.9 is a gamma

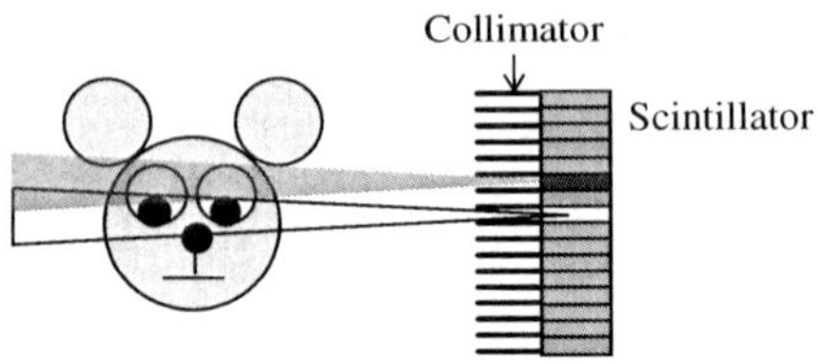

FIGURE 2.9 A scintillation camera equipped with a parallel hole collimator. Activity within the shaded and unshaded regions can be detected by the corresponding shaded and unshaded detector elements in the camera.

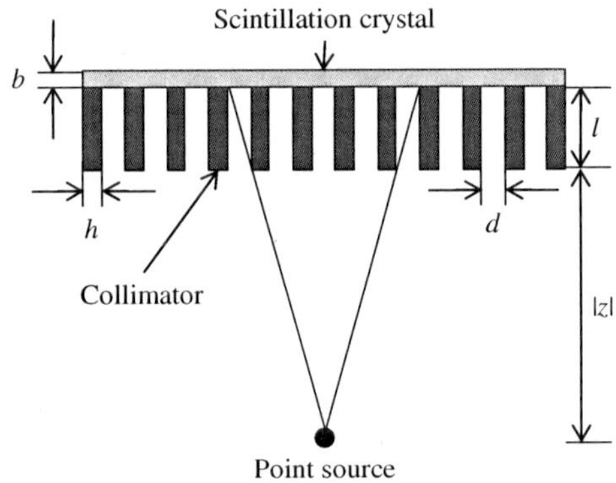

FIGURE 2.10 Camera with a parallel hole collimator.

camera with a parallel hole collimator that only allows gamma rays that travel in the directions perpendicular to the camera head to pass through and reach the scintillation detector. Events registered by a detector element (shown as solid grey in Fig. 2.9) are originated from the shaded area viewed by the detector element. The further away from the collimator surface, the broader the shaded region becomes due to the geometry. Radioactive sources that are far away from the camera surface may be detected by multiple detector elements and cannot be resolved by individual detectors. As a result, the resolution of the imaging system degrades as the distance between the object and the camera increases.

Image resolution of a scintillation camera is affected by both the detector intrinsic spatial resolution and the collimator resolution. The resolution of a parallel hole collimator and the corresponding system resolution can be expressed as Equations 2.5 and 2.6 where individual parameters are illustrated in Figure 2.10 [6]:

$$R_{coll} = \frac{d}{l}\left(l + b + |z|\right) \tag{2.5}$$

$$R_{sys} = \sqrt{R_{\text{int}}^2 + R_{coll}^2} \tag{2.6}$$

The sensitivity of a scintillation camera using a parallel hole collimator can be expressed as

$$\varepsilon = K^2 \left(\frac{d}{l_e}\right)^2 \left[\frac{d^2}{(d+h)^2}\right] \qquad K \approx 0.25 \tag{2.7}$$

From these equations, it can be seen that the resolution of a gamma camera using a parallel hole collimator depends on the dimension of the collimator opening (d), collimator thickness (l), and the distance from the object to the surface of the camera (z). The sensitivity of the system depends on the dimension of the collimator opening and collimator thickness but is independent of the location of the object.

The number of events registered by a detector element in Figure 2.9 is proportional to the integral of activity distribution within the shaded region (or volume if viewed in three dimensional (3D)) weighted by a spatially variant detection efficiency [52]. Gamma ray events registered by the camera are analogous to a projection image of the gamma-emitting radionuclides distribution within the object. A collection of the "projection images" measured from different angles allows one to reconstruct 3D tomographic images of the source distribution. One may use a simplified system model that ignores the spatially variant system resolution and sensitivity and assumes that events registered by a detector element are equivalent to the line integral of the activity distribution in the object. This will significantly reduce the complexity in image reconstruction algorithm, although the reconstructed images do not truly represent the distribution of radioactivity in the subject. More sophisticated image reconstruction algorithms that model the detector response accurately can provide more quantitative measurement of activity distribution, as the expense of longer computation time.

For a camera with a pinhole collimator (Fig. 2.11), the resolution depends on the collimator cone length (l), object-to-pinhole distance (z), pinhole aperture's diameter (d), and the collimator material's linear attenuation coefficient (μ). The collimator and system resolution can be expressed as

$$d_{eff} = \sqrt{d \cdot \left[d + 2\mu^{-1} \tan(\alpha/2)\right]} \tag{2.8}$$

$$R_{coll} \approx d_{eff} \cdot \left(\frac{l+z}{l}\right) \tag{2.9}$$

$$R_{sys} \approx \sqrt{R_{coll}{}^2 + \left(\frac{z}{l} \cdot R_{\text{int}}\right)^2} \tag{2.10}$$

The image resolution of the system can approach the collimator resolution if $z << l$ (i.e., if the object-to-pinhole distance is significantly smaller than the collimator cone length). To take advantage of this geometry, a gamma camera of large detection surface is coupled with pinhole collimator with large cone length. An animal is placed close to the pinhole such that the activity distribution in the subject is projected to the camera surface through the pinhole with a magnification.

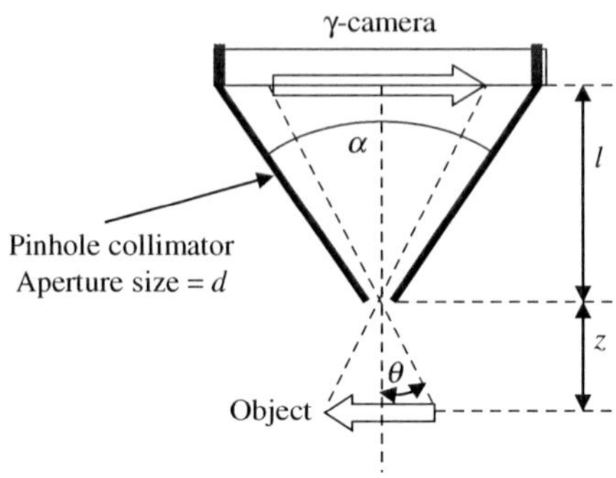

FIGURE 2.11 A camera with a pinhole collimator.

If the pinhole aperture (d) is small, the resolution of the image can be high despite the relatively low intrinsic spatial resolution of typical gamma cameras. This is commonly used in animal SPECT applications to achieve very high-resolution images [53].

The sensitivity of such a system can be expressed as

$$\varepsilon \approx \frac{d_{eff}^2 \cos^3\theta}{16 \cdot z^2} \tag{2.11}$$

Note that the sensitivity of the system is directly proportional to the dimension of the pinhole aperture but inversely proportional to the square of object-to-pinhole distance. For ultrahigh-resolution SPECT imaging applications, one would use a collimator with small aperture, which has limited sensitivity. To overcome this limitation, multi-pinhole collimator is commonly used to increase the overall system sensitivity. With a small object-to-pinhole distance, the sensitivity and the image resolution of the system are both improved. However, this limits the usable imaging FOV with the high magnification of the geometry. To overcome such limitation, one may image the subject with a step-and-shoot motion. As a result, the scan duration varies depending on whether one needs to image an entire animal or just an organ of interest locally. Current state-of-the-art animal SPECT systems can achieve submillimeter resolution images with a system sensitivity of a fraction of 1%.

2.5 CORRECTION TECHNIQUES

One of the major strengths of nuclear imaging techniques such as PET and SPECT is their ability to provide quantitative measurement of radiotracer distribution *in vivo*. This enables one to model biological system and to establish pharmacokinetics and pharmacodynamics of new drugs. In order to achieve the desirable quantitation, many physical effects such as random coincidences, attenuation and scatter of gamma rays, and systemic variation of detector efficiency must be corrected before or during the image reconstruction. Brief description of these physical factors and the corresponding correction technique is provided next.

2.5.1 Random Correction

Random coincidences can be estimated by Equation 2.1 if one knows the singles event rate of individual detectors. Alternatively, one may measure it using the "delayed coincidence window technique" that compares the time stamp of current singles event with previously occurred events that are kept in memory buffer to determine how many of the "current singles" are in coincidence with "previously occurred singles." These "delayed coincidences" are uncorrelated singles events that are off by a preselected fixed delay time, which occur purely by randomness. As a result, the "delayed coincidence" is a good estimate of the random coincidences registered by the coincidence circuits. Its outputs are flagged and stored separately from the "prompt coincidences" that are all events registered by the coincidence processor. "Delayed coincidences" are either directly subtracted from the prompt coincidence to remove the random coincidences in the data or used in statistical image reconstruction algorithms as an "estimate" of the random coincidences contribution to the measured events. In either case, the random coincidences need to be removed or compensated for.

In general, one can obtain higher counting statistics by injecting more radiotracer into a subject to be imaged. High counting statistics is critical to good image quality for nuclear imaging. However, excessive amount of radioactivity inside or near the imaging FOV will lead to rapid increase of random coincidences and may degrade the image quality. This is because the true coincidence rate increases linearly as one increases the amount of radioactivity employed, while the random coincidence rate increases quadratically because it is the product of singles event rates of two detectors in coincidence (as shown in Eq. 2.1). As a result, "inject more activity" does not always help the image quality of PET. For small animal imaging applications using mice, one may only need to inject radiotracers with activity as little as 1.85 MBq (50 uCi) if the targeting is highly specific. For many highly sensitive animal PET systems, 74 MBq (2 mCi) of radiotracer in a mouse or a rat is considered very high dose and should be avoided. Since every system is different, one needs to learn the count rate handling capability of a scanner and limit the total amount of radioactivity employed for PET imaging studies. This is often evaluated by the so-called noise equivalent count (NEC) rate [54] capability following the procedures defined by the NEMA standards that evaluate the performance of clinical and animal PET scanners [55, 56].

2.5.2 Attenuation Correction

For emission tomography, the signal is attenuated as gamma rays escape the object to reach the external detectors. The amount of attenuation generally depends on the dimension and the composition of the object. Figure 2.12 illustrates how the tissues in an object may attenuate the signal (annihilation gamma rays) in PET imaging. Assume that the radioactivity distribution function within the object is $A(x,y)$ and the linear attenuation coefficient function within the object is $\mu(x,y)$.

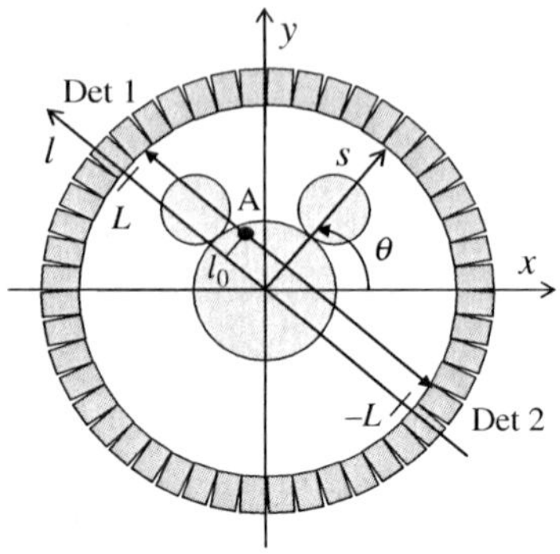

FIGURE 2.12 Attenuation of annihilation gamma rays in PET data acquisition.

For gamma rays traveling along a LOR between a pair of detectors and originated from the location "*A*" in Figure 2.12, the probability that the gamma ray will reach detector 1 is

$$\exp\left\{-\int_{l_0}^{L}\mu\left(x(l'),y(l')dl'\right)\right\}$$

That is, the probability depends on the integral of the linear attenuation coefficient function along the path from the origin of the gamma ray to the detector. Similarly, the probability that the second gamma ray travels along the opposite direction and reaches detector 2 is

$$\exp\left\{-\int_{-L}^{l_0}\mu(x(l'),y(l'))dl'\right\}$$

In order for this event to be detected as a valid coincidence event, both gamma rays have to be detected, and thus the probability is

$$\exp\left\{-\int_{l_0}^{L}\mu(x(l'),y(l'))dl'\right\}\cdot\exp\left\{-\int_{-L}^{l_0}\mu(x(l'),y(l'))dl'\right\}=\exp\left\{-\int_{-L}^{L}\mu(x(l'),y(l'))dl'\right\} \tag{2.12}$$

That is, the probability of a pair of gamma rays to escape the object and reach the detectors depends on the total length of the LOR through the object, as well as the linear attenuation coefficient function within the object, but independent of the depth of the source location along the LOR.

This property is unique to PET and is advantageous when it comes to correcting for the attenuation of the signal. One may use an external gamma ray source that emits 511 keV gamma rays to measure the attenuation along each LOR and use this correction factor for attenuation correction of the imaging data [57–63]. Alternatively, if one knows the linear attenuation coefficient distribution function of the tissues within the object, one can integrate the linear attenuation coefficient function along each LOR to estimate and correct for the attenuation of gamma ray signal [64–66]. The first approach factors in the actual scatter and attenuation of gamma rays by tissues and is generally the most accurate way to correct for attenuation of PET data. Unfortunately, this method is time consuming because one cannot use a very strong gamma ray source to speed up the data acquisition. As a result, it is not widely used, especially for clinical PET imaging. The second approach is commonly implemented using PET/CT scanners [67, 68]. CT images represent the linear attenuation coefficients function of the object at the corresponding X-ray energy (typically 80–100 keV on average). One can map the linear attenuation coefficients at CT energy to their corresponding values at 511 keV and then perform the integration in Equation 2.12 to estimate and correct for PET data.

In contrast to PET, attenuation correction of SPECT data is less straightforward. This is because the amount of attenuation that a gamma ray is subject to as it travels through an object is depth dependent. As a result, there are two spatially variant functions (radioactivity distribution and linear attenuation coefficient function) of the object affecting the measured events and their effects cannot be separated. For modern SPECT scanners that are equipped with CT (i.e., SPECT/CT scanners), the linear attenuation coefficient function can be estimated from CT images and subsequently used in the image reconstruction to compensate for the attenuation of the gamma rays through an iterative process [69–71]. As a result, SPECT images can be made quantitative using modern SPECT/CT scanners and sophisticated reconstruction algorithms.

2.5.3 Scatter Rejection and Correction

Energy discrimination is a technique commonly used in nuclear imaging system to reject scatter [11]. For a given radionuclide used for radiolabeling the tracer, the emitted gamma ray energy is known. As a result, one can choose the lower-level discriminator (LLD) and upper-level discriminator (ULD) to define an energy window (usually centered at the energy of the gamma rays emitted by the specific radionuclide). When an event is detected, it is accepted as a valid event only if its energy (or signal amplitude) falls within the selected energy window. Otherwise, the event is likely to have undergone Compton scatter in the object and lost part of its energy. In this case, the direction of the gamma ray is deviated from its original path. Such events are commonly rejected because they will lead to mispositioning of the origin of the gamma ray sources.

For SPECT cameras that employ NaI(Tl) detectors, the intrinsic energy resolution at 140 keV is typically <10% FWHM. As a result, an energy window of ±10–15% around the photopeak is common. Some radionuclides used for SPECT imaging emit gamma rays of different energies. For example, ^{111}In emits gamma rays of 171 and 245 keV. For this type of imaging studies, the system can be set up to employ two energy windows (e.g., windows 1 and 2 in Fig. 2.13 defined by the solid and dash lines, respectively), one centered at 171 keV and the other centered at 245 keV. Events that fall within either one of the two energy windows are considered valid events and used for image formation and reconstruction. The higher-energy gamma rays may be scattered by the object to lose part of their energy and end up with an

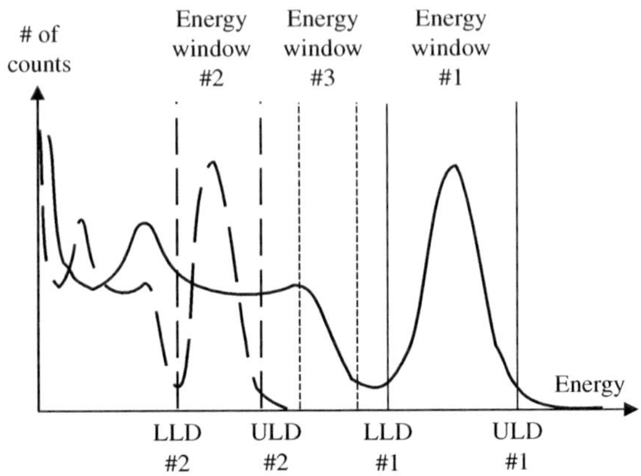

FIGURE 2.13 Scatter rejection using energy discrimination and windowing.

energy that falls within the lower-energy window. As a result, the accepted events in the lower-energy window contain both unscattered lower-energy gamma rays (valid events) and scattered higher-energy gamma rays (undesirable events, or "contamination"). The contamination can be estimated by multiplying the number of gamma rays accepted in yet another energy window (e.g., window no. 3 in Fig. 2.13 defined by dotted lines) whose LLD is above the ULD of the lower-energy window (window no. 2 of Fig. 2.13). All events in the energy window no. 3 are the results of scattered high-energy gamma rays. Therefore, the total number of events in window no. 3 multiplied by a precalibrated constant will be a reasonable estimate of the contamination in window no. 2. This allows one to correct for scatter contamination in tracers that emit multiple energy gamma rays.

For PET imaging, both annihilation gamma rays must fall within the energy window centered at 511 keV. Since PET detectors typically have an energy resolution between 10% and 20% FWHM, the energy window cannot be set too narrow. Otherwise, there is a good chance that unscattered gamma rays may be rejected with tight energy window. Furthermore, 511 keV gamma rays have a fairly high probability to interact with PET detectors through Compton scatter and only deposit partial energy. This is particularly true for high-resolution animal PET scanners that employ minute scintillation crystals that may not completely stop the secondary gamma rays resulted from Compton scatter of 511 keV gamma rays in a PET detector. As a result, energy window may range from ±10 to ±30% around the photopeak (511 keV), depending on the type of imaging studies. For clinical whole-body PET imaging, since a patient body may produce a lot of scatter coincidences, a tighter energy window is used to minimize the impact of Compton scatter. For animal PET studies, particularly for mouse imaging where scatter from the subject is minimal, the energy window may be widened to maximize the sensitivity of the system.

To further limit the impact of scatter to PET and SPECT images, scatter contamination can be estimated and corrected for during the image reconstruction process. It often requires knowledge of tissue density function and estimated activity distribution function within the object [72–78]. As a result, the scatter correction is often an iterative process during the reconstruction.

2.5.4 Normalization

Normalization corrects for the nonuniform response of detectors in a PET scanner [79, 80]. This nonuniformity may be caused by systematic factors such as radial offset of the detector LOR from the center of FOV, as well as the block detector design that leads to higher detection efficiency at the center of the block when compared to the edges and corners of a block. The nonuniformity may also be caused by drifting of electronics and deterioration of detectors over time. Therefore, it is critical to calibrate the detection efficiency of all LOR on a regular basis and normalize the emission data before or during the image reconstruction. This calibration often consists of two steps [81–83]. The first one is to measure the systemic error due to the geometry and block detectors. This is often done at the factory and the results saved as a "system file" that can be incorporated into routine normalization procedure. The second part of the normalization requires users to scan a uniform phantom that is positioned at the center of the FOV to gather coincidence events. This data is used to estimate the detection efficiency of all detectors and to establish a normalization file for properly normalizing data from regular imaging studies. To ensure the quantitative accuracy of data, a daily quality control procedure is often required. One would image the uniform phantom used for normalization procedure and compare the resulted images to that acquired right after the previous system calibration/normalization. If the mean and standard deviation of the images of the uniform phantom have not changed significantly, then the system is considered stable and there is not a need to perform a new normalization scan. Otherwise, the detector response and/or electronics performance may have drifted, and the system is due for a new normalization and calibration. Rigorous quality control routine of the imaging system is essential for achieving high level of quantitative accuracy that is necessary for characterizing pharmacokinetics and pharmacodynamics.

Similarly, a SPECT system needs to be calibrated in order to correct for the nonuniformity and nonlinearity of the gamma camera. Typically, a weak uniform radioactive source is placed in front of a camera head without a collimator to obtain a flood image that can be used to correct for the intrinsic nonuniformity of the camera. For each type of collimator, the camera also needs to be calibrated with the collimator attached in order to establish the uniformity correction file that factors in the nonuniform response caused by the collimator. If the nonuniformity of a gamma camera is not adequately corrected, the same high (or low) efficiency pattern of the detector system will appear in every projection angle as the camera is rotated around an object to acquire projection data for SPECT images. This consistent high/low efficiency pattern results in ring artifact in the reconstructed images.

2.5.5 Others

There are additional correction techniques that are necessary for PET and SPECT systems but often operate behind the scene and do not require users interaction. These include (i) baseline correction [84] for detector signals that may drift when event rate is high, (ii) dead-time correction to compensate for loss of events under high count rate condition [85–87], and (iii) decay correction to compensate for natural decay of the signal (radioactive decay) at different time points.

For human SPECT systems that typically have one or more large camera heads, center of rotation calibration is critical and needs to be checked daily. For small animal SPECT systems that have all detector panels mounted to the same gantry, the relative location of each detector is fixed and does not change over time. This type of calibration may only need to be done once in the factory to establish a correction setting and may not involve user intervention.

While PET and SPECT imaging techniques are inherently quantitative, rigorous quality control routine of the imaging system is essential for achieving high level of quantitative accuracy. This is particularly important if the systems are to be used to characterize pharmaceutical kinetics and pharmaceutical dynamics of new drugs.

2.6 APPLICATION NOTES

2.6.1 Common Types of Imaging Studies

Static, dynamic, whole-body, and gated imaging studies are the four most common types of studies employed for PET and SPECT systems [6, 52]. Static imaging study refers to the imaging of an object with a "snapshot" in time and space. That is, the imaging FOV is often a preselected region of the object, and the imaging time is also a preselected time point of the entire dynamic process. This is often used to see one target organ after the radiotracer reaches a steady state in the biological system. Alternatively, one may take a snapshot of the radiotracer distribution in an object at several time points if the biological process is slow and it is not feasible to image the object throughout the entire course of the process.

A dynamic imaging study refers to the imaging of an object with a snapshot in space but continuously in time. That is, one may select and place a region of interest in the imaging FOV and collect data continuously throughout the entire biological process. Data are often sorted into multiple frames of images. Playback of the image frames reveals the dynamics and kinetics of the radiotracer distribution within the imaging FOV. This is often used when the biological process under observation has a relatively short biological half-life and therefore requires the images to be collected continuously for a short period of time (an hour or so). For biological processes with longer halflives, a dynamic imaging study initially followed by multiple static imaging studies is often employed.

Wholebody imaging study refers to the imaging of an object with a snapshot in time but continuously in space. This is often done when the object is larger than the imaging FOV of the system and requires a step-and-shoot motion in order to cover the entire body in multiple steps. This is the most widely used protocol for clinical PET studies because the goal is to detect metastases for cancer staging. For research protocols, whole-body imaging is often used when the radiotracer distribution in all major organs throughout the body is needed. For example, one needs to measure the accumulation of the radiotracer in order to calculate radiation dose to individual organs.

Gated imaging study refers to the imaging of an object with a snapshot in space at different phases of a repeated cycle. For example, one may image different phases of cardiac cycle of an object by following the heartbeat of the object. An ECG signal is used as a trigger to PET or a SPECT system. Tag words are inserted into the data stream to indicate when each event occurs relatively to the cardiac cycle. Valid events are later on sorting into multiple subgroups to reveal the distribution of the radiotracer at different phases of the cardiac cycle. This type of study is often employed when the desired information may be blurred by the involuntary motion such as cardiac or pulmonary motions. While gated imaging study can reduce the blurring and motion artifact, it also divides valid events into multiple subgroups and thus reduces the counting statistics of individual gates of images.

The aforementioned four types of studies are not mutually exclusive of each other. Instead, they may be combined into one study. For example, if the scanner's imaging FOV is sufficient to cover the entire body of the subject (e.g., a mouse), dynamic whole-body imaging becomes possible and is very useful in establishing pharmacodynamics and pharmacokinetics of new drugs. Most PET scanners employ ring geometry that collects coincidence events from all angles simultaneously. Therefore, it is inherently capable of dynamic imaging study. In contrast, some SPECT systems require rotation of the camera to collect data from different angles for tomographic image reconstruction. As a result, the images represent measurement of the radiotracer distribution at selective time points. Some other SPECT systems employ multiple detector heads (or cameras) with multiple pinholes to collect projection image of the source distribution from multiple angles simultaneously [88–90]. This permits one to obtain tomographic images without any rotation and thus enables dynamic SPECT imaging. One potential limitation of such systems is that it often limits the imaging FOV in order to avoid excessive overlap of the projection measurement by the multiple pinhole camera. As a result, the dynamic images are limited to a small region instead of the entire body of a subject. This is a common challenge for any imaging system when the subject is bigger than the imaging FOV, and it will require careful experimental design in order to gather the necessary kinetics for studying new drugs.

2.6.2 Multiple-Tracer Imaging Capability

For SPECT imaging applications, radionuclides that emit gamma rays with distinct energy levels may be mixed and combined in the same study, as long as the scanner can resolve and identify individual gamma sources [91–94]. For example, one can label a small molecule that flows freely with blood using one radionuclide that emits gamma rays with energy E_1 and label another molecule that targets a specific type of receptor with a second radionuclide that emits gamma rays of energy E_2 [95]. When both radiolabeled molecules are injected, one can obtain two images, one centered on each energy window (E_1 and E_2). The first image reveals blood flow to the targeted area, while the second image reveals the receptor binding. Care needs to be taken to minimize or correct for the contamination of the scattered higher-energy gamma rays into the lower-energy window, similar to the issues illustrated in Figure 2.13 [96]. This dual- or multiple-tracer imaging capability of SPECT can potentially be very useful in studying how different molecules interact with the biological system simultaneously.

In contrast, dual- or multiple-tracer imaging capability for PET is significantly more difficult to implement because all PET radionuclides emit positrons that are converted to 511 keV gamma rays as signals. Therefore, the signal from different

PET radionuclides is not differentiable based on energy. Several groups are developing algorithms that can differentiate different PET tracers based on their kinetics [97, 98]. The accuracy of such techniques remains to be seen.

2.6.3 Multimodality Imaging

Multimodality imaging systems such as PET/CT and SPECT/CT are becoming increasingly popular because they provide both anatomical and functional information that complement each other [69, 99]. In addition to the anatomy, CT images also provide the linear attenuation coefficients of the tissues that can be used to correct for attenuation of gamma rays in PET and SPECT imaging. This allows one to reduce the scan duration for PET and SPECT imaging by eliminating transmission scans that use external radioactive source to measure tissue density. With adequate attenuation correction, PET and SPECT images can be made quantitative for establishing biological models and measuring pharmacokinetics. In recent years, MRI/PET and MRI/SPECT are also becoming possible despite their technological challenge [100–107]. Aside from the significant reduction of radiation dose to the object, MRI provides much better soft tissue contrast than CT and is particularly useful for imaging brain, breasts, and abdomen where density between different soft tissue types is similar [108]. In addition to the superb anatomical images from MRI, MR spectroscopy also provides chemical information that is not easily available from nuclear imaging techniques. As a result, MRI/PET and MRI/SPECT have the potential to further enhance the *in vivo* molecular imaging capability for studying biological processes and for accelerating drug development.

Another new trend in multimodality imaging technology is the combination of multiple "functional imaging" techniques such as PET and optical imaging [109, 110]. When multiple modalities that can each probe one or more types of biological processes are integrated, the combined system can be used to probe multiple processes simultaneously. As a result, an optical/PET or optical/SPECT system may lead to *in vivo* measurement of complex biological processes that involve the interplay of multiple components that need to be measured independently and simultaneously.

As multimodality imaging becomes the new standard in *in vivo* molecular imaging research, it is even more critical for users to understand the strengths and limitation of individual technologies and how to best combine them in study design. This chapter provides the fundamentals of PET and SPECT technologies. More complete information can be found in several of the excellent review papers listed in the following reference section.

REFERENCES

1. Huang SC, Carson RE, Hoffman EJ, et al. Quantitative measurement of local cerebral blood flow in humans by positron computed tomography and 15O-water. J Cereb Blood Flow Metab 1983;3:141–153.

2. Allport JR, Weissleder R. In vivo imaging of gene and cell therapies. Exp Hematol 2001;29:1237–1246.

3. Dubey P, Su H, Adonai N, et al. Quantitative imaging of the T cell antitumor response by positron-emission tomography. Proc Natl Acad Sci USA 2003;100:1232–1237.
4. Shah K, Jacobs A, Breakefield XO, et al. Molecular imaging of gene therapy for cancer. Gene Ther 2004;11:1175–1187.
5. Firestone RB, Shirley VS. *Table of Isotopes*. New York: John Wiley & Sons, Inc; 1996, pp. 3168.
6. Cherry SR, Sorenson JA, Phelps ME. *Physics in Nuclear Medicine*. 3rd ed. Philadelphia: Saunders; 2003.
7. Kak AC, Slaney M, IEEE/Engineering in Medicine and Biology Society. *Principles of Computerized Tomographic Imaging*. New York: IEEE Press; 1988.
8. Zeng G. *Medical Image Reconstruction: A Conceptual Tutorial*. 1st ed. Beijing: Springer; 2010.
9. Hoffman EJ, Huang S-C, Phelps ME, et al. Quantitation in positron emission computed tomography. IV. Effect of accidental coincidences. J Comput Assist Tomogr 1981;5:391–400.
10. Wernick MN, Aarsvold JN. *Emission Tomography the Fundamentals of PET and SPECT*. San Diego (CA): Elsevier Academic Press: 2004, xix, 576 pp.
11. Knoll GF. *Radiation Detection and Measurement*. 2nd ed. New York: Wiley; 1989.
12. Jeavons AP, Chandler RA, Dettmar CAR. A 3D HIDAC-PET camera with sub-millimetre resolution for imaging small animals. IEEE Trans Nucl Sci 1999;46:468–473.
13. Eisen Y, Shor A, Mardor I. CdTe and CdZnTe X-ray and gamma-ray detectors for imaging systems. IEEE Trans Nucl Sci 2004;51 (3 Pt 3):1191–1198.
14. Yin Y, Chen X, Wu H, et al. 3D spatial resolution of 350 m pitch pixelated CdZnTe detectors for imaging applications. IEEE Trans Nucl Sci 2013;60 (1):915.
15. Kim H, Furenlid LR, Crawford MJ, et al. SemiSPECT: a small-animal single-photon emission computed tomography (SPECT) imager based on eight cadmium zinc telluride (CZT) detector arrays. Med Phys 2006;33:465–474.
16. Mettivier G, Russo P, Cesarelli M, et al. Dedicated scanner for laboratory investigations on cone-beam CT/SPECT imaging of the breast. Nuclear Instruments & Methods in Physics Research Section a-Accelerators Spectrometers Detectors and Associated Equipment 2011;629:350–356.
17. Ogawa K, Muraishi M. Feasibility study on an ultra-high-resolution SPECT With CdTe detectors. IEEE Trans Nucl Sci 2010;57:17–24.
18. Cai L, Fu G, Tan JW, et al. A very-high resolution SPECT system based on the energy-resolved photon counting CdTe detectors. Nuclear Science Symposium and Medical Imaging Conference (NSS/MIC); IEEE 2010 Oct 30–Nov. 6; Knoxville, TN: 2788–2792.
19. Peng H, Levin CS. Design study of a high-resolution breast-dedicated PET system built from cadmium zinc telluride detectors. Phys Med Biol 2010;55:2761–2788.
20. Buzhan P, Dolgoshein B, Filatov L, et al. Silicon photomultiplier and its possible applications. Nucl Instrum Methods [A] 2003;504:48–52.
21. Georgievskya EA, Klemin SN, Filatov LA, et al. The solid state silicon photomultiplier for a wide range of applications. Proc. SPIE – Int Soc Opt Eng (USA) 2003;5126:37–42.
22. McClish M, Farrell R, Myers R, et al. Recent advances of planar silicon APD technology. Nucl Instrum Methods [A] 2006;567:36–40.
23. Pichler BJ, Boening G, Rafecas M, et al. Development of a compact PET detector module for DOI measurement based on LSO, monolithic APD arrays, and integrated preamplifiers. J Nucl Med 2001;42:102–102.

24. Shao Y, Meadors K, Silverman RW, et al. Dual APD array readout of LSO crystals: optimization of crystal surface treatment. IEEE Trans Nucl Sci 2002;49:649–654.
25. Shah KS, Grazioso R, Farrell R, et al. Position sensitive APDs for small animal PET imaging. IEEE Trans Nucl Sci 2004;51:91–95.
26. Song TY, Wu H, Komarov S, et al. A submillimeter resolution PET detector module using a multi-pixel photon counter array. Phys Med Biol 2010;55:2573–2587.
27. Casey ME, Nutt R. A multicrystal two dimensional BGO detector system for positron emission tomography. IEEE Trans Nucl Sci 1986;33:460–463.
28. Lecomte R, Martel C, Cadorette J. Study of the resolution performance of an array of discrete detectors with independent readouts for positron emission tomography. IEEE Trans Med Imaging 1991;10:347–357.
29. Siegel S, Silverman RW, Yiping S, et al. Simple charge division readouts for imaging scintillator arrays using a multi-channel PMT. IEEE Trans Nucl Sci 1996;43: 1634–1641.
30. Anger HO. Scintillation camera. Rev Sci Instr 1958;29:27–33.
31. Moses WW, Derenzo SE. Empirical observation of resolution degradation in positron emission tomographs utilizing block detectors. J Nucl Med 1993;34:101.
32. Tai YC, Wu H, Pal D, et al. Virtual-pinhole PET. J Nucl Med 2008;49:471–479.
33. Levin CS, Hoffman EJ. Calculation of positron range and its effect on the fundamental limit of positron emission tomography system spatial resolution. Phys Med Biol 1999;44:781–799.
34. DeBenedetti S, Cowan CE, Konneker WR, et al. On the angular distribution of two-photon annihilation radiation. Phys Rev X 1950;77:205–212.
35. Tomic N, Thompson CJ, Casey ME. Investigation of the "block effect" on spatial resolution in PET detectors. IEEE Trans Nucl Sci 2005;52:599–605.
36. Moisan C, Rogers JG, Buckley KR, et al. Design studies of a depth encoding large-aperture pet camera. IEEE Trans Nucl Sci 1995;42:1041–1050.
37. Tai Y-C, Laforest R. Instrumentation aspects of animal PET. Annu Rev Biomed Eng 2005;7:255–285.
38. Raylman RR. Positron emission tomography-guided biopsy with a dedicated breast scanner: initial evaluation. IEEE Trans Nucl Sci 2009;56:620–624.
39. Weinberg IN, Beylin D, Zavarzin V, et al. Positron emission mammography: high-resolution biochemical breast imaging. Technol Cancer Res Treat 2005;4:55–60.
40. Thompson CJ, Murthy K, Picard Y, et al. Positron emission mammography (PEM): a promising technique for detecting breast cancer. IEEE Trans Nucl Sci 1995;42:1012–1017.
41. Eriksson L, Wienhard K, Eriksson M, et al. The ECAT HRRT: NEMA NEC evaluation of the HRRT system, the new high-resolution research tomograph. IEEE Trans Nucl Sci 200249:2085–2088.
42. Huh SS, Clinthorne NH, Rogers WL. Investigation of an internal PET probe for prostate imaging. Nucl Instrum Methods [A] 2007;579:339–343.
43. Huber JS, Derenzo SE, Qi J, et al. Conceptual design of a compact positron tomograph for prostate imaging. IEEE Trans Nucl Sci 2001;48:1506–1511.
44. Komarov SA, Heyu W, Keesing DB, et al. Compton scattering in clinical PET/CT With high resolution half ring PET insert device. IEEE Trans Nucl Sci 2010;57:1045–1051.
45. Wu H, Pal D, Song TY, et al. Micro insert: a prototype full-ring PET Device for improving the image resolution of an animal PET scanner. J Nucl Med 2008;49:1668–1676.

46. Wu H, Pal D, O'Sullivan JA, et al. A feasibility study of a prototype PET insert device to convert a general purpose animal PET scanner to higher resolution. J Nucl Med 2008;49:79–87.
47. Pal D, O'Sullivan JA, Wu HY, et al. 2D linear and iterative reconstruction algorithms for a PET-insert scanner. Phys Med Biol 2007;52:4293–4310.
48. Tai YC, Wu HY, Janecek M. Initial study of an asymmetric PET system dedicated to breast cancer imaging. IEEE Trans Nucl Sci 2006;53:121–126.
49. Janecek M, Wu H, Tai YC. A simulation study for the design of a prototype insert for whole-body PET scanners. IEEE Trans Nucl Sci 2006;53:1143–1149.
50. Chatziioannou AF. PET scanners dedicated to molecular imaging of small animal models. Mol Imaging Biol 2002;4:47–63.
51. Larobina M, Brunetti A, Salvatore M. Small animal PET: a review of commercially available imaging systems. Curr Med Imaging Rev 2006;2:187–192.
52. Prince JL, Jonathan LM. *Medical Imaging Signals and Systems*. Upper Saddle River (NJ): Pearson Prentice Hall; 2006.
53. Rowland DJ, Cherry SR. Small-animal preclinical nuclear medicine instrumentation and methodology. Semin Nucl Med 2008;38:209–222.
54. Strother SC, Casey ME, Hoffman EJ. Measuring PET scanner sensitivity: relating countrates to image signal-to-noise ratios using noise equivalent counts. IEEE Trans Nucl Sci 1990;37:783–788.
55. NEMA, NEMA NU 2. *Performance Measurements of Positron Emission Tomographs*. Rosslyn (VA): National Electrical Manufacturers Association; 2001.
56. NEMA, NEMA NU 4. *Performance Measurements of Small Animal Positron Emission Tomographs*. Rosslyn (VA): National Electrical Manufacturers Association; 2008.
57. Huang SC, Hoffman EJ, Phelps ME, et al. Quantitation in positron emission computed tomography: 2. Effects of inaccurate attenuation correction. J Comput Assist Tomogr 1979;3:804–814.
58. Phelps ME, Hoffman EJ, Mullani NA, et al. Application of annihilation coincidence detection to transaxial reconstruction tomography. J Nucl Med 1975;16:210–224.
59. Dahlbom M, Hoffman EJ. Problems in signal-to-noise ratio for attenuation correction in high resolution PET. IEEE Trans Nucl Sci 1987;34:288–293.
60. Ostertag H, Kubler WK, Doll J, et al. Measured attenuation correction methods. Eur J Nucl Med 1989;15:722–726.
61. Ranger NT, Thompson CJ, Evans AC. The application of a masked orbiting transmission source for attenuation correction in PET. J Nucl Med 1989;30:1056–1068.
62. Meikle SR, Dahlbom M, Cherry SR. Attenuation correction using count-limited transmission data in positron emission tomography. J Nucl Med 1993;34:143–150.
63. Chatziioannou A, Dahlbom M, Hoh CK. Study on the use of transmission scans for whole body PET attenuation correction. IEEE Trans Nucl Sci 1994;41:1545–1550.
64. Huang SC, Carson RE, Phelps ME, et al. A boundary method for attenuation correction in positron computed tomography. J Nucl Med 1981;22:627–637.
65. Bergstrom M, Litton J, Eriksson L, et al. Determination of object contour from projections for attenuation correction in cranial positron emission tomography. J Comput Assist Tomogr 1982;6:365–372.
66. Tomitani T. An edge detection algorithm for attenuation correction in emission CT. IEEE Trans Nucl Sci 1987;34:309–312.
67. Beyer T, Kinahan P, Townsend D, et al. Attenuation correction for a combined 3D PET/CT scanner. Proceedings of the 1995 International Meeting on Fully Three-Dimensional Image Reconstruction in Radiology and Nuclear Medicine 1995:55–59.

68. Beyer T, Kinahan PE, Townsend DW, et al. The use of Xray CT for attenuation correction of PET data. Nuclear Science Symposium and Medical Imaging Conference. 1994 IEEE Conference Record (Cat. No.94CH35762) 1995;4:1573–1577.
69. Hasegawa B, Tang HR, Da Silva AJ, et al. Dual-modality imaging. Nucl Instrum Methods [A] 2001;471:140–144.
70. Kinahan PE, Hasegawa BH, Beyer T. X-ray-based attenuation correction for positron emission tomography/computed tomography scanners. Semin Nucl Med 2003;33: 166–179.
71. Hwang AB, Hasegawa BH. Attenuation correction for small animal SPECT imaging using x-ray CT data. Med Phys 2005;32:2799–2804.
72. Beekman FJ, den Harder JM, Viergever MA, et al. SPECT scatter modelling in non-uniform attenuating objects. Phys Med Biol 1997;42:1133–1142.
73. Beekman FJ, Kamphuis C, Frey EC. Scatter compensation methods in 3D iterative SPECT reconstruction: a simulation study. Phys Med Biol 1997;42:1619–1632.
74. Beekman FJ, de Jong HW, Slijpen ET. Efficient SPECT scatter calculation in non-uniform media using correlated Monte Carlo simulation. Phys Med Biol 1999;44:N183–N192.
75. Cherry SR, Meikle SR, Hoffman EJ. Correction and characterization of scattered events in three-dimensional PET. J Nucl Med 1993;34:671–678.
76. Bentourkia M, Msaki P, Cadorette J, et al. Object and detector scatter function dependence on energy and position in high-resolution pet. IEEE Trans Nucl Sci 1995;42:1162–1167.
77. Watson CC. New, faster, image-based scatter correction for 3D PET. IEEE Trans Nucl Sci 2000;47:1587–1594.
78. Watson CC, Newport D, Casey ME. A single scatter simulation technique for scatter correction in 3D PET. Proceedings of the 1995 International Meeting on Fully Three-Dimensional Image Reconstruction in Radiology and Nuclear Medicine 1995:215–219.
79. Hoffman EJ, Guerrero TM, Germano G, et al. PET system calibrations and corrections for quantitative and spatially accurate images. IEEE Trans Nucl Sci 1989;36:1108–1112.
80. Cutler PD, Hoffman EJ. Use of digital front-end electronics for optimization of a modular PET detector. IEEE Trans Med Imaging 1994;13:408–418.
81. Casey ME, Gadagkar H, Newport D. A component based method for normalization in volume PET. Proceedings of the 1995 International Meeting on Fully Three-Dimensional Image Reconstruction in Radiology and Nuclear Medicine 1995:67–71.
82. Badawi RD, Marsden PK. Developments in component-based normalization for 3D PET. Phys Med Biol 1999;44:571–594.
83. Bai B, Li Q, Holdsworth CH, et al. Model-based normalization for iterative 3D PET image reconstruction. Phys Med Biol 2002;47:2773–2784.
84. Binkley DM, Puckett BS, Swann BK, et al. A 10-Mc/s, 0.5-mu m CMOS constant-fraction discriminator having built-in pulse tail cancellation. IEEE Trans Nucl Sci 2002;49:1130–1140.
85. Eriksson L, Wienhard K, Dahlbom M. A simple data loss model for positron camera systems. IEEE Trans.Nucl Sci 1994;41:1566–1570.
86. Thompson CJ, Meyer E. The effect of live time in components of a positron tomograph on image quantification. IEEE Trans Nucl Sci 1987;34 (1):337–343.
87. Tai YC, Chatziioannou A, Dahlbom M, et al. Evaluation of a deadtime correction technique for simultaneous emission-transmission data acquisition in PET. J Nucl Med 1997;38:424–424.

88. Beekman FJ, Vastenhouw B. Design and simulation of a high-resolution stationary SPECT system for small animals. Phys Med Biol 2004;49:4579–4592.
89. Beekman FJ, van der Have F, Vastenhouw B, et al. U-SPECT-I: a novel system for submillimeter-resolution tomography with radiolabeled molecules in mice. J Nucl Med 2005;46:1194–1200.
90. Funk T, Despres P, Barber WC, et al. A multipinhole small animal SPECT system with submillimeter spatial resolution. Med Phys 2006;33:1259–1268.
91. Johnson LL, Seldin DW, Keller AM, et al. Dual isotope thallium and indium antimyosin spect imaging to identify acute infarct patients at further ischemic risk. Circulation 1990;81:37–45.
92. Neumann DR. Simultaneous dual-isotope spect imaging for the detection and characterization of parathyroid pathology. J Nucl Med 1992;33:131–134.
93. Zhang JJ, Park CH, Kim SK, et al. Dual isotope spect in the evaluation of recurrent brain-tumor. Clin Nucl Med 1992;17:663–664.
94. Schwartz RB, Carvalho PA, Alexander E, et al. Radiation necrosis vs high-grade recurrent glioma – differentiation by using dual-isotope spect with Tl-201 and Tc-99m-Hmpao. Am J Neuroradiol 1991;12:1187–1192.
95. Van Laere K, Casteels C, De Ceuninck L, et al. Dual-tracer dopamine transporter and perfusion SPECT in differential diagnosis of parkinsonism using template-based discriminant analysis. J Nucl Med 2006;47:384–392.
96. Tianyu M, Yiping S, Rutao Y, et al. PET and SPECT dual-tracer imaging: correction of 511 keV photon's down-scatter effect on SPECT. Nuclear Science Symposium Conference Record; IEEE 2007Oct. 26–Nov. 3; Honolulu, HI: 3404–3407.
97. Figueiras FP, Jimenez X, Pareto D, et al. Simultaneous dual-tracer PET imaging of the rat brain and its application in the study of cerebral ischemia. Mol Imaging Biol 2011;13:500–510.
98. Andreyev A, Celler A. Feasibility study of dual isotope PET. Nuclear Science Symposium and Medical Imaging Conference (NSS/MIC); IEEE 2010 Oct. 30-Nov. 6; Knoxville, TN: 2108–2111.
99. Beyer T, Townsend DW, Brun T, et al. A combined PET/CT scanner for clinical oncology. J Nucl Med 2000;41:1369–1379.
100. Shao Y, Cherry SR, Farahani K, et al. Simultaneous PET and MR imaging. Phys Med Biol 1997;42:1965–1970.
101. Yibao W, Catana C, Farrell R, et al. PET performance evaluation of an MR-compatible PET insert. IEEE Trans Nucl Sci 2009;56:574–580.
102. Schlemmer HP, Pichler BJP, Schmand M, et al. Simultaneous MR/PET imaging of the human brain: feasibility study. Radiology 2008;248:1028–1035.
103. Judenhofer MS, Catana C, Swann BK, Siegel SB, Jung WI, et al. PET/MR images acquired with a compact MR-compatible PET detector in a 7-T magnet. Radiology 2007;244:807–814.
104. Catana C, Wu Y, Judenhofer MS, et al. Simultaneous acquisition of multislice PET and MR images: initial results with a MR-compatible PET scanner. J Nucl Med 2006;47:1968–1976.
105. Schlemmer HP, Pichler BJ, Krieg R, Heiss WD. An integrated MR/PET system: prospective applications. Abdom Imaging 2009;34:668–674.
106. Cai JWTL, Cai L, Meng LJ. Experimental study of the response of 1–5 mm thick CdTe/CZT detectors inside strong magnetic field. Nuclear Science Symposium and Medical Imaging Conference (NSS/MIC); IEEE 2010 Oct. 30–Nov. 6; Knoxville, TN: 3403–3409.

107. Hugg J, Wagenaar D, Meier D, Benjamin MW, Nalcioglu O, Patt B. Simultaneous SPECT/MRI feasibility demonstration. J Nucl Med 2010;51:14.

108. Wehrl HF, Sauter AW, Judenhofer MS, Pichler BJ. Combined PET/MR imaging – technology and applications. Technol Cancer Res Treat 2010;9:5–20.

109. Culver J, Akers W, Achilefu S. Multimodality molecular imaging with combined optical and SPECT/PET modalities. J Nucl Med 2008;49:169–172.

110. Li C, Wang G, Qi J, et al. Three-dimensional fluorescence optical tomography in small-animal imaging using simultaneous positron-emission-tomography priors. Opt Lett 2009;34:2933–2935.

CHAPTER 3

PHOTOACOUSTIC TOMOGRAPHY AND ITS APPLICATIONS IN DRUG DELIVERY AND PHOTOTHERMAL THERAPY

LIANG SONG AND LIHONG V. WANG

3.1 INTRODUCTION TO PHOTOACOUSTIC TOMOGRAPHY (PAT)

Photoacoustic (or optoacoustic) tomography (PAT) is a novel hybrid imaging modality based on the photoacoustic effect, in which acoustic waves are generated upon absorption of electromagnetic waves (photons). Although Alexander Graham Bell discovered this effect as far back as 1880, it was not until the 1990s—with technological advances in lasers, computers, and computed tomography (CT)—that its biomedical applications started to become apparent [1–3]. In 2003, noninvasive *in vivo* functional PAT of the cerebral hemodynamics of a rat (with skin and skull intact) in response to whisker stimulations was reported [4], which initiated an exponential growth of the field of PAT; in 2006, *in vivo* functional PAT of hemoglobin oxygenation and melanoma was reported [5]. By now, PAT has demonstrated a wide array of potential biomedical applications, including the imaging of tumor angiogenesis [6–9], tumor hypoxia [10], gene expression [11], hemoglobin oxygen saturation [5, 12], breast cancer [13], sentinel lymph nodes [14–17], the eye [18–21], amyloid plaques [22], and atherosclerosis [23–25]. In addition, many promising nanocarriers for drug delivery and photothermal therapy—such as gold nanostructures and carbon nanotubes—also provide excellent optical absorption contrast for PAT [26]. As a result, PAT is now being actively explored to guide and evaluate drug delivery and photothermal therapy.

3.1.1 Principles of PAT

When a short laser (or radiofrequency or microwave) pulse is absorbed by structures (e.g., microvessels) in biological tissue, wideband ultrasonic (photoacoustic) waves are produced from a thermal- and stress-confined transient thermoelastic expansion [27].

Drug Delivery Applications of Noninvasive Imaging: Validation from Biodistribution to Sites of Action, First Edition. Edited by Chun Li and Mei Tian.

The photoacoustic emissions are then detected and reconstructed to form an image, whose amplitude reflects primarily the optical absorption properties and the Grüneisen parameter of tissue.

In PAT (except in optical-resolution photoacoustic microscopy, which will be described in detail later), both the ballistic and diffuse photons are utilized to generate photoacoustic signals, and the spatial resolution is determined primarily by ultrasonic rather than optical parameters. Since ultrasonic scattering in biological tissue is two to three orders of magnitude weaker than optical scattering, high spatial resolution can be achieved at depths beyond the optical transport mean free path (~1 mm in skin).

In general, if the laser excitation is in both thermal and stress confinement, the initial local photoacoustic pressure rise p_0 can be described by [28]

$$p_0 = \Gamma \eta_{th} A = \Gamma \eta_{th} \mu_a F, \tag{3.1}$$

where Γ denotes the dimensionless Grüneisen parameter, A the specific optical energy deposition (J/cm^3), η_{th} the percentage of A converted into heat, μ_a the optical absorption coefficient (cm^{-1}), and F the optical fluence (J/cm^2). It can be seen that p_0 depends on both the optical energy deposition and the thermal and mechanical properties of the tissue. In PAT of biological tissue, η_{th} is usually unchanged, while Γ is a function of temperature but normally does not change significantly during imaging (exceptions include, e.g., when the tissue is under photothermal therapy). Consequently, in many cases, the initial pressure rise reflects essentially the optical energy deposition ($\mu_a \times F$). However, the fact that Γ is a function of temperature has also been utilized to image tissue temperature with PAT [29–31], which can be useful for monitoring photothermal therapy.

In biological tissue, the primary endogenous chromophores include oxy- and deoxyhemoglobin (HbO_2 and HbR) and melanin. The strong intrinsic absorption of HbO_2 and HbR in the visible spectrum enables blood vessels to be readily photoacoustically imaged noninvasively, with high contrast and without using ionizing radiation. Using multiwavelength measurements, changes in both the concentration and oxygen saturation of hemoglobin can be quantified. Consequently, functional photoacoustic imaging of endogenous hemoglobin contrast is available with both high sensitivity and high spatial resolution at large depth.

3.1.2 Instrumentation of PAT

Photoacoustic tomography has two major forms: photoacoustic computed tomography (PACT) and focused scanning photoacoustic microscopy (PAM). In PACT, an array of unfocused ultrasonic transducers detects photoacoustic waves in parallel. A reconstruction algorithm is then used to produce a tomographic image. In PAM, hardware focusing (ultrasonic or optical) in conjunction with 2D raster mechanical or optical scanning produces a volumetric photoacoustic image. The lateral resolution of a PAM image comes from the ultrasonic or optical focusing, while the axial resolution comes from the time of flight of the photoacoustic waves. Within each form, there are various systems using different instruments and configurations. This section is not intended to give a complete review of PAT systems. Rather, we will briefly introduce representative systems within each of the two major forms. By doing so, we hope to give the readers a more vivid image of PAT before discussing its applications in drug delivery and photothermal therapy. More comprehensive reviews of PAT can be found elsewhere [27, 32, 33].

3.1.2.1 Photoacoustic Computed Tomography First-generation laser-based PACT was developed using a single unfocused ultrasonic transducer in conjunction with mechanical scanning [4, 34–36]. In such systems, a large number of scanning steps were needed to reconstruct a high-quality photoacoustic image. Hence, the imaging speed was usually slow. Recently, ultrasonic array transducers—including custom-made ring-shaped array transducers [37–40], hemispherical array transducers [41], and linear array transducers [14, 15, 42–46]—were used to develop next-generation PACT systems capable of imaging at a significantly higher speed.

Figure 3.1a and b shows a PACT system using a full-ring ultrasonic array [40]. For photoacoustic excitation, the system employed a tunable pulsed laser system (Vibrant HE 355 I, Opotek, Inc.). The laser pulse had a width of 5 ns and was fired at a 10-Hz repetition rate. The laser beam was collimated and homogenized before reaching the tissue surface. The full-ring ultrasonic array used for photoacoustic detection had 512 elements and a center frequency of 5 MHz. Each array element was cylindrically focused, forming a cross-sectional imaging slice ~2 mm thick. Due to an 8:1 multiplexing in data acquisition, a complete 2D cross-sectional scan required eight laser shots. Even with this multiplexing, a frame rate of 0.625 Hz without averaging was achieved. Using this PACT system, the cerebral cortexes of Swiss Webster mice were photoacoustically imaged *in vivo* with the scalp and skull intact (Fig. 3.1c and d). The photoacoustic images were validated with open skull anatomical photographs acquired after imaging (Fig. 3.1e and f).

Recently, using a half-ring ultrasonic array PACT system, photoacoustic whole-body small animal imaging has been achieved (Fig. 3.2a) [37]. This whole-body PACT system may prove particularly valuable in the development of new drugs and drug delivery systems. In addition, similar PACT systems capable of imaging the human breast have been developed [13, 47].

For clinical imaging, PACT systems using linear ultrasound arrays may possess several advantages: (i) they can be integrated with widely available commercial medical ultrasound imaging systems [14], (ii) they offer handheld operation [15] and can image many anatomical sites, and (iii) they offer high imaging speed and even real-time imaging and display [43, 44, 46]. For example, a handheld PACT imaging system based on a Philips iU22 clinical ultrasound imaging system (Philips Healthcare, Andover, MA) was developed [15]. This system, with at least 3–5-cm imaging depths, has demonstrated great potential in noninvasive mapping of sentinel lymph nodes for breast cancer staging (Fig. 3.2b–d).

3.1.2.2 Photoacoustic Microscopy In PAM, when the optical illumination is unfocused or only weakly focused, the lateral resolution is determined primarily by the ultrasonic focusing of detection. PAM of this type is usually termed acoustic-resolution photoacoustic microscopy (AR-PAM). When the optical illumination is tightly focused, confining the photoacoustic excitation into a

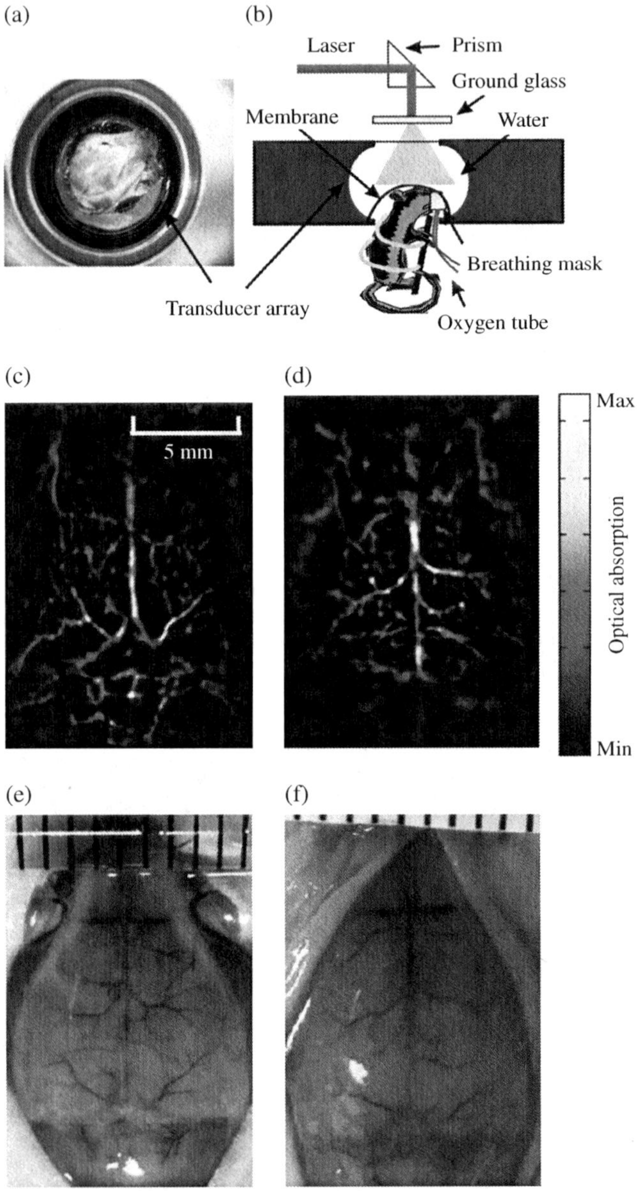

FIGURE 3.1 Full-ring ultrasonic array PACT system and brain cortex imaging. (a) Top view of the full-ring ultrasonic array and the animal head. (b) Side view of the system and the position of the animal. (c, d) *In vivo* photoacoustic cerebral cortex images acquired with scalp and skull intact. (e, f) Open-skull photographs of the cerebral cortexes corresponding to images in (c, d) (Reprinted with permission from reference [40]). For color details, please see color plate section.

significantly smaller region than the ultrasonic focus, the lateral resolution of PAM is determined primarily by the optical focus. PAM of this type has optical-diffraction-limited lateral resolution (which can be submicron) and thus is termed optical-resolution photoacoustic microscopy (OR-PAM).

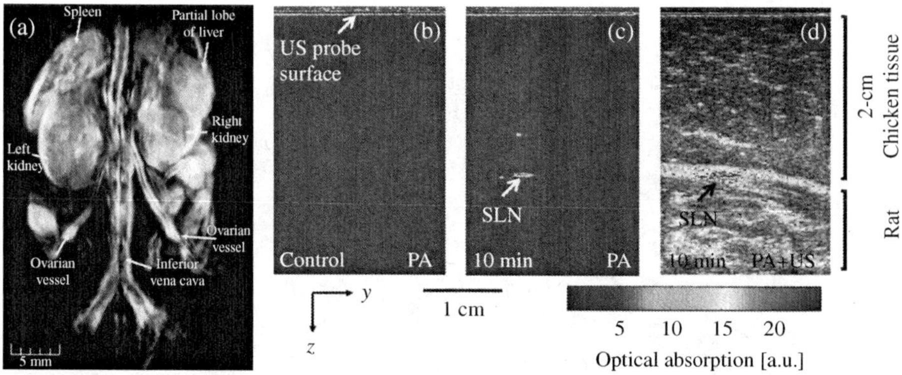

FIGURE 3.2 (a) *In vivo* whole-body image of a mouse acquired at 755 nm. (b–d) Noninvasive photoacoustic sentinel lymph node mapping *in vivo*, with ICG dye and a laser wavelength of 810 nm. Compared with the photoacoustic control image (b), the postinjection image (c) clearly shows the sentinel lymph node of a rat beneath a 2-cm thick layer of chicken tissue. The photoacoustic–ultrasonic co-registered image (d) shows both the tissue structure and sentinel lymph node (Reprinted with permission from reference [40]).

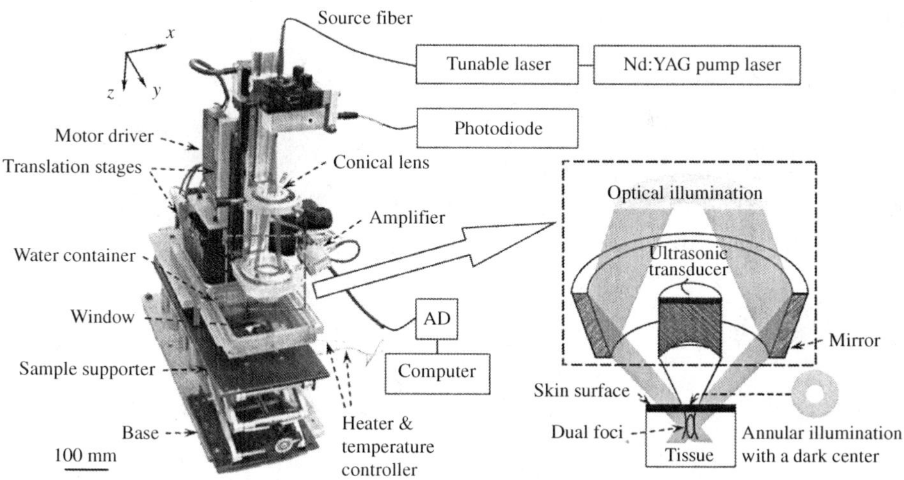

FIGURE 3.3 Acoustic-resolution photoacoustic microscopy system. Raster scanning of the components within the dashed box in the *x–y* plane produces a volumetric image. The window of the water container is sealed with a low-density polyethylene membrane. The sample is placed between the water container and the support for imaging (Reprinted with permission from reference [5]).

Figure 3.3 shows a representative AR-PAM system. For photoacoustic excitation, a tunable dye laser pumped by an Nd:YAG laser was used. With a conical lens, the delivered laser beam was focused into a "donut" shape on the surface of the imaging target. This donut-shaped (or dark-field) optical illumination reduced the photoacoustic background from the superficial region within the ultrasonic field of view. Furthermore, the optical focus and ultrasonic focus were confocal with each other, increasing the signal-to-noise ratio (SNR). In this system, the optical focus was 2 mm in diameter, much wider than the ultrasonic focus. To achieve high spatial resolution, an ultrasonic transducer

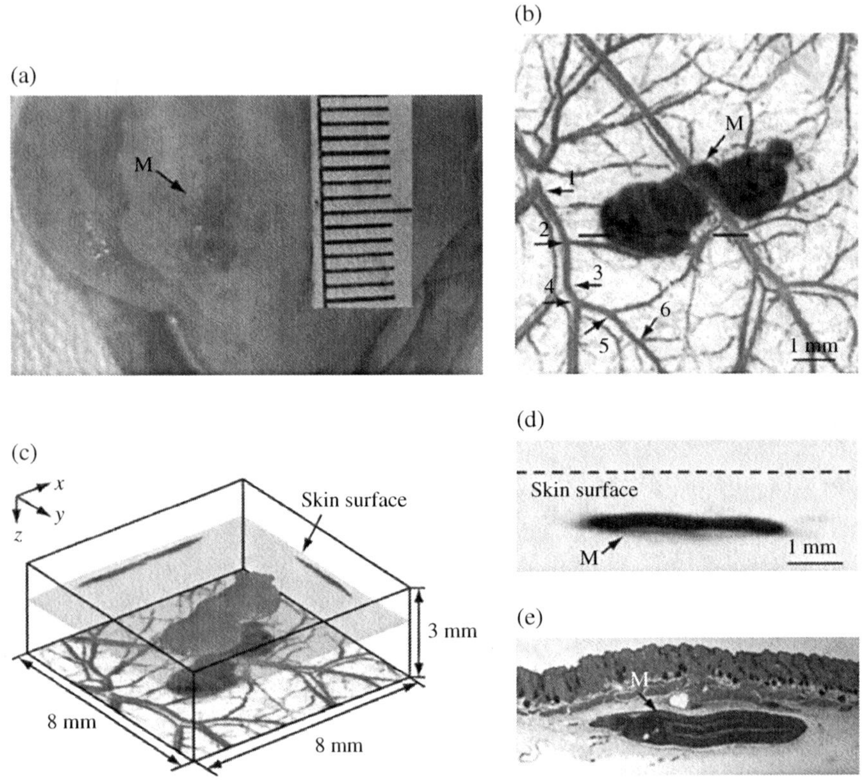

FIGURE 3.4 (a) Photograph of the melanoma, M. (b) Composite photoacoustic image. (c) 3D image of the melanoma acquired at 764 nm. The composite image in (a) is redrawn at the bottom. (d) Cross-sectional (B-scan) image of the melanoma corresponding to the dashed line in (b). (e) Hematoxylin and eosin (HE)-stained section corresponding to the region in (d) (Reprinted with permission from reference [5]). For color details, please see color plate section.

with a large numerical aperture (NA) (0.44), a high central frequency (50 MHz), and a wide bandwidth (70% nominal) was employed. The axial and lateral resolutions of the system were 15 and 45 μm, respectively, in the acoustic focal zone.

Using two laser wavelengths (764 and 584 nm), an inoculated melanoma and its surrounding blood vessels were imaged by AR-PAM with high contrast and resolution (Fig. 3.4). The blood vessels are pseudo-colored red in the 584-nm image and the melanoma brown in the 764-nm image. Up to six orders of vessel branching are observed in the image (indicated by the numbers 1–6). The thickness of the melanoma was photoacoustically measured to be 0.3 mm, which agreed with the results from histology (Fig. 3.4d and e).

The high imaging contrast and spatial resolution of PAM at large depths can potentially be used for treatment planning by imaging the tumor and its surrounding neovasculature before drug delivery. In addition, drugs delivered by absorbing nanocarriers can be imaged and tracked by PAM. Finally, the accumulation of drugs at the tumor region and the therapeutic efficacy can potentially be imaged and evaluated by PAM.

The principle of OR-PAM is similar to that of AR-PAM, except that the optical illumination is tightly focused to provide optical-diffraction-limited lateral resolution. Due to its high-spatial resolution, OR-PAM has been demonstrated to be a valuable tool for the study of microcirculation and angiogenesis [48]. Hence, OR-PAM may prove to be useful for the development of antiangiogenesis drugs and corresponding delivery systems.

3.2 PAT IN DRUG DELIVERY AND PHOTOTHERMAL THERAPY

3.2.1 Drug Delivery Using Multifunctional Nanocarriers

Since liposome was used as the first nanocarrier for drug delivery in the 1960s [49], various advanced drug delivery systems have emerged, generating a profound impact on many branches of medicine, including oncology, cardiology, ophthalmology, endocrinology, and immunology [50]. Advanced drug delivery systems have been shown to dramatically improve the pharmacological properties of free drugs [51–53]:

1. They can improve the solubility of hydrophobic drugs and protect drugs from premature degradation, reducing the required dosage.
2. They can enable sustained and even controlled drug release, enhancing the therapeutic activity and drug efficacy.
3. They can improve the localization of drugs with targeted delivery (either passive targeting through the enhanced permeability and retention (EPR) effect or active targeting through specific ligand–receptor interactions) and thus can reduce side effects and damage to normal tissue.
4. They can reduce unfavorable pharmacokinetics, for example, the undesired rapid renal clearance of drugs.
5. They can potentially deliver drugs through biological barriers (e.g., the blood–brain barrier) that are hard to cross by "free" drugs.
6. Multifunctional drug carriers for simultaneous drug delivery and imaging can potentially improve the planning, monitoring, and evaluation of disease treatment [50, 54].

Although clinically approved advanced drug delivery systems are based dominantly on liposomes and polymers [55, 56], many other nanocarriers—including polymeric nanoparticles, carbon nanotubes, and metal nanoparticles—are also being actively studied, and some of them have entered clinical trials [52]. In addition, to improve treatment planning, monitoring, and evaluation, there is a great interest in developing multifunctional nanocarriers that can simultaneously enable targeting, therapy, and *in vivo* imaging [57–60].

In general, a multifunctional nanocarrier is composed of two or more functional moieties. Figure 3.5 is a schematic of a typical multifunctional nanocarrier, which consists of a nanoparticle and multiple functional moieties. Some functional moieties, such as the targeting ligands and protecting polymers (e.g., polyethylene glycol), are usually conjugated on the surface of the nanoparticle, while others, including the

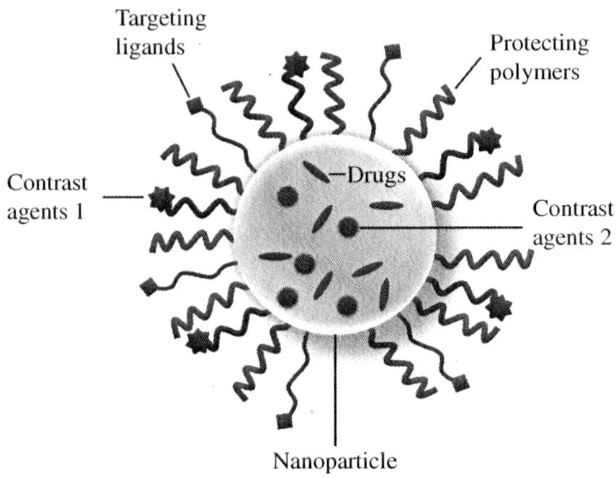

FIGURE 3.5 Schematic of a typical multifunctional nanocarrier.

therapeutic drugs and contrast agents, can be either entrapped in the cavities of the carrier or conjugated on its surface.

Targeting and protecting functional moieties, which aim to increase the localized drug concentration at targeted sites and the drug circulation lifetime, respectively, have been widely adopted in drug delivery systems [52]. Contrast agents have also been widely used to provide or enhance the contrast of various imaging technologies. There are multiple advantages to making the drug delivery carriers trackable and visible *in vitro* and *in vivo*. In clinical practice, predicting and detecting drug delivery to tumor sites are critically important. However, owing to the complex nature of the microenvironment of a tumor, the standard pharmacokinetic models are often inadequate for these purposes. In drug development, the immediate visualization of the delivery of drugs may enable the optimization of experimental treatment strategies and fast drug screening. Consequently, it is highly desirable that multifunctional nanocarriers and corresponding noninvasive imaging methods be developed.

As mentioned above, PAT provides extremely high sensitivity to optical absorption. In addition to hemoglobin and melanin, which provide intrinsic absorption contrast, a wide range of nanoparticles and organic dyes can offer high photoacoustic contrast. Moreover, PAT can image the tissue temperature during photothermal therapy. In conjunction with multifunctional nanocarriers, PAT can potentially be used to guide drug delivery or photothermal therapy, monitor the response to treatment, and assess the therapeutic efficacy.

3.2.2 PAT Using Gold Nanoparticles

Because of its inert biochemical properties and low toxicity, gold has long been used in medicine (e.g., for rheumatoid arthritis treatment) [61]. Gold nanostructures are versatile nanocarriers that have been reported in many preclinical drug delivery studies [62, 63]. Their surfaces can be functionalized with targeting ligands, protecting polymers, and fluorescent moieties. In addition, their sizes, shapes, and compositions can be easily controlled and tuned to cater to particular drug delivery and imaging applications. Indeed, a variety of gold nanostructures, including nanoshells, nanorods, and nanocages, have been developed and explored for drug delivery, therapy, and imaging uses. Of note, several classes of gold nanostructures have already been approved by the U.S. Food and Drug Administration (FDA) as drug delivery carriers or therapeutic agents in phase I clinical trials (e.g., http://www.cytimmune.com and http://www.nanospectra.com).

3.2.2.1 Nanoshells Since the first synthesis of Au-coated Au_2S nanoparticles [64, 65], gold nanoshells have attracted great attention because of their extraordinary optical properties. Their plasmon resonance optical absorption is not only strong but also highly tunable. By varying the ratio of the shell thickness and the core size, the absorption peak of gold nanoshells can be tuned to the near-infrared (NIR) region, where optical absorption and scattering in biological tissue are minimal and light penetration is relatively deep.

Compared with conventional thermal ablation of tumors by microwaves or focused ultrasound, photothermal ablation using gold nanoshells enables selective heating in the tumor region, which can reduce the damage to normal tissue. Following the initial demonstration of photothermal therapy of tumor using local injection of gold nanoshells [66], photothermal therapy using intravenous injection was also demonstrated in a murine tumor model [67]. Furthermore, immunotargeted gold nanoshells were developed to achieve a selective destruction of targeted carcinoma cells through photothermal therapy [68].

By conjugating gold nanoshells with thermally responsive polymer N-isopropylacrylamide-acrylamide (NIPAAm-co-AAm), the potential of gold nanoshell-based drug delivery and controlled release was also demonstrated [69]. When NIPAAm-co-AAm is heated to a critical temperature, it undergoes a reversible phase change, leading to the expulsion of water and dissolved drugs from the matrix. By mixing gold nanoshells into the polymer matrix, a pulsed NIR laser can be used to induce a local temperature rise, resulting in a phase change of the polymer and thus a pulsatile release of drugs. In addition, hollow gold nanoshells coupled with liposomes were also found capable of delivering and releasing drugs upon irradiation by NIR light [70].

In PAT, the first use of gold nanoshells was to enhance the imaging contrast of cerebral cortex vasculature in rats and mice [71, 72]. Recently, the progressive extravasation and accumulation of nanoshells within a solid tumor were imaged *in vivo* by PAT [73]. Immunocompetent BALB/c mice inoculated with CT26.wt murine colon carcinoma cells were used for *in vivo* experiments. Polyethylene glycol (PEGylated) gold nanoshells (gold-coated silica) with a peak optical absorption at 800 nm were intravenously administered. The tumor region was then photoacoustically monitored for 6 h. The experimental results revealed a heterogeneous distribution of nanoshells in the tumor—more concentrated in the tumor cortex and largely absent from the tumor core (Fig. 3.6). Furthermore, the tumor position was delineated, which can potentially be very useful for guidance of photothermal therapy.

In another study, the use of PAT to simultaneously image gold-coated iron oxide nanoshells and monitor the temperature during photothermal therapy was demonstrated [31]. During the photothermal therapy, ultrasound and photoacoustic signals were recorded every 10 s, and the temperatures (measured by both ultrasound imaging

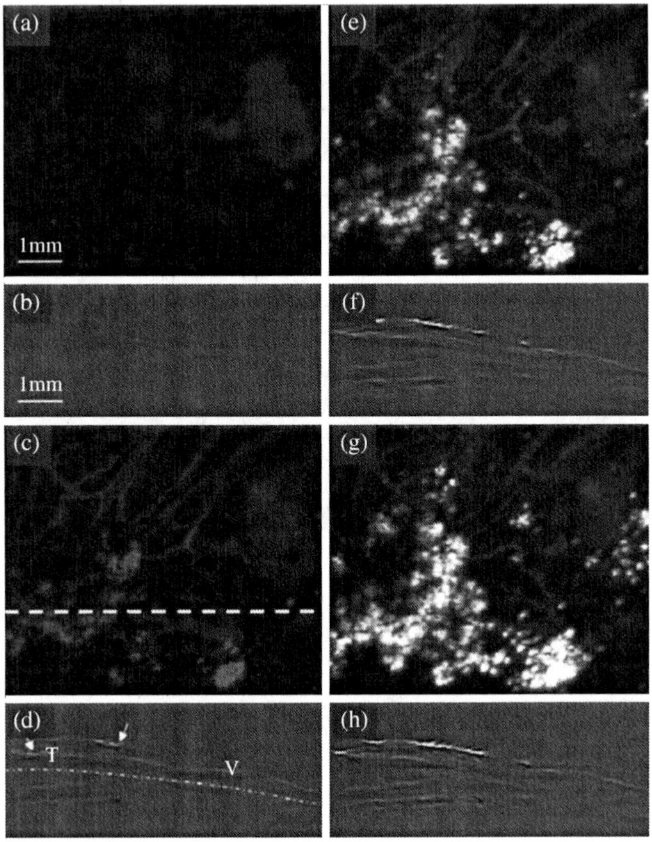

FIGURE 3.6 *In vivo* PAT of nanoshell extravasation from solid tumor vasculature. (a, c, e, and g) Photoacoustic images acquired prior to and 1.4, 3, and 5.8 h after nanoshell administration, respectively. The horizontal and vertical axes indicate the scanning directions of the acoustic transducer. (b, d, f, and h) B-scan images acquired before and 1.4, 3, and 5.8 h after nanoshell injection, respectively (corresponding to the scanning position indicated by the dashed line in (c)). The vertical and horizontal axes are the depth and lateral scanning directions, respectively. The dot-dashed line in (d) shows the contour of the skull; the arrows, nanoshells; T, tumor; V, vessel (Reprinted with permission from reference [73]).

and PAT) were computed. In this study, although both ultrasound imaging and PAT have demonstrated the capability to image tissue temperature, PAT showed a much higher sensitivity than ultrasound [31].

3.2.2.2 Nanorods Nanorods also have potential for both drug delivery and PAT. In fact, the synthesis of nanorods with various aspect ratios is quite simple and well established, and the size of nanorods that absorb NIR light is quite small and suitable for potential drug (gene) delivery use.

Although many other optical techniques such as OCT and two-photon microscopy have been used to image nanorods (based on scattering or photoluminescence), PAT is probably the only modality that can directly sense their high optical absorption

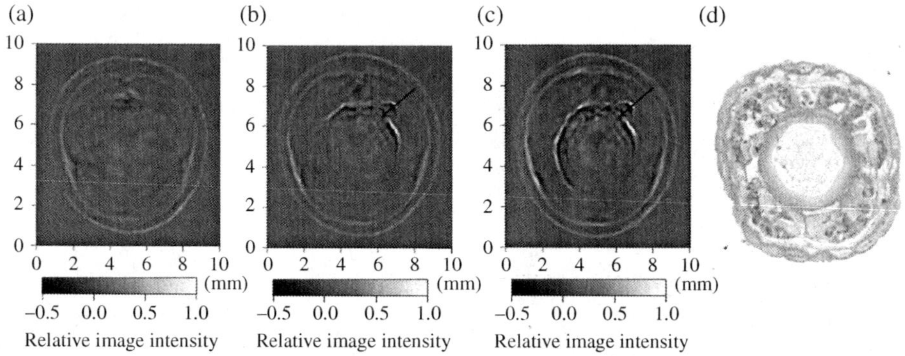

FIGURE 3.7 Two-dimensional cross-sectional PAT of a rat tail joint. (a) Image taken before the administration of Etanercept-conjugated gold nanorods. (b) Image taken after the first administration. (c) Image taken after the second administration. For each administration, 0.025 ml of agent with a 10-pM concentration was injected intra-articularly at positions indicated by the arrows in (b) and (c). (d) Histology of a similar cross section showing the morphological features, including the intra-articular tissue, vessels, and muscle (Reprinted with permission from reference [79]).

[74]. In fact, nanorods have been used in conjunction with PAT for a wide array of imaging applications, including targeted imaging of prostate cancer [75], imaging of the macrophage infiltration and inflammatory response of atherosclerosis [25, 76], noninvasive mapping of sentinel lymph nodes [77], and guidance of tumor ablation using high-intensity focused ultrasound (HIFU) [78].

PAT has been used to noninvasively image and monitor antirheumatic drug delivery using nanorods [79]. A PACT system using a single-element ultrasonic transducer of 10-MHz center frequency was employed. Gold nanorods with an aspect ratio of 3 (45 nm by 15 nm) and an absorption peak at 660 nm were synthesized. Etanercept (anti-TNF drug) molecules were then conjugated with the gold nanorods for antirheumatic therapy. *Ex vivo* experiments were conducted with rat tail joints, during which Etanercept-conjugated gold nanorods were injected intra-articularly using a needle. The results suggested that PAT had the potential to image both the major structures of joints and the nanorod-conjugated antirheumatic drugs (Fig. 3.7).

One potential issue associated with gold nanorods is that they melt and change shape easily upon pulsed laser irradiation [80]. As a result, their absorption peak may shift away from the desired wavelength selected for PAT. One attempt to address this issue was to coat gold nanorods with a thin layer of silica, which makes them more photothermally stable [80]. The shape changes of PEGylated and silica-coated nanorods under pulsed laser irradiation have been experimentally studied. Silica-coated gold nanorods showed a minimal shape change (Fig. 3.8), suggesting that silica coating can be a potential solution to the instability of gold nanorods under PAT.

3.2.2.3 Nanocages Nanocages are a new class of nanostructures that have compact sizes (typically 20–50 nm), precisely tunable surface plasmon resonance peaks, and large absorption cross-sections per volume that are particularly important for photothermal conversion [81]. Gold nanocages can be synthesized using a simple galvanic replacement reaction between a silver template and a gold salt [82]. In a

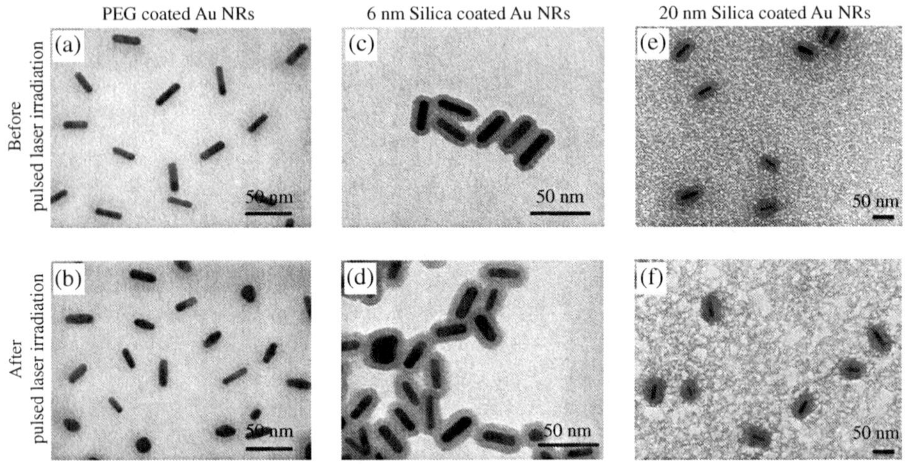

FIGURE 3.8 TEM images showing the morphology evolutions of various gold nanorods before and after 300 pulses of 20 mJ/cm^2 laser irradiation. (a, b) PEG-coated gold nanorods. (c, d) 6-nm silica-coated gold nanorods. (e, f) 20-nm silica-coated gold nanorods (Reprinted with permission from reference [80]).

typical reaction, a HAuCl4 solution is slowly titrated into a boiling suspension of silver nanocubes. Because of the electrochemical potential difference between gold and silver, silver will dissolve into the solution as ions, while gold is deposited on the outer surface of the cube. As the reaction continues, more silver from the interior of the cube will be removed through a small pit in the wall, forming a hollow gold–silver alloyed nanobox whose wall thickness can be precisely tuned by adjusting the amount of HAuCl4 added. In a later stage of the reaction when all pure silver has been removed, the HAuCl4 will begin to remove silver from the gold–silver alloy walls, forming highly porous structures known as gold nanocages. The plasmon resonance peaks of nanocages can be tuned by simply varying their wall thicknesses.

Like other gold nanoparticles, the surface of nanocages can be easily functionalized with tumor targeting moieties such as antibodies or polypeptides [83]. Unique to nanocages, however, are their hollow interiors and porous walls. Hence, nanocages can be loaded with drugs, dyes, or other functional agents. In addition, the pores of a gold nanocage can be formed at its corners, creating well-defined routes for drugs to enter and exit its interior. Furthermore, the pores of these nanocages can be controllably opened and closed by coating the nanocages with a thermosensitive smart polymer that expands and covers the pores at low temperatures and collapses to expose the pores at high temperatures [84]. As a result, with the use of a NIR laser, preloaded effectors such as drugs and dyes can be released in a controllable fashion (Fig. 3.9).

Photothermal therapy using gold nanocages has also been reported. In one *in vivo* study, PEGylated gold nanocages were used to photothermally treat inoculated bilateral tumors in mice [85]. The gold nanocages were intravenously administrated and delivered to the tumor via passive targeting (EPR effect). A CW diode laser was used as the photothermal therapy irradiation source. The temperature increase during photothermal treatment was monitored using an infrared camera, offering information for treatment planning. The therapeutic efficacy on tumor metabolism was evaluated using ^{18}F-fluorodeoxyglucose positron emission tomography

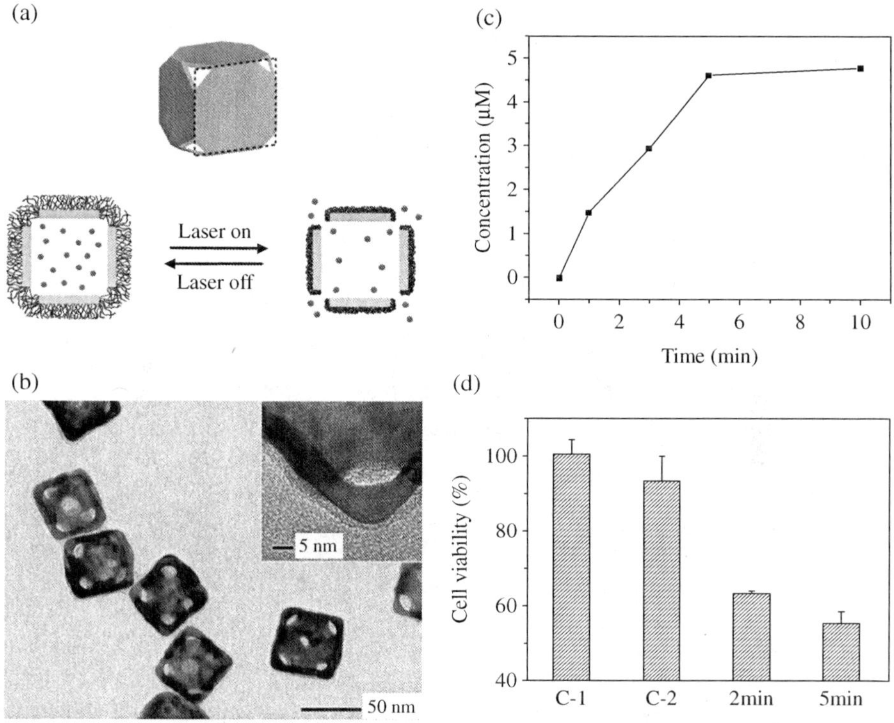

FIGURE 3.9 (a) Principles of smart polymer-coated nanocages. (b) TEM images of Au nanocages with surface covered by a pNIPAAm-*co*-pAAm copolymer of a critical temperature of 39°C. The inset shows a magnified TEM image of the corner of one nanocage. (c) A plot of the concentrations of Dox released from the Au nanocages on heating at 45°C for different periods of time. (d) Cell viability of samples after different treatments: (C-1) cells irradiated with a pulsed NIR laser for 2 min in the absence of Au nanocages; (C-2) cells irradiated with the laser for 2 min in the presence of Dox-free Au nanocages; and (2/5 min) cells irradiated with the laser for 2 and 5 min in the presence of Dox-loaded Au nanocages. A power density of 20 mW/cm^2 was used for these studies (Reprinted with permission from reference [84]).

(^{18}F-FDG PET). A decrease in tumor metabolic activity, an indication of effective therapy, was observed in tumors treated with a combination of gold nanocages and laser exposure (Fig. 3.10). Irreversible damage to the tumor cells was found upon histological examination. In addition, biodistribution studies showed that the uptake of the PEGylated gold nanocages by tumors was efficient, and the nanocages were distributed throughout the tumor, with the concentration in the tumor periphery being slightly higher than that in the tumor core.

In this study, the primary sensing and imaging tools used for treatment planning and evaluation are an infrared camera and a PET scanner. Although PET is an established technique to measure metabolic activity, it uses ionizing radiation and is quite expensive. The infrared camera is relatively inexpensive, but it cannot provide depth sectioning and high spatial resolution of temperature *in vivo*. In the future, a PAT system may be used to replace both the PET scanner and the infrared camera. PAT can noninvasively image both temperature and metabolic activity at high spatial resolution and is relatively inexpensive. In addition, PAT can directly image the

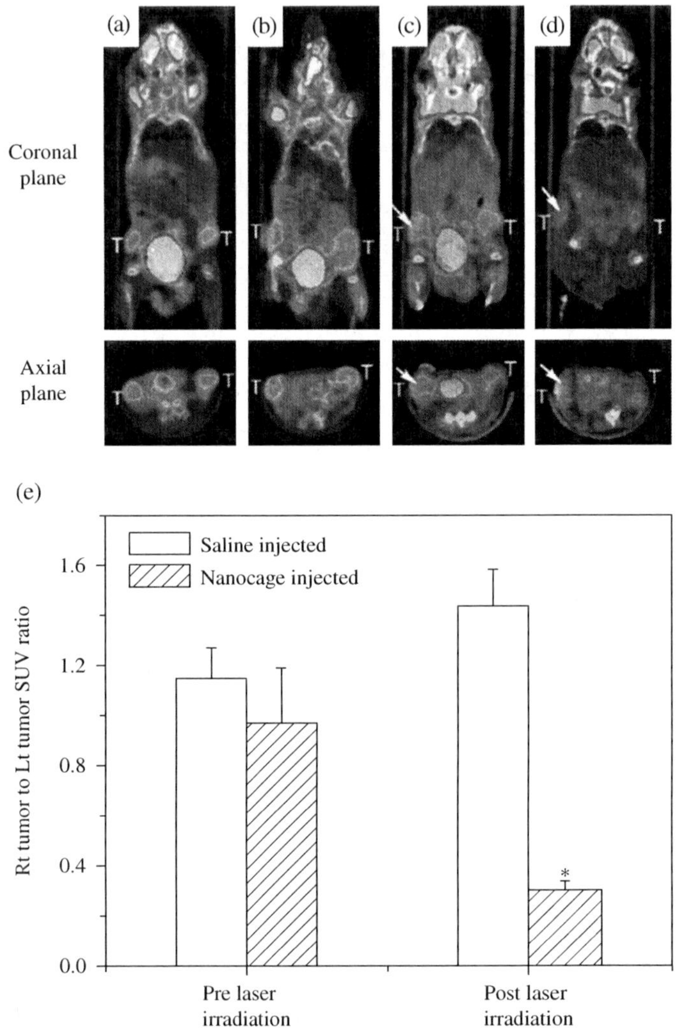

FIGURE 3.10 ^{18}F-FDG PET/CT co-registered images of mice intravenously administrated with either saline or Au nanocages, followed by laser treatment. (a) Saline-injected mouse prior to laser irradiation. (b) Nanocage-injected mouse prior to laser irradiation. (c) Saline-injected mouse after laser irradiation. (d) Nanocage-injected mouse after laser irradiation. The white arrows indicate the treated tumors. (e) Plot showing the ratios of the ^{18}F-FDG standardized uptake values (SUV, $p<0.001$) of laser-treated tumor (Rt tumor) to non-treated tumor (Lt tumor) (Reprinted with permission from reference [85]).

distribution of nanocages *in vivo* by sensing their strong optical absorption, which can be helpful for treatment planning.

PAT of gold nanocages has also been demonstrated for many other applications. In addition to *in vivo* vascular imaging in the brain cortex and noninvasive sentinel lymph node mapping [14, 15, 86], molecular PAT of melanoma using targeted bioconjugated nanocages has been reported [87]. Both active and passive targeting gold nanocages were used to enhance the contrast of PAT of melanoma *in vivo*. The surfaces of the gold nanocages were derivatized, with [Nle^4, d-Phe^7]-α-melanocyte-stimulating hormone ([Nle^4, d-Phe^7]-α-MSH) and polyethylene glycol (PEG), respectively. The former surface-coating molecules have been shown to be able to strongly bind to α-MSH receptors overexpressed on melanomas; the latter molecule was used simply to increase the circulation lifetime of the nanocages. The plasmon resonance absorption peak of the nanocages was at ~780 nm. A deep-penetration PAM system with a 10-MHz ultrasonic transducer was used to image the melanoma, and a high-resolution PAM system with a 50-MHz ultrasonic transducer was used to image the surrounding blood vessels. Two laser wavelengths, 778 nm and 570 nm, were used for the melanoma and vascular imaging, respectively. Figure 3.11 shows composite photoacoustic images of both the melanoma and blood vessels prior and after the injection of gold nanocages. It can be seen that, at 6 h post injection, the contrast enhancement with [Nle^4, d-Phe^7]-α-MSH-gold nanocages (38 ± 6%) was on average 300% greater than that with PEG-gold nanocages (13 ± 2%), suggesting that the [Nle^4, d-Phe^7]-α-MSH-gold nanocages had a higher probability of accumulating in the tumors. The results were further validated by quantifying the number of gold nanocages accumulated in the excised tumors using inductively coupled plasma mass spectrometry [87].

Eventually, to translate any of the aforementioned gold nanostructures into daily clinical use, their long-term toxicity and safety need to be better studied and addressed. Currently, the sizes of commonly used gold nanostructures for biomedical applications range from 20 to 150 nm, which is still not small enough for ultimate clearance. In fact, one study showed that particles smaller than ~6 nm in diameter could be cleared rapidly from the body [88]. However, nanoparticles in this size range are usually removed too rapidly from the body to be used effectively for imaging and therapy. Recently, one study trying to address this issue has led to the development of biodegradable plasmonic nanoclusters [89]. These nanoclusters consist of <5-nm primary gold particles stabilized by a weakly adsorbed biodegradable polymer. The small spacing between the primary gold nanoparticles within the clusters results in strong NIR absorption, which can be used for imaging and therapy. More importantly, the nanoclusters, stable at pH 7, can degrade into small primary gold nanoparticles at ~ pH 5 (the environment inside endosomes) within several days. The biodegraded primary gold nanoparticles can then be excreted from the body. Recently, the potential use of such biodegradable nanoclusters as contrast agents for PAT has been demonstrated [90].

3.2.3 PAT Using Carbon Nanotubes

One promising application of carbon nanotubes in drug delivery is the delivery of genes and siRNAs [91, 92]. Pristine carbon nanotubes are water insoluble and difficult to handle biologically. However, the surface of a carbon nanotube can be functionalized to enable biocompatibility [93]. Various moieties—such as drugs, contrast agents, and antibodies—can also be conjugated on the surface of carbon nanotubes to provide additional functions. For example, fluorescent molecules have been introduced to enable NIR fluorescent imaging of carbon nanotubes [94]. Due to their intrinsic optical absorption, carbon nanotubes have also been used for photothermal therapy [95]. Several articles have reviewed the drug delivery and therapeutic applications of carbon nanotubes [96, 97]. Below, we will mainly discuss the use of carbon nanotubes as a molecular contrast agent for PAT.

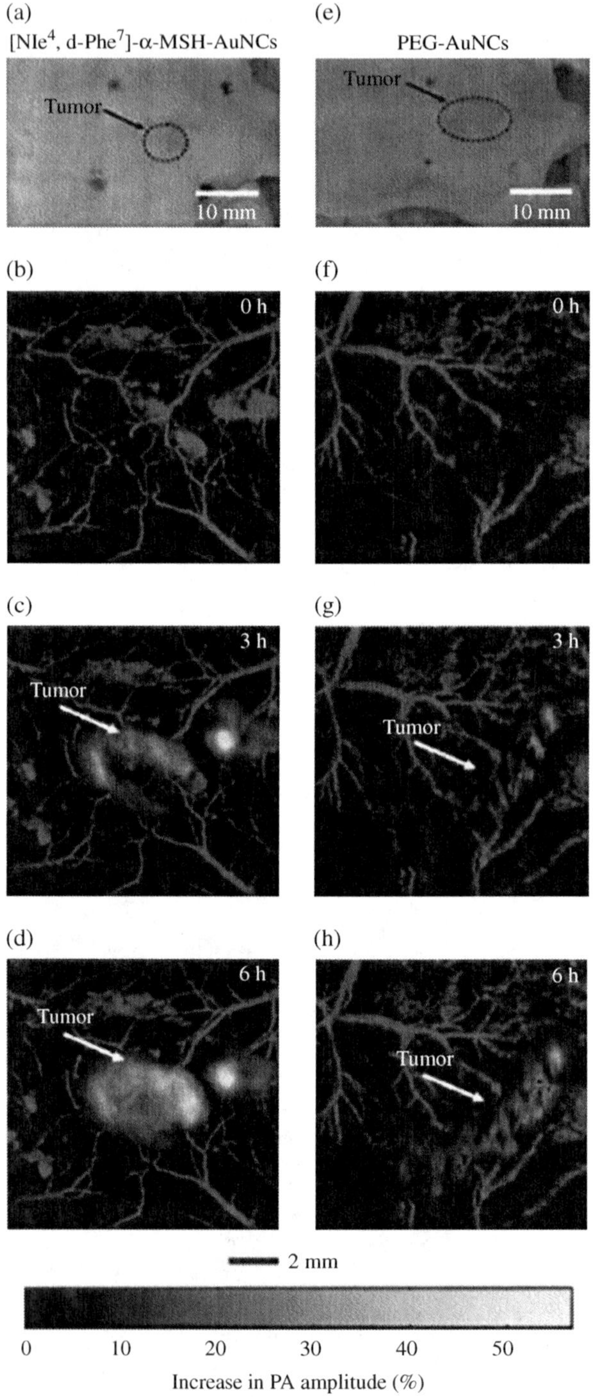

FIGURE 3.11 Noninvasive *in vivo* PAT of B16 melanomas using [Nle^4, d-Phe^7]-α-MSH- and PEG-gold nanocages. (a, e) Photographs of nude mice transplanted with B16 melanomas before nanocage injection. PA images of the B16 melanomas after intravenous injection of 100-μL 10 nM [Nle^4, d-Phe^7]-α-MSH- (b–d) and (f–h) PEG-AuNCs through the tail vein. The background vasculature images were obtained using PAM at 570 nm, and the melanoma images were obtained using PAM at 778 nm (Reprinted with permission from reference [87]). For color details, please see color plate section.

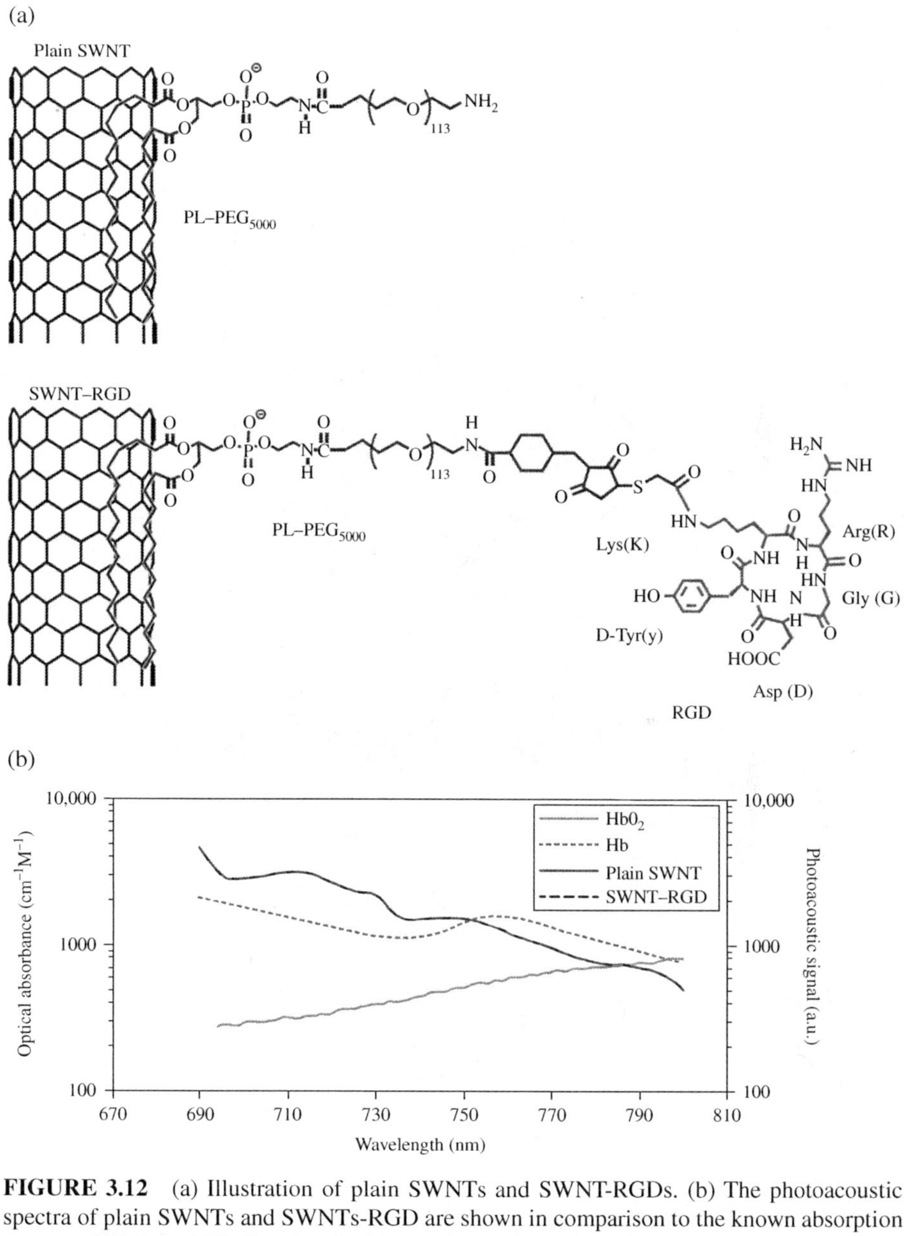

FIGURE 3.12 (a) Illustration of plain SWNTs and SWNT-RGDs. (b) The photoacoustic spectra of plain SWNTs and SWNTs-RGD are shown in comparison to the known absorption spectra of HbO_2 and Hb (Reprinted with permission from reference [98]).

Single-walled carbon nanotubes (SWNTs) were among the first type of carbon nanotubes used for PAT. In one study, both untargeted and targeted SWNTs were prepared for photoacoustic imaging [98] (Fig. 3.12). The untargeted SWNTs (plain SWNTs) were conjugated with PEG-5000-grafted phospholipid ($PL\text{-}PEG_{5000}$) to make them biocompatible; the targeted SWNTs were further coupled to RGD

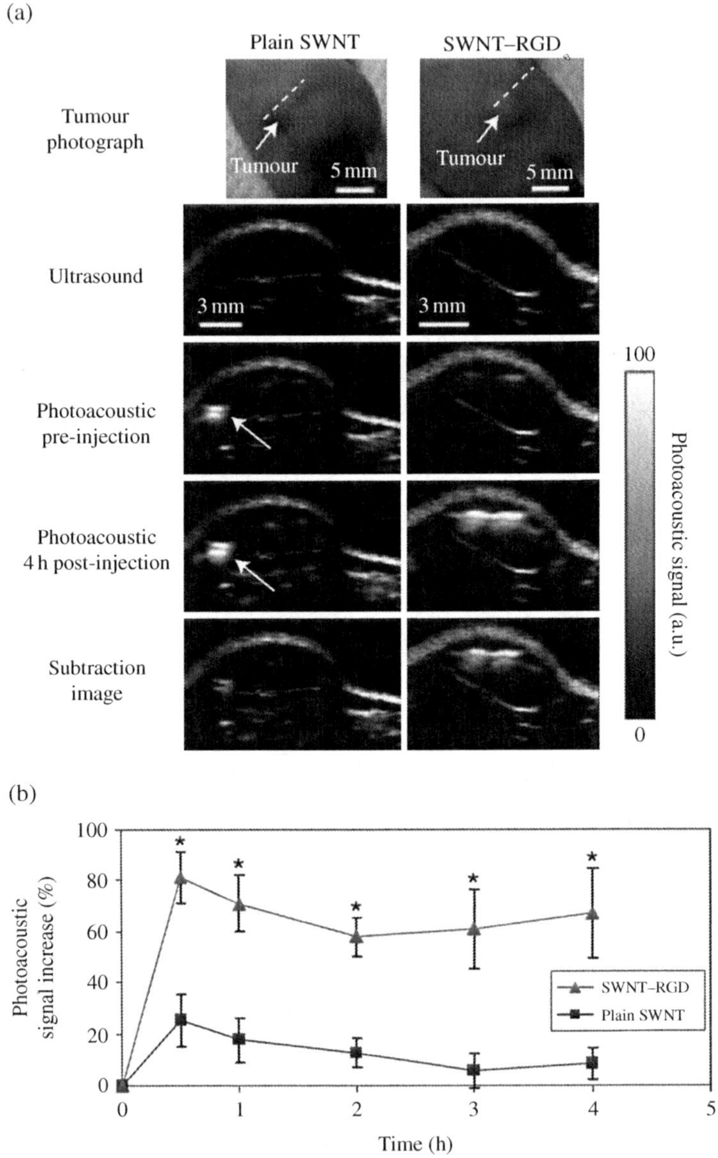

FIGURE 3.13 (a) Ultrasonic (grey) and photoacoustic (green) cross-sectional (along the dotted line) images of the tumor, before and 4 h after the injection of carbon nanotubes. (b) Mice injected with SWNT-RGDs showed a significantly higher photoacoustic signal than mice injected with plain SWNTs. The error bars represent standard error ($n = 4$) ($*p < 0.05$) (Reprinted with permission from reference [98]). For color details, please see color plate section.

peptides through $PL\text{-}PEG_{5000}$. The RGD-conjugated SWNTs can bind with high affinity to $\alpha_v\beta_3$ integrin, which is overexpressed in tumor neovasculature. The photoacoustic spectra of both types of SWNTs were measured and compared with the absorption of hemoglobin in the same spectral region (Fig. 3.12). The spectral overlap between plain SWNTs and SWNT-RGDs suggests that the RGD conjugation does not perturb the photoacoustic signal.

Two groups of mice bearing U87MG tumor xenografts (~100 mm^3) were used for *in vivo* experiments. Plain SWNTs and SWNT-RGDs of ~200 µl at a concentration of 1.2 µM were injected through the tail vein. Photoacoustic and ultrasonic images of the tumor were acquired before and 4 h after the injection (Fig. 3.13). While the ultrasonic images showed the skin and tumor boundaries, photoacoustic images showed the distribution of the carbon nanotubes. To remove the background signal, subtraction images were computed as the 4-h postinjection image minus the preinjection image. The high photoacoustic signal in the mouse injected with plain SWNTs (white arrows in Fig. 3.13) is not seen in the subtraction image, suggesting that it is due to blood vessels rather than SWNTs. Compared with the mice injected with plain SWNTs, those injected with SWNT-RGDs showed a significantly stronger photoacoustic signal in the tumor region. Further, the photoacoustic signal increases within the tumor region were quantified as a function of time (Fig. 3.13b). On average, 4 h after injection, the SWNT-RGDs resulted in an approximately eight times stronger photoacoustic signal than the plain SWNTs.

SWNTs have also been explored for noninvasive sentinel lymph node mapping using PAT and TAT [99, 100]. Compared with organic dyes that have been used for PAT of sentinel lymph nodes, carbon nanotubes can potentially be made targeted to enable molecular imaging of metastasis. Moreover, carbon nanotubes have a broad absorption spectrum, allowing a wide range of optical wavelengths for imaging. By contrast, most gold nanostructures have narrow plasmon resonance absorption spectra, restricting their use to fewer wavelengths.

However, there are also issues associated with carbon nanotubes. One drawback of using carbon nanotubes for PAT is that their intrinsic optical absorption is relatively low compared with gold nanostructures. Different attempts have been made to address this issue. In one study, gold-plated carbon nanotubes (golden carbon nanotubes) were synthesized and used as photoacoustic sensing and photothermal therapy agents. These golden carbon nanotubes showed ~100 times enhanced NIR contrast compared with conventional carbon nanotubes [101]. Indocyanine green dye-enhanced single-walled carbon nanotubes (SWNT-ICG) have also been developed [102]. Compared with conventional SWNTs, SWNT-ICGs showed an ~300 times increase in photoacoustic signal strength, leading to subnanomolar sensitivities.

As with gold nanostructures, the safety and long-term toxicity of carbon nanotubes also need to be addressed. In fact, there is still an ongoing debate on the risk-to-benefit assessments for carbon nanotubes [97]. In the future, more systematic and extensive studies will be needed to reach a consensus.

3.3 SUMMARY

Photoacoustic tomography (PAT) is a hybrid imaging modality combining optical absorption contrast and ultrasonic spatial resolution for deep tissue imaging beyond the quasi-ballistic regime. We consider PAT to have two major forms: PACT and PAM. PACT utilizes an ultrasonic array, or an unfocused ultrasonic transducer in conjunction with scanning, to detect the photoacoustic signals and uses a reconstruction algorithm to form a 3D image. In contrast, PAM utilizes a tightly focused ultrasonic transducer in conjunction with raster scanning to form a 3D photoacoustic image. Depending on whether the optical illumination is focused or not, PAM can be further categorized into OR-PAM and AR-PAM. In all PAT forms, except in OR-PAM, both the ballistic and diffuse photons contribute to the photoacoustic signals used for image formation. As a result, PAT breaks through the depth limit of the optical transport mean free path, providing rich optical contrast and high ultrasonic resolution. The rich optical absorption contrast enables PAT to image intrinsic physiological functions and exogenous drug delivery and therapeutic agents with high sensitivity. The deep penetration permits potential clinical translation of PAT. In addition, PAT can monitor the temperature during photothermal therapy. Along with the development of multifunctional nanocarriers, PAT is emerging as a promising imaging tool for guiding and evaluating drug delivery and photothermal therapy. In this chapter, the recent development of PAT-compatible multifunctional nanocarriers has been reviewed. The use or potential use of PAT with these nanocarriers for various applications in drug delivery and photothermal therapy has been discussed. Even though the application of PAT in drug delivery and photothermal therapy is still in its infancy, it is a rapidly emerging field that can potentially have a fundamental impact on clinical practice.

ACKNOWLEDGMENTS

This work was sponsored in part by National Institutes of Health grants R01 EB000712, R01 EB008085, R01 CA134539, U54 CA136398, R01 EB010049, and 5P60 DK02057933 to L.W., as well as the Shenzhen Development and Reform Commission Fund [2010] 1599, the Guangdong Innovation Research Team Fund for Low-cost Healthcare Technologies (GIRTF-LCHT), and the Low-cost Health Engineering Research Program of the Chinese Academy of Sciences [2011] to L.S.. L.W. has a financial interest in Microphotoacoustics, Inc., and Endra, Inc., which, however, did not support this work.

REFERENCES

1. Kruger RA, Liu PY, Fang YR, et al. Photoacoustic ultrasound (Paus) – reconstruction tomography. Med Phys 1995;22:1605–1609.
2. Oraevsky AA, Jacques SL, Tittel FK. Measurement of tissue optical properties by time-resolved detection of laser-induced transient stress. Appl Opt 1997;36:402–415.
3. Wang LHV, Zhao XM, Sun HT, et al. Microwave-induced acoustic imaging of biological tissues. Rev Sci Instr 1999;70:3744–3748.
4. Wang XD, Pang YJ, Ku G, et al. Noninvasive laser-induced photoacoustic tomography for structural and functional in vivo imaging of the brain. Nat Biotechnol 2003;21:803–806.
5. Zhang HF, Maslov K, Stoica G, et al. Functional photoacoustic microscopy for high-resolution and noninvasive in vivo imaging. Nat Biotechnol 2006;24:848–851.
6. Ku G, Wang XD, Xie XY, et al. Imaging of tumor angiogenesis in rat brains in vivo by photoacoustic tomography. Appl Opt 2005;44:770–775.

7. Lu W, Melancon M, Xiong C-Y, et al. Brain tumor theranostics: photoacoustic tomography-guided thermal ablation mediated with targeted HAuNS. World Molecular Imaging Congress. Kyoto, Japan; 2010 Sept 9.
8. Jose J, Manohar S, Kolkman RGM, et al. Imaging of tumor vasculature using Twente photoacoustic systems. J Biophotonics 2009;2:701–717.
9. Lao YQ, Xing D, Yang SH, et al. Noninvasive photoacoustic imaging of the developing vasculature during early tumor growth. Phys Med Biol 2008;53:4203–4212.
10. Li ML, Oh JT, Xie XY, et al. Simultaneous molecular and hypoxia imaging of brain tumors in vivo using spectroscopic photoacoustic tomography. IEEE 2008;96:481–489.
11. Li L, Zemp RJ, Lungu G, et al. Photoacoustic imaging of lacZ gene expression in vivo. J Biomed Opt 2007;12:020504.
12. Zhang HF, Maslov K, Sivaramakrishnan M, et al. Imaging of hemoglobin oxygen saturation variations in single vessels in vivo using photoacoustic microscopy. Appl Phys Lett 2007;90:053901.
13. Ermilov SA, Khamapirad T, Conjusteau A, et al. Laser optoacoustic imaging system for detection of breast cancer. J Biomed Opt 2009;14:024007.
14. Erpelding TN, Kim C, Pramanik M, et al. Sentinel lymph nodes in the rat: noninvasive photoacoustic and US imaging with a clinical US system. Radiology 2010;256:102–110.
15. Kim C, Erpelding TN, Maslov K, et al. Handheld array-based photoacoustic probe for guiding needle biopsy of sentinel lymph nodes. J Biomed Opt 2010;15:046010.
16. Song KH, Stein EW, Margenthaler JA, et al. Noninvasive photoacoustic identification of sentinel lymph nodes containing methylene blue in vivo in a rat model. J Biomed Opt 2008;13:054033.
17. Song L, Kim C, Maslov K, et al. High-speed dynamic 3D photoacoustic imaging of sentinel lymph node in a murine model using an ultrasound array. Med Phys 2009;36:3724–3729.
18. de la Zerda A, Paulus YM, Teed R, et al. Photoacoustic ocular imaging. Opt Lett 2010;35:270–272.
19. Hu S, Rao B, Maslov K, et al. Label-free photoacoustic ophthalmic angiography. Opt Lett 2010;35:1–3.
20. Jiao SL, Jiang MS, Hu JM, et al. Photoacoustic ophthalmoscopy for in vivo retinal imaging. Opt Express 2010;18:3967–3972.
21. Silverman RH, Kong FT, Chen YC, et al. High-resolution photoacoustic imaging of ocular tissues. Ultrasound Med Biol 2010;36:733–742.
22. Hu S, Yan P, Maslov K, et al. Intravital imaging of amyloid plaques in a transgenic mouse model using optical-resolution photoacoustic microscopy. Opt Lett 2009;34:3899–3901.
23. Sethuraman S, Amirian JH, Litovsky SH, et al. Spectroscopic intravascular photoacoustic imaging to differentiate atherosclerotic plaques. Opt Express 2008;16:3362–3367.
24. Wang B, Su JL, Karpiouk AB, et al. Intravascular photoacoustic imaging. IEEE J Quantum Electron 2010;16:588–599.
25. Wang B, Yantsen E, Larson T, et al. Plasmonic intravascular photoacoustic imaging for detection of macrophages in atherosclerotic plaques. Nano Lett 2009;9:2212–2217.
26. Kim C, Favazza C, Wang LHV. In vivo photoacoustic tomography of chemicals: high-resolution functional and molecular optical imaging at new depths. Chem Rev 2010; 110:2756–2782.
27. Xu MH. Wang LHV. Photoacoustic imaging in biomedicine. Rev Sci Instr 2006;77:041101.
28. Wang LV, Wu H-I. *Biomedical Optics: Principles and Imaging*. Hoboken: John Wiley & Sons, Inc.; 2007.

29. Pramanik M, Wang LV. Thermoacoustic and photoacoustic sensing of temperature. J Biomed Opt 2009;14:054024.
30. Larina IV, Larin KV, Esenaliev RO. Real-time optoacoustic monitoring of temperature in tissues. J Phys-D-Appl Phys 2005;38:2633–2639.
31. Shah J, Park S, Aglyamov S, et al. Photoacoustic imaging and temperature measurement for photothermal cancer therapy. J Biomed Opt 2008;13:034024.
32. Li CH, Wang LHV. Photoacoustic tomography and sensing in biomedicine. Phys Med Biol 2009;54:R59–R97.
33. Wang LV. Multiscale photoacoustic microscopy and computed tomography. Nat Photonics 2009;3:503–509.
34. Hoelen CGA, de Mul FFM, Pongers R, et al. Three-dimensional photoacoustic imaging of blood vessels in tissue. Opt Lett 1998;23:648–650.
35. Kostli KP, Frauchiger D, Niederhauser JJ, et al. Optoacoustic imaging using a three-dimensional reconstruction algorithm. IEEE J Quantum Electron 2001;7:918–923.
36. Wang XD, Xu Y, Xu MH, et al. Photoacoustic tomography of biological tissues with high cross-section resolution: reconstruction and experiment. Med Phys 2002;29:2799–2805.
37. Brecht HP, Su R, Fronheiser M, et al. Whole-body three-dimensional optoacoustic tomography system for small animals. J of Biomed Opt 2009;14:064007.
38. Gamelin J, Maurudis A, Aguirre A, et al. A real-time photoacoustic tomography system for small animals. Opt Express 2009;17:10489–10498.
39. Yang XM, Maurudis A, Gamelin J, et al. Photoacoustic tomography of small animal brain with a curved array transducer. J Biomed Opt 2009;14:054007.
40. Li CH, Aguirre A, Gamelin J, et al. Real-time photoacoustic tomography of cortical hemodynamics in small animals. J Biomed Opt 2010;15:010509.
41. Kruger RA, Lam RB, Reinecke DR, et al. Photoacoustic angiography of the breast. Med Phys 2010;37:6096–6100.
42. Niederhauser JJ, Jaeger M, Lemor R, et al. Combined ultrasound and optoacoustic system for real-time high-contrast vascular imaging in vivo. IEEE Trans Med Imaging 2005;24:436–440.
43. Song L, Maslov K, Bitton R, et al. Fast 3-D dark-field reflection-mode photoacoustic microscopy in vivo with a 30-MHz ultrasound linear array. J Biomed Opt 2008;13:054028.
44. Song L, Maslov K, Shung KK, et al. Ultrasound-array-based real-time photoacoustic microscopy of human pulsatile dynamics in vivo. J Biomed Opt 2010;15:021303.
45. Yang DW, Xing D, Yang SH, et al. Fast full-view photoacoustic imaging by combined scanning with a linear transducer array. Opt Express 2007;15:15566–15575.
46. Zemp RJ, Song L, Bitton R, et al. Realtime photoacoustic microscopy in vivo with a 30-MHz ultrasound array transducer. Opt Express 2008;16:7915–7928.
47. Oraevsky AA, Ermilov SA, Conjusteau A, et al. Initial clinical evaluation of laser optoacoustic imaging system for diagnostic imaging of breast cancer. Breast Cancer Research and Treatment 2007;106:S47.
48. Hu S, Wang LV. Photoacoustic imaging and characterization of the microvasculature. J Biomed Opt 2010;15:011101.
49. Bangham AD, Standish MM, Watkins JC. Diffusion of univalent ions across the lamellae of swollen phospholipids. J Mol Biol 1965;13:238–252.
50. Langer R. Drug delivery and targeting. Nature 1998;392:5–10.
51. Allen TM, Cullis PR. Drug delivery systems: entering the mainstream. Science 2004;303:1818–1822.

52. Peer D, Karp JM, Hong S, et al. Nanocarriers as an emerging platform for cancer therapy. Nat Nanotechnol 2007;2:751–760.

53. Bhaskar S, Tian FR, Stoeger T, et al. Multifunctional Nanocarriers for diagnostics, drug delivery and targeted treatment across blood-brain barrier: perspectives on tracking and neuroimaging. Part Fibre Toxicol 2010;7:3.

54. Langer R. Perspectives: drug delivery – drugs on target. Science 2001;293:58–59.

55. Wagner V, Dullaart A, Bock AK, et al. The emerging nanomedicine landscape. Nat Biotechnol 2006;24:1211–1217.

56. Farokhzad OC, Langer R. Impact of nanotechnology on drug delivery. ACS Nano 2009;3:16–20.

57. Guo R, Zhang LY, Qian HQ, et al. Multifunctional nanocarriers for cell imaging, drug delivery, and near-IR photothermal therapy. Langmuir 2010;26:5428–5434.

58. Liong M, Lu J, Kovochich M, et al. Multifunctional inorganic nanoparticles for imaging, targeting, and drug delivery. ACS Nano 2008;2:889–896.

59. Torchilin VP. Multifunctional nanocarriers. Adv Drug Deliv Rev 2006;58:1532–1555.

60. Torchilin VP. Multifunctional nanocarriers for delivery of drugs, genes, and diagnosticals in the body. Febs J 2009;276:85.

61. Kean WF, Kean IR. Clinical pharmacology of gold. Inflammopharmacology 2008; 16:112–125.

62. Han G, Ghosh P, Rotello VM. Functionalized gold nanoparticles for drug delivery. Nanomedicine (Lond) 2007;2:113–123.

63. Paciotti GF, Myer L, Weinreich D, et al. Colloidal gold: a novel nanoparticle vector for tumor directed drug delivery. Drug Deliv 2004;11:169–183.

64. Averitt RD, Sarkar D, Halas NJ. Plasmon resonance shifts of Au-coated Au2S nanoshells: insight into multicomponent nanoparticle growth. Phys Rev Lett 1997;78:4217–4220.

65. Zhou HS, Honma I, Komiyama H, et al. Controlled synthesis and quantum-size effect in gold-coated nanoparticles. Phys Rev B 1994;50:12052–12056.

66. Hirsch LR, Stafford RJ, Bankson JA, et al. Nanoshell-mediated near-infrared thermal therapy of tumors under magnetic resonance guidance. Proc Natl Acad Sci USA 2003;100:13549–13554.

67. O'Neal DP, Hirsch LR, Halas NJ, et al. Photo-thermal tumor ablation in mice using near infrared-absorbing nanoparticles. Cancer Lett 2004;209:171–176.

68. Loo C, Lowery A, Halas NJ, et al. Immunotargeted nanoshells for integrated cancer imaging and therapy. Nano Lett 2005;5:709–711.

69. Sershen SR, Westcott SL, West JL, et al. An opto-mechanical nanoshell-polymer composite. Appl Phys B 2001;73:379–381.

70. Wu GH, Milkhailovsky A, Khant HA, et al. Remotely triggered liposome release by near-infrared light absorption via hollow gold nanoshells. J Am Chem Soc 2008; 130:8175–8177.

71. Lu W, Huang Q, Geng KB, et al. Photoacoustic imaging of living mouse brain vasculature using hollow gold nanospheres. Biomaterials 2010;31:2617–2626.

72. Wang YW, Xie XY, Wang XD, et al. Photoacoustic tomography of a nanoshell contrast agent in the in vivo rat brain. Nano Lett 2004;4:1689–1692.

73. Li ML, Wang JC, Schwartz JA, et al. In-vivo photoacoustic microscopy of nanoshell extravasation from solid tumor vasculature. J Biomed Opt 2009;14:010507.

74. Eghtedari M, Oraevsky A, Copland JA, et al. High sensitivity of in vivo detection of gold nanorods using a laser optoacoustic imaging system. Nano Lett 2007;7:1914–1918.

75. Agarwal A, Huang SW, O'Donnell M, et al. Targeted gold nanorod contrast agent for prostate cancer detection by photoacoustic imaging. J Appl Phys 2007;102:064701.
76. Kim K, Huang SW, Ashkenazi S, et al. Photoacoustic imaging of early inflammatory response using gold nanorods. Appl Phys Lett 2007;90:223901.
77. Song KH, Kim C, Maslov K, et al. Noninvasive in vivo spectroscopic nanorod-contrast photoacoustic mapping of sentinel lymph nodes. Eur J Radiol 2009;70:227–231.
78. Cui HZ, Yang XM. In vivo imaging and treatment of solid tumor using integrated photoacoustic imaging and high intensity focused ultrasound system. Med Phys 2010; 37:4777–4781.
79. Chamberland DL, Agarwal A, Kotov N, et al. Photoacoustic tomography of joints aided by an Etanercept-conjugated gold nanoparticle contrast agent—an ex vivo preliminary rat study. Nanotechnology 2008;19:095101.
80. Chen YS, Frey W, Kim S, et al. Enhanced thermal stability of silica-coated gold nanorods for photoacoustic imaging and image-guided therapy. Opt Express 2010;18:8867–8878.
81. Chen J, Saeki F, Wiley BJ, et al. Gold nanocages: bioconjugation and their potential use as optical imaging contrast agents. Nano Lett 2005;5:473–477.
82. Cobley CM, Au L, Chen J, et al. Targeting gold nanocages to cancer cells for photothermal destruction and drug delivery. Expert Opin Drug Deliv 2010;7:577–587.
83. Chen J, Wang D, Xi J, et al. Immuno gold nanocages with tailored optical properties for targeted photothermal destruction of cancer cells. Nano Lett 2007;7:1318–1322.
84. Yavuz MS, Cheng Y, Chen J, et al. Gold nanocages covered by smart polymers for controlled release with near-infrared light. Nat Mater 2009;8:935–939.
85. Chen JY, Glaus C, Laforest R, et al. Gold nanocages as photothermal transducers for cancer treatment. Small 2010;6:811–817.
86. Yang X, Skrabalak SE, Li ZY, et al. Photoacoustic tomography of a rat cerebral cortex in vivo with au nanocages as an optical contrast agent. Nano Lett 2007;7:3798–3802.
87. Kim C, Cho EC, Chen J, et al. In vivo molecular photoacoustic tomography of melanomas targeted by bioconjugated gold nanocages. ACS Nano 2010;4:4559–4564.
88. Choi HS, Liu W, Misra P, et al. Renal clearance of quantum dots. Nat Biotechnol 2007; 25:1165–1170.
89. Tam JM, Tam JO, Murthy A, et al. Controlled assembly of biodegradable plasmonic nanoclusters for near-infrared imaging and therapeutic applications. ACS Nano 2010; 4:2178–2184.
90. Yoon SJ, Mallidi S, Tam JM, et al. Utility of biodegradable plasmonic nanoclusters in photoacoustic imaging. Opt Lett 2010;35:3751–3753.
91. Pantarotto D, Singh R, McCarthy D, . Functionalized carbon nanotubes for plasmid DNA gene delivery. Angew Chem Int Ed Engl 2004;43:5242–5246.
92. Liu Z, Winters M, Holodniy M, et al. siRNA delivery into human T cells and primary cells with carbon-nanotube transporters. Angew Chem Int Ed Engl 2007;46:2023–2027.
93. Liu Z, Cai W, He L, et al. In vivo biodistribution and highly efficient tumour targeting of carbon nanotubes in mice. Nat Nanotechnol 2007;2:47–52.
94. Welsher K, Liu Z, Sherlock SP, et al. A route to brightly fluorescent carbon nanotubes for near-infrared imaging in mice. Nat Nanotechnol 2009;4:773–780.
95. Kam NWS, O'Connell M, Wisdom JA, et al. Carbon nanotubes as multifunctional biological transporters and near-infrared agents for selective cancer cell destruction. Proc Nat Acad Sci USA 2005;102:11600–11605.
96. Liu Z, Tabakman S, Welsher K, et al. Carbon nanotubes in biology and medicine: in vitro and in vivo detection, imaging and drug delivery. Nano Res 2009;2:85–120.

97. Kostarelos K, Bianco A, Prato M. Promises, facts and challenges for carbon nanotubes in imaging and therapeutics. Nat Nanotechnol 2009;4:627–633.

98. de la Zerda A, Zavaleta C, Keren S, et al. Carbon nanotubes as photoacoustic molecular imaging agents in living mice. Nat Nanotechnol 2008;3:557–562.

99. Pramanik M, Swierczewska M, Green D, et al. Single-walled carbon nanotubes as a multimodal-thermoacoustic and photoacoustic-contrast agent. J Biomed Opt 2009; 14:034018.

100. Pramanik M, Song KH, Swierczewska M, et al. In vivo carbon nanotube-enhanced non-invasive photoacoustic mapping of the sentinel lymph node. Phys Med Biol 2009;54: 3291–3301.

101. Kim JW, Galanzha EI, Shashkov EV, et al. Golden carbon nanotubes as multimodal photoacoustic and photothermal high-contrast molecular agents. Nat Nanotechnol 2009;4:688–694.

102. de la Zerda A, Liu ZA, Bodapati S, et al. Ultrahigh sensitivity carbon nanotube agents for photoacoustic molecular imaging in living mice. Nano Lett 2010;10:2168–2172.

CHAPTER 4

RAMAN MICROSPECTRAL IMAGING FOR LABEL-FREE DETECTION OF NANOPARTICLE-MEDIATED CELLULAR AND SUBCELLULAR DRUG DELIVERY

TATYANA CHERNENKO, LARA MILANE, CHRISTIAN MATTHÄUS, MAX DIEM, AND MANSOOR AMIJI

4.1 INTRODUCTION

Advances in molecular biology, genetic engineering, and subsequent understanding of diseases at the molecular level have resulted in tremendous opportunity to develop novel therapeutic strategies that target cellular and subcellular structures. The United States National Institutes of Health has defined "nanomedicine" as molecular-scale intervention for prevention, diagnosis, and treatment of diseases [1]. Since nanomedicine encompasses a vast spectrum of research, from the development of nanodelivery systems to instrumentation needed for their visualization, synergistic progress in all the fields is of importance.

Pharmacological aspects of nanomedicine are concerned with improving the therapeutic effects of existing chemotherapeutics with delivery systems, or vehicles. Such systems are needed to solubilize hydrophobic agents and/or increase the concentration of the therapeutic payload at the tumor interstitium. This occurs if the delivery system can avoid recognition by the immune system, bypass excretion from the body, and remain circulating within the blood for sustained amount of time to allow population of the tissue of interest. The physical characteristics of the designed nanocarrier must thus be optimized with respect to the toxicity levels. Once the nanoparticle has reached the tissue, it must traverse the cell and subsequently avoid intracellular degradation, in order to deliver the drug to the site of action [2–4].

Cellular targeting is thus another parameter where nanomedicine has been involved extensively [5]. Nano-drug delivery systems must have the capability to not

Drug Delivery Applications of Noninvasive Imaging: Validation from Biodistribution to Sites of Action, First Edition. Edited by Chun Li and Mei Tian.

only traverse the tissue of interest but also penetrate the cell, in order to deliver the therapeutic payload to the subcellular site of action such as the nucleus, mitochondria, or lysosomes. This can be achieved by designing the nanocarrier with an active target, which will be taken up into the cell via facilitated or active transport, avoiding nonspecific endocytotic uptake pathways. Following successful intracellular delivery of the active pharmaceutical, the carrier must ultimately be easily degraded within the cell without inducing undesirable side effects. Thus, another requirement for the drug-encapsulating system must be its *in vivo* biocompatibility [6].

Apart from optimal fulfillment of the above requirements for drug delivery carrier formulations, there is a need for noninvasive imaging modalities, which have the capacity to gather information on the delivery system as well as its effect on the biological environment. There are a number of imaging techniques for diverse biological samples based on light absorption, emission, and scattering effects [7]. Since optical microscopy does not offer great contrast to minute subcellular biochemical changes, most imaging methods employ some staining procedure for sample visualization. These techniques include fluorescence imaging epifluorescence and confocal fluorescence microscopy. With the introduction of lasers, biological imaging has grown to include such techniques as two-photon microscopy (TPM), second- and third-generation imaging (SHIM, THIM), or fluorescence lifetime imaging (FCM, FRET, FLIM), among many others, for increased contrast and resolution.

Fluorescence microscopy, however, has inherent drawbacks such as low contrast, autofluorescence, and photobleaching, which compromise the continuous monitoring of a sample. It also cannot detect all the minute and subtle biochemical changes without the use of contrast-enhancing imaging agents, such as diverse chemical, biological, and/or immunohistochemical stains. Introduction of such fluorescent labels, however, may alter the physical and biochemical properties of the molecules of interest. Thus, alternate biochemical imaging modalities, which do not compromise the molecular integrity of the system under investigation, such as Raman microspectroscopy, have been gaining recognition [8, 9].

Raman microspectroscopy is a spectral imaging method, which couples Raman spectroscopy and optical microscopy to offer biochemical mapping of a sample, rendering a map based on inherent chemical signatures. Vibrational spectroscopy has the capability to image the sample based on its inherent biochemical signatures, consequently bypassing extensive staining procedures. The technique is based on the identification of molecular vibrations that are characteristic of distinct functional groups comprised within all molecules. For cells and tissues, these vibrational fingerprints arise from the functional groups of proteins, nucleic acids, lipids, phospholipids, and carbohydrates—the basic building blocks of mammalian cells. Such a biochemical map can provide information regarding a sample's chemical composition, protein conformation, and various subcellular metabolic activities, which are all reflected in the spectral features of cellular compartments [10, 11].

Using Raman spectroscopy, we have scrutinized subcellular architecture as well as intracellular dynamics of drug-delivery systems. Since it offers a label-free biochemical contrast of the sample, it is possible to probe the intracellular environment simultaneously with the delivery system [8, 12, 13]. With the noninvasive character of the technique, we are able to obtain biochemical information of the sample, which offers an advantage over many invasive imaging modalities.

Raman microspectroscopy, just as fluorescence microspectroscopy, can be carried out confocally. This allows the collection of a Raman image of a confocal slice of the sample deep within the sample, up to a depth of about 100 μm. This optical sectioning capability is necessary when following the uptake of drugs or drug delivery systems into cells, since one must assure that the particles are located within the cell and not on the cell surface.

In this chapter, we will concentrate on the applications of Raman spectral imaging to three parts. First, the sensitivity of Raman imaging to subcellular organization for an unstained cell. Next, Raman images of cells incubated with nanoparticles will be discussed as well as the potential for using Raman imaging to monitor the uptake and fate of the nanoparticle carrier systems. Finally, results will be introduced that demonstrate the capability of Raman spectral imaging to follow the release of drugs from nanoparticle carrier systems after intracellular penetration.

4.2 RAMAN MICROSPECTRAL IMAGING

4.2.1 Instrumentation Setup

The cellular images presented in this chapter were recorded using a CRM200 Confocal Raman Microscope (WITec GmbH, Ulm, Germany) employing 488-nm (c. 30 mW at 488 nm) excitation wavelength and a water immersion objective (60x/NA = 1.00, WD = 2.0 mm), with a spatial resolution or step size of 500-nm and 250-ms pixel exposure time. Excitation is provided by an air-cooled Ar ion laser (Melles Griot, model 532). The exciting laser radiation is coupled into a Zeiss microscope through a wavelength-specific single-mode optical fiber. The incident laser beam is collimated via an achromatic lens and passes a holographic band-pass filter to remove Raman signatures of the optical fiber before it is focused onto the sample through the water immersion objective (60x/NA = 1.00, WD = 2.0 mm). The spot of the exciting radiation is determined by the diffraction limit, which is approximately 300 nm for 488-nm excitation wavelength.

The sample is located on a piezo-electrically driven microscope scanning stage with an x,y resolution of c. 3 nm and a repeatability of ±5 nm and z resolution of c. 0.3 nm and ±2 nm repeatability. The sample is scanned through the laser focus in a raster pattern at a constant stage speed of fractions of one micrometer per second. The continuous motion prevents sample degradation at the focal point of the laser beam. Spectra are collected at a 0.5-μm grid, with a dwell time of 0.5 s.

Raman backscattered radiation was collected through the microscope objective and passed through a holographic edge filter to block Rayleigh scattering and reflected laser light before being focused into a multimode optical fiber. The 50-μm-diameter single-mode input fiber and the 50-μm-diameter multimode output fiber provided the optical apertures for the confocal measurement. The light emerging from the output optical fiber was dispersed by a 30-cm focal length, f/4 Czerny–Turner monochromator, incorporating interchangeable gratings (1800/mm, blazed at 500 nm and 600/mm, blazed at 500 nm). The light was finally detected by a back-illuminated deep-depletion, 1024 × 128 pixel charge-coupled device camera operating at −82 °C.

Fluorescence measurements were performed on the same instrument also using 488-nm excitation. For confirmation of mitochondrial distribution observed in Raman imaging, the cells were first scanned to collect the Raman spectra. Subsequently, MitoTracker® stain was carefully added and the fluorescence emission of the cell was rescanned at significantly lower laser power with a dwell time of 50 ms per data point.

Data were exported from the WITec software as ASCII files into our offline data processing software. Typical data sets (referred to as spectral hypercubes) contain c. 100×100 individual Raman spectra collected from individual voxels of the cell. Each spectrum consists of 1024 intensity points.

4.2.2 Methods of Analysis

Raw Raman spectral data sets of cells may consist of upward of 10,000 individual spectra that are very rich in chemical information. Thus, visual analysis of raw spectra would be an extremely arduous task, and hence we apply univariate and multivariate analyses to view and analyze the data, respectively. Among the multivariate methods used are hierarchical cluster analysis and vertex component analysis, as described next.

4.2.2.1 Hierarchical Cluster Analysis (HCA) Hierarchical cluster analysis (HCA) is a method that segments spectra in a data set by similarity. HCA searches for the smallest "distance," or highest similarity between two spectra. The similarity of the spectra is expressed as a matrix of similarity measures, or of correlation coefficients C_{LM}:

$$C_{LM} = \frac{\sum_{i=1}^{N}\left(S_i^L\right)\left(S_i^M\right)}{\sqrt{\sum_{i=1}^{N}\left(S_i^L\right)^2}\sqrt{\sum_{i=1}^{N}\left(S_i^M\right)^2}}$$

S^L and S^M are spectra from individual pixels within the data set, each comprised of N data points. The correlation matrix C_{LM} consists of $n \times n$ entries, where n is the number of spectra in a data set. Each element in this matrix is the projection of spectral vector S^L onto spectral vector S^M. Since the vectors are normalized, their dot products can range from one (identical spectra) to zero for completely different spectra.

$$\begin{pmatrix} - & C_{12} & C_{13} & C_{14} & \cdots & C_{19} \\ & - & C_{23} & C_{24} & \cdots & C_{29} \\ & & - & C_{34} & \cdots & C_{39} \\ & & & & & \\ & & & & & - \end{pmatrix} \rightarrow \begin{pmatrix} - & C_{n2} & C_{13} & C_{14} & \cdots & C_{18} \\ & - & C_{n3} & C_{n4} & \cdots & C_{n8} \\ & & - & C_{34} & \cdots & C_{38} \\ & & & & & \end{pmatrix} \rightarrow \begin{pmatrix} - & C_{12} & C_{13} & \cdots & C_{n7} \\ & - & C_{23} & \cdots & C_{n7} \\ & & & & \\ & & & & - \end{pmatrix}$$

After calculation of the correlation matrix of all spectra in the data set, spectra are grouped together in an iterative process. In brief, the correlation matrix is searched for the two most similar spectra. These are merged into a single object and its similarity measure is recalculated by averaging the scores for the two original spectra. A new matrix is then calculated by comparing this new item to all other spectra in the data set. Again the two most similar items are merged. This process is iterated until all spectra in the data set have been grouped into a single item.

Spectra in a group are assigned the same color code to create a pseudo-color image that is based on spectral similarity and does not require any pure component spectra. The mean spectra for each cluster reflect the different chemical composition of each cluster.

4.2.2.2 *Vertex Component Analysis (VCA)* Vertex component analysis (VCA) is a multivariate color decomposition method that decomposes the spectrum of each pixel into contributions from the most dissimilar spectra of the data set, which are considered the "pure component" or "end-member" (EM) spectra. VCA proceeds from the original dimensionality and reduces the dimensionality step-by-step by a process known as "orthogonal subspace projection." In each of these steps, the problem is reduced from a p-simplex to a $(p - 1)$-simplex. Thus, VCA decomposes the spectra into a few basis spectra, or EMs, which represent the degree to which spectra can be unmixed.

Subsequently, all spectra S in a data set are expressed as linear combinations of the EM spectra E:

$$S = EA + W$$

where E represents the matrix that is constructed from the spectra of the EMs, A is a matrix that determines the fractional contribution α of each EM to each spectrum, and W accounts for arbitrary fluctuations such as noise. The fractional abundances α are normalized to 1:

$$\sum_{i=1}^{n} a_{ij} = 1$$

for a given spectrum.

The unmixing solution is now to find E and A with the constraint that both matrices contain only positive contributions, since negative EMs or abundances do not have a physical meaning. Mathematically, the problem can be solved by positive matrix factorization approaches, and several algorithms have been described in the literature.

The abundance values for a given EM are scaled and converted to a monochrome intensity map of the imaged object. An overlay of these RGB color maps yields a pseudo-color map with mixed colors indicating the intensity values of each of three EMs.

4.2.3 Raman Imaging of Cellular and Subcellular Structures

Cells are composed of membrane phospholipids, proteins, as well as intracellular organelles such as mitochondria, ribosomes, and nucleus. From a chemical perspective, cells are mainly composed of proteins, carbohydrates, nucleic acids, lipids/phospholipids, and cholesterol as the main component. A microenvironmental imbalance within cells

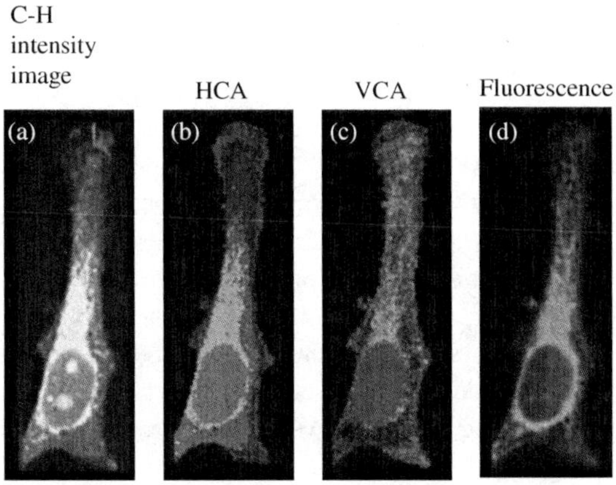

FIGURE 4.1 Univariate (C–H stretching intensity) image of a HeLa cell grown on CaF_2. (a) C–H intensity image, (b) image following HCA, (c) image following VCA, and (d) fluorescence image.

or any biochemical inconsistence therein in any of these cellular features may result in disease. In order to visualize any of the possible changes within a cell, fluorescence microscopy has been widely utilized with an aim of detecting specific changes. However, the biochemical imbalance must first be known in order to be assigned a particular label.

Raman microspectroscopy offers the capability to image noninvasively a biological sample, obtaining a biochemical functional map. Since this technique is based on recording inherent vibrational patterns of biomolecules, no label is required. Thus Raman microspectroscopy is very sensitive to chemical changes of subcellular constituents [14, 15].

In Raman spectroscopy, a very small volume (c. $0.3 \times 0.3 \times 1.3$ μm^3), or a voxel, is sampled at one point in time. The spectral information thus acquired within the diffraction-limited volume reflects the biochemical composition of the particular volume sampled. However, within that small volume, minute biochemical variations are present. For example, if we scrutinize voxel spectra recorded in the nucleus of a cell, the observed spectrum describes contributions due to nucleic acids and proteins; if the voxel is in the perinuclear region, the protein signal is overlaid with signals originating from membrane-rich organelles, such as the endoplasmic reticulum (ER), Golgi, or mitochondria. Due to the complex biochemical information contained within one Raman spectrum, inter-pixel variations within the sample are difficult to discriminate without the use of multivariate methods of analysis.

Biochemical information obtained from a cell is comprised of a superposition of molecular vibrational bands, over a large spectral range (500–4000/cm). These chemical signatures have been previously interpreted and will not be addressed here [16, 17]. The simplest way to visualize the obtained data set is by viewing the abundance of intensity over one region (or *univariate* image). The intensity region usually plotted is the C-H stretching region of 2800–3100/cm. An example of this univariate analysis is shown in Figure 4.1a. However, as can be seen from the image, only topography of the cell is represented, apart from a prominent nucleus. The nucleus is rendered visible, yet the chemical contrast of the diverse subcellular compartments is not easily discernible.

Thus to enhance biochemical contrast and characterize components within the intracellular matrix, we utilize multivariate analysis methods, which sort the entire spectral hypercube. These methods classify spectra and subsequently group them into biochemically distinct clusters based either by similarity (as done by HCA) or dissimilarity (as done by VCA).

Applications of multivariate methods of analysis to a data set collected from a HeLa cell are shown in Figure 4.1b and c. Both algorithms distinguish quite easily the spectral contributions of the nucleus (pseudo-colored blue) as well as gradations within the cytoplasm. These gradations are due to distinct biochemical environments, which are reflected in the vibrational spectra collected.

HCA and VCA have the capability to also discern the subcellular environments, as is shown in Figure 4.1b and c, where the variations in the green regions are in fact due to mitochondrial aggregations. This is corroborated by staining the mitochondria with a MitoTracker Green® dye in Figure 4.1d. Mitochondrial aggregations are seen to be highest in the perinuclear region, as is seen in white in Figure 4.1d, where the signal overfilled the detector. HCA rendered the cluster for high mitochondrial aggregation in the same area. VCA, on the other hand, not only distinguished regions of high mitochondrial aggregation but also generated a higher spatially resolved biochemical map of the subcellular heterogeneity.

Diverse unmixing algorithms have been previously applied to tissues [18], in order to distinguish cancerous or metastatic regions from normal tissue, as well as cells [8, 12, 13]. Algorithms of such a nature have shown great potential in distinguishing, within thousands of spectral data points, an underlying biochemical theme that either unites biological regions within cells or tissues or separates them into distinct clusters.

4.3 ILLUSTRATIVE EXAMPLES OF INTRACELLULAR DELIVERY

4.3.1 Polymeric Nanoparticle Systems

Preparation of poly(epsilon caprolactone) (PCL) and poly(D,L-lactide-co-glycolide) (PLGA) nanoparticles was performed as described below. Pluronic® F-108 is an ABA triblock copolymer of poly(ethylene oxide) (PEO) and poly(propylene oxide) (PPO). PCL/Pluronic® F-108 and PLGA/Pluronic® F-108 blend nanoparticles were prepared using solvent displacement in an acetone/water system as previously described [19]. In short, a solution of PCL or PLGA was prepared in acetone or ethanol, respectively, and was introduced into an aqueous solution containing known concentrations of Pluronic® (F-68 or F-108) under magnetic stirring. Typically, a 1% (w/v) solution of PCL in 50 ml of acetone was introduced into an aqueous solution of Pluronic®. The rate of addition was controlled through a burette at about 1 ml/min, and stirring was maintained at a speed that facilitated formation of a vortex in the solution. The nanoparticles were either freeze-dried at this point (with surfactant remaining in the bulk) or after removal of the excess (unadsorbed) stabilizer either by centrifugation (10,000 rpm for 20 min followed by washing twice with 50 ml of water) and/or equilibrium dialysis (against water using dialysis membrane having 15,000 Da cutoff—from Spectrum Labs, United States, which results in selective leaching of free Pluronics in solution through the membrane).

N-palmitoyl-(D31)-D-*erythro*-sphingosine (C_{16}-d_{31} ceramide; deuterated) was purchased from Avanti Polar Lipids (Alabaster, AL). The PCL nanoparticles were loaded with C_{16}-d_{31} ceramide at 20% w/w.

4.3.2 Cell Culture

Human HeLa cells (cell line CCL-2, American Type Culture Collection, Manassas, VA) were grown in 75-cm^3 culture flasks (Fisher Scientific) with 7.0 ml of Dulbecco's Modified Eagle's medium (DMEM, ATCC) and 10% fetal bovine serum (FBS, ATCC). Cells were incubated at 37 °C and 5% CO_2. Cells were grown on CaF_2 windows to avoid scattering from the regular glass windows. Prior to data collection, cells were fixed in a 10% phosphate-buffered formalin solution (Sigma-Aldrich, St. Louis, MO) and washed in phosphate-buffered saline. Incubations with nanoparticle systems were carried out at 37 °C. All Raman measurements were performed at room temperature.

4.3.3 Raman Imaging of Drug Delivery Systems

Chemotherapeutic payload delivery to a specific site of action is vital in anticancer treatments or gene therapy. Several different types of therapeutics do not reach their specific tissue or cellular target due to being prematurely metabolized or excreted from the body. Others, which do reach the desired site, encounter cellular barriers, such as the plasma membrane, which further prevents its subsequent biochemical function. In order to circumvent the problems of degradation and/or excretion of the administered pharmaceuticals, diverse delivery systems have been increasingly utilized. Biodegradable and biocompatible materials are the most desirable for the formulations of the drug delivery systems [4, 20, 21].

Biodegradable polymers and their blends have been utilized in the medicinal field prior to their use as platforms for nanodelivery systems. Mostly, their employment is seen in cardiovascular stent coatings, or as the largest ingredient in aerosol particles for oral drug delivery [22, 23]. These polymers have been further adapted for formulations in nanodelivery methods due to their ease of degradation without cytotoxic effects *in vivo* [24].

The most extensively utilized technique for monitoring drug internalization into cells is fluorescence microscopy, for which a fluorophore must be incorporated into the delivery system alongside the drug of interest. Thus, it is not necessarily the therapeutic itself that is monitored, but rather the fluorescing dye. Another way to visualize the distribution pattern of the drug is to label it specifically with a fluorophore. This, however, may change its physicochemical properties, rendering it fruitless for a therapeutic effect.

In order to circumvent these problems, we employ Raman microspectroscopy to provide a noninvasive and nondestructive method for cellular imaging. Since the technique is based upon the identification of molecular vibrations that are characteristic of different functional groups contained within molecules, it is a label-free method and hence offers an advantage over extensive fluorescence staining procedures. We have utilized this method to study two biocompatible and

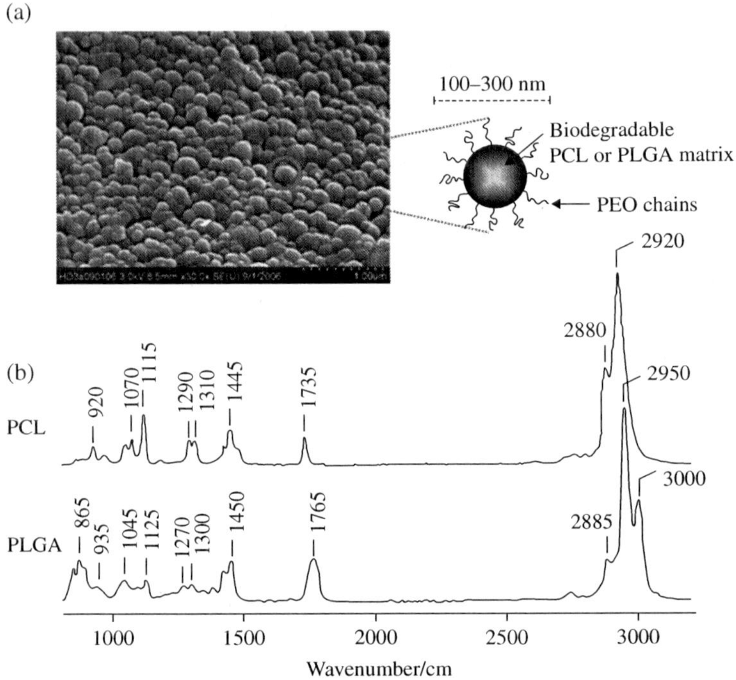

FIGURE 4.2 (a) Scanning electron microscopy image of PEO-modified biodegradable polymeric nanoparticles. The schematic on the right illustrates the basic architecture of the 100–300-nm particle, which contains either PCL or PLGA biodegradable matrix with PEO chains extending outside. (b) Raman spectra of the PCL or PLGA nanoparticles, demonstrating their characteristic peaks at 1730 and 2920/cm for PCL and 1760 and 2950/cm for PLGA [13].

biodegradable polymer-based nanodelivery systems made of PCL and PLGA. The two systems have been modified with a PEO (using Pluronic® F-108), because such modifications *in vivo* prolong the nanoparticles' circulation time, decrease their aggregations, as well as prevent their clearance from blood by opsonization. This process involves recognition of the exogenous nanoparticles as pathogens, which marks them for digestion by macrophages [25].

The two raw materials were first analyzed using Raman spectroscopy in order to establish their vibrational signatures for subsequent characterization of their intracellular behavior, kinetics, and dynamics upon internalization inside the cell body. The chemical spectra obtained are shown in Figure 4.2. PCL and PLGA both exhibit their highest peak intensities in the C-H stretching region (between 2800 and 3100/cm) at 2920 and 2950/cm, respectively. The C=O stretches specific to the ester linkages are located at 1735 and 1765/cm. Other Raman peaks of less intensity can be observed between 1000 and 1500/cm. We have thus been able to follow the uptake of the nanoparticles without any extraneous labels or dyes, detecting the signal of the nanoparticles biochemically.

In order to establish the internalization patterns of the two biocompatible nanoparticle systems, HeLa cell cultures were incubated with either system for different time

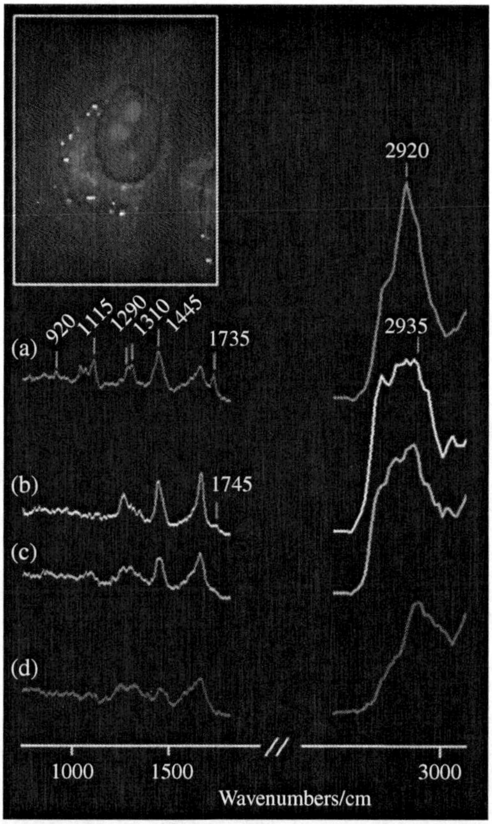

FIGURE 4.3 *Inset*: Pseudo-color Raman image (VCA analysis) of a HeLa cell incubated with PEO-modified PCL nanoparticle-containing medium for 7 h. Red regions: PCL nanoparticle aggregates, yellow regions: lipid/phospholipid inclusions, green: membrane-rich organelles such as ER, Golgi apparatus, and mitochondria; blue: proteins of the cell body. Traces (a–d) represent the Raman spectra of the corresponding regions. For color details, please see color plate section.

intervals. Cells were then either fixed in formalin solution or grown in nanoparticle-free cell culture medium for additional time periods in order to follow nanoparticles' internal degradation pattern. An incubation time for pure PCL nanoparticles of 7 h was chosen as a starting point since nanoparticle internalization seems to reach a plateau at this time point due to their intracellular saturation and accumulation. This was further verified with longer incubation times, which did not result in any increase in concentration of the nanoparticles within the cell. One such HeLa cell is shown in Figure 4.3.

Intracellular inclusions became consistently visible in the HeLa cells after 7-h incubation of cell culture with nanoparticle-containing medium. Since the HeLa cell line does not exhibit internal lipid/phospholipid inclusions, they appear only after nanoparticle internalization. In Figure 4.3, intracellular inclusions are clearly seen and pseudo-colored red and yellow. Red inclusions exhibit spectral features prevalent to the PCL nanoparticles themselves, while the yellow inclusions underscore features specific to phospholipid aggregates. [12] This is further seen from the corresponding spectra, where the red trace has peaks correlating with the pure PCL-nanoparticle spectrum shown in Figure 4.2. The C-H stretching intensity region still retains the sharpest peak at 2920/cm, while the peak due to ester linkage is seen at 1735/cm. The peaks identified in the yellow trace, namely, the ester peak at 1745/cm and the shifting and rise of a number of peaks in the C-H stretching region, are prevalent in phospholipid inclusions. These inclusions are seen to be overlapping with the nanoparticle aggregates in red, demonstrating the formation of endocytotic vesicles within the cells, which surround the nanoparticles.

In order to establish an intracellular pattern of degradation of the PCL nanoparticles, a time-course experiment was performed. After 7-h incubation, the PCL-containing medium was removed and replaced with fresh nanoparticle-free medium, in order to prevent any further nanoparticle internalization and solely follow their degradation. As can be seen from Figure 4.4, PCL nanoparticles do not get fully degraded even after a 24-h incubation period. This is clearly seen from the persistence of red inclusions within the cell body, which is further seen from the corresponding EM spectrum below the image. The two PCL-characteristic peaks are still easily identified (at 1735/cm and 2920/cm) from the signatures of the diverse subcellular compartments. As the internal translocation and sorting of the nanoparticles persists, their spectral features merge with the spectral characteristics of the cellular components. For example, the 1735/cm signal is replaced by a shallower and broader peak at 1745/cm. In addition, the peak at 2920/cm decreases and is accompanied by an increase of several additional peaks in the C-H stretching region. This can be seen from the green EM spectrum, where the peak shapes and ratios in the C-H stretching region clearly vary with incubation time. This can be explained by an increasing density of phospholipids encompassing the nanoparticles, detected via an increase in the aliphatic side-chain vibrational bands. Such integration of the nanocarriers into vesicles of high phospholipid content, or secondary endosomes, is consistent with previous studies, where early and secondary endosome formation was detected with nanoparticle internalization [26, 27], as part of their metabolic pathway.

With the increase of incubation time of the cell cultures with nanoparticle-free medium proceeding their incubation with the nanoparticle-containing medium, a degradation pathway may be deduced. This is further elucidated upon the use of multivariate analyses methods, which aid in data visualization. Diverse metabolic pathways may thus be followed via spectral changes. First, the degradation of the nanoparticles occurs via hydrolysis of the ester bonds, which is manifested in the 1735/cm peak. Furthermore, subcellular sorting of the nanocarriers can be monitored by the variations of the C-H stretching region, which changes from one prominent peak at 2920/cm to a broader and more jagged peak due to engulfment of the nanoparticles and subsequent integration of the PCL signal with that of the intracellular vesicle formations.

Another biocompatible material utilized in nanoparticle preparation is PLGA polymer. It has faster kinetics when compared with internalization of PCL nanoparticles. This is underscored in Figure 4.5 and Figure 4.6, where the PLGA nanoparticles are seen in the cell body after 2-h incubation. As with PCL nanoparticles, PLGA internalization may also be followed by the ester linkage peak centered

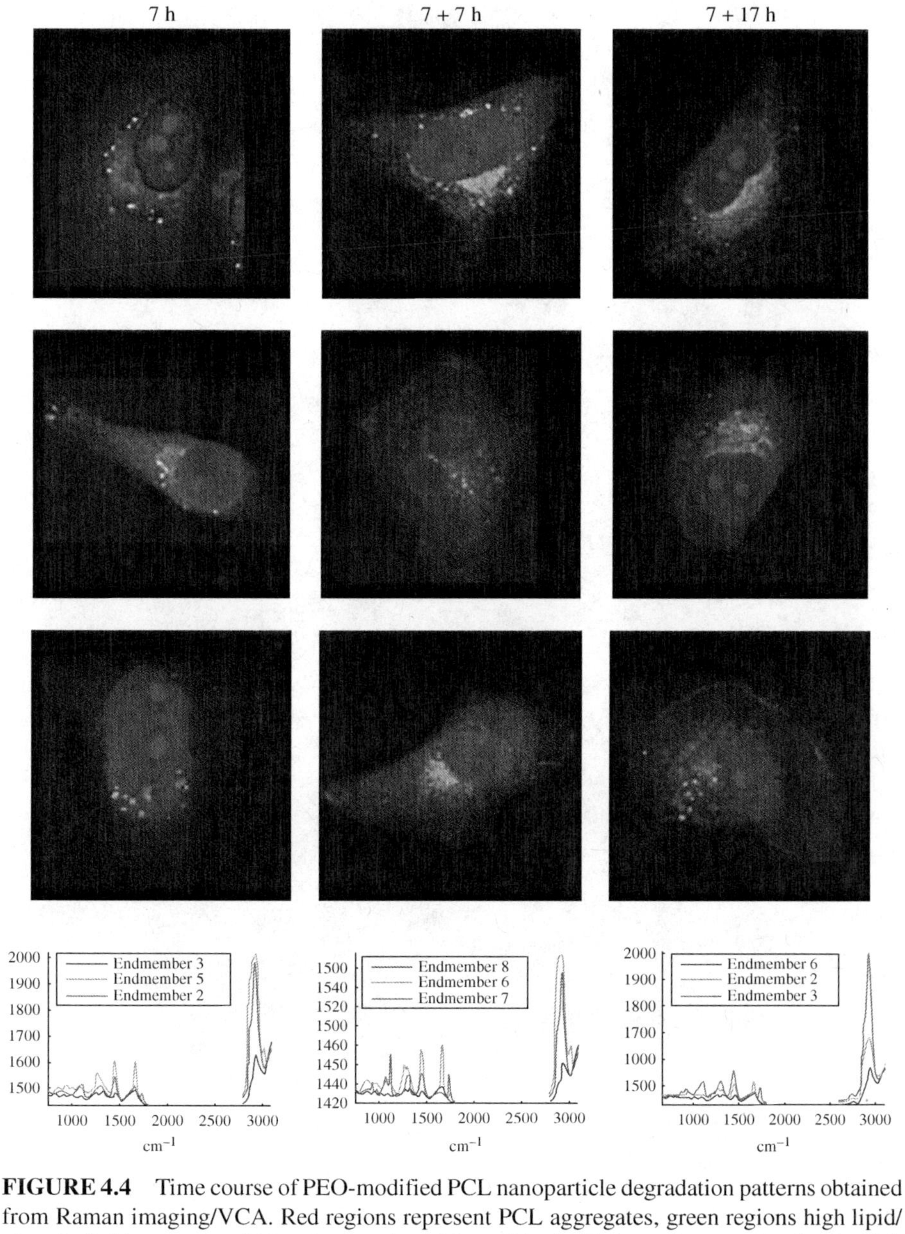

FIGURE 4.4 Time course of PEO-modified PCL nanoparticle degradation patterns obtained from Raman imaging/VCA. Red regions represent PCL aggregates, green regions high lipid/phospholipid content, and blue the proteins of the cell body. Top row: images of cells after 7-h incubation with PCL-containing medium. Middle row: images of cells incubated with PCL-containing medium for 7 h followed by incubation with fresh medium for another 7 h. Bottom row: images of cells incubated in PCL-containing medium for 7 h followed by incubation for another 17 h in fresh medium. EM spectra for the three incubation times are shown on left, corresponding to cells of the last row. For color details, please see color plate section.

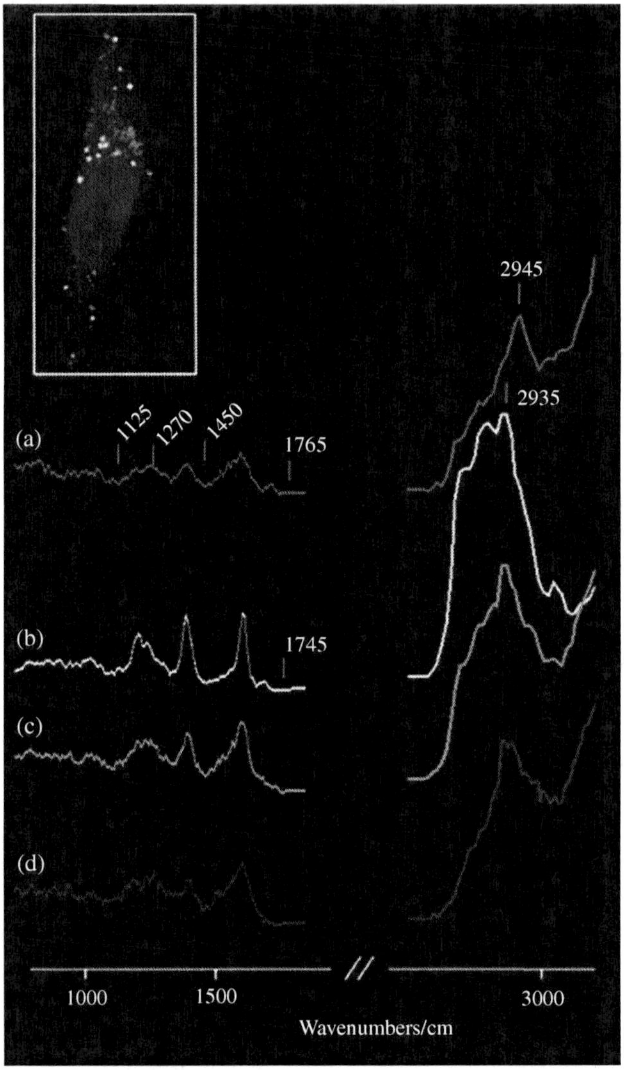

FIGURE 4.5 *Inset*: Pseudo-color Raman image (VCA analysis) of a HeLa cell incubated with PEO-modified PLGA-containing medium for 2 h. Red regions: PLGA aggregates, yellow regions: lipid/phospholipid inclusions, green regions: membrane-rich organelles such as ER, Golgi apparatus, and mitochondria; blue regions: proteins of cell body. Traces (a–d) represent the Raman spectra of the corresponding regions. For color details, please see color plate section.

at 1765/cm, as well as a prominent peak at 2945/cm in the C-H stretching region. Unlike the haphazard aggregations of the PCL particles within the cell body, the PLGA particulates tend to aggregate in the perinuclear region, which are seen as pseudo-colored red inclusions in Figure 4.5. This demonstrates a possible integration of the PLGA nanoparticles with the subcellular structures such as the ER and Golgi apparatus, and has been previously observed utilizing opossum kidney (OK) renal tubule and human bronchial epithelial (HBE) cell lines [26].

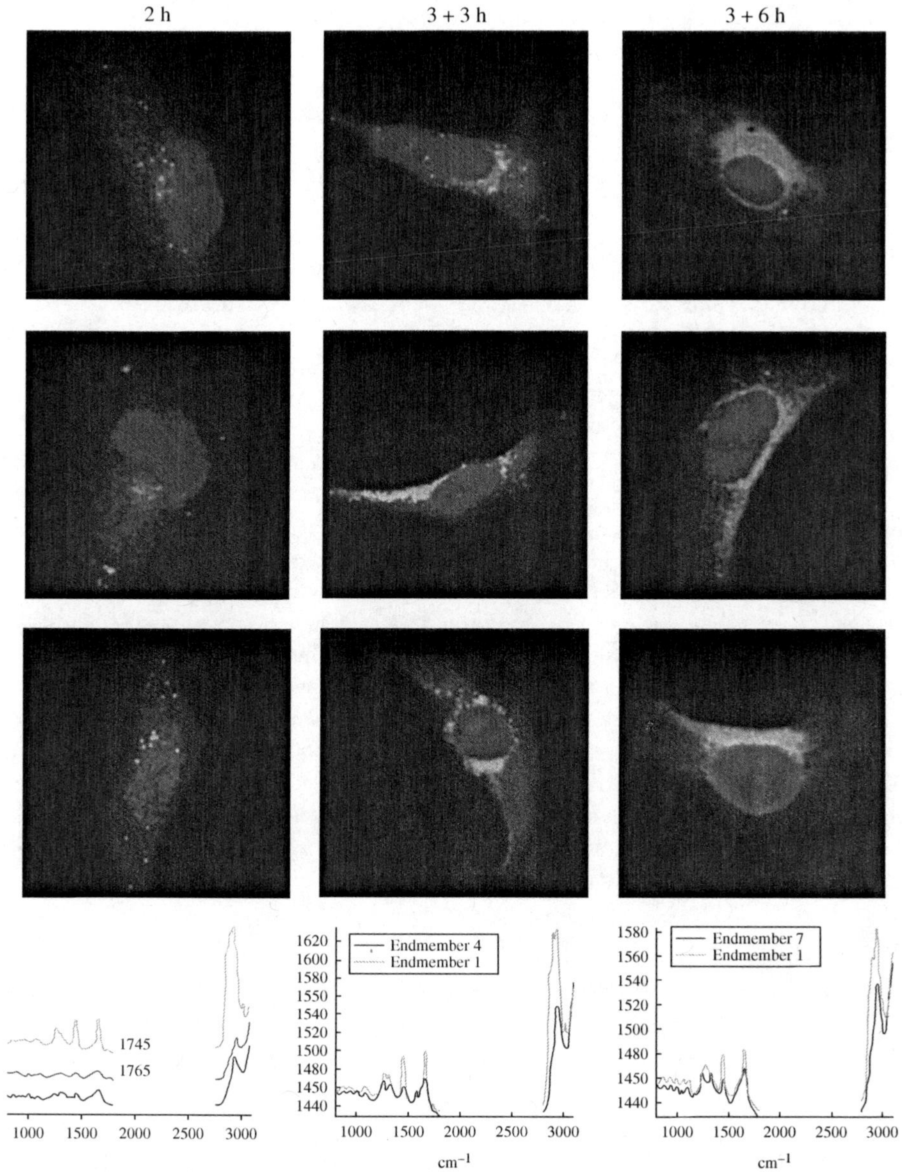

FIGURE 4.6 Time course of PEO-modified PLGA nanoparticle degradation patterns obtained from Raman imaging/VCA. Red regions represent PLGA aggregates, green regions high lipid/phospholipid content, and blue the proteins of the cell body. Top row: images of cells after 2-h incubation with PLGA-containing medium. Middle row: images of cells incubated with PLGA-containing medium for 3 h followed by incubation with fresh medium for another 3 h. Bottom row: images of cells incubated in PLGA-containing medium for 3 h followed by incubation for another 6 h in fresh medium. EM spectra for the three incubation times are shown on left, corresponding to cells of the last row. For color details, please see color plate section.

Incubation of the cell cultures with fresh medium, after incubation with nanoparticle-containing medium, a degradation pattern may be established of the PLGA system, shown in Figure 4.6. Upon a 3-h incubation in fresh medium, PLGA-specific signal was not detected as a separate EM following VCA analysis. Rather, a signal was detected rich in phospholipid character, delineating distributed inclusions, which is pseudo-colored green (Fig.4.6 middle column). This rendered abundance corresponds to the overlay of the ER and Golgi apparatus with the compartments digesting or retaining residual PLGA nanoparticles. This is further seen from the EM spectra, where the green trace has features attributed to high phospholipid content, which is prevalent in vesicle formations. As the incubation times in fresh medium are increased, the PLGA-specific ester peak disappears and is followed in synergy with an increase of a peak at 1745/cm, which is characteristic of diverse lipid and phospholipid formations. This is consistent with the previous studies of PLGA nanoparticles being sorted into Golgi-associated late endosomes [26]. After 6-h incubation in the fresh medium, the nanoparticle-characteristic peaks are no longer visible, along with distinct phospholipid inclusions that encompassed the residual PLGA particulates. This shows a complete digestion of the introduced PLGA nanoparticles, followed by a label-free method.

Our results closely parallel previous studies that show PCL degradation patterns to be much slower than those of PLGA. Furthermore, PCL particles remain intact particles in the cell body for a longer period [6, 26, 28, 29]. The exact intracellular pathway of the nanoparticle digestion is still not fully known. However, it has been suggested that nanoparticle degradation may follow a number of pathways such as mere entrapment in the sorting vesicles, excretion from recycling endocytotic vesicles, or lysosome degradation [27]. As the particles are incorporated into the cell body, their chemical signatures are easily distinguished from the rest of the cell; however, as the catabolic pathways proceed, shifting, and decreasing/increasing peak heights and ratios in the rendered Raman spectra delineate the system's degradation paths.

4.3.4 Raman Imaging of Drug Release from Polymeric Nanoparticles and Intracellular Distribution

Raman microspectroscopy offers great advantage over diverse staining methods for distinguishing the internalization of the pharmaceutical carriers as well as giving insight into the dissociation patterns of the payload. Research on intracellular drug uptake, its fate, effects, and metabolism inside the cell is of vital consequence. Thus, apart from imaging cellular features, we have utilized Raman microspectroscopy for observations of uptake and metabolic activities of biologically active substances. One such pharmaceutical agent is the drug ceramide. We have utilized PCL nanoparticles as the loading platform for delivering ceramide to cells.

Ceramide is a second messenger signaling molecule involved in differentiation, proliferation, and apoptosis [30]. It has also been found to reduce the apoptotic threshold level via activation of pro-apoptotic inducers. In order to allow the drug from the intracellular milieu and the PCL nanoparticles spectroscopically, we have exploited isotopic labeling, namely, deuteration. Ceramide utilized in this experiment is a C_{16} ceramide derivative, where one of the aliphatic side chains was constructed with deuterium covalently bound to the carbon backbone. Raman spectra of the raw

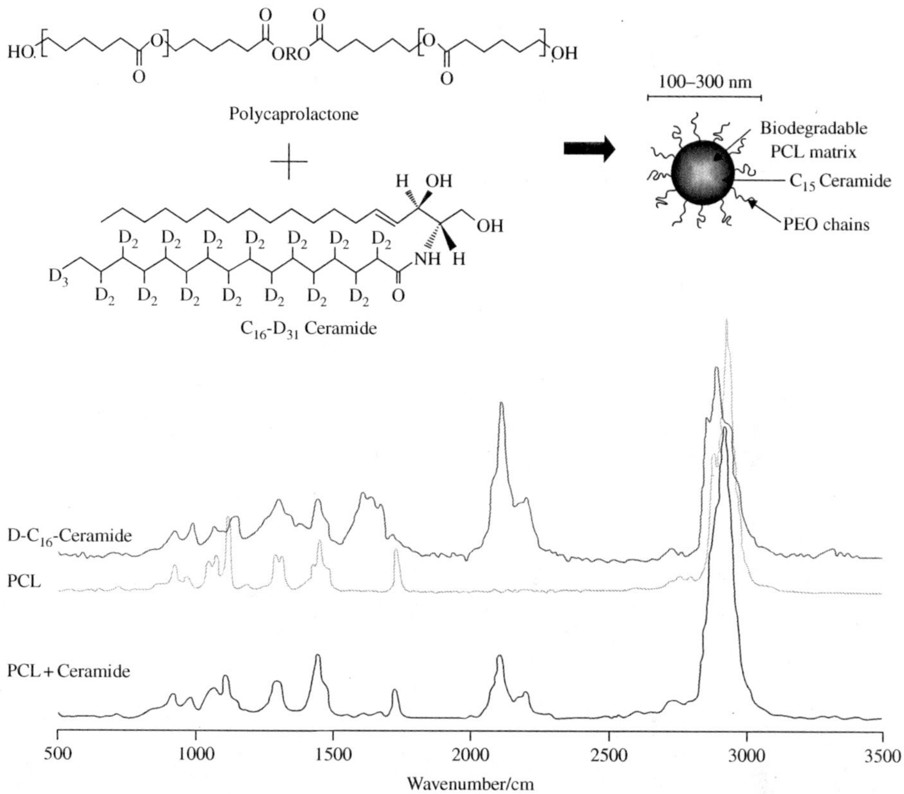

FIGURE 4.7 Raman spectra of the deuterated C_{16} ceramide derivative (light gray trace), PCL polymer (dark gray trace), as well as the ceramide-loaded nanoparticle (black trace). Molecular structures for the two systems as well as a cartoon of the loaded nanoparticle are shown above the spectra.

materials are shown in Figure 4.7. The most prominent peak in the spectrum of the loaded nanoparticle is still within the C-H stretching region, centered at 2920/cm, which corresponds to the PCL polymer. The spectrum characteristic of the C_{16}-d_{31} ceramide exhibits high-intensity C-D stretching bands at 2105 and 2203/cm. As Raman spectroscopy is very sensitive to small chemical differences, the C-D stretching vibrations are easily distinguished from the rest of the biological signatures, spectra of which are barren in the region of 1800–2800/cm.

A time-course experiment was performed in order to follow the dissociation patterns of the ceramide payload from the PCL nanoparticles. Seven-h and 14-h incubation points were taken based on the reproducibility of the results. Abundances of three EM spectra were again utilized in order to reconstruct the biochemical functional map of the cells. The blue abundance correlates with the protein-rich spectrum, delineating the cell body; green intracellular areas depict membrane-rich organelles such as the ER and mitochondria; while the red inclusions within the images are specific to EM containing high C-D stretching intensities due to deuterated alkyl chains of the ceramide.

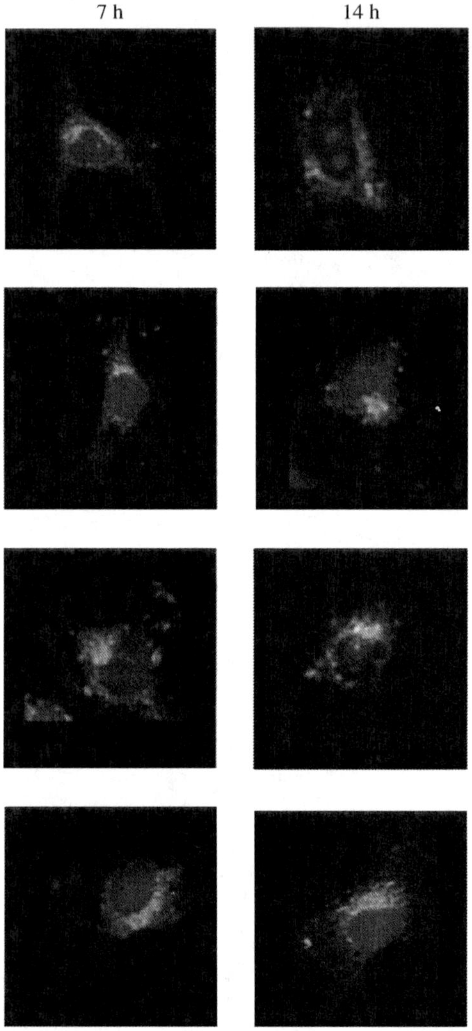

FIGURE 4.8 Time-course experiment of ceramide-loaded nanoparticle distribution within a HeLa cell culture. Top row: 7-h incubation period of HeLa cells. Bottom row: 14-h incubation period of HeLa cells. All cell data analyzed via VCA shows cell body: blue intensity, membrane-rich organelles: green intensity, internalized nanoparticle: red intensity. For color details, please see color plate section.

As with the non-loaded PCL nanoparticles, the loaded nanoparticles exercised similar kinetics, thus establishing a 7-h time point as the starting incubation time, since prior to 7 h cells did not contain detectable concentrations of the nanoparticles. At 14-h incubation, PCL nanoparticles were still found detectable within the cell body, consistent with the non-loaded PCL carriers. This is further seen in Figure 4.8.

After 7-h incubation period, the C_{16}-d_{31} ceramide derivative is still mostly associated with the nanoparticles. This is seen from the red inclusions in the figure, spectral

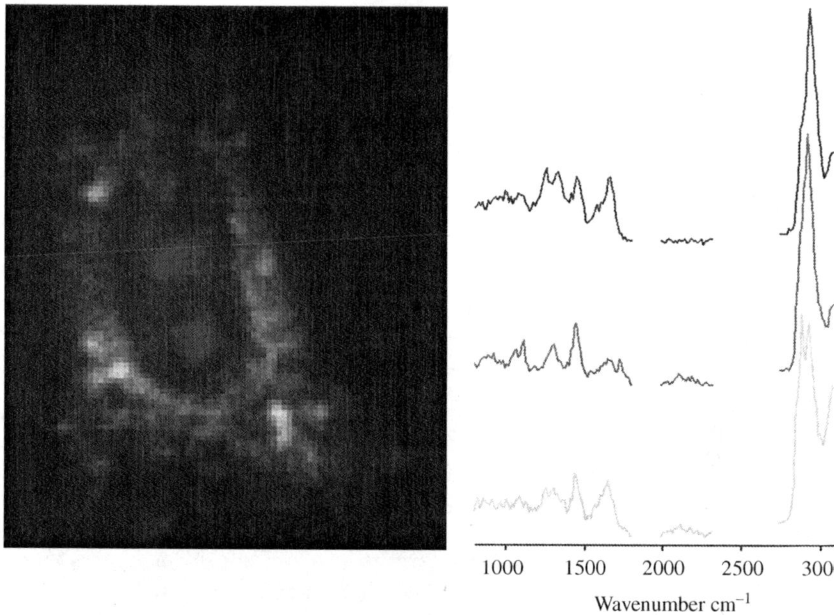

FIGURE 4.9 HeLa cell analyzed via VCA of a 14-h incubation period. Blue region denotes highest protein content; green abundance denotes membrane-rich regions, while the red clusters delineate nanoparticle aggregations. Spectra on the right of the image are of the three overlaid abundances rendered by the VCA. For color details, please see color plate section.

signatures of which demonstrate presence of the PCL nanoparticles along with the signatures of the deuterated drug (EM spectra not shown). VCA-analyzed images of all cells incubated for 7 h with the loaded nanoparticles exhibit similar spectral features where the nucleus (pseudo-colored blue) and the membrane-rich subcellular organelles (green) are easily differentiated from the inclusions of encapsulated PCL nanoparticles along with its ceramide cargo (red).

After incubation of the cell cultures with the loaded nanoparticles for 14 h, the ceramide payload begins to dissociate from the nanoparticle and leach out into the cell body. Viewing Figure 4.8 does not give insight into this particular phenomenon, where the same three EM abundances are overlaid, namely, delineating the nucleus, membrane-rich organelles, and the nanoparticles. However, the EM spectra provide more information, as is shown in Figure 4.9, where one of the cells is analyzed in greater detail.

VCA-analyzed image of a HeLa cell grown in culture for 14 h with the loaded nanoparticles is shown in Figure 4.8. Three most distinct biochemical abundances are overlaid to create the RGB map of the cell. Nucleus and nucleoli are shown pseudo-colored dark and light blue, respectively, while the red inclusions exhibit a signal that correlates to the aggregations of C_{16}-d_{31} ceramide-loaded PCL nanoparticles. The green intracellular region, on the other hand, along with portraying the character of membrane-rich organelles, also contains spectral signatures that can be attributed to signal of the C-D stretching vibrations of C_{16}-d_{31} ceramide at 2105 and 2203/cm. Also, a prominent peak in the C-H stretching region has become visible at 2890/cm that also corresponds to the presence of

C_{16}-d_{31} ceramide. This demonstrates that along with the biological information obtained of the subcellular architecture, namely, the localization of the mitochondria and other membrane-rich cellular compartments, Raman microspectral imaging has the capability to evaluate drug destination, which in this particular case is in the perinuclear region.

4.4 CONCLUSIONS

Nanoparticulate drug carriers are useful for achieving target-specific delivery of chemotherapeutics and for enhancing the intracellular availability of a therapeutic payload. The use of drug carriers can transform exogenous molecules with undesirable pharmacokinetics into therapeutic agents. Delivery systems are also employed in order to solubilize hydrophobic therapeutics that otherwise once administered, would be cleared from the body. Immunotherapeutic as well as chemotherapeutic strategies may also benefit from nanodelivery systems due to their potential use in vaccine administration.

With the increased interest in nanocarriers of cytologically active substances in the pharmaceutical field, research on their intracellular fate, effects, and metabolism has vital consequence. It is thus critical to develop in parallel the diverse nanodevices and methods to establish their biocompatibility, intracellular trafficking, and dynamics.

Vibrational spectroscopic techniques, such as Raman microspectroscopy, coupled with optical microscopy to provide a noninvasive and nondestructive method for cellular imaging. Cellular signatures obtained utilizing Raman imaging method are able to describe the subcellular pathway of the distribution of nanoparticles via decreases, increases, and shifting of molecule-characteristic peaks.

Following these chemical subcellular variations can portray the fate of the diverse particles, their metabolism, and excretion. Raman microspectroscopy has been employed for observations of metabolic activities and uptake of biologically active substances, not only their carrier systems. We can thus follow the uptake and the distribution of nanoparticles as well as the payload, via their characteristic chemical signatures within the cell, noninvasively mapping the distribution and degradation patterns.

ACKNOWLEDGMENTS

The research described is this chapter was partially supported by the National Cancer Institute (NCI) of the National Institutes of Health through grants CA-135594 (to M.A.) and CA-090346 (to M.D.). Additionally, both Tatyana Chernenko and Lara Milane were fellows of the Interdisciplinary Graduate Education and Research Training (IGERT) grant on Nanomedicine Science and Technology co-funded by the NCI and the National Science Foundation. Scanning electron microscopy was performed by William Fowle at the Electron Microscopy Center at Northeastern University.

REFERENCES

1. Saha M. Nanomedicine: promising tiny machine for the healthcare in future – a review. Oman Med J 2009;24 (4):242–247.
2. Jabr-Milane L, van Vlerken L, Devalapally H, et al. Multi-functional nanocarriers for targeted delivery of drugs and genes. J Control Release 2008;130 (2):121–128.
3. Brown R, Links M. Clinical relevance of the molecular mechanisms of resistance to anti-cancer drugs. Expert Rev Mol Med 1999;1 (15):1–21.
4. Kaul G, Amiji M. Long-circulating poly(ethylene glycol)-modified gelatin nanoparticles for intracellular delivery. Pharm Res 2002;19 (7):1061–1067.
5. Jabr-Milane LS, van Vlerken LE, Yadav S, et al. Multi-functional nanocarriers to overcome tumor drug resistance. Cancer Treat Rev 2008;34 (7):592–602.
6. Chawla JS, Amiji MM. Biodegradable poly([var epsilon]-caprolactone) nanoparticles for tumor-targeted delivery of tamoxifen. Int J Pharm 2002;249 (1–2):127–138.
7. White NS, Errington RJ. Fluorescence techniques for drug delivery research: theory and practice. Adv Drug Deliv Rev 2005;57 (1):17–42.
8. Matthäus C, Chernenko T, Judith A. Newmark, et al. Label-free detection of mitochondrial distribution in cells by nonresonant Raman microspectroscopy. Biophys J 2007;93 (2):668–673.
9. Christian M. Tatyana C; Luis Q, et al. Raman microscopic imaging of cells and applications monitoring the uptake of drug delivery systems. Proc. SPIE 2008;6991:699106.
10. Matthäus C, Bird B, Miljkovic´ M, et al. Infrared and Raman microscopy in cell biology. Methods Cell Biol 2009;89:275–308.
11. Chernenko T. *Biochemical Imaging of Drug-Delivery Systems Using Raman Micro-Spectroscopy to Probe Cell Architecture and Behaviors of Nanoparticles*. Saarbrucken: LAP LAMBERT Academic Publishing AG & Co.KG; 2010.
12. Matthäus C, Kale A, Chernenko T, et al. New ways of imaging uptake and intracellular fate of liposomal drug carrier systems inside individual cells, based on Raman microscopy. Mol Pharm 2008;5 (2):287–293.
13. Chernenko T, Matthäus C, Milane L, et al. Label-free Raman spectral imaging of intracellular delivery and degradation of polymeric nanoparticle systems. ACS Nano 2009;3 (11):3552–3559.
14. Krafft C, Knetschke T, Salzer R, et al. Identification of organelles and vesicles in single cells by Raman microspectroscopic mapping. Vib Spectrosc 2005;38 (1–2):85–93.
15. Krafft C, Knetschke T, Funk RH, et al. Studies on stress-induced changes at the subcellular level by Raman microspectroscopic mapping. Anal chem 2006;78 (13):4424–4429.
16. Romeo MJ, Boydston-White S, Matthäaus C, et al. Infrared and Raman microspectroscopic studies of individual human cells. In: Diem MG, Griffith P, Chalmers J, editors. *Vibrational Spectroscopy for Medical Diagnosis*. Hoboken: John-Wiley & Sons Ltd; 2008. p 325–333.
17. Naumann D. FT-infrared and FT-Raman spectroscopy in biomedical research. In: Gremlich HU, Yan B, editors. *Infrared and Raman Spectroscopy of Biological Materials*. New York: Marcel Dekker, Inc.; 2001. p 323–377.
18. Wood BR, Chiriboga L, Yee H, et al. Fourier transform infrared (FTIR) spectral mapping of the cervical transformation zone, and dysplastic squamous epithelium. Gynecol Oncol 2004;93 (1):59–68.

19. Shenoy DB, Amiji MM. Poly(ethylene oxide)-modified poly(varepsilon-caprolactone) nanoparticles for targeted delivery of tamoxifen in breast cancer. Int J Pharm 2005;293 (1–2):261–270.

20. Hutmacher, DW. Polymers for medical applications. In: Buschow KHJ, et al., editors. *Encyclopedia of Materials: Science and Technology*. Oxford: Elsevier; 2001. p 7664–7673.

21. Kaul G, Amiji M. Tumor-targeted gene delivery using poly(ethylene glycol)-modified gelatin nanoparticles: in vitro and in vivo studies. Pharm Res 2005;22 (6):951–961.

22. Garg S, Serruys PW. Coronary stents: looking forward. J American Coll Cardiol 2010;56 (10 Suppl):43S–78S.

23. Lockwood NA, Hergenrother RW, Patrick LM, et al. In vitro and in vivo characterization of novel biodegradable polymers for application as drug-eluting stent coatings. J Biomater Sci Plym Ed 2010;21 (4):529–552.

24. Torchilin VP. *Nanoparticulates as Drug Carriers*. London: Imperial College Press; 2006.

25. Moghimi S, Vega E, Garcia M, et al. Polymeric nanoparticles as drug carriers and controlled release implant devices. In: Torchilin VP, editor. *Nanoparticulates as Drug Carriers*. Boston: Imperial College Press; 2006. p 29–42.

26. Cartiera MS, Johnson KM, Rajendran V, et al. The uptake and intracellular fate of PLGA nanoparticles in epithelial cells. Biomaterials 2009;30 (14):2790–2798.

27. Panyam J, Labhasetwar V. Targeting intracellular targets. Curr Drug Deliv 2004;1 (3):235–247.

28. Pitt CG. Poly-epsilon-caprolactone and its copolymers. In: Langer R, Chasin M, editors. *Biodegradable Polymers as Drug Delivery Systems*. New York: Marcel Dekker, Inc; 1990. p 71–120.

29. Sung HJ, Meredith C, Johnson C, et al. The effect of scaffold degradation rate on three-dimensional cell growth and angiogenesis. Biomaterials 2004;25 (26):5735–5742.

30. Siskind LJ. Mitochondrial ceramide and the induction of apoptosis. J Bioener Biomembr 2005;37 (3):143–153.

CHAPTER 5

ANATOMICAL AND FUNCTIONAL MRI

EDWARD F. JACKSON

5.1 INTRODUCTION

The soft tissue contrast provided by magnetic resonance imaging (MRI) often makes it the modality of choice in oncological imaging, particularly in neuroimaging and musculoskeletal applications. The excellent sensitivity of MRI for detecting lesions is due to the dependence of the image contrast and signal-to-noise ratio (SNR) on a wide range of both intrinsic and extrinsic parameters. Intrinsic parameters, which depend upon the individual tissue characteristics, include, but are not limited to, the spin–lattice (longitudinal) relaxation time (T_1), spin–spin (transverse) relaxation time (T_2), proton density, temperature, and the velocity and local chemical environment of the nuclei of interest. Extrinsic parameters that affect image contrast and SNR are those chosen by the individual performing the examination. A partial list of such extrinsic parameters includes the particular type of image acquisition sequence, the echo time (TE); repetition time (TR); excitation flip angle; various types of specialized magnetization preparation pulses, for example, saturation and inversion pulses; specialized gradient field pulses, for example, diffusion sensitization pulses, etc. By appropriate manipulation of the extrinsic parameters, a very wide range of image contrast can be obtained and can be tailored to provide excellent visualization of anatomy, pathology, and, in some cases, function. In this chapter, the physical principles of MRI will be briefly introduced, and current state-of-the-art imaging hardware and acquisition techniques will be reviewed. By necessity, the review of the basics of magnetic resonance (MR) physics will be brief, and for more detailed discussions, the reader is referred to additional introductory references [1–6]. In addition, this chapter only addresses MR techniques that are available on current commercially available state-of-the-art clinical MR scanners. Therefore, exciting new techniques, such as hyperpolarized ^{13}C-pyruvate MR applications [7, 8], are not included.

Drug Delivery Applications of Noninvasive Imaging: Validation from Biodistribution to Sites of Action, First Edition. Edited by Chun Li and Mei Tian.

5.2 PHYSICAL PRINCIPLES OF MRI

5.2.1 Basic Concepts

There are three major requirements for obtaining images using MRI. The first is the presence of a nucleus with a nonzero nuclear magnetic moment. Such nuclei have an odd number of protons and/or neutrons. Some of the more commonly utilized nuclei in MRI and magnetic resonance spectroscopy are given in Table 5.1. Virtually all clinical imaging studies are based on proton MRI due to the large natural abundance of protons in human tissue and the high natural MR sensitivity of the proton due to its large gyromagnetic ratio.

Nuclei that have nonzero magnetic moments are, in a classical physics sense, like very weak bar magnets (Fig. 5.1). However, in the absence of an applied magnetic field, the nuclear magnetic moments point in random directions, and there is no measurable *net* magnetization (Fig. 5.1). To generate net magnetization, the nuclear magnetic moments must be polarized. Therefore, given nuclei with nonzero magnetic moments, the second component required in order to generate an MR image is a relatively large and highly homogeneous magnetic field, B_0. In clinical imaging studies, the most common magnetic field strengths currently employed are in the range of 0.5–3.0 tesla (T). However, there are currently a limited number of whole-body research scanners with magnetic field strengths in the 4.0–9.4 T range.

The proton is a "spin-1/2" nucleus. Therefore, when placed in a magnetic field, there are only two allowed states: aligned with the external magnetic field (parallel, lower energy state) or aligned opposed to the magnetic field (antiparallel, higher energy state). If a patient is placed in such a field, more of the magnetic moments tend to align parallel with the applied magnetic field than antiparallel (Fig. 5.2). The result is a nonzero net magnetization, that is, an equilibrium magnetization, M_0. The population of the nuclei in each of the two states is governed by a Boltzmann distribution [5], and for a sample at room temperature with a 1.5-T magnet, only about 10 out of a million protons are preferentially aligned with the external field. Therefore, the inherent sensitivity of MRI is quite low, and it is quite fortunate that the human body has an abundance of water and fat, so sufficient numbers of protons are available from which to form an image in a reasonable acquisition time.

TABLE 5.1 Properties of Some Common Nuclei with Potential Use in MR Studies

Nucleus	Spin	Gyromagnetic ratio (MHz/T)	Natural isotopic abundance (%)	Sensitivity relative to ^{1}H (%)[a]
^{1}H	1/2	42.57	99.98	100
^{13}C	1/2	10.71	1.11	1.6
^{19}F	1/2	40.06	100	83.4
^{23}Na	3/2	11.26	100	9.3
^{31}P	1/2	17.24	100	6.6

The higher the value of the gyromagnetic ratio, which is a constant for a given nucleus, the higher the native MR sensitivity.

[a]At constant field for equal number of nuclei.

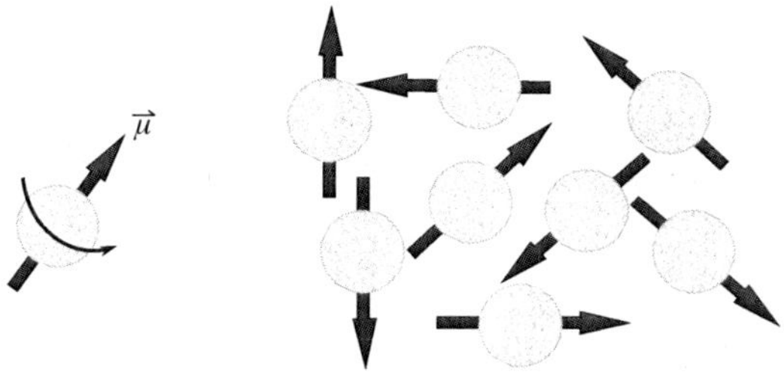

FIGURE 5.1 A nucleus with nonzero spin has an associated magnetic moment, μ, and behaves, in a classical physics view, similar to a very small bar magnet (left). A collection of spins in the absence of an applied magnetic field has no *net* magnetization since the magnetic moments are randomly oriented (right).

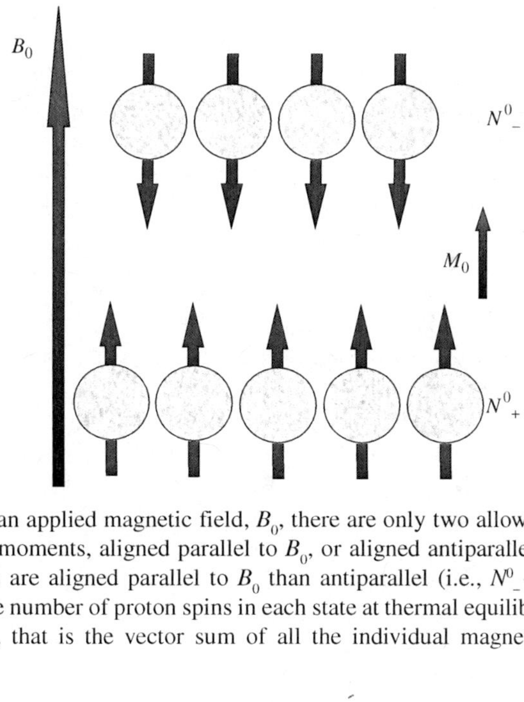

FIGURE 5.2 In an applied magnetic field, B_0, there are only two allowed directions of the spin-1/2 magnetic moments, aligned parallel to B_0, or aligned antiparallel to B_0. Since more magnetic moments are aligned parallel to B_0 than antiparallel (i.e., $N^0_- < N^0_+$, where the N^0 values represent the number of proton spins in each state at thermal equilibrium), there is a *net* magnetization, M_0, that is the vector sum of all the individual magnetic moments in the ensemble average.

At this point, net magnetization has been created that is aligned parallel to the applied magnetic field. However, to detect the magnetization requires that it be perpendicular to the applied magnetic field rather than parallel to it. To accomplish this, an additional magnetic field is applied to the nuclei. This applied field, B_1, is a time-varying field that is applied at a frequency proportional to the energy separation of the two allowed states,

$$\upsilon_0 = \gamma B_0 \tag{5.1}$$

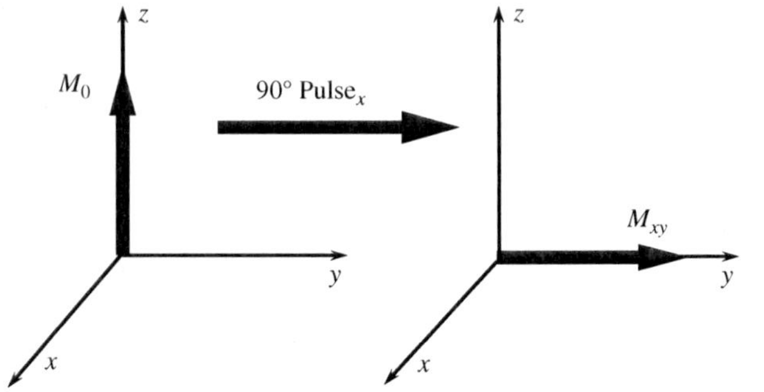

FIGURE 5.3 Evolution of the net magnetization, M_0, from the longitudinal (z)-direction to the transverse (x,y) plane under the influence of a 90° pulse along the x-direction in the rotating frame of reference.

where γ is the gyromagnetic ratio, a constant for a given nucleus, for example, 42.57 MHz/T for protons, and B_0 is the applied static magnetic field amplitude. This frequency is known as the Larmor frequency and is a key concept in MRI. If the radiofrequency (RF) field, B_1, is applied at this frequency and in a direction perpendicular to the static field, B_0, then the net magnetization can be made to rotate by an arbitrary angle with respect to the applied static field. For a rectangular RF pulse, the angle of rotation, commonly called the "flip angle," α, is given by

$$\alpha = \gamma B_1 t_p \tag{5.2}$$

where B_1 is the amplitude of the pulse and t_p is the duration of the pulse. (This equation can be generalized for any arbitrary B_1 pulse shape.) If the values of B_1 and t_p are chosen such that α is 90°, the pulse is commonly called a "90° pulse," and the net magnetization is left perpendicular to the applied static field (in the "transverse plane"). The combination of the RF pulse and the precession of the spins make visualizing the result of a 90° pulse on the net magnetization somewhat difficult. However, if one considers the motion of the spins from a frame of reference that is rotating at the precession rate, that is, the Larmor frequency, then the complication of the precessional motion is removed (Fig. 5.3). This choice of reference frames is analogous to choosing to view a child riding on a merry-go-round while standing on the merry-go-round rather than standing on the ground nearby; changing to the reference frame of the merry-go-round removes the rotational component of the motion and greatly simplifies describing the motion of the child. This frame is known as the "rotating frame of reference."

After the pulse is terminated, the net magnetization precesses about the applied static field at the Larmor frequency. The time-varying magnetic field resulting from the precession of the magnetization induces a time-varying voltage, that is, the "free induction decay" (FID) (Fig. 5.4), in an appropriately tuned and oriented RF coil, and this is the manner by which the MR signal is measured. In some cases, the RF coil used to produce the B_1 pulse and the RF coil used to detect the MR signal are one and the same. This is true for some older head and extremity coils and for whole-body coils. In other cases, a large RF coil is used to transmit the B_1 pulses, and a smaller

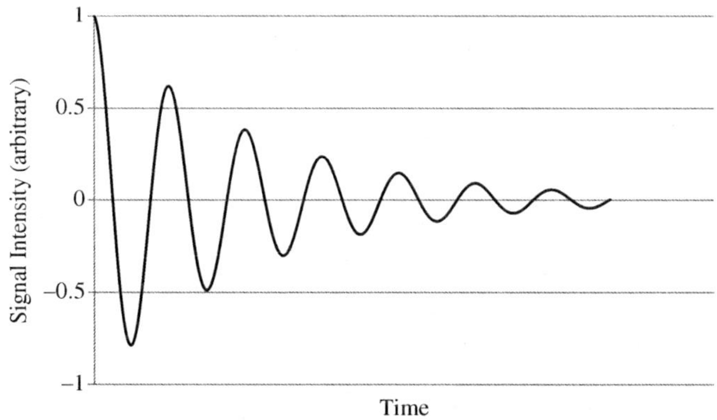

FIGURE 5.4 Following a 90° RF (B_1) pulse, the net magnetization precesses and the time-varying magnetic field generates a time-varying voltage signal, known as the FID, in a receiver coil.

RF coil is used to detect the signal. In general, the receive coil should be matched in size and shape to the anatomy of interest as closely as possible to optimize the SNR of the resulting images, and, as a result, a wide variety of RF receive coils, including many phased array coils, are available. For further details of RF coil design and integration of the coils in the overall MR scanner hardware, the reader is referred to the references [1, 6].

Ultimately, the net magnetization returns to the equilibrium condition parallel to the applied static field, and in the process energy is transferred to the surroundings or, in the solid-state physics terminology, "the lattice." The rate at which this occurs is determined by the intrinsic parameter T_1, which is the "spin–lattice" or longitudinal relaxation time. Fortunately, different tissue types exhibit different T_1 times, and this becomes an important parameter in determining image contrast (Fig. 5.4). At the same time that the net magnetization is recovering, there is a dephasing of the net transverse magnetization due to "spin–spin" interactions. The rate at which the net transverse magnetization dephases is determined by the intrinsic parameter T_2, which is the "spin–spin" or transverse relaxation time. As with the T_1 relaxation times, T_2 relaxation times vary with tissue type and can have an important effect on image contrast (Fig. 5.5). Some representative T_1 and T_2 relaxation times are given in Table 5.2 [9].

In general, T_1 relaxation times are always greater than T_2 relaxation times, and T_1 and T_2 relaxation times are both long in pure fluids, for example, the orbits and cerebrospinal fluid (CSF). The general dependence of T_1 and T_2 upon the size and rate of translational/rotational motion of a molecule is shown in Figure 5.6.

Spin–spin dephasing at a rate determined by T_2 occurs from the moment the net magnetization is "flipped" into the transverse plane. This dephasing is irreversible but gives rise to useful image contrast. Unfortunately, there are other contributions to spin dephasing that, uncorrected, yield an overall loss of SNR. With the exception of applications to perfusion imaging and other susceptibility mapping techniques (discussed next), such dephasing phenomena are detrimental to MR imaging as they result in decreased SNR. One of the most common sources of dephasing is local magnetic field

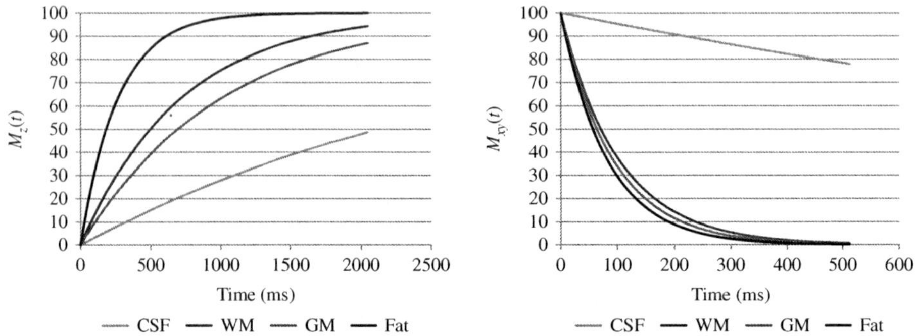

FIGURE 5.5 A 90° pulse converts the longitudinal magnetization, M_0, into the measurable transverse magnetization, M_{xy}. Following the pulse, the longitudinal magnetization recovers at a rate determined by the spin–lattice, or T_1, relaxation time (left). Independent of the spin–lattice relaxation, the transverse magnetization dephases at a rate determined by the spin–spin, or T_2, relaxation time (right). Like the T_1 relaxation times, the T_2 relaxation times vary from tissue to tissue. CSF: cerebrospinal fluid, WM: white matter, GM: gray matter, Fat: long-chain fatty acids (primarily methyl, methylene, methane protons).

TABLE 5.2 Representative Relaxation Times[a]

Tissue	T_1 at 0.5T (ms)	T_1 and 1.5T (ms)	T_2 (ms)
Gray matter	656	920	101
White matter	539	790	92
Cerebrospinal fluid	>2000	>4000	>4000
Skeletal muscle	600	870	47
Fat	215	260	84
Liver	323	490	47

[a] *Source*: From reference 9.

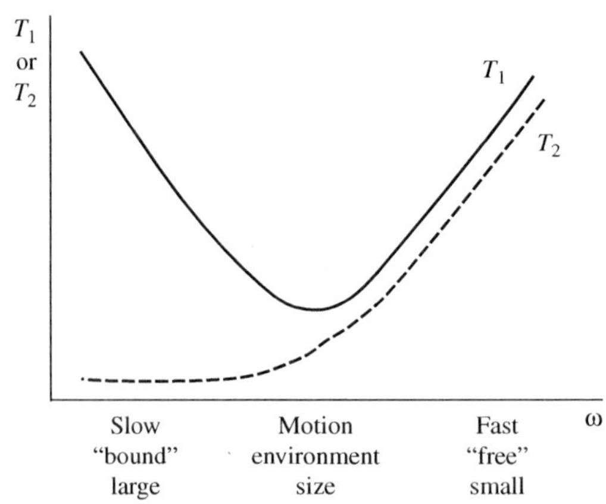

FIGURE 5.6 Relative proton T_1 and T_2 relaxation times as a function of molecular size, environment, and motion.

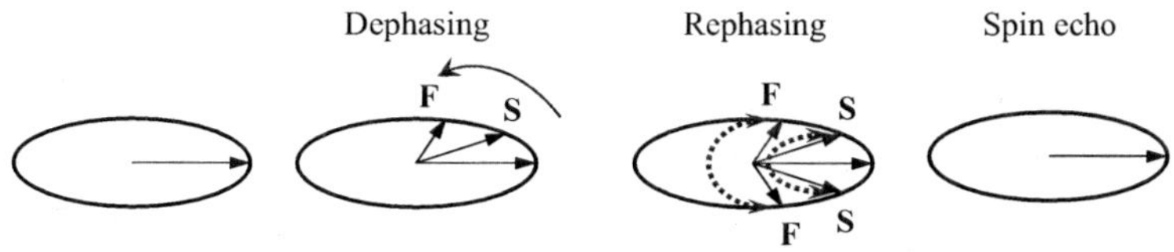

FIGURE 5.7 Spin-echo formation. Following the 90° pulse, the transverse magnetization dephases due to both spin–spin interactions and local magnetic field inhomogeneities, with some spins precessing faster (F) than others (S). Although spin–spin dephasing is irreversible, the dephasing due to the inhomogeneities can be reversed by applying a 180° pulse that "flips" the spins in the transverse plane. Then the "faster" precessing spins rephase with the "slower" precessing spins at the TE, and the effects of the magnetic field inhomogeneities are compensated.

inhomogeneities due to imperfections in the applied static field homogeneity or, more commonly, due to magnetic susceptibility changes in the human body. These susceptibility changes are most evident at tissue/air or tissue/bone interfaces and give rise to apparent intravoxel inhomogeneities in the applied field. Such inhomogeneities cause a decrease in the "apparent T_2," or T_2*, and the larger the inhomogeneity the shorter the value of T_2* and the faster the rate of dephasing, that is,

$$\frac{1}{T_2^*} = \frac{1}{T_2} + k\Delta B_0 \tag{5.3}$$

where ΔB_0 represents the field inhomogeneity and k is a constant. Fortunately, many of the effects of spin dephasing due to magnetic field inhomogeneities can be reversed using *spin echoes*. The concept of spin echoes and how they compensate for local field inhomogeneities is illustrated in Figure 5.7.

5.2.2 Spatial Localization

The application of a static field, B_0, and a RF field, B_1, results in the formation of net magnetization that can be detected. However, there is as yet no spatial localization of the detected signal. To encode position typically requires the application of three additional spatially varying, or gradient, magnetic fields, each one orthogonal to the other two. For each of the three magnetic gradient fields, the component of the magnetic field parallel to the static field increases (ideally) linearly with position, that is, in either the x-, y-, or z-direction. Figure 5.8 shows a generic representation of the required RF and other fields required to generate a two-dimensional spin-echo (SE) image. Such a diagram is often referred to as a pulse sequence and generally shows the RF pulses (B_1) and the pulses on the three gradient fields required for spatial localization (G_i, where i corresponds to x, y, or z).

Spatial localization occurs in three steps. For two-dimensional imaging sequences, the MR signal is first localized to a single slice in the patient using a gradient field pulse that is applied simultaneously with a band-limited RF pulse. The thickness of the slice selected by the RF/gradient pulse combination is given by

$$\Delta z = \frac{\Delta \upsilon_{\mathrm{RF}}}{\gamma G_{\mathrm{slice}}} \tag{5.4}$$

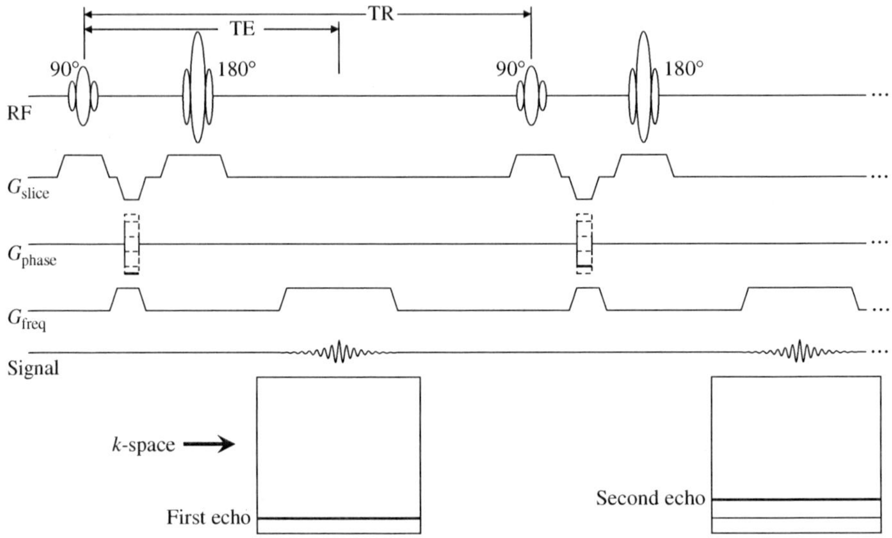

FIGURE 5.8 A generic SE pulse sequence timing diagram. RF represents the B_1 RF pulses, G_{slice} represents the slice-selecting gradient, G_{phase} represents the phase-encoding gradient, G_{freq} represents the frequency-encoding gradient, signal represents the detected spin echo, TE is the echo time, and TR is the repetition time. Two repetitions of the sequence are shown, with each repetition filling in one line of the necessary 128–512 lines of *k*-space.

where $\Delta\upsilon_{RF}$ is the bandwidth of the RF pulse and G_{slice} is the amplitude of the slice-selecting gradient field.

The use of the slice-selecting gradient/RF pulse combination localizes the detected MR signal to a slice of thickness Δz. However, assuming there is perfect magnetic field homogeneity within the slice, all of the nuclei resonate at a single frequency, that is, there is no localization within the slice. To encode position in the remaining two dimensions, two additional gradient pulses are used. The first is known as the *frequency-encoding* gradient and is applied at the same time the MR signal is detected (Fig. 5.8). Under the influence of this additional gradient, assumed for the moment to be applied along the *x*-direction, the Larmor frequency is modified such that

$$\upsilon_{freq} = \gamma(B_0 + xG_{freq}) \tag{5.5}$$

where x is the position along the x-direction, G_{freq} is the amplitude of the frequency-encoding gradient field, and υ_{freq} is the Larmor frequency modified by the addition of the gradient field. Equation 5.5 indicates there is a linear, one-to-one correspondence between the frequency of precession of the nuclear spins and the position of the spins along the x-direction. Therefore, the action of the frequency-encoding gradient field is to encode position in terms of frequency, and if we can measure the precessional frequencies of groups of nuclear spins, we can determine their position. Taking the Fourier transform of the detected signal accomplishes this task.

At this point, the signal received from the nuclear spins has been localized to a slice, and the location within that slice has been encoded in one of the two remaining directions. The location of nuclear spins along the second direction is encoded using a *phase-encoding* gradient (Fig. 5.8). The phase-encoding gradient is typically applied 128–512 times, once per TR, in conventional single-echo SE imaging, and at each repetition, the amplitude of the phase-encoding gradient pulse is changed by a fixed amount. The result of this action is that each frequency-encoded echo that is acquired has a slightly different phase than did the previous echo. Therefore, there is a one-to-one correspondence of phase to position in a fashion similar to the one-to-one correspondence of frequency to position in the frequency-encoding direction. Taking a second Fourier transform of the detected signal in the phase-encoding direction allows for the complete reconstruction of the image of the anatomy within the slice of thickness Δz.

It should be noted that the gradient magnetic field coil performance has been, and continues to be, the limiting factor in how fast MR images can be obtained. When the gradient coils are activated, the changing magnetic field induces currents, known as eddy currents, in nearby conductive materials. These induced eddy currents give rise to magnetic fields that oppose the applied gradient fields (Lenz's law) and generally limit the rate at which the gradient fields can be switched. This, in turn, limits the rate at which images can be acquired. In the last decade, the gradient coil designs and the associated current amplifiers have been highly optimized for rapid switching and have allowed commercial scanners with specialized fast receivers and fast gradient subsystems to acquire images in as short a time as 50–100 ms. Such systems are typically characterized by the maximum gradient strength, in millitesla per meter (mT/m), as well as their slew rate, in T/m/s. While systems in the early 1990s had slew rates of around 10 T/m/s, many systems are currently commercially available with slew rates of 150–200 T/m/s and above. Such systems are required for ultrafast MRI studies.

5.2.3 Basic Image Contrast Mechanisms

The advantage of MRI over other modalities used for imaging soft tissue is the wide range of image contrasts that can be obtained by modifying the basic pulse sequence timings or by the use of additional RF pulses. Historically, the three basic types of image contrasts obtained using MRI were proton density-weighted (PDW), T_1-weighted (T1W), and T_2-weighted (T2W) images. In general, if TE is short such that T_2 decay is not significant, and TR is long to allow nearly complete recovery of the longitudinal magnetization, then T_1 and T_2 processes are not significant, and the image will be proton density-weighted. If, on the other hand, TE is short such that T_2 decay is minimal and TR is short to enhance differences in T_1 relaxation times (Fig. 5.5), the dominant contributor to image contrast is T_1 weighting. Finally, if TE is increased to emphasize differences in T_2 relaxation times (Fig. 5.5) and TR is long such that all tissues achieve nearly complete recovery of the longitudinal magnetization, the image contrast is primarily determined by T_2 weighting. PDW, T1W, and T2W images have been common image contrast options for many years.

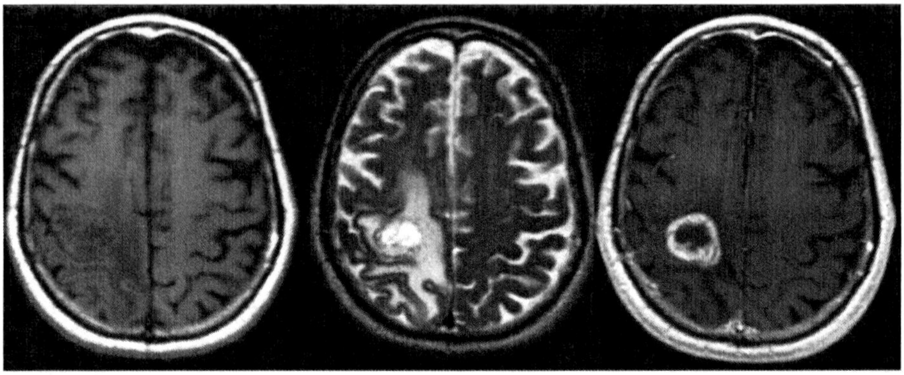

FIGURE 5.9 SE images from a patient with a glioblastoma: T1W, TE/TR = 12/500 ms (left), T2W, TE/TR = 100/3000 ms (center), and T1W post-contrast agent infusion, TE/TR = 12/500 ms (right).

In oncological imaging, an additional contrast mechanism is commonly exploited by the administration of contrast agents based on the paramagnetic atom gadolinium (Gd). Such a paramagnetic substance, if in close proximity to water, will cause a shortening of both the T_1 and T_2 relaxation times of the water protons (Eq. 5.6), with the shortening of T_1 being the dominant effect for the concentrations of contrast agent typically administered. In these equations, $T_{1,0}$ and $T_{2,0}$ are the relaxation values of the native tissue, that is, before the contrast agent is administered:

$$\frac{1}{T_1} = \frac{1}{T_{1,0}} + r_1[\text{Gd}] \quad \text{and} \quad \frac{1}{T_2} = \frac{1}{T_{2,0}} + r_2[\text{Gd}] \tag{5.6}$$

In the case of contrast agent-enhanced T1W imaging, the shortened T_1 relaxation times in areas where Gd atoms are in close proximity to water protons cause these areas to become hyperintense on T1W images as compared to identical images without the administration of Gd. Unfortunately, free Gd is toxic and, therefore, must be tightly chelated to a readily excreted substance. The first Food and Drug Administration (FDA)-approved MR contrast agent used DTPA as the chelate, but other chelates are now in clinical use [1].

Gadolinium-based contrast agents do not cross an intact blood–brain barrier (BBB). Therefore, enhancement on T1W images acquired postinfusion of contrast agent demonstrates breakdown of the BBB and/or increased local microvascular density due to a pathological condition. Brain neoplasms commonly, but not always, cause such a BBB breakdown and/or increased vascularity, as do certain other disease states. Also, neoplasms in other areas of the body may enhance following Gd infusion due to lesion-induced increased permeability and/or vascularity. Other contrast agents based on paramagnetic substances, superparamagnetic iron oxides (SPIOs), and ultrasmall SPIOs (USPIOs) are also in use and in clinical trials. Examples of T1W, T2W, and T1W post-Gd-DTPA infusion images are given in Figure 5.9.

5.2.4 Other Imaging Sequences

The preceding discussions have been developed around the SE pulse sequence. However, there are several other major classes of pulse sequences that have been introduced primarily to increase the speed of image acquisition. These sequences include gradient-echo pulse sequences, fast SE pulse sequences, and echo-planar imaging (EPI) sequences.

5.2.4.1 Gradient-Echo Sequences Gradient-echo pulse sequences are similar to the SE sequence but do not use the 180° refocusing pulse. Instead, the MR signal is refocused solely by the use of appropriate gradient pulses. The advantage of the gradient-echo sequences is that the omission of the 180° pulse and associated delays significantly reduces the image acquisition time and allows for very rapid acquisition of T1W images in particular. Unfortunately for newcomers to MRI, there are several varieties of gradient-echo pulse sequences, and the problem is exacerbated by the fact that each manufacturer has its own lengthy list of names for the sequences. Convenient sources of information for sorting through the "alphabet soup" of gradient-echo sequences from various manufacturers are found in references 3 and 10.

The most common applications of gradient-echo imaging are in T1W "volume" acquisitions of typically 64–192 1–2-mm "slices," which can be displayed in three-dimensional (3D) fashion, and in breath-hold abdominal imaging and other body imaging applications (Fig. 5.10). Another application of gradient-echo-based sequences is in time-of-flight (TOF) and phase contrast (PC) MR angiography (discussed in section 5.2.4.4). Although gradient-echo sequences allow for faster image acquisition rates, there are inevitable trade-offs that come with the omission of the 180° pulse. The primary downside is the loss of the compensation for local field inhomogeneities that the spin echo provides. Therefore, gradient-echo images demonstrate increased artifacts due to dephasing of the MR signal in areas of inhomogeneities, particularly those inhomogeneities caused by magnetic susceptibility changes at air/tissue and tissue/bone interfaces, for example, near the skull base and sinuses [5]. Since the effects of this dephasing become more prominent as TE increases, gradient-echo-based T2W images (more correctly T_2*W images) are prone to significant artifacts at such interfaces. Therefore, gradient-echo sequences are commonly used for T1W imaging and MR angiography, both of which require short TEs, but are less routinely used for T2*W imaging. Gradient-echo T2*W images are, however, used for certain special imaging applications, particularly in the imaging of nerve roots and for the detection of areas of intracranial hemorrhage.

5.2.4.2 Fast Spin-Echo Sequences As fast acquisition of T2*W images using gradient echoes can be problematic in terms of artifacts, other fast imaging techniques were developed that allowed for the acquisition of T2W images of similar quality to SE images, but at a much faster rate. The most commonly used technique is the fast spin-echo (FSE) (or turbo spin-echo, TSE) sequence based on a method originally proposed by Hennig et al. [11]. In this technique, the improved acquisition rate is provided by uniquely phase-encoding several echoes during a given TR interval rather than just encoding a single echo as in the conventional SE sequence shown in Figure 5.8. The number of echoes phase-encoded in a given TR interval is frequently known as the echo-train length

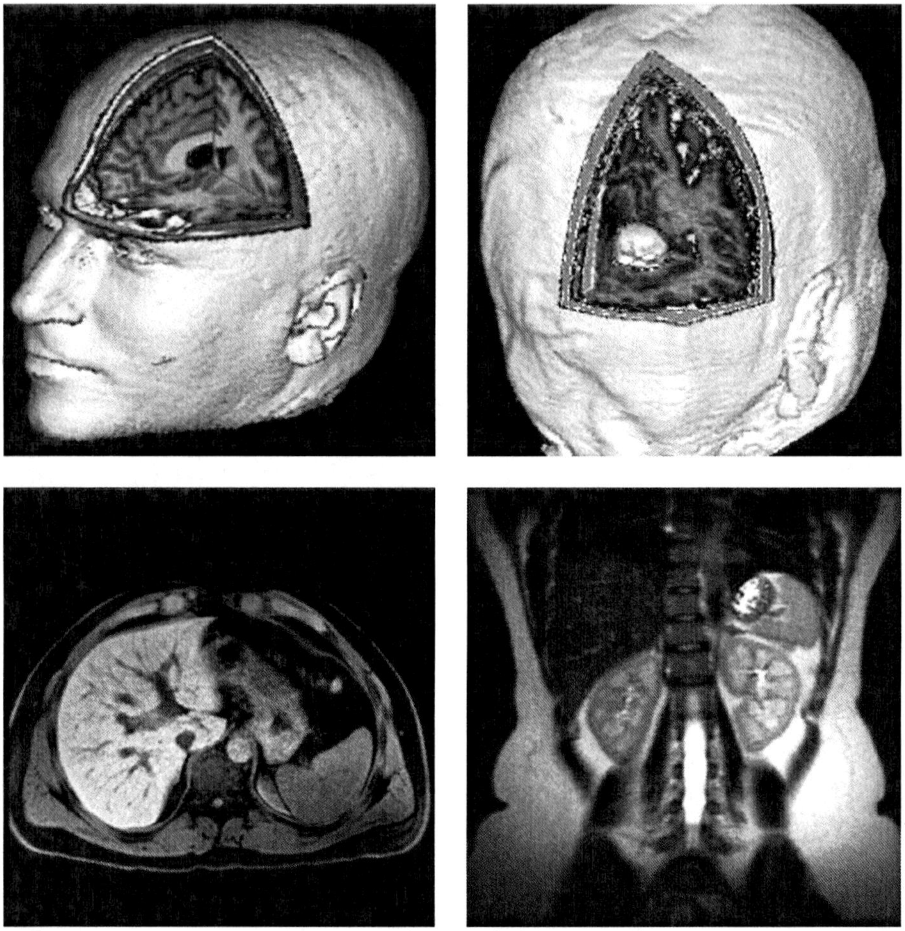

FIGURE 5.10 Accelerated image acquisition rates provided by fast gradient-echo techniques allows for the acquisition of high spatial resolution 3D data in neuroimaging (top frames) and breath-hold applications in abdominal imaging to allow excellent soft tissue contrast acquisitions in moving organ systems (bottom frames).

(ETL). For a given ETL, the image acquisition time is theoretically reduced by a factor of ETL as compared to the conventional SE sequence. Most commonly, ETLs of 8–32 are used for T2W FSE imaging, although larger values of ETL may be used for heavily T2W images, such as those used in MR cholangiopancreatography, or for ultrahigh-speed T2W imaging techniques such as SS-FSE or TurboHASTE [12, 13]. The dramatic savings in acquisition time can be used to improve the SNR, the spatial resolution, or simply to reduce the acquisition time. There is a practical limit to how large ETL can be, of course, without resulting in significant artifacts. This is particularly true for PDW or T1W FSE imaging where the use of large values of ETL at short TEs can result in blurring and loss of contrast for small lesions [14, 15]. Fortunately, this blurring issue is much less significant in T2W imaging where TE values are typically long, and FSE T2W imaging has become a standard in MR imaging.

Although T2W images acquired using FSE are generally similar to those obtained using conventional SE sequences, there are some noticeable differences [14–19]. First, fat is significantly brighter on FSE T2W images as compared to SE images acquired with the same TE and TR. This is primarily due to decreased *J*-coupling effects on the signal intensity of fat in the FSE sequence compared to the SE sequence. The increased intensity of fat on FSE T2W images frequently results in the need to use fat suppression techniques (discussed in section 5.2.5). A second difference in T2W images acquired with FSE versus SE is the increased T_2 weighting on FSE images due to the contribution of multiple echoes to the "effective echo time" rather than a single echo at a well-defined TE as is obtained in the SE sequence. Finally, the amount of RF power that is absorbed by the patient is significantly higher for FSE scans than for SE scans due to the large number of rapidly applied 180° pulses in the FSE acquisition. This becomes more problematic at higher magnetic field strengths, for example, 3.0T.

5.2.4.3 Echo-Planar Sequences Currently, the fastest widely available commercial image acquisition technique is EPI [2, 5]. The basic idea behind the EPI sequence is similar to the FSE sequence in that multiple phase- and frequency-encoding gradients are applied in a given TR interval, but in EPI, this concept is pushed to the extreme limit. In "snapshot" or "single-shot" SE EPI acquisitions, for example, *every* echo is obtained following a single 90°–180° pulse pair. This results in image acquisition times as short as 50–100 ms and makes EPI ideal for imaging dynamic processes. The acquisition of snapshot EPI images, however, requires an MR scanner that has very fast acquisition rate capability and extremely high-quality gradient field subsystems. Otherwise, high-quality single-shot EPI images are difficult to obtain reproducibly. To ease the requirements of single-shot EPI somewhat, multishot or "mosaic" EPI techniques have also been intsroduced that are not as fast as single-shot scans but have decreased geometric distortions and other artifacts. Clinical applications of EPI in routine oncological MRI include diffusion MRI, T_2*-based perfusion MRI, and blood oxygen level dependent functional MRI (BOLD fMRI) (discussed in subsequent sections).

5.2.4.4 Magnetic Resonance Angiography (MRA) Sequences The most common MRA techniques are based upon gradient-echo sequences and time-of-flight (TOF) or phase-contast (PC) techniques [1, 20]. The physical basis for TOF MRA techniques is known as "flow-related enhancement." In TOF MRA, T1W images are acquired with very short TE and TR times. As the images are being acquired from a volume of tissue, the spins in that volume become partially saturated due to the rapidly repeated RF pulses, that is, the longitudinal magnetization never recovers completely during the TR. Therefore, the signal from the volume of tissue being imaged is decreased due to this partial saturation. Blood flowing into the imaging volume, however, is not partially saturated. Therefore, the signal from the blood is typically significantly larger than the signals from the surrounding partially saturated tissues. As a result, signal from the blood in the vessels appears bright relative to that from the surrounding stationary tissue. The TOF technique works best for rapid flow that is perpendicular to the slices in the imaging volume as this maximizes the flow-related enhancement. However, it is frequently necessary to spatially saturate the vessels flowing in the opposite direction to the vessels of interest in order to minimize contribution to the MRA image from these vessels. In this case, "spatially saturate" means to apply a 90° pulse outside of the imaging volume to the blood in the vessels that are to be suppressed. If the saturated blood then flows into the imaging volume and a 90° pulse is applied, the resulting magnetization will primarily be in the longitudinal direction, that is, parallel to the applied static field, at the time of signal acquisition. As no signal is detected from magnetization that is parallel or antiparallel to the static field, the signal from the saturated blood will be suppressed. If such spatial saturation techniques are not employed, signals from both arterial and venous vessels may appear in the resulting image, and overlap of the two can cause confusion.

The TOF techniques are the most commonly utilized sequences for MRA, particularly for studies of relatively fast flow. They work well with very short TE values, which minimize loss of signal secondary to dephasing of the spins by turbulent flow. Such dephasing can cause significant overestimation of stenosis as the TE times increase. The same MR system improvements that have resulted in faster and faster imaging, especially the improved gradient subsystems, are also resulting in continuous improvement in the quality of MRA by decreasing acquisition times (decreased patient motion artifacts, better contrast due to emphasized flow-related enhancement effects, and better throughput) and decreasing TEs (decreased dephasing and less overestimation of stenosis). Examples of typical TOF MRA images of the circle of Willis and carotid arteries are given in Figure 5.11a and b.

While TOF techniques are the most commonly used noncontrast agent MRA option, PC techniques have some particular benefits. The basic mechanism behind PC MRA is quite different from TOF techniques. Whereas TOF MRA relies upon flow-related enhancement and appropriate use of spatial saturation pulses, PC MRA capitalizes on the fact that the signal intensity from nuclear spins can be made to depend upon the velocity of the spins. During a PC MRA pulse sequence, the moving nuclear spins accrue phase that is directly proportional to their velocity, and this phase can be measured. Therefore, PC MRA can depict the velocity (direction *and* speed) of the nuclear spins and, if the vessel diameter is also measured, the flow rate. Furthermore, such PC MRA acquisitions can be gated to the cardiac cycle to allow for the display of flow velocity as a function of cardiac cycle to examine multidirectional flow (biphasic or triphasic flow patterns).

Because of the way PC MRA images are formed, they have two main advantages compared to TOF MRA images. First, PC MRA is more useful when imaging slow flow. This is because the image contrast does not depend on flow-related enhancement but rather on the actual velocity of the moving protons. This also means that careful placement of spatial saturation bands to suppress flow opposed to the direction of interest and preferential placement of the scan plane perpendicular to the direction of the flow are not necessary. Second, quantitative

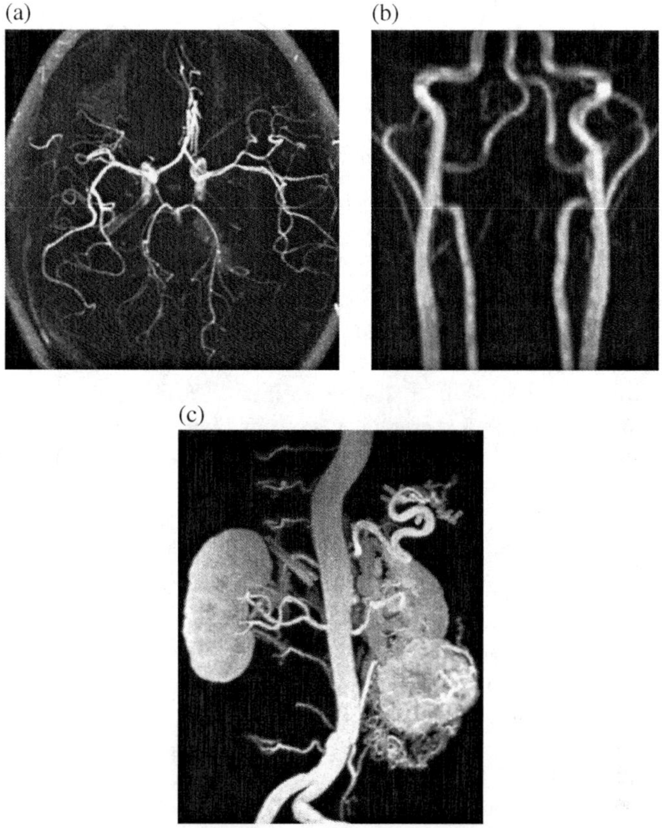

FIGURE 5.11 (a) 3D TOF MRA of the circle of Willis presented as an axial plane maximum intensity projection (top left). (b) A maximum intensity projection view of a two-dimensional TOF MRA of the carotid and vertebral arteries (top right). Each of these datasets were obtained without the injection of any exogenous contrast agent. (c) A maximum intensity projection view of coronal plane MRA data acquired in a single breath-hold following bolus infusion of Gd-DTPA (bottom). Notice the excellent distal coverage of the renal arteries as well as the visualization of the small lumbar arteries.

measures of flow velocities and flow rates can be obtained using PC MRA. There are, of course, some disadvantages of PC MRA relative to TOF techniques. First, for the same spatial resolution, the image acquisition time for a PC MRA study is longer than is required for a TOF MRA study. Second, the user must specify the expected maximum flow velocity (the *v*elocity *enc*oding (VENC) value) before the scan is initiated. If the actual flow velocity exceeds the specified VENC value, the flow will be "aliased" in the MRA images, typically resulting in the displayed flow velocity underestimating, sometimes substantially, the actual flow velocity. On the other hand, if the specified VENC is significantly greater than the actual flow velocity, the signal from the vessel will be small, and the SNR in the resulting MR angiogram will be poor. The dependence of the quality of PC MRA on the VENC value sometimes requires the acquisition of quick, low-resolution test scans with a range of VENC values, and then using the value that gives optimum SNR for the actual clinical PC MRA data acquisition.

Neither of the MRA techniques discussed thus far requires the infusion of contrast agent, and that is one of the chief advantages of MRA. However, the use of bolus Gd-based contrast agent infusion immediately prior to the acquisition of very rapid MRA images can substantially improved the visualization of distal and/or small vessels, particularly in body applications [20, 21]. Improvements in gradient subsystems have allowed for the acquisition of such 3D bolus contrast agent-enhanced scans in a single breath-hold (Fig. 5.11c). In addition to improving the visualization of smaller vessels, the infusion of the contrast agent minimizes the dependence of TOF techniques on flow-related enhancement, and the scan plane can be made independent of flow direction. Unfortunately, the timing of the image acquisition with respect to the initiation of bolus infusion now becomes critical in order to achieve good arterial phase MRA since spatial saturation bands, used heavily in nonbolus contrast agent-enhanced TOF MRA, are now ineffective in suppressing the venous flow. Due to the short acquisition time, however, several single breath-hold 3D image sets can be acquired following the bolus infusion. This allows the acquisition of various phases of contrast agent distribution, from nearly pure arterial phase to a later mixed arterial–venous phase. The use of such bolus contrast agent-enhanced scans has dramatically increased the applications of MRA in vascular imaging of the body.

5.2.5 Fat Suppression and Phase-Sensitive MRI Techniques

As mentioned earlier in the discussion of the FSE sequence, it is sometimes useful to suppress the signal from fat in order to improve the conspicuity of nearby lesions. This is particularly true on FSE T2W images or on some postcontrast T1W images. Currently, the most commonly utilized fat suppression technique is the application of frequency-selective saturation pulses ("fat-sat" or "chem-sat") [1]. Saturation, in this case, refers to the application of a narrow bandwidth 90° pulse that affects only the fat protons. Fortunately, this is possible since the fat and water protons resonate at frequencies that are separated by about 215 Hz (at 1.5 T, the separation scales linearly with B_0). If such a pulse is applied to the fat protons, and then is followed immediately by a SE or FSE sequence, the magnetization of the fat protons is left parallel or antiparallel to the applied static field at the TE and is not detected. The water protons, on the other hand, are not affected. Although this is the most common technique for fat suppression and does not, ideally, result in partial loss of water signal like inversion recovery (IR) techniques, there can be areas of suboptimal fat suppression. This is particularly true for large field-of-view images where the variable magnetic field homogeneity can cause shifts of the water and fat resonance frequencies with the result being uneven suppression of fat within the field of view. (Fat suppression efficiency decreases as the fat resonance frequency shifts away from the center frequency of the narrow bandwidth saturation pulse.) Other areas of susceptibility-induced inhomogeneities, such as skull base and neck, may also demonstrate variable suppression efficiency.

On large field-of-view scans, where fat-saturation techniques yield suboptimal suppression, short T1 IR (STIR) techniques are sometimes used for suppressing the fat signal [1–3]. The STIR sequence is based on a conventional SE sequence but has a 180° "inversion pulse" that precedes the 90° slice selective pulse by an amount of time known as the inversion time (TI). The inversion pulse inverts the net longitudinal magnetization and immediately after the pulse the magnetization begins to recover, at a rate determined by the T_1 relaxation time, to the equilibrium condition parallel to the applied static magnetic field. If the value of TI is chosen such that the longitudinal magnetization for a given tissue is zero when the 90° pulse of the SE sequence is applied, the magnetization at the time of the spin echo will be in the longitudinal direction and will not be detected. Therefore, one use of IR sequences is to suppress certain tissues of interest by taking advantage of the tissue-dependent T_1 relaxation times. In STIR, a short TI value is used, and the value is selected so that the longitudinal magnetization of fat is zero when the 90° pulse is applied. This results in the desired fat suppression in the resulting image. Since the T_1 relaxation times depend on the static magnetic field strength, the value of TI for a given tissue changes with field strength as well. At 1.5 T, the value of TI that yields this suppression is around 155 ms.

The advantage of STIR for fat suppression lies in the fact that the inversion pulse is not frequency selective like the fat-sat pulse. Therefore, the quality of fat suppression is not sensitive to magnetic field homogeneities. This is why STIR sequences are sometimes preferred over fat-sat sequences for large field-of-view scans or scans centered away from isocenter, for example, wrist imaging. However, this advantage does not come without a price, which is the unavoidable partial suppression of the water signal in addition to the suppression of the fat signal, with a concomitant decrease in SNR of the resulting image. Therefore, to achieve SNR that is comparable to the fat-sat case, longer image acquisition times are required. This is compounded by the fact that fewer anatomical slices can be acquired per TR in the case of STIR due to the time required to apply the inversion pulse and TI interval compared to the time required to apply the frequency-selective saturation pulse.

A final means of fat suppression is the use of phase-sensitive techniques. These techniques, like the frequency-selective saturation technique, take advantage of the differing resonance frequencies of fat and water (~215-Hz difference at 1.5 T). Because of this frequency difference, fat protons precess at a different rate than the water protons. If the water and fat protons are allowed to dephase by exactly 180° at the time the echo is detected, the fat and water signals will partially cancel one another in voxels that contain both fat and water. This phenomenon can also be used to generate separate images of fat and water using a technique originally proposed by Dixon [22]. In this technique, two SE images are obtained. In the first image, TE is chosen such that the water and fat protons are in phase at the TE, that is, the $S \sim W + F$, where S is the total signal intensity, F is the fat signal intensity, and W is the water signal intensity. In the second image, the TE of the sequence is modified very slightly such that the fat and water protons are out of phase by 180° at the time the echo is detected, that is, $S \sim W - F$. If the raw data (real and imaginary channels, not the magnitude data) from the two sets of images are added appropriately, the water image results, that

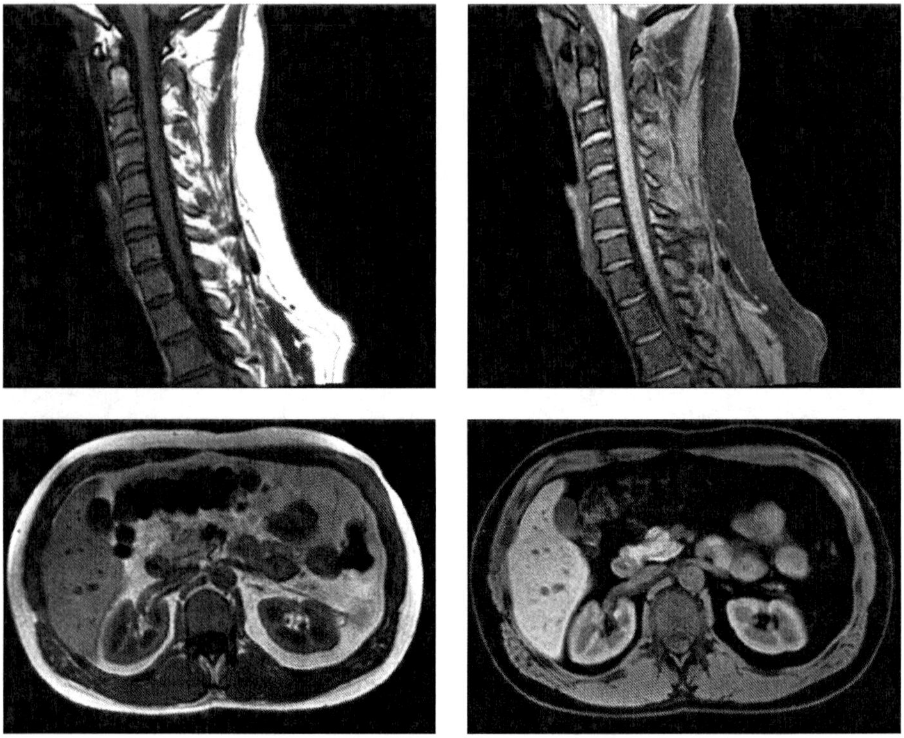

FIGURE 5.12 T1W SE images obtained without (top left) and with (top right) frequency-selective saturation of the fat signal ("fat-sat" pulse). Similar applications in abdominal imaging are shown in the lower left (unsuppressed) and lower right (fat-suppressed) images.

is, $(W+F)+(W-F)=2W$. On the other hand, if they are subtracted, the fat image is obtained, that is, $(W+F)-(W-F)=2F$. Several commercial implementations of phase-sensitive imaging, all based on the Dixon technique, are now available and are becoming increasingly popular. Examples of images obtained without and with fat suppression are shown in Figure 5.12.

5.2.6 Fluid-Attenuated Inversion Recovery (FLAIR) Pulse Sequences

T_2-FLAIR sequences are based on IR techniques discussed earlier with respect to fat suppression. However, in T_2-FLAIR, the value of TI is chosen such that signals from fluids with long T_1 relaxation times, like CSF, vitreous humor, and nonproteinaceous cysts, are strongly suppressed, while signals from brain and other tissues with shorter T_1 relaxation times are only minimally affected. A very common use of this technique is in the acquisition of fluid-suppressed T2W images [23, 24]. This can be very beneficial when evaluating, for example, low conspicuity periventricular lesions (e.g., multiple sclerosis plaques) or lesions adjacent to sulci (e.g., cortical infarcts), which can otherwise be masked by the adjacent CSF. Examples of the increase in lesion conspicuity in T_2-FLAIR images as compared to conventional FSE T2W images are shown in Figure 5.13.

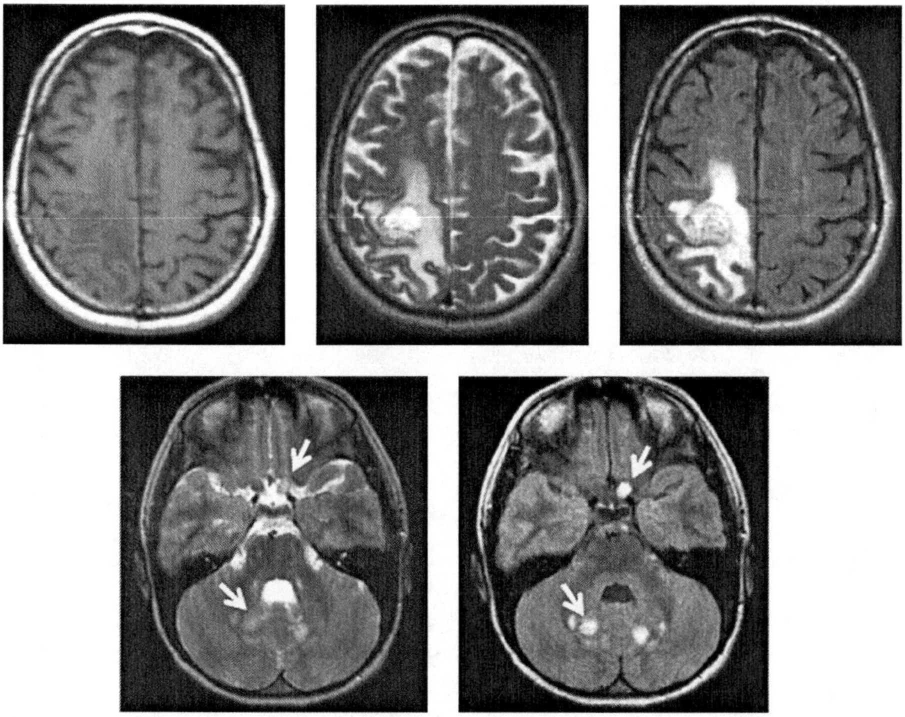

FIGURE 5.13 Typical T1W (top left), T2W (top center), and T2-FLAIR (top right) images of a patient with a high-grade neoplasm. Note the significant vasogenic edema, best seen on the T2-FLAIR images. FSE T2W (lower left) and fast T2-FLAIR (lower right) images of a patient with neurofibromatosis type 1. Notice the dramatic improvement in lesion conspicuity in the T2-FLAIR images compared with the T2W images (arrows).

5.3 CURRENT STATE-OF-THE-ART CLINICAL AND CLINICAL RESEARCH APPLICATIONS IN ONCOLOGY

Historically, PDW, T1W, T2W, and T1W postcontrast agent images, with or without fat suppression, formed the primary foundation for oncological imaging using MR. Currently, T_2-FLAIR images are standard for neuroimaging and have replaced the previously ubiquitous PDW images in brain imaging. However, several additional types of image-weighting techniques have recently become increasingly common. This section will introduce some of these techniques.

5.3.1 Diffusion-Weighted Pulse Sequences

The addition of appropriate gradient pulses within the TE allows for the generation of image contrast that depends heavily on the rate of diffusion of the nuclear spins [25–27]. In general, images are acquired with and without diffusion-weighting gradient pulses. In the diffusion-weighted images, tissues with freely diffusing water protons are *hypo*intense relative to tissues with water protons that

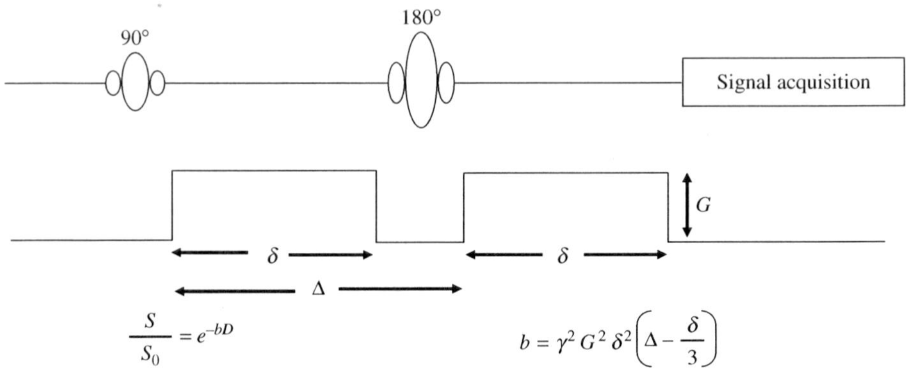

FIGURE 5.14 Generic diffusion imaging pulse sequence (RF and diffusion-sensitizing gradients only). S_0 is the signal obtained with the diffusion-sensitizing gradients turned off, S is the signal obtained with the gradients switched on, and b is the b-value, which is a user-selectable parameter.

have restricted motion. The contrast between the freely diffusing and restricted water protons increases as the area and separation of the diffusion-sensitizing gradient pulses increase, and these parameters are directly related to the often quoted user-selectable "*b*-factor" value [26, 28]. More specifically, the attenuation of the signal from a given tissue in a diffusion-weighted image (S) relative to the signal from a matched image with no diffusion weighting (S_0), that is, $b=0$, is given by $S/S_0 = e^{-bD}$, where b is the b-factor and D is the "apparent" diffusion coefficient (ADC) (Fig. 5.14). Therefore, the larger the diffusion coefficient, the greater the signal loss under the influence of diffusion-sensitizing gradients with a given b-factor. Appropriate postprocessing of the images also allows the calculation of an "ADC" image in which the pixel intensity is approximately proportional to the diffusion coefficient [26, 28]. In the ADC images, therefore, regions of freely diffusing spins are *hyper*intense relative to regions where diffusion is restricted.

If the diffusion-sensitizing gradients are applied appropriately along multiple directions, anisotropic diffusion images can be generated in addition to an isotropic diffusion image that is obtained from the trace of the diffusion tensor [2, 6, 29]. There are advantages and disadvantages of each. For example, water diffuses more freely along white matter tracts than it does transverse to the tracts. Therefore, the intensity of normal white matter in anisotropic ADC images depends upon the direction of the applied diffusion-sensitizing gradients (Fig. 5.15). This could cause confusion in the characterization of a lesion if the directionality of the white matter tracts is not fully appreciated. Therefore, it is frequently useful to examine the isotropic, or average, diffusion-weighted or ADC images, where the directional information is removed, rather than the anisotropic images. On the other hand, the anisotropic images may be quite useful in evaluating disease processes that disrupt the normal white matter tracts. By computing quantities such as the fractional anisotropy (FA) in each pixel, images that highlight the white matter tracts can be computed [30, 31], and fiber-tracking techniques can be used to visualize the larger

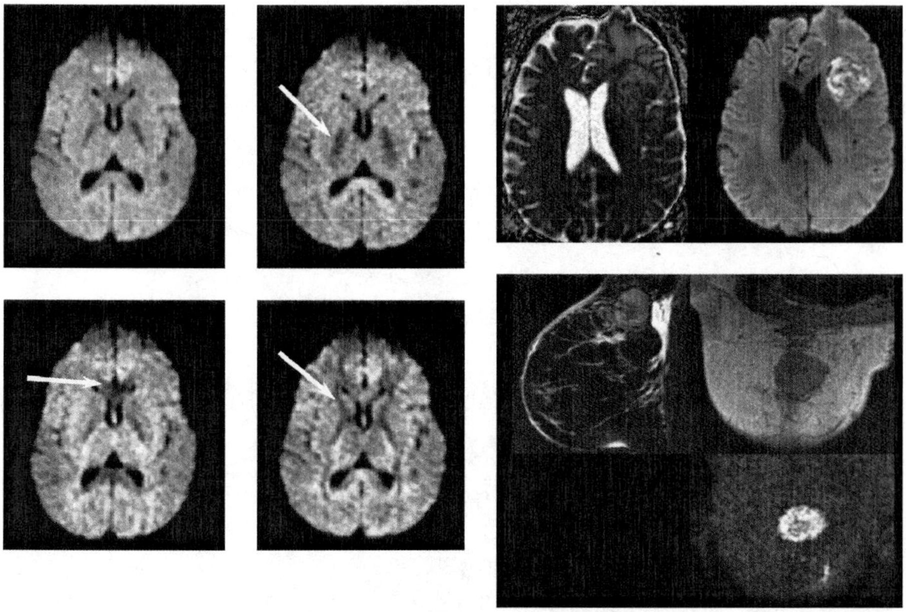

FIGURE 5.15 Left panel: Isotropic and anisotropic diffusion images of normal human brain. The top left image is an isotropic ("trace") diffusion-weighted image acquired using snapshot SE EPI sequence. The remaining images are diffusion-weighted images with the diffusion-sensitizing gradients applied in the superior–inferior (top right), right–left (bottom left), and anterior–posterior (bottom right) directions. The arrows indicate regions of anisotropic diffusion due to the orientation of the white matter tracts. Right panel: ADC (top left) and diffusion-weighted (top right) images from a glioma patient. T2W, T1W, and diffusion-weighted images from a patient with breast cancer (bottom). On diffusion-weighted images, areas of restricted diffusion, common in the case of increased cell density and proliferation, are hyperintense. On ADC images, such areas are hypointense. Successful tumor therapy, whether causing necrotic or apoptotic cell death, *ultimately* causes ADC values to increase and diffusion-weighted values to decrease.

white matter tracts to better appreciate the effects of tumors and other lesions on the white matter tracts (Fig. 5.16).

In general, diffusion restriction increases with increased cell density and cell volume fraction. Therefore, rapid cell growth, as seen in high-grade neoplasms, generally results in diffusion restriction and hyperintense signal on diffusion-weighted images (and hypointense signal on ADC images). Successful therapeutic intervention, causing either necrotic or apoptotic cell death, will ultimately result in a decreased signal intensity on diffusion-weighted images (decrease cell density) and an increased signal intensity on ADC images [32–36].

While diffusion imaging was initially only available for brain imaging, recent advances in scanner capabilities, including higher gradient performance and improvements in diffusion imaging pulse sequence implementations, have resulted in diffusion imaging applications becoming increasingly common in head and neck, breast, prostate, liver, and even whole-body studies [37].

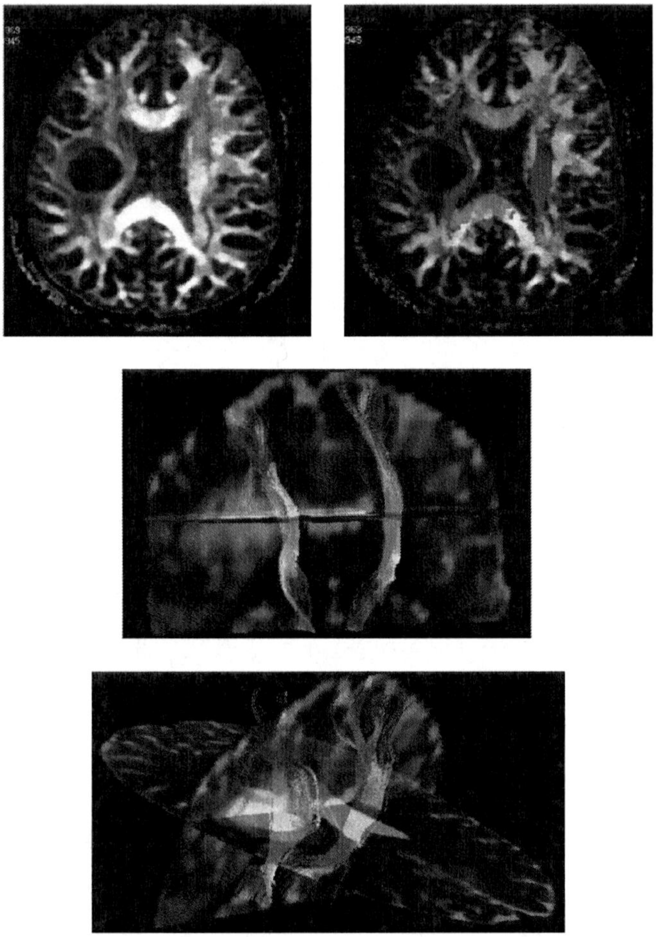

FIGURE 5.16 Top left: FA index image in which the pixel intensity is proportional to the degree of diffusion anisotropy providing a means of imaging white matter tracts. Top right: A color FA map in which white matter tracts running left to right are red, those running anterior–posterior are green, and those running superior–inferior are blue. Middle and bottom: Fiber-tracking images obtained by placing seed pixels in the primary motor cortex in a patient with a high-grade glioma causing a medial shift of the corticospinal white matter tract. For color details, please see color plate section.

5.3.2 Dynamic Contrast-Enhanced MRI (DCE-MRI)

Dynamic contrast-enhanced MRI (DCE-MRI) relies on rapid sampling (typically every 4–10 s) of T1W signal intensity changes from the vascular space and from the ROI before, during, and following contrast agent injection to allow the computation of vascular volume fraction (v_p), endothelial volume transfer constant (K^{trans}), contrast agent reflux rate (k_{ep}), and extravascular, extracellular volume fraction (v_e) (Fig. 5.17). Commonly applied analysis strategies in DCE-MRI applications are based on a two-compartment pharmacokinetic model [6, 38, 39].

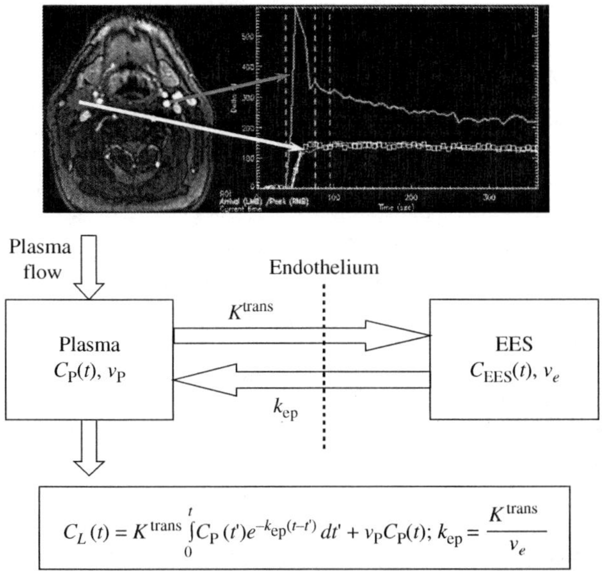

FIGURE 5.17 Basic DCE-MRI definitions. Top panel: A single T1W frame from a DCE-MRI acquisition with ROIs in a node with metastatic disease (from a squamous cell carcinoma of the tongue base) and in a reference vessel. The corresponding signal intensity versus time curves are shown on the right. Bottom panel: A common two-compartment pharmacokinetic model used to analyze the DCE-MRI data to extract the physiological parameters of interest. The parameters are defined in the text.

Differences in the details of the analysis strategies are most commonly (i) whether or not the vascular volume fraction is included in the calculation of the total contrast agent concentration in a given ROI (C_{ROI}), that is,

$$C_{ROI}(t) = K^{trans}\int_0^t C_P(t')e^{-k_{ep}(t-t')}dt' + v_P C_P(t) \tag{5.7}$$

(three-parameter generalized kinetic model)

versus

$$C_{ROI}(t) = K^{trans}\int_0^t C_P(t')\,e^{-k_{ep}(t-t')}dt' \tag{5.8}$$

(two-parameter kinetic model)

and (ii) whether the plasma compartment contrast agent concentration, $C_p(t)$, is measured (patient-derived arterial input function) or is assumed based on prior reported data (population-derived arterial input function). In the earlier equations, the remaining parameter of interest, v_e, can be obtained as K^{trans}/k_{ep}. It should be again noted that the conversion of signal intensity change data to contrast agent concentration data requires a nonlinear transformation that is based on the native tissue (pre-contrast agent administration) T_1 relaxation values, the specific pulse sequence used, and the acquisition parameters used for that pulse sequence. As a result of the complexities of these conversions, some investigators perform the DCE-MRI pharmacokinetic analysis using the measured signal intensity change data directly without conversion to contrast agent concentration. In this case, it is important that the flip angle used for the DCE-MRI acquisition be relatively large, that is, 25–30°, in order to provide a quasilinear signal response as a function of Gd concentration, particularly in the case of the vascular input function ROI. Details of these models and data analysis strategies are described in Chapter 6 of the text edited by Jackson et al. [40].

One problem with MR-based perfusion imaging techniques using commonly administered contrast agents relates to the size/molecular weight of the agent itself. Most commonly administered contrast agents have molecular weights in the ~500–700 da range and, as such, extravasate with relatively high rates for all tissues outside an intact blood brain barrier (BBB). In DCE-MRI applications, for example, this results in the K^{trans} parameter reflecting both the delivery of the agent (flow) and the permeability (extraction) in lesions associated with high "leakage" of the contrast agent [39]. If the flow is not the limiting factor, then K^{trans} is a good approximation of the permeability-surface area product. If flow is limiting, however, then K^{trans} is most generally a measure of the extraction-flow product. In this regard, both DCE-MRI and DCE-CT would benefit from the availability of approved contrast agents with higher molecular weight.

5.3.3 Dynamic Susceptibility Change (DSC) Perfusion Imaging

Two major MRI techniques have been proposed for direct mapping of regional cerebral blood volume (rCBV) and regional cerebral blood flow (rCBF). The first technique takes advantage of the transient susceptibility change that occurs when a bolus of paramagnetic contrast agent passes through the capillary bed. The susceptibility change causes enhanced dephasing of nearby spins due to the temporary decrease in T_2*, and this results in a transient decrease in signal intensity on T2W or T2*W images (Fig. 5.18). Therefore, areas of increased perfusion will appear hypointense relative to regions of decreased perfusion on T2W or T2*W images. Under the assumption that the contrast agent remains intravascular, that is, no fenestration of the BBB, such images can be used to construct maps of rCBV and/or rCBF [6, 41, 42]. If, however, the BBB is fenestrated and contrast agent "leaks" from the vascular space into the interstitial space, the value of rCBV obtained by bolus contrast infusion perfusion mapping will be underestimated unless appropriate deconvolution model analysis with leakage correction is used [43, 44]. This correction is particularly important given the low molecular weight of the clinically available MR contrast agents, as mentioned earlier. An advantage of this technique, as opposed to nonbolus contrast agent techniques, is the favorable SNR of the perfusion images, particularly if double-dose bolus injections of contrast agent are used or high-field, for example, 3.0T, scanners are utilized.

5.3.4 Arterial Spin Labeling Perfusion Imaging

The second method for mapping microvascular blood flow does not require the administration of an exogenous contrast agent. In arterial spin labeling MRI (ASL-MRI), labeled arterial blood water is used as an endogenous contrast agent

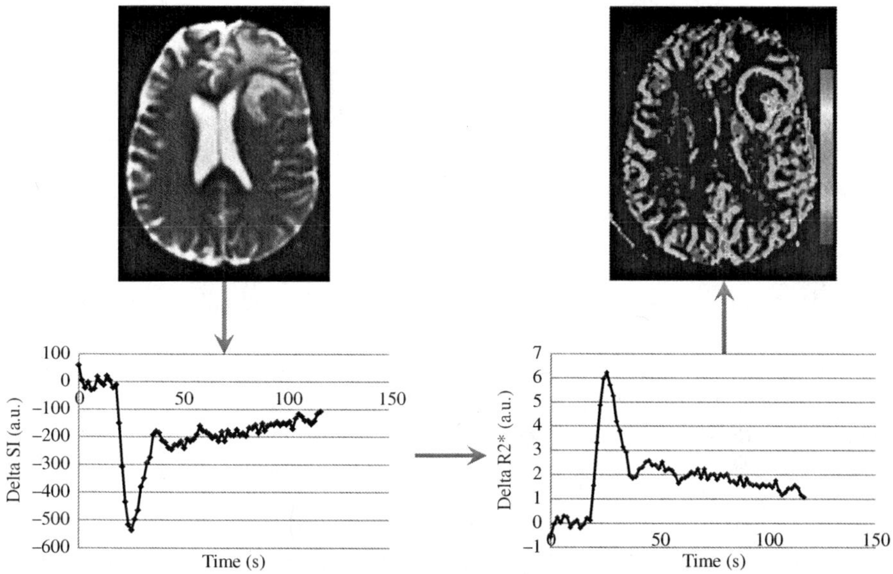

FIGURE 5.18 A representative SE EPI T2W image (top left), extracted signal intensity versus time curve (bottom left) converted to a change in T_2* relaxation rate versus time curve (bottom right) and the resulting perfusion map (top right) obtained using the bolus contrast agent infusion DSC technique (0.1 mmol/kg of Gd-DTPA). Note the expected contrast between gray matter and white matter in the perfusion map due to higher microvascular blood volume in the gray matter.

to allow for the computation of rCBV and rCBF maps. While a variety of specific implementations of ASL-MRI have been proposed and are in use in research centers and, more recently, available on commercial MR systems, the basic principle is that the longitudinal magnetization state of the blood is inverted prior to entering the imaging volume and this "tagging" causes the magnetization from the blood entering the imaging volume to differ from that of the tissues within the imaging volume. Two sets of images are obtained, one without this tagging (M_T^{Control}) and one with the tagging (M_T^{Label}). The image sets are subtracted, resulting in perfusion-weighted images. To convert the perfusion-weighted images to quantitative blood flow images requires additional information, including the T_1 relaxation rate for the tissues of interest (R_{1T}) and arterial blood (R_{1A}), as well as the magnetization labeling efficiency of the blood (α), the blood–brain partition coefficient (λ), and labeled blood transit time (δ), that is, CBF can be calculated from the following equation:

$$\frac{M_T^{\text{Control}} - M_T^{\text{Label}}}{M_T^{\text{Control}}} = \frac{\text{CBF}2\alpha}{\lambda R_{1T}} e^{-R_{1A}\delta} \tag{5.9}$$

The required parameters are either measured directly, for example, patient-specific R_{1T} measurements, or estimated from prior experiments or literature values, for example, literature values of λ. Details of the various implementations of ASL-MRI are provided in the references [6, 45–48].

The obvious advantage of ASL-MRI lies in the absence of any need for exogenous contrast agent. The disadvantages relative to contrast agent-enhanced techniques, however, are less sensitivity (requiring multiple signal averages with the concomitant longer scan time) and challenges of obtaining multiple acquisition sections with comparable measurement accuracy. However, with the recent increased scrutiny on Gd contrast agents due to the purported association of such contrast agents with nephrogenic systemic fibrosis in a small number of patients with severe renal insufficiency and the ability to obtain repeated rCBF measures in a single imaging session, commercial implementations of ASL-MRI are now available, and applications are expanding, particularly on 3.0-T systems where the added SNR is advantageous in such applications. With this expansion of commercially available ASL options, the applications have extended from neuroimaging-centric applications to those outside the brain, including applications in renal [49] and breast [50] studies.

It is commonly accepted that tumors frequently induce vascular neogenesis and vascular permeability changes, and highly vascular tumors are often more malignant than those do not have increased local vascularity. Treatment-related changes and benign lesions, on the other hand, do not typically exhibit the same level of perturbed angiogenesis. Therefore, the ability to assess vascularity should improve the characterization of lesions and bolus-infusion perfusion imaging has been proposed as a tool for aiding in the differential diagnosis of tumor progression versus treatment-related changes and for assessing tumor grade [51–56]. Such quantitative measures of vascularity also allow the assessment of the efficacy of antiangiogenic drug therapy. There are, in fact, numerous examples of DCE- and DSC-MRI applications in the assessment of targeted therapies, particularly with antiangiogenic/antivascular agents and radiation therapy [53, 57–62]. In antiangiogenic/antivascular therapy applications, profound decreases (35–65%) in DCE-MRI K^{trans} measures, for example, are frequently reported, often as early as 24–48 h following initial dose [62–67]. As a result, DCE-MRI imaging biomarker measures have been increasingly used as secondary/exploratory end points in phase I/II clinical trials of novel antiangiogenic/antivascular agents. In radiation oncology applications, both DCE-MRI and DSC-MRI applications have demonstrated that areas of tumors that exhibit poor perfusion measures at baseline, for example, low values of blood volume, blood flow, v_p, and/or K^{trans}, or demonstrate no or minimal increase during radiation therapy are typically associated with poor outcome, for example, local failure [53, 57, 58, 61, 67–69]. Finally, while ASL applications currently are fewer in number than contrast-enhanced applications, reported studies include those addressing the management of cerebrovascular disease, central nervous system neoplasms, and epilepsy [70–75]. As ASL capabilities have recently been made available on high-field MR scanners from all major vendors, however, it is anticipated that such applications will increase in number rather quickly. Figure 5.19 illustrates applications of DCE- and DSC-MRI in the case of a patient with a grade IV glioblastoma.

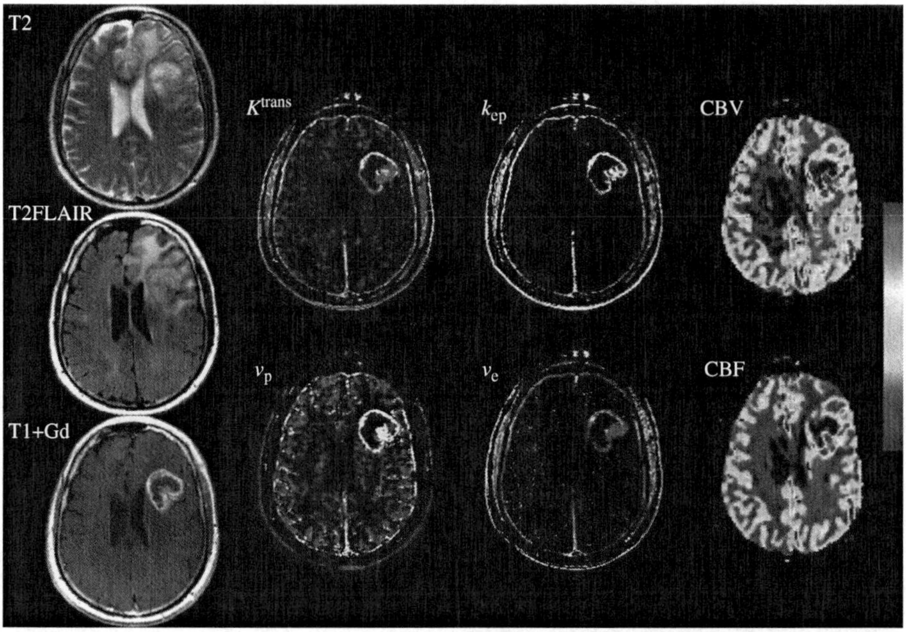

FIGURE 5.19 Various anatomic images and corresponding functional parameter maps in a patient with pathologically proven glioblastoma. The images on the left are standard anatomic images (T2W, T2-FLAIR, and T1W post-Gd contrast agent images). The functional parametric maps obtained from a DCE-MRI acquisition (K^{trans}, k_{ep}, v_p, and v_e) and obtained from a DSC-MRI acquisition (rCBV and rCBF) are shown. For color details, please see color plate section.

5.3.5 BOLD Functional MRI Pulse Sequences

The underlying physiological principle for the blood oxygen level dependent (BOLD) technique for indirectly assessing neuronal activation is that such activation in response to a stimulus, for example, a motor, cognitive, or vision task, results in local vasodilation and a concomitant increase in CBF and CBV. The completely noninvasive BOLD fMRI technique takes advantage of the fact that deoxyhemoglobin is a paramagnetic substance, which causes increased local dephasing of the spins due to T_2* (susceptibility) effects, while oxyhemoglobin is a diamagnetic substance, which has minimal T_2* effects. When neuronal activation occurs, local vasodilation yields a local increase in the concentration of oxyhemoglobin compared to the control (resting) state. It has been shown from positron emission tomography studies that the rate of oxygen extraction is less than the rate of oxyhemoglobin delivery [76], and thus there is a local decrease in the deoxyhemoglobin-to-oxyhemoglobin ratio [77, 78]. This results in less T_2* dephasing in areas near neuronal activation. Therefore, on T2*W images, areas of activation are hyperintense relative to the control state. Unfortunately, this change in signal intensity is relatively small (~1–5% at 1.5T and ~20% at 4T) [79], and relatively sophisticated signal analysis must be performed to determine statistically significant regions of activation [80]. However, the BOLD contrast phenomenon does provide a completely noninvasive means of indirectly mapping areas of activation by measuring the

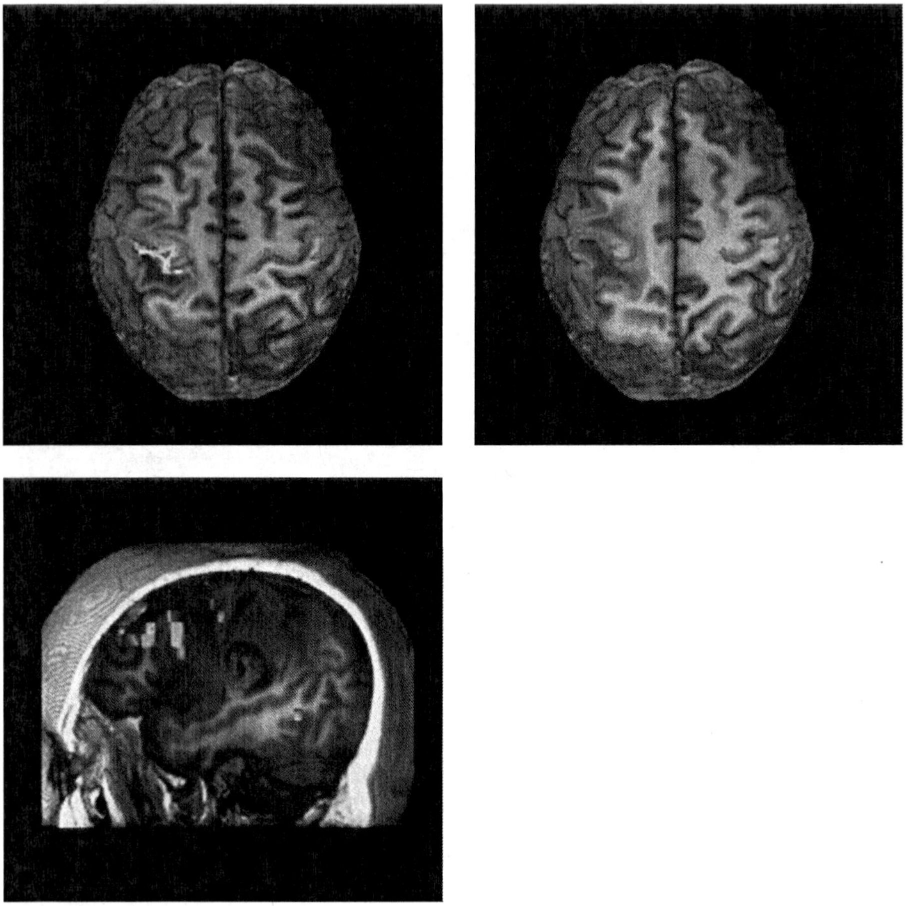

FIGURE 5.20 BOLD fMRI activation maps obtained by cross-correlation of the known stimulus pattern (a simple left-hand finger-tapping task) with the signal intensity time course on a pixel-by-pixel basis in a patient with a neoplasm (two different section locations, top left and top right, are shown). The yellow/blue/green pixels represent the measured BOLD fMRI response, while the orange pixels represent the Brodmann areas 1, 2, and 3 based on a standard stereotactic atlas. The fMRI BOLD response obtained during an expressive speech task (silently producing as many words as possible beginning with a specified letter in 20-s intervals) is shown on the bottom left. For color details, please see color plate section.

hemodynamic response changes secondary to the neuronal activation. Since the first BOLD studies were published in 1992, a large number of BOLD studies have been reported with widely varying applications including visual stimulation, auditory stimulation, memory, language, sensory and motor stimulation, psychiatric disorders, and pain research. In the oncology setting, one of the most common applications is in presurgical planning and surgical navigation. Examples of BOLD fMRI activation maps are given in Figure 5.20 for motor and expressive speech tasks. Figure 5.21 illustrates the combination of BOLD fMRI and diffusion tensor imaging. It is important to note that BOLD fMRI acquisition strategies have also been used to evaluate the oxygen status of tissue, which is of particular importance in radiation therapy applications [81, 82].

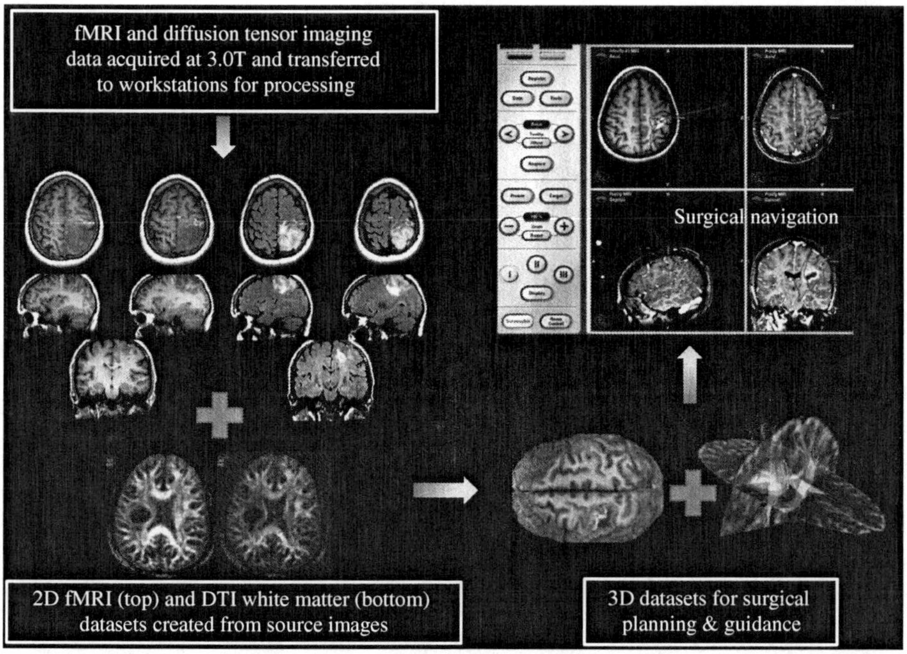

FIGURE 5.21 Combination of BOLD fMRI and diffusion tensor imaging (FA and corticospinal fiber-tracking maps) results and transfer to image-guided surgery navigation system. While this application is for identification of regions of motor "eloquent cortex" and associated white matter tracts, any of the parameter maps discussed, for example, K^{trans} and rCBV, can be similarly used for targeting of optimal biopsy targets or for targeting local delivery of drug. For color details, please see color plate section.

5.3.6 Current Status of Quantitative Imaging Biomarkers

Each of the parameters described earlier is an example of a quantitative imaging biomarker (QIB). As noted earlier, several of the aforementioned MR QIBs have been used in drug development, for example, in preclinical mechanism of action studies and in phase I/II clinical trials of new drugs, particularly targeted antiangiogenic/antivascular drugs. Such MR QIBs have also been used in drug delivery research, for example, convective delivery of injected agent [83, 84]. To transition such QIBs from secondary/exploratory end points to primary end points will require completion of additional efforts necessary to qualify such measures [85]. This is not only true for MR-based QIBs but for QIBs from all imaging modalities, including CT volumetrics, FDG-PET SUV measures, etc. To aid in this process, several organizations, including the Radiological Society of North America (RSNA) through its Quantitative Imaging Biomarker Alliance (QIBA) efforts (http://qibawiki.rsna.org) [86], are actively working with other scientific organizations, equipment vendors, end users, etc. to identify and promulgate methodologies that will, per the goal of QIBA, "improve the value and practicality of quantitative biomarkers by reducing variability across devices, patients, and time" and to work with the FDA, for example, to qualify such QIBs and extend their consistent and reproducible applications beyond individual centers of excellence to inter-vendor, inter-center applications.

5.4 CONCLUSIONS

The goals of this chapter have been to briefly introduce the physical principles of MRI, to briefly outline some current state-of-the-art applications of MRI to oncological imaging, and, hopefully, to pique the interest of the reader sufficiently to send him or her to the literature to gain further insight into the rapidly expanding role of MRI in oncology. In the roughly 30-year history of clinical MRI, the number of applications of MRI, and the progress in discovering and implementing new means of manipulating image contrast based on a wide range of chemical and physiological bases, has been incredibly impressive. This rate of development does not show any signs of slowing in the near future.

REFERENCES

1. Edelman RR, Hesselink JR, Zlatkin MB, et al. Physics, Instrumentation, and Advanced Techniques. In: Edelman RR, Hesselink JR, Zlatkin MB, Crues JV, editors. *Clinical Magnetic Resonance Imaging*. Volume 1, Philadelphia: Elsevier Health Sciences; 2005.
2. Bernstein M, King KF, Zhou XJ. *Handbook of MRI Pulse Sequences*. Philadelphia: Elsevier Science; 2004.
3. Elster AD, Burdette JH. *Questions and Answers in Magnetic Resonance Imaging*. 2nd ed. Philadelphia: Elsevier Health Sciences; 2000.
4. Hashemi RH, Bradley WG, Lisanti CJ. *MRI: The Basics*. 3rd ed. Philadelphia: Lippincott, Williams & Wilkins; 2010.
5. Haacke EM, Brown RW, Thompson MR, et al. *Magnetic Resonance Imaging. Physical Principles and Sequence Design*. New York: Wiley-Liss; 1999.
6. Yankeelov TE, Pickens DR, Price RR. *Quantitative MRI in Cancer*. Boca Raton: CRC Press; 2011.
7. Golman K, Olsson LE, Axelsson O, et al. Molecular imaging using hyperpolarized 13C. Br J Radiol 2003;76 Spec No 2:S118–S127.
8. Ross BD, Bhattacharya P, Wagner S, et al. Hyperpolarized MR imaging: neurologic applications of hyperpolarized metabolism. Am J Neuroradiol 2010;31:24–33.
9. Wood ML, Bronskill MJ, Mulkern RV, et al. Physical MR desktop data. J Magn Reson Imaging 1993;3 (S):19–26.
10. Siemens Healthcare. 2010. MRI Acronyms. Available at http://www.medical.siemens.com/siemens/it_IT/gg_mr_FBAs/files/MAGNETOM_World/Application_Tips/MRI_Acronyms.pdf. Accessed 2013 April 17.
11. Hennig J, Nauerth A, Friedburg H. RARE imaging: a fast imaging method for clinical MR. Magn Reson Med 1986;3:823–833.
12. Yamashita Y, Abe Y, Tang Y, et al. In vivo and clinical studies of image acquisition in breath-hold MR cholangiopancreatography: single-shot projection techniques versus multislice technique. Am J Roentgenol 1997;168:1449–1454.
13. Barish M, Soto J, Yucel E. Magnetic resonance cholangiopancreatography of the biliary ducts: techniques, clinical applications, and limitations. Top Magn Reson Imaging 1996;8:302–311.
14. Constable RT, Anderson AW, Zhong J, et al. Factors influencing contrast in fast spin-echo MR imaging. J Magn Reson Imaging 1992;10:497–511.
15. Constable RT, Gore JC. The loss of small objects in variable TE imaging: implications for FSE, RARE, and EPI. Magn Reson Med 1992;28:9–24.

16. Atlas SW, Hackney DB, Listerud J. Fast spin-echo imaging of the brain and spine. Magn Reson Q 1993;9:61–83.
17. Constable RT, Smith RC, Gore JC. Signal-to-noise and contrast in fast spin echo (FSE) and inversion recovery FSE imaging. J Comput Assist Tomogr 1992;16:41–47.
18. Henkelman RM, Hardy PA, Bishop JE, et al. Why fat is bright in RARE and fast spin-echo imaging. J Magn Reson Imaging 1992;2:533–540.
19. Norbash AM, Glover GH, Enzmann DR. Intracerebral lesion contrast with spin-echo and fast spin-echo pulse sequences. Radiology 1992;185:661–665.
20. Potchen EJ, Haacke EM, Siebert JE, et al. *Magnetic Resonance Angiography*. St Louis: Mosby-Year Book, Inc: 1993.
21. Prince MR, Grist TM, Debatin JF. *3D Contrast MR Angiography*. New York: Springer-Verlag; 1997.
22. Dixon WT. Simple proton spectroscopic imaging. Radiology 1984;153:189–194.
23. DeCoene B, Hajnal JV, Gatehouse P. MR of the brain using fluid-attenuated inversion recovery (FLAIR) pulse sequences. Am J Neuroradiol 1992;13:1555–1564.
24. Hajnal JV, DeCoene B, Lewis PD. High signal regions in normal white matter shown by heavily T_2-weighted CSF nulled IR sequences. J Comput Assist Tomogr 1992;16:506–513.
25. Le Bihan D. Clinical intravoxel incoherent motion imaging. In: Potchen EJ, Haacke EM, Siebert JE, Gottschalk A, editors. *Magnetic Resonance Angiography Concepts & Applications*. St Louis: Mosby-Year Book, Inc.; 1993. p 485–497.
26. Le Bihan D, Turner R, Douek P, et al. Diffusion MR imaging: clinical applications. Am J Roentgenol 1992;159:591–599.
27. Le Bihan D, Turner R, Moonen CTW, et al. Imaging of diffusion and microcirculation with gradient sensitization: design, strategy, and significance. J Magn Reson Imaging 1991;1:7–28.
28. Le Bihan D. Theoretical principles of perfusion imaging. Application to magnetic resonance imaging. Invest Radiol 1992;27:S6–S11.
29. Moseley ME, Cohen Y, Kucharczyk J, et al. Diffusion-weighted MR imaging of anisotropic water diffusion in cat central nervous system. Radiology 1990;176:439–445.
30. Basser PJ, Mattiello J, LeBihan D. MR diffusion tensor spectroscopy and imaging. Biophys J 1994;66:259–267.
31. Basser PJ, Mattiello J, Turner R, et al. Diffusion tensor echo-planar imaging of human brain. Twelfth Annual Meeting of the Society of Magnetic Resonance in Medicine: New York: Society of Magnetic Resonance in Medicine; 1993. pp 584.
32. Chenevert TL, McKeever PE, Ross BD. Monitoring early response of experimental brain tumors to therapy using diffusion magnetic resonance imaging. Clin Cancer Res 1997;3:1457–1466.
33. Chenevert TL, Stegman LD, Taylor JM, et al. Diffusion magnetic resonance imaging: an early surrogate marker of therapeutic efficacy in brain tumors. J Natl Cancer Inst 2000;92:2029–2036.
34. Gerstner ER, Sorensen AG. Diffusion and diffusion tensor imaging in brain cancer. Semin Radiat Oncol 2011;21:141–146.
35. Padhani AR. Diffusion magnetic resonance imaging in cancer patient management. Semin Radiat Oncol 2011;21:119–140.
36. Pope WB, Lai A, Mehta R, et al. Apparent diffusion coefficient histogram analysis stratifies progression-free survival in newly diagnosed bevacizumab-treated glioblastoma. Am J Neuroradiol 2011;32:882–889.

37. Koh DM, Blackledge M, Padhani AR, et al. Whole-body diffusion-weighted MRI: tips, tricks, and pitfalls. Am J Roentgenol 2012;199:252–262.

38. Tofts PS. Modeling tracer kinetics in dynamic Gd-DTPA MR imaging. J Magn Reson Imaging 1997;7:91–101.

39. Tofts PS, Brix G, Buckley DL, et al. Estimating kinetic parameters from dynamic contrast-enhanced T(1)-weighted MRI of a diffusable tracer: standardized quantities and symbols. J Magn Reson Imaging 1999;10:223–232.

40. Jackson A, Buckley DL, Parker GJM. *Dynamic Contrast-Enhanced Magnetic Resonance Imaging in Oncology*. Berlin: Springer-Verlag; 2005.

41. Rosen BR, Belliveau JW, Aronen HJ, et al. Susceptibility contrast imaging of cerebral blood volume: human experience. Magn Reson Med 1991;22:293–299.

42. Rosen BR, Belliveau JW, Vevea JM, et al. Perfusion imaging with NMR contrast agents. Magn Reson Med 1990;14:249–265.

43. Ostergaard L, Sorensen AG, Kwong KK, et al. High resolution measurement of cerebral blood flow using intravascular tracer bolus passages. Part II: Experimental comparison and preliminary results. Magn Reson Med 1996;36:726–736.

44. Ostergaard L, Weisskoff RM, Chesler DA, et al. High resolution measurement of cerebral blood flow using intravascular tracer bolus passages. Part I: Mathematical approach and statistical analysis. Magn Reson Med 1996;36:715–725.

45. Steger TR, White RA, Jackson EF. Input parameter sensitivity analysis and comparison of quantification models for continuous arterial spin labeling. Magn Reson Med 2005;53:895–903.

46. Liu TT, Brown GG. Measurement of cerebral perfusion with arterial spin labeling: Part 1. Methods. J Int Neuropsychol Soc 2007;13:517–525.

47. Deibler AR, Pollock JM, Kraft RA, et al. Arterial spin-labeling in routine clinical practice. Part 1: Technique and artifacts. Am J Neuroradiol 2008;29:1228–1234.

48. Petersen ET, Zimine I, Ho YC, et al. Non-invasive measurement of perfusion: a critical review of arterial spin labelling techniques. Br J Radiol 2006;79:688–701.

49. Artz NS, Sadowski EA, Wentland AL, et al. Reproducibility of renal perfusion MR imaging in native and transplanted kidneys using non-contrast arterial spin labeling. J Magn Reson Imaging 2011;33:1414–1421.

50. Kawashima M, Katada Y, Shukuya T, et al. MR perfusion imaging using the arterial spin labeling technique for breast cancer. J Magn Reson Imaging 2012;35:436–440.

51. Aronen HJ, Gazit IE, Louis DN, et al. Cerebral blood volume maps of gliomas: comparison with tumor grade and histologic findings. Radiology 1994;191:41–51.

52. Aronen HJ, Glass J, Pardo FS, et al. Echo-planar MR cerebral blood volume mapping of gliomas. Clinical utility. Acta Radiol 1995;36:520–528.

53. Voglein J, Tuttenberg J, Weimer M, et al. Treatment monitoring in gliomas: comparison of dynamic susceptibility-weighted contrast-enhanced and spectroscopic MRI techniques for identifying treatment failure. Invest Radiol 2011;46:390–400.

54. Barajas RF, Jr., Chang JS, Segal MR, et al. Differentiation of recurrent glioblastoma multiforme from radiation necrosis after external beam radiation therapy with dynamic susceptibility-weighted contrast-enhanced perfusion MR imaging. Radiology 2009;253:486–496.

55. Jackson A, Kassner A, Annesley-Williams D, et al. Abnormalities in the recirculation phase of contrast agent bolus passage in cerebral gliomas: comparison with relative blood volume and tumor grade. Am J Neuroradiol 2002;23:7–14.

56. Nagesh V, Chenevert TL, Tsien CI, et al. Quantitative characterization of hemodynamic properties and vasculature dysfunction of high-grade gliomas. NMR Biomed 2007;20:566–577.

57. Zahra MA, Hollingsworth KG, Sala E, et al. Dynamic contrast-enhanced MRI as a predictor of tumour response to radiotherapy. Lancet Oncol 2007;8:63–74.
58. Matsusue E, Fink JR, Rockhill JK, et al. Distinction between glioma progression and post-radiation change by combined physiologic MR imaging. Neuroradiology 2010;52:297–306.
59. Mayr NA, Yuh WT, Jajoura D, et al. Ultra-early predictive assay for treatment failure using functional magnetic resonance imaging and clinical prognostic parameters in cervical cancer. Cancer 2010;116:903–912.
60. Cao Y. The promise of dynamic contrast-enhanced imaging in radiation therapy. Semin Radiat Oncol 2011;21:147–156.
61. Franiel T, Hamm B, Hricak H. Dynamic contrast-enhanced magnetic resonance imaging and pharmacokinetic models in prostate cancer. Eur Radiol 2011;21:616–626.
62. Morgan B, Thomas AL, Drevs J, et al. Dynamic contrast-enhanced magnetic resonance imaging as a biomarker for the pharmacological response of PTK787/ZK 222584, an inhibitor of the vascular endothelial growth factor receptor tyrosine kinases, in patients with advanced colorectal cancer and liver metastases: results from two phase I studies. J Clin Oncol 2003;21:3955–3964.
63. Liu G, Rugo HS, Wilding G, et al. Dynamic contrast-enhanced magnetic resonance imaging as a pharmacodynamic measure of response after acute dosing of AG-013736, an oral angiogenesis inhibitor, in patients with advanced solid tumors: results from a phase I study. J Clin Oncol 2005;23:5464–5473.
64. Herbst RS, Hong D, Chap L, et al. Safety, pharmacokinetics, and antitumor activity of AMG 386, a selective angiopoietin inhibitor, in adult patients with advanced solid tumors. J Clin Oncol 2009;27:3557–3565.
65. Batchelor TT, Sorensen AG, di Tomaso E, et al. AZD2171, a pan-VEGF receptor tyrosine kinase inhibitor, normalizes tumor vasculature and alleviates edema in glioblastoma patients. Cancer Cell 2007;11:83–95.
66. Levin VA, Bidaut L, Hou P, et al. Randomized double-blind placebo-controlled trial of bevacizumab therapy for radiation necrosis of the central nervous system. Int J Radiat Oncol Biol Phys 2011;79:1487–1495.
67. Essock-Burns E, Lupo JM, Cha S, et al. Assessment of perfusion MRI-derived parameters in evaluating and predicting response to antiangiogenic therapy in patients with newly diagnosed glioblastoma. Neuro Oncol 2011;13:119–131.
68. Cao Y, Popovtzer A, Li D, et al. Early prediction of outcome in advanced head-and-neck cancer based on tumor blood volume alterations during therapy: a prospective study. Int J Radiat Oncol Biol Phys 2008;72:1287–1290.
69. Loiselle CR, Eby PR, DeMartini WB, et al. Dynamic contrast-enhanced MRI kinetics of invasive breast cancer: a potential prognostic marker for radiation therapy. Int J Radiat Oncol Biol Phys 2010;76:1314–1319.
70. Wolf RL, Detre JA. Clinical neuroimaging using arterial spin-labeled perfusion magnetic resonance imaging. Neurotherapeutics 2007;4:346–359.
71. Ludemann L, Warmuth C, Plotkin M, et al. Brain tumor perfusion: comparison of dynamic contrast enhanced magnetic resonance imaging using T1, T2, and T2* contrast, pulsed arterial spin labeling, and H2(15)O positron emission tomography. Eur J Radiol 2009;70:465–474.
72. Detre JA, Rao H, Wang DJ, et al. Applications of arterial spin labeled MRI in the brain. J Magn Reson Imaging 2012;35:1026–1037.
73. Hartkamp NS, Petersen ET, De Vis JB, et al. Mapping of cerebral perfusion territories using territorial arterial spin labeling: techniques and clinical application. NMR Biomed 2012. Epub ahead of print. DOI: 10.1002/nbm.2836.

74. Watts JM, Whitlow CT, Maldjian JA. Clinical applications of arterial spin labeling. NMR Biomed 2013. Epub ahead of print. DOI: 10.1002/nbm.2904.
75. Chawla S, Wang S, Wolf RL, et al. Arterial spin-labeling and MR spectroscopy in the differentiation of gliomas. Am J Neuroradiol 2007;28:1683–1689.
76. Fox P, Raichle M. Focal physiological uncoupling of cerebral blood flow and oxidative metabolism during somatosensory stimulation in human subjects. Proc Natl Acad Sci USA 1986;83:1140–1144.
77. Kwong KK, Belliveau JW, Chesler DA, et al. Dynamic magnetic resonance imaging of human brain activity during primary sensory stimulation. Proc Natl Acad Sci USA 1992;89:5675–5679.
78. Sorenson AG, Rosen BR. Functional MRI of the brain. In: Atlas SW, editor. *Magnetic Resonance Imaging of the Brain and Spine*. 2nd ed. Philadelphia: Lippincott-Raven Publishers; 1996. p 1501–1545.
79. Kwong KK. Functional magnetic resonance imaging with echo planar imaging. Magn Reson Q 1995;11:1–20.
80. Bandettini PA, Jesmanowicz A, Wong EC, et al. Processing strategies for time-course datasets in functional MRI of the human brain. Magn Reson Med 1993;30:161–173.
81. Mendichovszky I, Jackson A. Imaging hypoxia in gliomas. Br J Radiol 2011;84 Spec No 2:S145–S158.
82. Padhani AR, Krohn KA, Lewis JS, et al. Imaging oxygenation of human tumours. Eur Radiol 2007;17:861–872.
83. Mardor Y, Rahav O, Zauberman Y, et al. Convection-enhanced drug delivery: increased efficacy and magnetic resonance image monitoring. Cancer Res 2005;65:6858–6863.
84. Linninger AA, Somayaji MR, Mekarski M, et al. Prediction of convection-enhanced drug delivery to the human brain. J Theor Biol 2008;250:125–138.
85. Buckler AJ, Bresolin L, Dunnick NR, et al. Quantitative imaging test approval and biomarker qualification: interrelated but distinct activities. Radiology 2011;259:875–884.
86. Buckler AJ, Bresolin L, Dunnick NR, et al. A collaborative enterprise for multi-stakeholder participation in the advancement of quantitative imaging. Radiology 2011;258:906–914.

CHAPTER 6

QUANTITATIVE IMAGING IN DRUG DELIVERY

KOORESH I. SHOGHI

The increasing availability of animal models of disease (e.g., transgenic) has brought about opportunities to utilize small animal positron emission tomography (PET) in assessing efficacy of therapeutic interventions and, in parallel, to validate the use of current and novel imaging agents and targets for translation to the clinic. One of the many advantages of PET imaging is that it offers quantifiable measures of the underlying biology. In most instances, PET images are quantified by normalizing average values in regions of interest (ROIs). This "normalization" scheme takes the form of the "standardized uptake value" (SUV) or related to a reference region presumably devoid of the underlying biological process under investigation. The application of SUV has been discussed in detail; we will also discuss quantifying the kinetics of the imaging agents through compartmental modeling. Recent examples of kinetic modeling in small animals span many fields, including oncologic imaging [1, 2], cardiac imaging [3, 4], and neuroimaging [5] among others. By far, the predominant choice of tracer is ^{18}FDG largely owing to its availability. This chapter is meant as a primer on quantification of PET and single photon emission computed tomography (SPECT) data. Where appropriate, references will be provided for further reading. We will start with a primer on image quantification followed by a brief discussion on factors that may affect image quantification.

6.1 RADIOPHARMACEUTICALS AND TARGETS

The availability of radiopharmaceuticals is growing rapidly owing to a large extent by the availability of -omic data. The NIH has compiled a database tabulating potential imaging agents [6]. Broadly, imaging agents can be categorized into four groups, namely, metabolic agents, "trapped" agents, perfusion agents, and agents targeting receptors [7]. Metabolic agents, as the name implies, are metabolized much like endogenous substrates.

Drug Delivery Applications of Noninvasive Imaging: Validation from Biodistribution to Sites of Action, First Edition. Edited by Chun Li and Mei Tian.

These include ^{11}C-labeled glucose or palmitate, for example, which are widely used to quantify glucose and fatty acid metabolism, respectively. "Trapped" agents are those that have been modified chemically to be trapped within a cell following an enzymatic reaction, and to a large extent, there is some overlap between metabolic and trapped agents. The best-known example of an trapped agent is ^{18}FDG in which the 2'-OH has been replaced with F-18, thereby blocking further downstream metabolism (although there are reports disputing this assertion [8]); another is ^{18}F-3′-fluoro-3′-deoxy-L-thymidine (^{18}FLT) [9], a pyrimidine analog that is phosphorylated by the enzyme thymidine kinase 1 (TK1), leading to intracellular trapping, thus providing a measure of cellular proliferation. Perfusion agents as the name implies measure perfusion/blood flow in tissue, for example, myocardial blood flow (MBF) using [^{15}O]H_2O. Finally, the broadest category of imaging agents comprises radiolabeled ligands targeting receptors with a broad spectrum of radiopharmaceuticals, including nanoparticles. To a large extent, the quantitative approaches described later are dependent on the radiopharmaceutical in question, its target, and the *in vivo* assay of interest.

6.2 SEMIQUANTITATIVE ANALYSIS

The simplest semiquantitative approach for characterization of image data is to characterize the activity of the tracer in a target ROI, denoted by $C_T(t)$, at a given time. It is important to note that the activity of the tracer in tissue is per unit volume (e.g., nCi/ml). Given the dynamic nature of tracer kinetics, the observed activity is a function of time, and thus an optimal time-point should be selected to report on image data in the context of the biological assay of interest. When comparing between various study groups (such as treatment groups and disease models), however, the activity at a given time point may be of little value due to potential variability in imaging protocols. For this reason, ROI activity is often normalized relative to a "reference." The choice of reference can be relative to the activity in an "inert" tissue, the injected dose (ID), or standardized for weight to account for the available distribution volume.

6.2.1 Target-to-Reference Measures

When the activity in a target ROI is normalized to a reference ROI, the reference tissue should be inert in the sense that the radiopharmaceutical's kinetics in the reference tissue is not influenced by study in question (disease model, treatment, etc.). For example, in calculating the uptake of ^{18}FLT in tumors, the muscle is often used as a reference region since differentiated myocytes do not proliferate.

6.2.2 Target Activity Normalized to ID

The activity in the target ROI can be also normalized to ID represented as percent injected dose (%ID). In Equation 6.1, it is assumed that the activity in tissue, C_T, is in units of mCi per cc (cm^3) as a function of ID, D_{inj}, in units of mCi and tissue density, ρ, in units of grams per cc. The %ID/cc is a crude semiquantitative measure that does not account for differences in body mass and routes of tracer extraction and elimination, all of which can affect quantitative comparison across groups:

$$\frac{\%\text{ID}}{\text{cc}} = \frac{C_\text{T}[\text{mCi / cc}] \cdot 100}{D_\text{inj}[\text{mCi}]}. \tag{6.1}$$

In many cases, it is further assumed that the tissue density is unity, and thus the measure is represented as %ID/g tissue:

$$\frac{\%\text{ID}}{\text{g}} = \frac{\%\text{ID / cc}}{\rho[\text{g / cc}]}. \tag{6.2}$$

The assumption on the density in tissue, however, is tissue dependent. For example, lung tissue density is significantly less than unity. In addition, heterogeneous tumors may exhibit varying tissue densities.

6.2.3 Standardized Uptake Value

In an effort to account for variability in the bioavailability of the radiopharmaceutical between subjects, in SUV analysis the %ID/g is normalized to subject mass (M_S). Considerations and variations in SUV analysis have been discussed extensively elsewhere [10–15]. Much like measures of %ID/g, SUV measures are subject to bias if not all the activity is available for imaging at target region due to variability in tracer extraction at site of administration:

$$\text{SUV} = \frac{\%\text{ID}}{\text{g}} \cdot \frac{M_\text{S}[\text{g}]}{100}. \tag{6.3}$$

6.2.4 Factors Affecting Semiquantitative Analysis

As alluded to earlier, given the dynamic nature of tracer kinetics in tissue, all semiquantitative measures described earlier can be temporal in dynamic imaging, or static, representing a single snapshot of the measure at a given time post injection of radiopharmaceutical. In case of the latter, it is imperative to characterize the optimal time point to report a measure. Ideally, semiquantitative measures should be reported under steady-state conditions when tissue and blood are in equilibrium. Other factors that affect semiquantitative analysis include ROI definition, scanner calibration and cross-calibration, scanner resolution and partial volume effects, as well as competition with endogenous substrates among others, all of which are discussed in Section 6.5.

6.3 QUANTITATIVE ANALYSIS

6.3.1 Noncompartmental Modeling

Noncompartmental modeling is a model-independent approach that does not require explicit knowledge of the various compartments and metabolites of the tracer. As such, the body or tissue of interest is viewed as one compartment with an input and output, irrespective of any potential compartments or pools of metabolites within the tissue. The underlying assumptions in applications involving noncompartmental modeling are (i) conservation of mass, (ii) linearity of system, and (iii) steady-state conditions in deriving outcome measures of interest, such as clearance rate defined as the ratio of the dose and the area under curve of the plasma time-activity curve (TAC) and the mean residence time (MRT):

$$\mathrm{MRT} = \frac{\int_0^\infty t \cdot c(t)dt}{\int_0^\infty c(t)}. \tag{6.4}$$

6.3.2 Compartmental Modeling (Tracer Kinetic Analysis)

Kinetic modeling offers several advantages in comparison to "normalization" schemes, in that it offers measures of transport/perfusion rates that can be used to tease out delivery issues from metabolism/binding of ligands. Kinetic analysis also offers means to measure metabolic rates and receptor density, as well as fracture of underlying vasculature that can be used to discern "true" tissue activity especially in situations when one expects changes in vasculature, for example, due to injury. Kinetic analysis offers means to separate specific from nonspecific signal, for example, when a ligand is found to have a secondary nonspecific binding site. Finally, with more advanced application of kinetic analysis, spatial information on the kinetics in tissue may be characterized.

Generally, once an imaging agent is administered into the subject, be it human, mouse, rat, etc., it will distribute throughout the whole body and will be extracted by various tissues. Let F denote the blood flow (ml/min/g), $C_a(t)$ (nCi/ml) the arterial concentration of the imaging agent in blood at time t, and $C_v(t)$ the venous concentration of the imaging agent at time t; then the *flux* (nCi/min) per *gram* of tissue, J_T, is defined by

$$J_T = J_{in} - J_{out} = FC_a - FC_v. \tag{6.5}$$

The extraction fraction, E, can be calculated by the ratio of the arterial–venous differential relative to the arterial concentration (Fig. 6.1):

$$E = \frac{C_a - C_v}{C_a}. \tag{6.6}$$

Assuming low arterial–venous differential, first-pass extraction by tissue can be approximated by

$$E = 1 - e^{-P \cdot S/F} \tag{6.7}$$

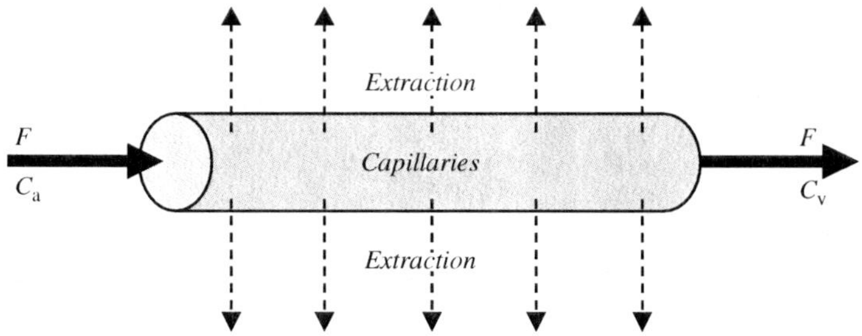

FIGURE 6.1 Extraction of tracer at tissue from capillaries. The blood flows at a rate F with the concentration of the radiopharmaceutical on the arterial side provide by C_a while on the venous the concentration is C_v.

where P denotes the permeability (cm/min), S denotes the surface area (cm^2/g), and F denotes blood flow. Thus, the unidirectional flux into tissue is approximated by

$$J = F \cdot E \cdot C_a. \tag{6.8}$$

Traditionally, in tracer compartmental modeling, the product of F and E is denoted by the rate constant K_1 (ml/min/g) representing the uptake rate constant of the tracer into tissue, as discussed later.

The application of compartmental modeling involves a certain workflow. Since, in most cases, there is some prior knowledge about the radiopharmaceutical's mode of interaction with the target, the general workflow involves the optimization of a "skeleton" model against the data. If the quality of the fit is adequate, as described later, then further biological correlation is necessary to confirm that image-derived outcome measures correlate with biological assays. Alternatively, if the fit is not adequate, the model needs to be reconfigured to account for additional processes. In many respects, the workflow in identifying a compartmental model to match a biological correlate is an iterative process.

6.3.2.1 *Assumptions of Compartmental Modeling* Typical application of compartmental modeling to quantification of PET and SPECT images carries several assumptions. First, compartmental models, by nature, assume that each compartmental represented a well-mixed pool of a metabolite. In other words, compartmental modeling ignores the presence of concentration gradients, which undoubtedly exist. Moreover, given the spatial resolution of the scanner, the activity in a given voxel may represent a "mixed" pool of multiple metabolites. Second, in invoking compartmental model analysis, it is assumed that underlying physiological processes are at steady state. Steady state implies that the rate constant depicting the underlying physiological process is constant, that is, independent of time, or time invariant. If steady-state conditions are not met, the system is considered time varying and, mathematically, nonlinear in nature. With the two assumptions described earlier, the flux of tracer from compartment A to B is assumed linearly proportional to the concentration in compartment A; in other words,

$$J_{A \to B} = kC_A, \tag{6.9}$$

where k denotes the rate constant of proportionality (in units of inverse time). Equation 6.9 forms the basis for the application of compartmental modeling to quantify PET data using linear time-invariant ordinarily differential equations (ODEs).

6.3.2.2 *Ordinarily Differential Equations* In setting up the system of ODE describing the kinetics of a radiopharmaceutical in tissue, the net flux into each compartment is defined as the sum of inflow minus outflow of the metabolites in question. In practical terms, inflows and outflows may represent conversion of the parent compound or a metabolite from one species to another. Alternatively, they may present transport from one physical pool to another. The general equation describing the net change in flux in first tissue compartment (denoted by $C_1(t)$) is given by

$$\frac{dC_1}{dt} = K_1 C_P(t) - \sum_{j \neq 1} k_{1j} C_{1j}(t) + \sum_{k \neq 1} k_{k_1} C_{k_1}(t) \tag{6.10}$$

where the net change in flux (in units of concentration per units time) is a sum of all fluxes outward from compartment 1 to j and from inward flux from compartment k ($k = 1 \ldots N$) to compartment 1 with N denoting the number of compartments. As discussed later, $C_P(t)$ denotes the plasma input function representing the forcing function of the parent radiopharmaceutical (or its metabolites in case of multiple input functions) to tissue. Ideally, the plasma input function should represent a metabolite-corrected TAC of the radiopharmaceutical in plasma. The system of ODEs for a given radiopharmaceutical is dependent on the nature of reaction or the radiopharmaceutical's target application. Generally, since the total activity in the ROI is bound to include some underlying vasculature, a vascular fraction, V_f, is added to the total activity in the ROI:

$$\text{ROI}(t) = \sum_k C_k(t) + V_f C_B(T) \tag{6.11}$$

where k denotes summation over all tissue compartments and $C_B(t)$ denotes total blood radioactivity.

Models for Perfusion Kinetics In modeling perfusion of a radiopharmaceutical into tissue, one considers the flux of the radiopharmaceutical from blood to tissue (Fig. 6.2). Generally, this is accomplished by setting a single compartment with bidirectional parameters for the uptake and clearance of the radiopharmaceutical to/from tissue. Mathematically, the ODE for perfusion kinetics is provided by

$$J_T = \frac{dC_T(t)}{dt} = K_1 C_P(t) - k_2 C_T(t). \tag{6.12}$$

The analytical solution of Equation 6.12 is

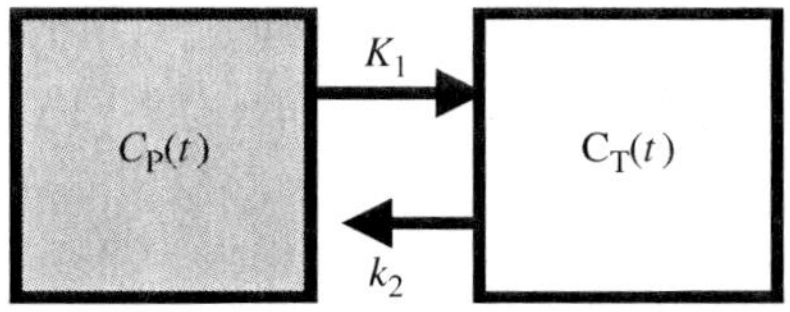

FIGURE 6.2 The perfusion compartmental model for [^{15}O]H_2O kinetics. The input function is denoted by $C_p(t)$, while the dashed lines denote tissue compartment. $C_p(t)$ denotes the input function, which in case of [^{15}O]H_2O is the arterial curve. $C_T(t)$ denotes the activity in tissue.

$$C_T(t) = C_P(t) \otimes K_1 e^{-k_2 t}. \tag{6.13}$$

At steady state, $dC_T/dt=0$; thus the ratio

$$\frac{K_1}{k_2} = \frac{C_T}{C_P} \tag{6.14}$$

represents the partition coefficient, λ, between tissue and blood. For an inert perfusion radiopharmaceutical such as [^{15}O]H_2O, the partition coefficient is known to be $\lambda = 0.92$ ml/g. For example, [^{15}O]H_2O is routinely used to quantify MBF in human and animal studies [3, 4, 16–18]. If we denote K_1 = MBF, then the total ROI activity is given by

$$\text{ROI}(t) = C_P(t) \otimes \text{MBF} \cdot e^{-\text{MBF}\cdot t/0.92} + V_f C_B(t). \tag{6.15}$$

To obtain estimates of MBF, the simulated ROI data is optimized against image-derived ROI data as described in subsequent sections.

Models for Enzymatic Kinetics Radiopharmaceuticals are widely used to characterize *in vivo* metabolic processes in a wide range of application including brain imaging [19, 20], cardiac substrate metabolism [3, 4, 18, 21], and imaging systems for gene expression [22, 23] among others. As an example of modeling the kinetics of a radiopharmaceutical involved in metabolism, we will focus our attention to the metabolism of ^{18}FDG. Endogenous glucose is transported into tissue by one of several glucose transporters (GLUTs) in a tissue-specific manner. For example, in the heart, GLUT1 and GLUT4 dominate. As a glucose analog, ^{18}FDG is similarly transported into tissue via GLUT (Fig. 6.3). Its transport is denoted by the rate constant K_1 (ml/g/min), while ^{18}FDG clears tissue with a rate constant k_2 (/min). Once in tissue, ^{18}FDG is phosphorylated by hexokinase/glucokinase with a rate constant k_3 (/min) to generate ^{18}FDG-6-phosphate, which as discussed earlier is presumably trapped within the cell. It is generally assumed that dephosphorylation of ^{18}FDG-6-phosphate is negligible, that is, $k_4 = 0$. Thus, ^{18}FDG-6-phosphate represents the trapped metabolite of ^{18}FDG allowing visualization of total ^{18}FDG utilization.

FIGURE 6.3 The model describes the kinetics of a radiopharmaceutical that enters the tissue to be converted to a trapped metabolite, which is in fact the generally accepted model for FDG imaging in the brain. The bidirectional transport across the BBB is denoted by K_1 (ml/g/min) and k_2 (/min). The phosphorylation of FDG to FDG-6-phosphate (F6P) is given the rate constant k_3 (/min). Dephosphorylation of FDG-6-phosphate, which may exist in some tissue, such as liver, is denoted by the rate constant k_4 (/min).

Assuming that the processes described earlier are linear, the compartmental model can be represented by a system of ODEs,

$$\begin{aligned}\frac{dC_{T_1}(t)}{dt} &= K_1 C_P(t) - (k_2 + k_3) C_{T_1}(t).\\ \frac{dC_{T_2}(t)}{dt} &= k_3 C_{T_1}(t)\end{aligned} \tag{6.16}$$

The total radioactivity in an ROI as a function of time is given by

$$\mathrm{ROI}(t) = [C_{\mathrm{FDG}}(t) + C_{\mathrm{FDG6P}}(t)] + V_f C_B(t) \tag{6.17}$$

where V_f denotes the fraction of underlying vasculature in tissue. Optimization of the earlier equation against PET data yields estimates of K_1, k_2, k_3, and V_f. The tissue metabolic rate of glucose, $\mathrm{MR_{glu}}$, is provided by

$$\mathrm{MR_{glu}} = \frac{K_1 k_3}{k_2 + k_3}\frac{[\mathrm{glu}]_P}{\mathrm{LC}} = K\frac{[\mathrm{glu}]_P}{\mathrm{LC}} \tag{6.18}$$

where $[\mathrm{glu}]_P$ denotes the plasma concentration of glucose and LC denoted the much-debated lumped constant for the differential metabolism of ^{18}FDG and glucose [24]. In a similar fashion, compartmental models for more elaborate kinetics such as [^{11}C]glucose [18], [^{11}C]palmitate [25], and ^{18}FDOPA [26], to name a few, are devised.

Models of Ligand–Receptor Kinetics Kinetic analysis of PET ligand–receptor interactions is a rich subject in PET imaging, largely due to the wide range of targeted radiopharmaceuticals and applications. In quantifying PET ligand–receptor interactions, the fundamental concept includes the notion of a *free* pool of PET radioligands (denoted by "C_F") and a *bound* pool (denoted by "C_B") in tissue. The availability of the PET radioligand in plasma is denoted by $C_P(t)$ that serves as the input function to the model. Mathematically, the system of ODEs describing the kinetics of the ligand in tissue is given by

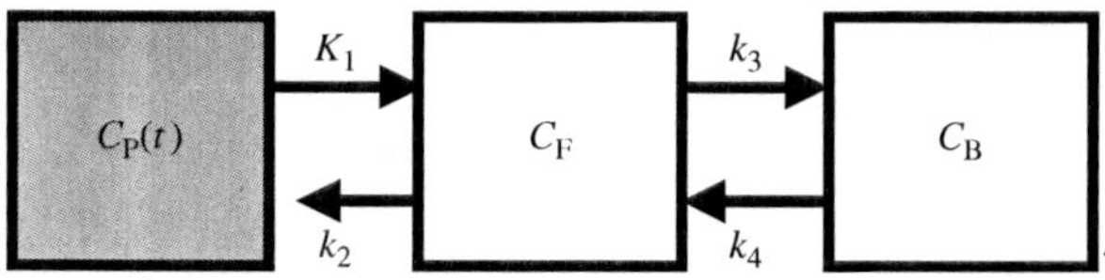

FIGURE 6.4 The kinetics of ligand–receptor interaction in the absence of nonspecific binding. The bidirectional transport of the ligand across into tissue is denoted by K_1 (ml/g/min) and k_2 (/min). The "free" pool of ligand is denoted by "F," and the concentration of bound receptors is denoted by "B," the association rate constant of the ligand that is proportional to the density of the receptors $k_3 = k_{on}{*}B_{max}$. The dissociation rate constant is provided by $k_4 = k_{off}$.

$$\begin{aligned}\frac{dC_F}{dt} &= K_1C_P(t) - k_2C_F - k_{on}\left(B_{max} - C_B\right)C_F + k_{off}C_B \\ \frac{dC_B}{dt} &= k_{on}\left(B_{max} - C_B\right)C_F - k_{off}C_B\end{aligned} \tag{6.19}$$

where the bidirectional transport from blood to tissue is represented by K_1 (ml/g/min) and k_2 (/min), respectively; B_{max} denotes the maximal density of target receptors (nM); k_{on} denotes the forward rate constant for the association of the ligand to the receptor; and k_{off} denotes the dissociation rate constant of the ligand from the receptor. Assuming that the ligand concentration is significantly less than the available receptor density, that is, $C_B \ll B_{max}$, the system of ODEs is linear with $k_3 = k_{on}B_{max}$ and $k_4 = k_{off}$ (Fig. 6.4),

$$\begin{aligned}\frac{dC_F}{dt} &= K_1C_P(t) - \left(k_2 + k_3\right)C_F + k_4C_B \\ \frac{dC_B}{dt} &= k_3C_F - k_4C_B\end{aligned} \tag{6.20}$$

The total activity in the ROI is thus

$$\mathrm{ROI}(t) = C_T(t) + V_fC_P(t) \tag{6.21}$$

where $C_T(t) = C_F(t) + C_B(t)$. The system of ODEs can be solved analytically or numerically. The analytical solution is obtained following Laplace transform:

$$\begin{aligned}C_T(t) &= C_F(t) + C_B(t) \\ &= \frac{K_1}{\phi_1 - \phi_2}\left[\left(k_4 + \phi_1\right)e^{\phi_1 t} - \left(k_4 + \phi_2\right)e^{\phi_2 t}\right] \\ &\quad + \frac{K_1k_3}{\phi_1 - \phi_2}\left[e^{\phi_1 t} - e^{\phi_2 t}\right]\end{aligned} \tag{6.22}$$

where ϕ denotes the eigenvalues

$$\phi_{1,2}=\frac{-(k_2+k_3+k_4)\pm\sqrt{(k_2+k_3+k_4)^2-4k_2k_4}}{2}.$$

When the simulated ROI kinetics in Equation 6.21 is optimized against PET data, one obtains kinetic estimates of K_1 through k_4 and V_f. At steady state, $dC_F/dt = dC_B/dt = 0$. The binding potential (BP) is a steady-state concept written as

$$\frac{C_B}{C_F}=\frac{k_3}{k_4}=\frac{B_{max}}{K_D}=\text{BP}. \tag{6.23}$$

where K_D denotes the dissociation constant $K_D = k_{off}/k_{on}$. The BP is typically used as a measure of receptor density in PET imaging of targeted radiopharmaceuticals with the underlying assumption that K_D is constant. Another useful steady-state concept is the total uptake of the tracer, characterized by the distribution volume (DV) between the concentrations of bound compartment to plasma provided by

$$\text{DV}_B=\frac{K_1}{k_2}\frac{k_3}{k_4}. \tag{6.24}$$

The DV for the total uptake of the tracer including steady-state concentration in the free pool and concentration of bound ligand is provided by

$$\text{DV}=\frac{K_1}{k_2}\left(1+\frac{k_3}{k_4}\right). \tag{6.25}$$

The earlier measure can be similarly obtained using the graphical Logan approach described in Section 6.3.3.

Several variations of the compartmental model in Figure 6.4 include a simplified ligand–receptor model in which the kinetics of uptake and binding are indistinguishable, thus reducing the number of compartments from two to one [27, 28]. In addition, the compartmental model can be extended to include nonspecific binding such that the free pool of ligand can bind to either a specific (target) receptor or a nonspecific receptor denoted by C_S and C_{NS}, respectively (Fig. 6.5). The inclusion of a nonspecific compartment adds another ODE and two parameters to the system of the ODEs in Equation 6.20. Various formulations of nonspecific binding are reported in the literature (see description in reference 29). For example, if the kinetics between the nonspecific pool and the free pool is sufficiently rapid, the two compartments are at pseudo-equilibrium such that only a fraction, θ, of the free pool is available for binding to the target receptor. In other words,

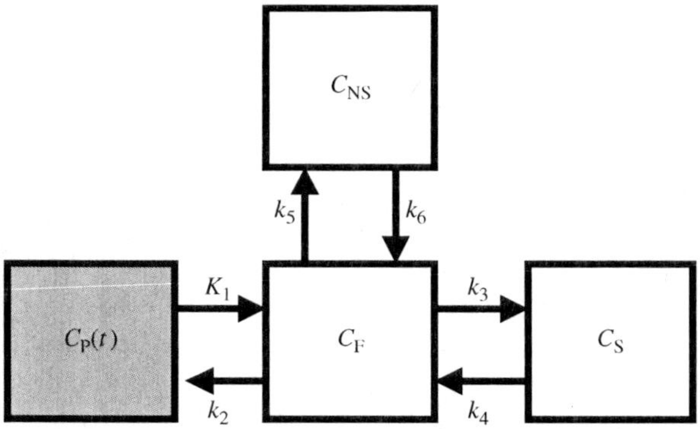

FIGURE 6.5 The kinetics of ligand–receptor interaction in the presence of nonspecific binding or two binding targets. The ligand has two pools of bound receptors—specific (C_S) and nonspecific (C_{NS}) with differing kinetics. In comparison to Figure 6.3, the models include two additional parameters for the association and dissociation of the ligand from the nonspecific receptors with rate constants k_5 (/min) and k_6 (/min).

$$\mathrm{BP} = \frac{C_\mathrm{B}}{\theta \cdot C_\mathrm{F}} = \frac{B_\mathrm{max}}{K_\mathrm{D}} \tag{6.26}$$

Overall, the kinetic estimates through compartmental modeling provide measures of uptake (transport), clearance, and binding of the radiopharmaceutical to its target. In an effort to simplify the analysis of PET ligand–receptor data, a graphical approach may be used to obtain estimates of the DV of the radiopharmaceutical (discussed in Section 6.3.3).

6.3.2.3 Spectral Analysis In spectral analysis, the input function for a given tissue is convolved with a sum of Gaussians representing the impulse response function (IRF) of the tissue. The advantage of such a preliminary analysis is that it does not require a priori knowledge of the compartmental model's configuration. The general form for spectral analysis formation is provided by

$$C_\mathrm{T}(t) = C_\mathrm{b}(t) \otimes \mathrm{IRF}(t) = C_\mathrm{P}(t) \otimes \sum_{i=1}^{N} e^{-\lambda_i t} \tag{6.27}$$

where $C_T(t)$ denotes the activity in a given tissue, $C_p(t)$ denotes the input function, N denotes the order of the spectral model, and λ_i denotes the eigenvalues of the IRF. The number of Gaussians (and hence the number of parameters) needed to describe the IRF can be judged by a nested F-test for the number of parameters [30], Akaike information criterion (AIC) [31], and the Schwarz criteria [32] as described in Section 6.3.2.5. The total observed activity in tissue is thus provided by

$$C_{\mathrm{PET}}(t) = C_{\mathrm{b}}(t) \otimes \sum_{i=1}^{N} e^{-\lambda_i t} + V_{\mathrm{f}} C_{\mathrm{b}}(t) \tag{6.28}$$

where V_{f} denotes the vascular fraction in a given tissue. Optimization of Equation 6.28 against PET image data provides estimates of the eigenvalues and the order of the spectral model N. Using the order of the spectral model, representative compartmental model configurations can be constructed to derive measures of imaging agent's kinetic estimates. In general, the macro-parameters derived from spectral analysis of order $N=2$ can be mapped to the microparameters of a two-compartmental model.

6.3.2.4 Metabolite-Corrected Input Function Accurate quantification of PET images requires the knowledge of the plasma TAC of the radiopharmaceutical [24, 33], as alluded to earlier. The input function commonly known as the plasma input function, denoted by $C_{\mathrm{p}}(t)$ in the earlier equations, serves as a forcing function to a kinetic model. The gold standard for the determination of the input function is an invasive blood sampling procedure [34, 35], in which activity concentrations of arterial blood samples are measured directly at select intervals and subsequently corrected for metabolites (if any). However, arterial sampling is plagued by several problems, such as technical handling and dispersion of the TAC due to delays in sampling and voiding of blood for catheters. More importantly, as an invasive technique, the procedure is not desirable. In small animal imaging, the procedure is challenging because of the small size of animal blood vessels and the potential perturbation to the physiology due to the loss of blood [35]. Therefore, less invasive, image-based estimation techniques are desirable. The literature is ripe in image-based methods for human applications [36–39]. In preclinical imaging, we have recently introduced a hybrid image and blood sampling (HIBS) algorithm whereby the peak of the image is derived from recovery-corrected image LV ROIs and the tail is derived from five to seven blood samples, both of which are linked by a Bezier interpolation algorithm [40]. The HIBS algorithm is ideal for kinetic analysis of tracers using short-lived radionuclides such as ^{11}C in ^{11}C-palmitate or ^{11}C-glucose where blood samples are readily available from metabolite ($^{11}CO_2$) correction analysis. For tracers where metabolites are not an issue, such as 2-deoxy-2-[^{18}F]fluoro-D-glucose (FDG), it would be ideal to minimize the number of blood samples for the reconstruction of the input function. To that end, we have recently proposed a hybrid method that extracts the peak of the input function from the image while the tail of the input function is derived by fitting a parametric model. In addition, recovery and spillover values are fitted simultaneously [41]. In Table 6.1, we survey methods for derivation of the blood activity profile in small animals for subsequent metabolite correction if necessary.

6.3.2.5 Parameter Estimation Parameters (p) are generally estimated by minimizing the weighted residual sum of squares (WRSS) between observed PET

TABLE 6.1 Survey of Methods to Reconstruct the Input Function in Small Animal Imaging

Number	Description	Ref.
1.	Hybrid image–blood sampling (nonparametric)	40
2.	Application of maximum likelihood factor analysis (MLFA) to heart images (Fig. 6.1)	42
3.	Hybrid single-input–dual-output (HSIDO) semiparametric reconstruction	41
4.	Estimation of ^{18}FDG input function by multitissue kinetic modeling	43
5.	Spillover and partial volume correction for image-derived input function	34
6.	Microfluidic based blood sampling	44

data ($C_{PET}(t)$) and predicted PET data $\hat{C}_{PET}(t)$ based on the kinetic model using the following relation:

$$\mathrm{WRSS}(t_k,\mathrm{p}) = \sum_{k=i}^{N} w_k \left(C_{PET}(t_k) - \hat{C}_{PET}(t_k,\mathrm{p})\right)^2 \tag{6.29}$$

where w_k denotes the weight at frame $k(k = 1 \ldots N)$. Typically, the weighting at each time point k with scan duration Δt_k is assumed to be pseudo-Poisson, $w_k = \sqrt{C_{PET}(t_k)/\Delta t_k}$, although it is not a strict. To minimize the WRSS, various numerical algorithms are used such as the Levenberg–Marquardt nonlinear least-squares algorithm implemented in MATLAB (MathWorks, Inc.). To assess the quality of the fit, various qualitative and quantitative approaches can be used, including visual inspection and parsimony criteria, minimizing the AIC [31] and the Schwarz criteria [32] with regard to minimizing the WRSS with the fewest number of parameters. The number of parameters needed to justify the models can also be evaluated using a nested F-test statistics [30], as applied previously [26]. One caveat in minimizing the WRSS using any minimization algorithms is that the convergence of the algorithm is iterative and requires initial conditions. Depending on the landscape of the cost-function WRSS, which may include many local minima, the convergence of the algorithm, and hence the estimated parameters, p, may differ. It is therefore suggested to try various initial conditions in the estimation process to ensure convergence to the global minima.

6.3.3 Graphical Approaches

In an attempt to simplify quantification of PET data, various approaches have been proposed to characterize the uptake of a "trapped" radiopharmaceutical and receptor-targeting ligand.

6.3.3.1 Patlak Graphical Analysis The Patlak graphical approach [45, 46] is applicable when the radiopharmaceutical is designed to be trapped with the cell. The system of ODEs describing such as system is provided in Equation 6.16 with total activity in the ROI given in Equation 6.17. For simplicity, let C_1 denote the concentration of FDG in tissue and C_2 denote the concentration of FDG-6-phospate in tissue (corresponding to C_{FDG} and C_{FDG6P}, respectively). The graphical approach proposed by Patlak and Blasberg amounts to normalization of the tissue and blood data as follows:

$$\frac{\mathrm{ROI}(t)}{C_{\mathrm{P}}(t)} = \left(\frac{K_1 k_3}{k_2 + k_3}\right)\frac{\int_0^{\mathrm{T}} C_{\mathrm{P}}(t)dt}{C_{\mathrm{P}}(t)} + \left[\left(\frac{K_1 k_3}{k_2 + k_3}\right)\frac{C_1(t)}{C_{\mathrm{P}}(t)} + \frac{V_{\mathrm{f}} C_{\mathrm{B}}(t)}{C_{\mathrm{P}}(t)}\right]. \tag{6.30}$$

After some time $t > t^*$, the slope of the linear segment of the plot $\mathrm{ROI}(t)/C_{\mathrm{P}}(t)$ versus $\int_0^{\mathrm{T}} C_{\mathrm{P}}(t)dt \,/\, C_{\mathrm{P}}(t)$ equals the net uptake of the tracer denoted by K_i:

$$K_i = \frac{K_1 k_3}{k_2 + k_3}. \tag{6.31}$$

The slope is determined by fitting a polynomial of order one through the linear segment of the plot. Note that the net uptake of the tracer K_i is comparable to that compartmental derives estimate. The condition for linearity is that the intercept is constant. To meet this condition, after $t > t^*$, the reversible compartment concentration follows the plasma concentration so that $C_1(t)/C_{\mathrm{P}}(t)$ and $C_{\mathrm{B}}(t)/C_{\mathrm{P}}(t)$ are constant, ensuring that the intercept is constant [45]. The Patlak graphical approach can be also used with a reference tissue as input in place of the plasma activity curve [45]. The underlying assumption in using reference region Patlak analysis is that the partition coefficient of target and reference regions is approximately the same. Thus, to assess the validity of the assumption, it is suggested that compartmental modeling will be performed on target and several potential reference regions to characterize a suitable reference region.

6.3.3.2 Logan Graphical Analysis The Logan graphical analysis [47] is applicable in case of reversible ligand–receptor kinetics described through Equation 6.20 with total activity in the ROI provided by Equation 6.21. The graphical approach proposed by Logan amounts to the normalization of the ROI and plasma activity curve $C_{\mathrm{P}}(t)$ as follows:

$$\frac{\int_0^{\mathrm{T}} \mathrm{ROI}(t)dt}{\mathrm{ROI(T)}} = \left[\frac{K_1}{k_2}\left(1 + \frac{k_3}{k_4}\right)\right]\frac{\int_0^{\mathrm{T}} C_{\mathrm{P}}(t)dt}{\mathrm{ROI(T)}} + \beta \tag{6.32}$$

where β denotes the intercept (detailed derivation of the intercept is provided in [47]). After some time $t > t^*$, the slope of the linear segment of the plot $\int_0^T \mathrm{ROI}(t)dt / \mathrm{ROI(T)}$ versus $\int_0^T C_P(t) / \mathrm{ROI(T)}$ equals the DV of the tracer

$$\mathrm{DV} = \frac{K_1}{k_2}\left(1 + \frac{k_3}{k_4}\right). \tag{6.33}$$

The Logan analysis-derived DV is comparable to the compartmental derived DV in Equation 6.25. Similar to the Patlak analysis approach, a reference tissue may be used in lieu of the plasma TAC. The two main assumptions in using a reference tissue Logan analysis are that (i) the reference region is devoid of specific binding (i.e., $k_3 = k_4 = 0$ in the reference region) and (ii) the ratio of partition coefficients K_1/k_2 in the target and $K_1^{\mathrm{REF}} / k_2^{\mathrm{REF}}$ reference regions is close to unity. The reference tissue Logan analysis amounts to

$$\frac{\int_0^T \mathrm{ROI}(t)\,dt}{\mathrm{ROI(T)}} = \mathrm{DVR}\left(\frac{\int_0^T \mathrm{REF}(t)dt + \mathrm{REF}(t) / k_2^{\mathrm{REF}}}{\mathrm{ROI(T)}}\right) + \beta' \tag{6.34}$$

where β' denotes the intercept and k_2^{REF} the clearance rate constant in the reference region. The term k_2^{REF} can represent a population mean but is often neglected. The distribution volume ratio (DVR) yields the following relation:

$$\begin{aligned}\mathrm{DVR} &= \frac{\mathrm{DV}}{\mathrm{DV^{REF}}} = \frac{\frac{K_1}{k_2}\left(1 + (k_3 / k_4)\right)}{K_1^{\mathrm{REF}} / k_2^{\mathrm{REF}}}. \\ &= 1 + \frac{k_3}{k_4} = 1 + \frac{B_{\max}}{K_{\mathrm{D}}} = 1 + \mathrm{BP}\end{aligned} \tag{6.35}$$

Note that the BP can be obtained through this analysis (mindful of the earlier assumption) using the relation $\mathrm{BP} = \mathrm{DVR} - 1$.

6.4 APPLICATION OF QUANTITATIVE IMAGING

To illustrate an application of quantitative imaging, we utilize an example of metabolic imaging in diabetes. There is increased evidence suggesting that diabetic patients have a predisposition to heart failure resulting from impairment in heart muscle contraction, particularly abnormalities in diastolic function [48]. Although several theories have been put forth to explain diabetic cardiomyopathy, evidence has emerged that diabetic cardiomyopathy is at least partly a consequence of severe alterations in myocardial energy metabolism [49, 50]. In particular, insulin resistance

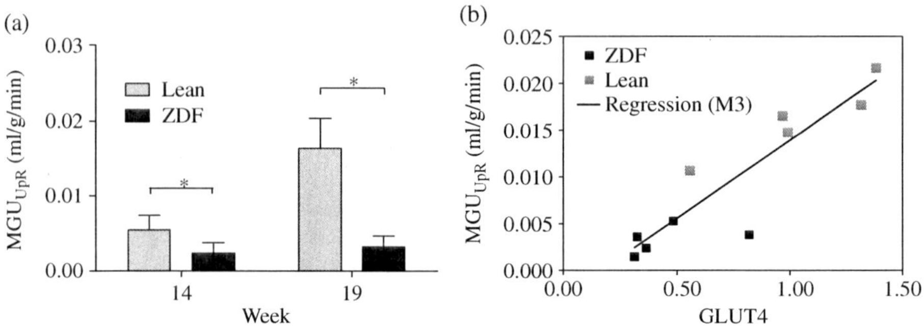

FIGURE 6.6 (a) Myocardial glucose utilization uptake rate (MGU_{UpR}) of lean and ZDF rats at 14 and 19 weeks of age. (b) Correlation between MGU_{UpR} and GLUT4 expression. *denotes significant at $p < 0.05$.

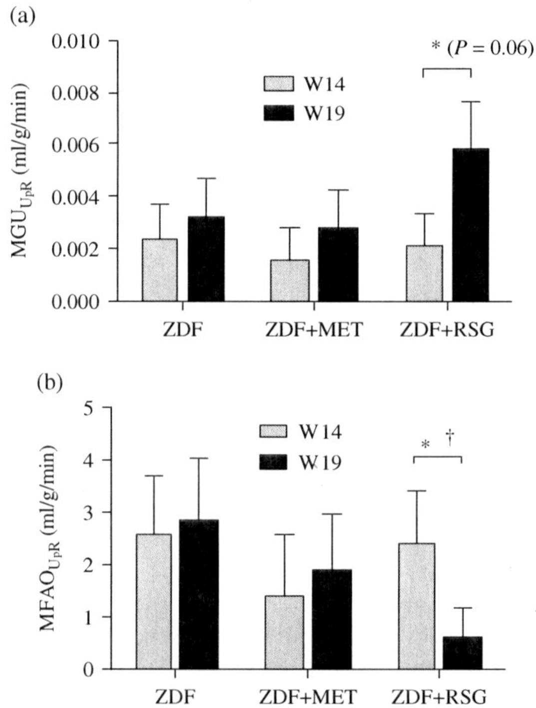

FIGURE 6.7 (a) Myocardial fatty acid utilization uptake rate ($MFAU_{UpR}$). (b) Myocardial fatty acid oxidation uptake rate ($MFAO_{UpR}$) in untreated (UT), metformin-treated rats (MET), and rosiglitazone-treated (RSG) ZDF rats at week 14 (W14) and week 19 (W19). *denotes that the treatment is significantly better than no treatment; † denotes that the ROSI treatment is significantly better than MET treatment.

in type 2 diabetes mellitus (T2D) shifts the balance of substrate utilization away from glucose utilization to almost exclusively on fatty acids for its energy needs [51]. To that end, the ZDF rat is a well-characterized model for T2D with obesity resulting from loss-of-function mutation in the leptin receptor [52]. Preclinical PET-FDG imaging was used to characterize myocardial glucose utilization (MGU) and myocardial fatty acid utilization (MFAU) in 14- and 19-week-old ZDF rats [3, 4]. Glucose uptake rate and utilization were characterized by performing a Patlak graphical analysis, as described in Section 6.3.3.1. ZDF rats exhibited significant reduction in MGU compared to nondiabetic rats (Fig. 6.6a). Moreover, image-derived measures of MGU uptake rate correlated with gene expression of GLUT4 following imaging (Fig. 6.6b). As an application of quantitative imaging in assessing therapeutic efficacy, we highlight the use of preclinical imaging in assessing the effects of antidiabetic therapies on myocardial substrate metabolism. Antidiabetic therapies such as metformin and rosiglitazone attempt to normalize substrate availability and to restore glycemic control by enhancing insulin sensitivity in the periphery [53]. Little is known, however, about the *in vivo* efficacy of either metformin or rosiglitazone on MFAU. To that end, preclinical PET imaging was used to assess MGU and myocardial fatty acid oxidation (MFAO) in the diabetic heart of ZDF rats longitudinally following treatment with metformin and rosiglitazone [3]. Using a kinetic model for [^{11}C]palmitate, quantitative analysis was used to delineate oxidative and esterification pathways of fatty acid metabolism. Overall, the results indicated increased MGU (Fig. 6.7a) and reduced MFAO (Fig. 6.7b) following rosiglitazone treatment, which correlated with gene expression analysis of relevant pathways [3].

6.5 FACTORS AFFECTING QUANTITATIVE ACCURACY

Quantification of PET and SPECT images is a function of the quality of the data extracted from the images. To that end, variability in quantified images is common between imaging sites as well as within sites due to interoperator variability. Thus, it is essential to be mindful of the following factors that may affect quantification of images.

6.5.1 Region of Interest

In most cases, the ROI drawn on images is the first step in the quantitative process. Typically, ROI values are reported as the mean in the ROI with a measure corresponding to the variability in the activity within the ROI, such as the standard deviation (SD). As the ROI mean value is a function of the ROI boundaries, the mean value may change significantly depending on the boundaries of the ROI. Moreover, considering that both SPECT and PET data are subject to partial volume effects, boundaries are typically not obvious and to a large extent are subject to interpretation by the operator. Taken together, variability in definition of ROI is expected, but efforts should be made to minimize variability by establishing a consistent protocol for drawing ROIs, for example, by using the top 10% of voxels. Alternatively, one may resort to automated and semiautomated methods to drawing ROIs, although at this stage such methods are mostly experimental [54].

6.5.2 Competitive Kinetics and Mass Effect

The underlying assumption in PET and SPECT imaging is that the introduction of the tracer does not perturb the physiology or its target, for example, by inhibiting or activating a receptor. In addition, since the radiopharmaceutical is in trace amounts, it is believed that it does not compete with endogenous substrates or ligands. In fact, while these assumptions generally hold, in practice their validity depends on endogenous conditions. For example, radiopharmaceuticals may be in competition for transport or enzymatic process or in competition with endogenous ligands [55–57]. Moreover, in translational imaging studies where most of the initial characterization is performed in small animals, the mass effect of the ID needs to be considered [58]. For example, as reported in [58], an ID of 10 mCi [^{11}C]raclopride with a specific activity of about 1000 Ci/mmol in human studies will result in less than 1% of receptor occupancy. In contrast, in mice studies, an ID of 1 mCi will result in an upward of 70% receptor occupancy. Thus, the concentration of endogenous substrates and potential mass effect of the ID need to be considered in designing translational radiopharmaceutical imaging studies.

6.5.3 Partial Volume Effects Correction

The in-plane resolution of both PET and SPECT imaging devices is finite resulting in image blurring and partial volume effects. Clinical PET scanners, for example, have an in-plane resolution of 5–6 mm full-width-at-half-maximum (FWHM), while preclinical imaging scanners have an in-plane resolution ranging from 1.3 to 2 mm FWHM radically. This limitation in spatial resolution is a combination of multiple factors including instrument geometry, positron range, scatter, and reconstruction. Individual treatment of aforementioned factors is beyond the scope of this chapter (the reader is referred to [59, 60] for additional detail). A consequence of partial volume effects is loss of recovery and spillover. In practice, the extent of partial volume effects is a nonlinear function of the ratio of dimension of the structure and the FWHM of the scanner [61]. With the increasing ratio of dimension to FWHM, there is improved recovery of counts in the voxel (and less spillover). Two scenarios where considerations of partial volume effects are essential are in longitudinal imaging where the dimension of the structure is expected to change over time and in response to therapy where, for example, tumor shrinkage may occur. A rule of thumb is that the object dimension should be at least twice the FWHM of the scanner to recover ~90% of the counts. Various schemes have been developed to correct for partial volume effects ranging from physical phantom studies [62] to anatomical correction schemes using MR or CT images [63] to name a few. The application of any of the corrections schemes is dependent on the study design and resources available.

6.5.4 Calibration and Cross-Calibration

As quantitative imaging capabilities are one of the hallmarks of PET imaging and to some extent SPECT, imaging scanners and counting instruments need to be calibrated and cross-calibrated to produce reliable quantities output. Typically the imaging workflow requires the use of various instruments, such as a gamma-counter and a dose calibrator to support the use of the scanner. The dose calibrator is used to record the radioactivity in the ID, for example. The gamma-counter may be used to count the radioactivity in blood samples. Thus, calibration of scanners and cross-calibration between the various instruments is essential. Numerous calibration and cross-calibration schemes can be devised. Typically, the dose calibrator is calibrated using a standard that may be obtained through the National Institute of Standards and Technology (NIST). In theory, the scanner and the gamma-counter can be calibrated if the counting efficiency ε of the gamma-counter is known

$$\varepsilon = \frac{\text{cpm}}{\text{dpm}}$$

to relate image counts (counts per minute or cpm) to gamma-counter disintegrations (disintegrations per minute or dpm). The counting efficiency is a function of the geometry and radionuclide among other factors. Thus, while calibration based on counting efficiency may be applicable in some centers, it may not be ideal in other imaging centers where multiple radionuclides are used. In case of the latter, to calibrate the gamma-counter and a scanner, one may choose to prepare a standard of known activity diluted at multiple concentrations in vials, from which a standard curve is prepared. The vials can be imaged in the scanner and the ROI is matched to the gamma-counter counts. Alternatively, a standard source can be used to calibrate the scanners. In either scenario, calibration should be performed on a regular basis. Moreover, since both PET and SPECT data are compared across multiple centers, cross-calibration of imaging scanners should be part of any multicenter imaging protocol [64, 65].

REFERENCES

1. Kim SJ, Lee JS, Im KC, et al. Kinetic modeling of 3′-deoxy-3′-^{18}F-fluorothymidine for quantitative cell proliferation imaging in subcutaneous tumor models in mice. J Nucl Med 2008;49:2057–2066.
2. Pollok KE, Lahn M, Enas N, et al. In vivo measurements of tumor metabolism and growth after administration of enzastaurin using small animal FDG positron emission tomography. J Oncol 2009;2009:596560.
3. Shoghi KI, Finck BN, Schechtman KB, et al. In vivo metabolic phenotyping of myocardial substrate metabolism in rodents: differential efficacy of metformin and rosiglitazone monotherapy. Circ Cardiovasc Imaging 2009;2 (5):373–381.
4. Shoghi KI, Gropler RJ, Sharp T, et al. Time course of alterations in myocardial glucose utilization in the Zucker diabetic fatty rat with correlation to gene expression of glucose transporters: a small-animal PET investigation. J Nucl Med 2008;49: 1320–1327.
5. Guo Y, Gao F, Wang S, et al. In vivo mapping of temporospatial changes in glucose utilization in rat brain during epileptogenesis: an ^{18}F-fluorodeoxyglucose-small animal positron emission tomography study. Neuroscience 2009;162:972–979.
6. Cheng KT, Menkens A, Bryant S, et al. NIH MICAD initiative and guest author program opportunities. J Nucl Med 2007;48:19N.
7. Shoghi KI. Quantitative small animal PET. QJ Nucl Med Mol Imaging 2009;53: 365–373.

8. Southworth R, Parry CR, Parkes HG, et al. Tissue-specific differences in 2-fluoro-2-deoxyglucose metabolism beyond FDG-6-P: a 19F NMR spectroscopy study in the rat. NMR Biomed 2003;16:494–502.
9. Shields AF, Grierson JR, Dohmen BM, et al. Imaging proliferation in vivo with [F-18]FLT and positron emission tomography. Nat Med 1998;4:1334–1336.
10. Boellaard R. Standards for PET image acquisition and quantitative data analysis. J Nucl Med 2009;50 (Suppl 1):11S–20S.
11. Ito K, Kubota K, Morooka M, et al. Diagnostic usefulness of (18)F-FDG PET/CT in the differentiation of pulmonary artery sarcoma and pulmonary embolism. Ann Nucl Med 2009;23 (7):671–676.
12. Lucignani G, Paganelli G, Bombardieri E. The use of standardized uptake values for assessing FDG uptake with PET in oncology: a clinical perspective. Nucl Med Commun 2004;25:651–656.
13. Ning XH, Meng QL, Wang YZ, et al. Application of ^{18}F-fluorodeoxyglucose positron emission tomography in diagnosis of malignant diseases. Chin Med Sci J 2009; 24:117–121.
14. Suzuki K, Nishioka T, Homma A, et al. Value of fluorodeoxyglucose positron emission tomography before radiotherapy for head and neck cancer: does the standardized uptake value predict treatment outcome? Jpn J Radiol. 2009;27:237–242.
15. Wahl RL, Jacene H, Kasamon Y, et al. From RECIST to PERCIST: evolving considerations for PET response criteria in solid tumors. J Nucl Med 2009;50 (Suppl 1):122S–150S.
16. Herrero P, Kim J, Sharp TL, et al. Assessment of myocardial blood flow using 15O-water and 1-11C-acetate in rats with small-animal PET. J Nucl Med 2006;47:477–485.
17. Herrero P, Staudenherz A, Walsh JF, et al. Heterogeneity of myocardial perfusion provides the physiological basis of perfusable tissue index. J Nucl Med 1995;36:320–327.
18. Herrero P, Weinheimer CJ, Dence C, et al. Quantification of myocardial glucose utilization by PET and 1-carbon-11-glucose. J Nucl Cardiol 2002;9:5–14.
19. Martinez D, Narendran R. Imaging neurotransmitter release by drugs of abuse. Curr Top Behav Neurosci 2010;3:219–245.
20. Urban N, Abi-Dargham A. Neurochemical imaging in schizophrenia. Curr Top Behav Neurosci 2010;4:215–242.
21. Herrero P, Dence CS, Coggan AR, et al. L-3-11C-lactate as a PET tracer of myocardial lactate metabolism: a feasibility study. J Nucl Med 2007;48:2046–2055.
22. Brader P, Wong RJ, Horowitz G, et al. Combination of pet imaging with viral vectors for identification of cancer metastases. Adv Drug Deliv Rev 2012;64 (8):749–55.
23. Gambhir SS, Barrio JR, Herschman HR, et al. Assays for noninvasive imaging of reporter gene expression. Nucl Med Biol 1999;26:481–490.
24. Wienhard K. Measurement of glucose consumption using [(18)F]fluorodeoxyglucose. Methods 2002;27:218–225.
25. Bergmann SR, Weinheimer CJ, Markham J, et al. Quantitation of myocardial fatty acid metabolism using PET. J Nucl Med 1996;37:1723–1730.
26. Shoghi-Jadid K, Huang SC, Stout DB, et al. Striatal kinetic modeling of FDOPA with a cerebellar-derived constraint on the distribution of volume of 30MFD: a PET investigation using non-human primates. J Cereb Blood Flow Metab 2000;20:1134–1148.
27. Gallezot JD, Nabulsi N, Neumeister A, et al. Kinetic modeling of the serotonin 5-HT(1B) receptor radioligand [(11)C]P943 in humans. J Cereb Blood Flow Metab 2010; 30 :196–210.

28. Wu Y, Carson RE. Noise reduction in the simplified reference tissue model for neuroreceptor functional imaging. J Cereb Blood Flow Metab 2002;22:1440–1452.
29. Meyer JH, Ichise M. Modeling of receptor ligand data in PET and SPECT imaging: a review of major approaches. J Neuroimaging 2001;11:30–39.
30. Landaw EM, DiStefano JJ 3rd.. Multiexponential, multicompartmental, and noncompartmental modeling. II. Data analysis and statistical considerations. Am J Physiol. 1984;246: R665–R677.
31. Akaike H. Citation classic – a new look at the statistical-model identification. CC/Eng Tech Appl Sci 1981;51:22–22.
32. Schwarz G. Estimating the dimension of a model. Annals of Statistics 1978;6:461–464.
33. Acton PD, Zhuang H, Alavi A. Quantification in PET. Radiol Clin North Am 2004;42: 1055–1062, viii.
34. Fang YH, Muzic RF Jr. Spillover and partial-volume correction for image-derived input functions for small-animal 18F-FDG PET studies. J Nucl Med 2008;49:606–614.
35. Laforest R, Sharp TL, Engelbach JA, et al. Measurement of input functions in rodents: challenges and solutions. Nucl Med Biol 2005;32:679–685.
36. Chen BC, Germano G, Huang SC, et al. A new noninvasive quantification of renal blood flow with N-13 ammonia, dynamic positron emission tomography, and a two-compartment model. J Am Soc Nephrol 1992;3:1295–1306.
37. Germano G, Chen BC, Huang SC, et al. Use of the abdominal aorta for arterial input function determination in hepatic and renal PET studies. J Nucl Med 1992;33:613–620.
38. Huang SC, Barrio JR, Yu DC, et al. Modelling approach for separating blood time-activity curves in positron emission tomographic studies. Phys Med Biol 1991;36:749–761.
39. Kuhle WG, Porenta G, Huang SC, et al. Quantification of regional myocardial blood flow using 13N-ammonia and reoriented dynamic positron emission tomographic imaging. Circulation 1992;86:1004–1017.
40. Shoghi KI, Welch MJ. Hybrid image and blood sampling input function for quantification of small animal dynamic PET data. Nucl Med Biol 2007;34:989–994.
41. Su Y, Shoghi KI. Single-input-dual-output modeling of image-based input function estimation. Mol Imaging Biol 2010;12 (3):286–294.
42. Su Y, Welch MJ, Shoghi KI. The application of maximum likelihood factor analysis (MLFA) with uniqueness constraints on dynamic cardiac microPET data. Phys Med Biol 2007;52:2313–2334.
43. Ferl GZ, Zhang X, Wu HM, et al. Estimation of the 18F-FDG input function in mice by use of dynamic small-animal PET and minimal blood sample data. J Nucl Med 2007;48:2037–2045.
44. Wu HM, Sui G, Lee CC, et al. In vivo quantitation of glucose metabolism in mice using small-animal PET and a microfluidic device. J Nucl Med 2007;48:837–845.
45. Patlak CS, Blasberg RG. Graphical evaluation of blood-to-brain transfer constants from multiple-time uptake data. Generalizations. J Cereb Blood Flow Metab 1985;5:584–590.
46. Patlak CS, Blasberg RG, Fenstermacher JD. Graphical evaluation of blood-to-brain transfer constants from multiple-time uptake data. J Cereb Blood Flow Metab 1983; 3:1–7.
47. Logan J, Fowler JS, Volkow ND, et al. Graphical analysis of reversible radioligand binding from time-activity measurements applied to [N-11C-methyl]-(-)-cocaine PET studies in human subjects. J Cereb Blood Flow Metab 1990;10:740–747.

48. Fang ZY, Prins JB, Marwick TH. Diabetic cardiomyopathy: evidence, mechanisms, and therapeutic implications. Endocr Rev 2004;25:543–567.

49. Lopaschuk GD, Belke DD, Gamble J, et al. Regulation of fatty acid oxidation in the mammalian heart in health and disease. Biochim Biophys Acta 1994;1213:263–276.

50. Lopaschuk GD. Metabolic abnormalities in the diabetic heart. Heart Fail Rev 2002;7: 149–159.

51. Saddik M, Lopaschuk GD. Triacylglycerol turnover in isolated working hearts of acutely diabetic rats. Can J Physiol Pharmacol 1994;72:1110–1119.

52. Phillips MS, Liu Q, Hammond HA, et al. Leptin receptor missense mutation in the fatty Zucker rat. Nat Genet 1996;13:18–19.

53. Inzucchi SE. Oral antihyperglycemic therapy for type 2 diabetes: scientific review. Jama 2002;287:360–372.

54. Kim J, Cai W, Feng D, et al. Segmentation of VOI from multidimensional dynamic PET images by integrating spatial and temporal features. IEEE Trans Inf Technol Biomed 2006;10:637–646.

55. Elsinga PH, Hendrikse NH, Bart J, et al. Positron emission tomography studies on binding of central nervous system drugs and P-glycoprotein function in the rodent brain. Mol Imaging Biol 2005;7:37–44.

56. Narendran R, Hwang DR, Slifstein M, et al. In vivo vulnerability to competition by endogenous dopamine: comparison of the D2 receptor agonist radiotracer (-)-N-[11C] propyl-norapomorphine ([11C]NPA) with the D2 receptor antagonist radiotracer [11C]-raclopride. Synapse 2004;52:188–208.

57. Rominger A, Wagner E, Mille E, et al. Endogenous competition against binding of [(18) F]DMFP and [(18)F]fallypride to dopamine D(2/3) receptors in brain of living mouse. Synapse 2010;64:313–322.

58. Kung MP, Kung HF. Mass effect of injected dose in small rodent imaging by SPECT and PET. Nucl Med Biol 2005;32:673–678.

59. Soret M, Bacharach SL, Buvat I. Partial-volume effect in PET tumor imaging. J Nucl Med 2007;48:932–945.

60. Thomas BA, Erlandsson K, Modat M, et al. The importance of appropriate partial volume correction for PET quantification in Alzheimer's disease. Eur J Nucl Med Mol Imaging 2011;38:1104–1119.

61. Cherry SR, Sorenson J, Phelps ME. Physics in Nuclear Medicine. 3rd ed. Philadelphia: Saunders; 2003.

62. Hoetjes NJ, van Velden FH, Hoekstra OS, et al. Partial volume correction strategies for quantitative FDG PET in oncology. Eur J Nucl Med Mol Imaging 2010;37:1679–1687.

63. Hofmann M, Pichler B, Scholkopf B, et al. Towards quantitative PET/MRI: a review of MR-based attenuation correction techniques. Eur J Nucl Med Mol Imaging 2009;36 (Suppl 1):S93–S104.

64. Geworski L, Knoop BO, de Wit M, et al. Multicenter comparison of calibration and cross calibration of PET scanners. J Nucl Med 2002;43:635–639.

65. Hasegawa T, Oda K, Wada Y, et al. A practical method of determining cross-calibration factors of PET scanners by moving a point-like (22)Na radioactive source. Ann Nucl Med 2010;24:655–661.

CHAPTER 7

PET IMAGING OF THE PHARMACOKINETICS OF SMALL MOLECULAR-WEIGHT DRUGS

STEPHEN M. MOERLEIN AND ZHUDE TU

7.1 INTRODUCTION

A key aspect of drug action is its pharmacokinetic behavior *in vivo*. Pharmacokinetics (PK) describes the fate of the drug following administration to an organism and includes the classical parameters absorption, distribution, metabolism, and excretion (ADME) [1]. The importance of PK to the clinical utility of a drug is based on the premise that adequate concentrations of a drug must be achieved at the target in order for the desired therapeutic effect to occur. Traditionally, most human PK estimates have relied upon on spectroscopic analysis of plasma samples acquired in a serial fashion.

Because drug action depends on the drug reaching the appropriate target *in vivo*, positron emission tomography (PET) measurement of the tissue concentrations as a function of time in human subjects has immediate relevance to pharmacotherapy. The principal utility of PET in this situation is the direct measurement of tissue accumulation of tracer on a regional basis within organs. This is especially valuable where blood levels of drug show a poor correlation with tissue accumulation at target sites and therefore do not give a good index of pharmacological effect. PET imaging of these drugs may therefore be of use in estimating the concentrations, and thus pharmacological effects, in target tissues. In addition, PET can reveal tissue accumulation of drug that may lead to unwanted toxicities that are independent of the desired pharmacological action in the target tissue. Such accumulation also may not be apparent from traditional measurements of drug in the blood. Unexpected therapeutic response may be explained by abnormal accumulation of drug at target sites under altered conditions, such as aging, disease, or when the subject is undergoing therapy.

This chapter will focus on the application of PET for the evaluation of the PK of small molecular weight drugs in human subjects. Substantial progress has occurred

Drug Delivery Applications of Noninvasive Imaging: Validation from Biodistribution to Sites of Action, First Edition. Edited by Chun Li and Mei Tian.

in this field, and these achievements will be presented as an overview with specific examples from the literature.

7.2 PET METHODOLOGY

The rationale for the application of PET to drug PK is that the spatial distribution of a labeled tracer within a subject can be quantified in a regional and noninvasive manner as a function of time. The capability to noninvasively measure tissue time-activity curves makes PET a powerful tool for the analysis of the physiological fate of labeled drugs in humans. The radioactive drug is administered to the subject, and serial measurements of the distribution of radioactivity are acquired serially. Because PET instrumentation provides regional data as μCi/cc, dividing the quantified image data by the tracer specific activity (μCi/nmol) gives a direct measurement of the biodistribution of the labeled drug in nmol/cc.

Positron emission tomography is uniquely qualified as a tool for measurement of the PK of drugs in human subjects. PET methodology generates high-resolution images that quantify the regional distribution of radioactivity within a subject in a temporal fashion [2]. Image data can be treated with appropriate mathematical models to derive key descriptive parameters of PK. Note that PET images are corrected for attenuation, so accurate PK data can be acquired for deep organs, and reliance on plasma kinetics as a surrogate measurement is unnecessary. The mass of positron-emitting tracers is very low, associating very little risk of toxicity with the imaging procedure, and careful evaluation of dosimetry minimizes the radiation exposure to the subject. Since PET is noninvasive, repeat imaging of human subjects is feasible, which makes longitudinal or interventional studies possible. This aspect of PET is extremely valuable for clinical investigation of drug action.

PET instrumentation detects annihilation radiation with high sensitivity, so it has good temporal resolution and can be used for dynamic imaging [2]. Quantification of high-resolution PET images is enhanced by scatter and attenuation correction. These characteristics make PET a powerful research tool for the noninvasive assessment of drug distribution *in vivo*. The major disadvantages of PET for PK measurements are that there is no discrimination of the chemical source of signals in the images, and studies are limited by the relatively short half-lives of positron-emitting nuclides. Despite these limitations, PET technology offers special opportunities for noninvasive evaluation of drug behavior in living human subjects.

It is important to note that PET is virtually the only means for acquiring quantitative PK data from human subjects on a regional basis. Although similar data may be derived from invasive methods in animal subjects, or postmortem examination of human samples, the translation of these data to living human subject is often ambiguous. Direct PET measurement of drug concentrations within human tissues avoids such ambiguity and thus gives greater precision to PK measurements.

Analysis of image data with tracer kinetic models enhances the impact of PET studies of PK. In the modeling of tracer kinetics, radioactivity in tissue and plasma is compared, and complicated physiological processes are simplified into mathematical descriptions of tracer movement between a framework of compartments. Such mathematical treatment derives objective PK parameters that are relevant to the subject and drug under investigation. Such mathematical modeling is similar to what is done in classical pharmacokinetics [1] but differs in that data is obtained directly from tissues on a regional basis, and PET images represent a composite of the original labeled drug as well as any radiometabolites that are formed.

It should be noted that similar imaging techniques have been used for many years by the nuclear medicine community to determine tracer behavior *in vivo* for estimation of the internal radiation dosimetry of radiopharmaceuticals. In these applications, time-activity curves are measured in the various tissues of the body, and this information is analyzed using mathematical models to calculate intra- and interorgan radiation exposure [3, 4]. The similarity in PET methods for measuring tracer tissue distribution for dosimetry calculations, and tissue distribution of labeled drugs for evaluation of PK, is apparent. In fact, many PET PK studies have been combined with radiation dose estimation.

Although it is beyond the scope of this chapter to discuss tracer kinetic modeling in detail, a few comments will be made to give the reader a general understanding of the topic. The ultimate goal of tracer kinetic models is to quantify a pharmacologically relevant parameter, such as drug tissue concentration or receptor site occupancy. Such models may originally be based on the *in vitro* assays used in tracer development, but *in vivo* tracer kinetics are of necessity more complicated, since in the physiological environment influences like perfusion, metabolism, and plasma protein binding come into play. This greater complexity requires a larger number of compartments for adequate mathematical treatment.

Many types of mathematical models have been devised to describe tracer kinetics *in vivo*. The appropriateness of the particular model depends on the administration rate of the tracer (bolus or constant infusion), whether the drug binds to its target in a reversible or irreversible fashion, whether arterial blood sampling or reference region data is used, and if linear or nonlinear regression analysis is employed. Graphic or nongraphic models can be used to describe the tracer kinetics from PET imaging. The reader is referred to several excellent review articles dealing with the subject of tracer kinetic models [5–12].

7.3 PHARMACOKINETICS

In pharmaceutical research, a major decision point for further development of a lead compound concerns its *in vivo* tissue distribution and its PK. For a drug to be effective, it must reach suitable concentrations at the site of action for an adequate period of time for the desired effect on the target to be elicited. This is key to the ultimate success of a drug entity, because if an agent that is active *in vitro* does not physically reach the biological site of action, pharmacological effect will not occur. Besides drug development, measurement of the PK of existing drugs is sometimes needed in routine clinical practice. This is because individual variation in the classical ADME parameters may impact therapeutic effectiveness in any particular patient being treated.

Application of PET techniques for determination of drug PK is relatively straightforward and derives several important PK parameters. The major caveat to the use of PET methodology is that the image must be corrected for tracer metabolism to give

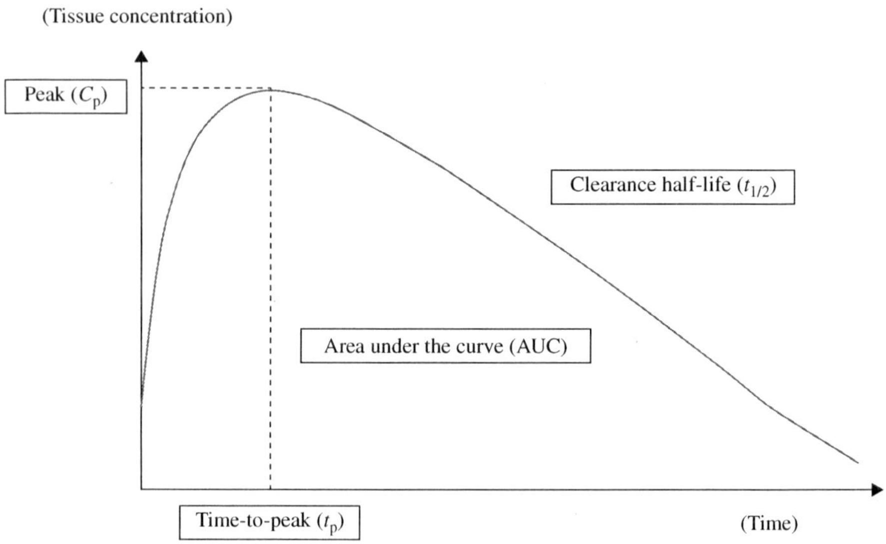

FIGURE 7.1 PK parameters that can be measured noninvasively on a regional basis by PET. C_p, peak concentration; t_p, time-to-peak concentration; $t_{1/2}$, biological half-life; AUC, area under the curve.

precise estimation of intact drug distribution. However, there are a variety of methods for correcting PET data for radiometabolite formation, so this should not be interpreted as a major impediment.

Some important PK parameters that can be measured by PET are illustrated in Figure 7.1. Note that these values can be determined either for an entire organ or for specific regions within an organ. The latter more selective measurements are especially valuable for assessing the mechanism of drug action, where the target distribution within an organ might be heterogeneous. In either case, PET analysis is much more comprehensive than traditional methods for PK studies, which are predominantly limited to chemical analysis of blood samples and extrapolation to tissue concentrations. With PET, the maximum concentration (C_{max}) and the time at which this concentration is achieved (t_p) can be measured for any human tissue noninvasively. In addition, the clearance (CL) half-life and the area under the curve (AUC) can also be derived from PET data. Determination of these PK parameters for a drug is very useful, because it helps determine optimum dosing regimens for therapeutic effect.

On a more fundamental level, other PK issues that can easily be addressed with PET imaging is whether or not the drug reaches the target site and the major routes of elimination (renal, hepatic, pulmonary). These measurements are particularly useful for drug development, where go-no-go decisions must be made concerning further research on lead compounds. These less mathematically refined descriptions of drug distribution are more useful in this case because they are relatively rapid and less expensive than comprehensive PK determinations. Simplified measurements of relative tissue enrichment such as standardized uptake values (SUVs) are often made for this purpose.

TABLE 7.1 PET Imaging of Positron-Emitting Drugs in Human Subjects: Infectious Disease, Oncology, and Cardiology

Labeled drug	Pharmacological class	Medical indications [13, 14]
Infectious disease		
[^{11}C]Erythromycin [15]	Macrolide antibiotic	Bacterial endocarditis; respiratory infections; skin infections; sexually transmitted disease
[^{18}F]Ciprofloxacin [16–18]	Fluoroquinolone antibiotic	Respiratory, abdominal, skin, bone, urinary tract infections
[^{18}F]Trovafloxacin [19]	Fluoroquinolone antibiotic	Meningitis; respiratory, skin, abdominal infections; sexually transmitted disease
[^{18}F]Lomefloxacin [20]	Fluoroquinolone antibiotic	Bronchitis; skin, postoperative, urinary tract infections
[^{18}F]Fleroxacin [21]	Fluoroquinolone antibiotic	Sexually transmitted disease; bone, skin, intestinal, urinary tract infections
[^{18}F]Fluconazole [22]	Triazole antifungal	Candidiasis, candidemia, cryptococcal meningitis
Oncology		
5-[^{18}F]Fluorouracil [23–29]	Chemotherapeutic drug	Breast, colorectal, skin, gastric, pancreatic cancer
[^{11}C]Docetaxel [30]	Chemotherapeutic drug	Breast, gastric, prostate, lung cancer
[^{11}C]Temozolomide [31]	Chemotherapeutic drug	Glioblastoma; astrocytoma
[^{11}C]Vinblastine [32]	Chemotherapeutic drug	Lymphoma; Hodgkin's disease; melanoma; breast, bladder, cervical, ovarian, lung, testicular cancer
Cardiology		
[^{11}C]Verapamil [33]	Calcium channel antagonist	Cardiac arrhythmia; hypertension
[^{11}C]Epinephrine [34]	Sympathomimetic drug	Cardiac arrest, anaphylaxis, glaucoma
[^{11}C]Phenylephrine [35, 36]	Sympathomimetic drug	Hypotension, shock, nasal congestion

Tables 7.1 and 7.2 list the clinically used drugs that have been labeled with positron-emitting nuclides and imaged in human subjects using PET methodology. There are a large number of drugs tabulated, and they have been labeled with either ^{11}C ($t_{1/2}$ = 20 min) or ^{18}F ($t_{1/2}$ = 110 min) for PET detection. The list is diverse, as is the pharmacopeia of clinically used drugs. The various drugs are categorized into the various therapeutic applications for the specific drugs. Representative drugs for the major medical specialties are shown in the tables, including infectious disease, oncology, cardiology, neurology, psychiatry, and surgery.

For all of the drugs in Tables 7.1 and 7.2, the respective molecular structure enables radiolabeling with ^{11}C or ^{18}F without alteration of the drug. Figure 7.2 gives examples of how positron-emitting nuclides can be incorporated into drug structures

TABLE 7.2 PET Imaging of Clinically-Used Drugs in Human Subjects: Neurology, Psychiatry and Other Specialties

Labeled drug	Pharmacological class	Medical indications [13, 14]
Neurology		
[^{11}C]Vinpocetine [37, 38]	Neuroprotectant	Cerebrovascular disease; Alzheimer's disease
[^{11}C]Flumazenil [39]	BZD antagonist	BZD reversal
[^{11}C]Diphenylhydantoin [40]	Anticonvulsant	Seizure
[^{11}C]Donepezil [41]	AChE inhibitor	Alzheimer's disease
[^{11}C]Physostigmine [42]	AChE inhibitor	Anticholinergic toxicity
Psychiatry		
[^{11}C]Methylphenidate [43]	CNS stimulant	ADHD; narcolepsy
[^{11}C]Doxepin [44]	Histamine H_1 antagonist	Depression; pain syndromes
[^{11}C]Mirtazapine [45]	Antidepressant	Depression
[^{11}C]Clorgyline [46, 47]	MAO-A inhibitor	ADHD; bipolar disorder; depression
[^{11}C]Deprenyl [47, 48]	MAO-B inhibitor	Parkinson's disease; depression
[^{11}C]Befloxatone [49]	MAO-A inhibitor	Depression
[^{11}C]Chlorpromazine [50]	Phenothiazine psychotropic	Behavioral syndrome; nausea; vomiting
[^{11}C]Rolipram [51]	Phosphodiesterase inhibitor	Depression
[^{11}C]Nicotine [52, 53]	nACh agonist	Smoking cessation
Surgery		
[^{11}C]Cocaine [54]	Anesthetic	Surgical procedures
Pulmonary		
[^{11}C]Triamcinolone [55–57]	β-Adrenergic agonist	Asthma, inflammatory disorders
Veterinary		
[^{11}C]Carfentanil [58–60]	Opioid agonist	Tranquilizer
[^{11}C]Diprenorphine [61, 62]	Opioid antagonist	Tranquilizer reversal
[^{11}C]Metomidate [63]	Anesthetic	Surgical procedures

without chemical modification of the original compound. Note that not every drug can be labeled with a positron-emitting nuclide. Not only must the appropriate element already be present in the molecular structure of the drug, but its location must be amenable to facile synthesis within the constraints created by the half-life of the radionuclide. When such radiosynthesis of a positron-emitting drug is feasible, the PET data of its distribution *in vivo* correlate precisely to the PK of the clinically used drug.

As discussed earlier, any radioactive metabolites generated within the data acquisition period are also detected in the field of view of the PET scanner. The commingling of native labeled drug and radiometabolite biodistribution is an unavoidable disadvantage to the use of PET for PK studies, since it is impossible to distinguish annihilation radiation by the chemical structure from which it arises. However, the impact of this effect is minimized for drugs that are metabolized slowly relative to the time span of the imaging session. It is also possible to correct image data using tracer kinetic models that incorporate plasma metabolite measurements.

(a) **$[^{18}F]$Ciprofloxacin**

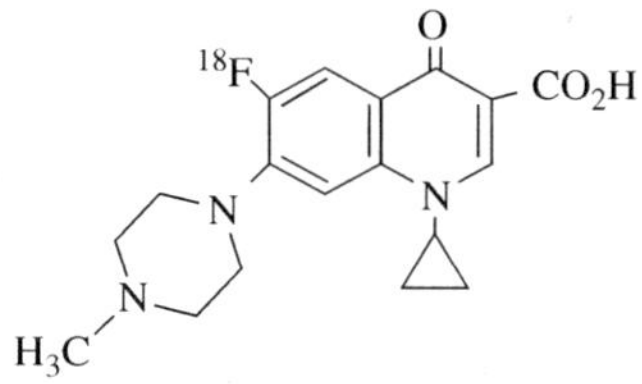

(b) **$[^{18}F]$Fluorouracil**

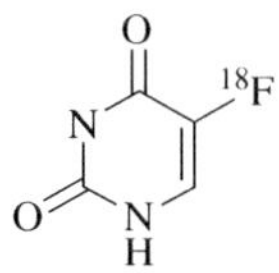

(c) **$[^{11}C]$Verapamil**

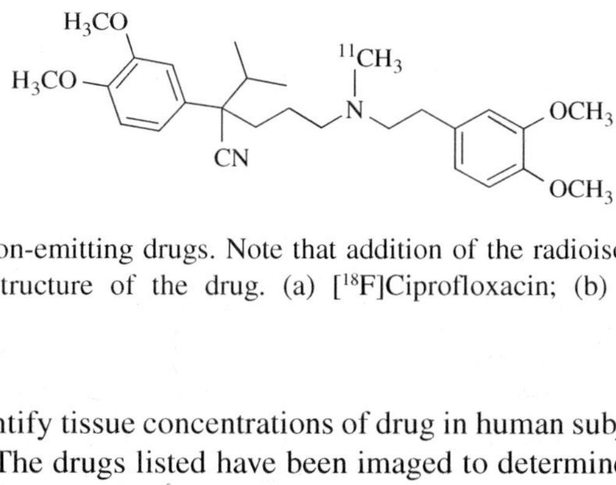

FIGURE 7.2 Examples of positron-emitting drugs. Note that addition of the radioisotope has not changed the molecular structure of the drug. (a) $[^{18}F]$Ciprofloxacin; (b) $[^{18}F]$Fluorouracil; (c) $[^{11}C]$Verapamil.

The capability of PET to quantify tissue concentrations of drug in human subjects on a regional basis is valuable. The drugs listed have been imaged to determine the rate of accumulation of drugs within tissues of interest, together with the retention, CL, and regional distribution within organs. This data is informative for assessing the temporal relationship between drug administration and drug concentrations needed to elicit pharmacological effect. In this way, the onset of action and the duration of action of drugs can be estimated from the tissue time-activity curves in the target tissue. Organs of elimination such as kidneys and liver can be imaged in a similar fashion, for insight into the rate at which the drugs leave the body. The latter information may be important in cases of impaired excretion, leading to increased drug toxicity.

The type of PK studies performed with the drugs listed in Tables 7.1 and 7.2 depends to a large extent on the medical specialty in which the drug is used. For specialties like infectious disease and oncology, the pathology is widespread, and whole-body PK studies are routinely performed. For medical specialties like cardiology, neurology, or psychiatry, disease is more focused, and limited organ PK studies are more common. PK studies with drugs used in infectious disease and oncology (Table 7.1) will be discussed in detail, since they involve whole-body measurements. Cardiovascular drugs (Table 7.1) will also be discussed, as PK studies with these drugs demonstrate how pharmacological mechanisms of action can be assessed using PET. Drugs used in neurology, psychiatry, and other medical specialties are listed in Table 7.2, but these will not be individually discussed because many of these tracers will be described in the section on biomarkers.

7.3.1 Infectious Disease

The tissue concentration of antimicrobials is critical for therapeutic effect. This is because concentration exceeding the minimal inhibitory concentration (MIC) must be attained to stop the growth of the infectious microbe. The MIC depends on the particular bacterium and the specific antibiotic, with higher MIC required for more resistant infectious organisms or weaker antimicrobials. It is important to establish optimal dosage regimens to assure that sufficiently high antibiotic concentrations are achieved to be therapeutic, particularly at infected sites. The tissue concentrations of drug are often difficult to predict based on blood, sputum, or urine levels, so the use of PET for direct measurement of the tissue PK of antimicrobials is ideally suited for evaluation of the pharmacotherapy of infectious disease.

7.3.1.1 [^{11}C]Erythromycin Erythromycin is a bacteriostatic antibiotic that is active against dividing organisms by interfering with bacterial tRNA and preventing protein synthesis. It is effective against several organisms and has multiple Food and Drug Administration (FDA)-approved indications, including treatment of endocarditis, respiratory infection, skin and soft tissue infection, pertussis, and sexually transmitted diseases. Erythromycin was the first antibiotic examined in humans by PET. In one of the early PK studies in humans using PET, [^{11}C]erythromycin was administered intravenously to patients with lobar pneumonia [13]. In this study, the unaffected lung served as a control measurement to the diseased lung. The radiolabeled drug rapidly entered the lung, and by 10 min, drug concentrations greatly exceeded the MIC for most sensitive organisms. The extravascular concentrations in the lungs ranged from about 5.5 to 6.6 µg/g. It was also noted that there was no significant difference between the erythromycin concentration in normal and pneumonic lungs, demonstrating that the disease process did not alter delivery of drug to the target site and that therapeutic antimicrobial levels were attained.

7.3.1.2 [^{18}F]Ciprofloxacin Fluoroquinolones are unique among antibiotics in that they are the only direct inhibitors of bacterial DNA synthesis. They are active against a wide variety of gram-negative and gram-positive bacteria and are used to treat several types of infection, including urinary tract, respiratory tract, skin, bone, and abdominal infection. From the imaging perspective, fluoroquinolones have the advantage of being metabolically stable and of having a fluorine in their molecular structure that can be labeled with fluorine-18.

Ciprofloxacin is a fluoroquinolone that is dosed twice a day. The drug has FDA approval for treatment of upper and lower respiratory infection (including pneumonia) and infection of the bone, skin, abdomen, and urinary tract. The PK of [^{18}F]ciprofloxacin has been evaluated by PET imaging of healthy volunteers that had been pretreated with a therapeutic ciprofloxacin dosing regimen for 5 days [14]. Time-activity curves were measured by PET in several organs for up to 6 h. It was found that the drug kinetics could be described by a two-compartment model. One compartment includes kidneys, heart, and spleen and has a CL half-life of 60–90 min. The second compartment includes the liver, muscle, and lungs and has much slower CL. The greatest tissue accumulation was in the kidneys and liver, and approximately two-thirds of the dose is renally excreted. The distribution to the brain was minimal, which may be explained by ciprofloxacin being a substrate for the p-glycoprotein (Pgp) efflux pump. No bone uptake nor plasma metabolite formation was seen on HPLC analysis of plasma samples, adding support that the PET-reconstructed biodistribution represents unchanged [^{18}F]ciprofloxacin and is therefore a valid measurement of ciprofloxacin PK.

The concentration of ciprofloxacin in peripheral tissues ranged from 1.3 μg/nL (skeletal muscle) to 9.5 μg/mL in kidney [14]. Because of the high drug concentration in the kidneys, the MIC was exceeded in this organ for all susceptible pathogens, which underlies the drug's great effectiveness in treating urinary tract infections. However, tenfold lower concentrations are attained in muscle and lung. The concentrations in these tissues are still much greater than the MIC for gram-negative bacteria like *Haemophilus influenzae*, *Moraxella catarrhalis*, and *Klebsiella* sp., so ciprofloxacin is effective for respiratory or soft tissue infection by these causative agents. At the dose evaluated, ciprofloxacin would not be adequate for treatment of skin or respiratory infection induced by microbes with higher MIC, such as *Pseudomonas aeruginosa*, or the gram-positive *Staphylococcus* sp. and *Streptococcus* sp. This prediction of efficacy by comparison of the bacterial pathogen susceptibilities with specific tissue PK measured by PET demonstrates the unique value of this noninvasive technique.

7.3.1.3 [^{18}F]Trovofloxacin Trovofloxacin is a newer broad-spectrum fluoroquinolone that has a longer duration of action than ciprofloxacin and requires only once-daily dosing. The tissue PK of this drug was evaluated by PET imaging of [^{18}F]trovofloxacin [15]. The tracer was administered to healthy adults that had received 200 mg daily for 5–8 days prior to imaging. Serial PET images and blood samples were obtained for 6–8 h. Peak drug concentrations 5–10 times the MIC for *Enterobacteriaceae* and anaerobes were attained in virtually all organs. In peripheral tissues, plateau concentration (measured at 2–8 h) ranged from a low of 1.5 μg/g in bone to 11.3 μg/g in the liver. Unlike ciprofloxacin, trovofloxacin crosses the blood–brain barrier to give a plateau concentration of 0.9 μg/g, which exceeds the MIC for CNS infectious agents like *Streptococcus pneumoniae*, *Neisseria meningitidis*, and *H. influenzae*. These data suggest that this fluoroquinolone is an effective antibiotic for treatment of a variety of infections at diverse anatomical sites.

7.3.1.4 [^{18}F]Lomefloxacin Lomefloxacin is another fluoroquinolone antibiotic that is approved for human use in countries other than the United States. It has been used to treat exacerbation of chronic bronchitis, urinary tract infection, and surgical prophylaxis. The drug, like trovofloxacin, is used with once-daily dosing. [^{18}F]Lomefloxacin was imaged rectilinearly with a PET scanner to assess the biodistribution of the drug in humans [16]. Unlike other PET investigations of fluoroquinolones, this study evaluated oral (rather than intravenous (i.v.)) administration of the drug. Imaging occurred for 7 h post ingestion. There was little bone uptake, suggesting that metabolism was minimal during the imaging session. The drug was found to move from the stomach to the small intestine, where it was absorbed into the circulation and excreted via the kidneys into the bladder. Blood and urine time-activity curves were also determined. Liver activity peaked at about 35% of the ingested dose, with a tissue concentration of ~65 μg/mL [16]. The lung concentration of lomefloxacin peaked at about 25 μg/mL. Comparison of these drug concentrations to the MIC of susceptible infectious agents was not made.

7.3.1.5 [^{18}F]Fluconazole Another class of antibiotics is the antifungal drugs. Fluconazole is a synthetic antifungal drug that is effective for the treatment of *Candida* and *Cryptococcus* species. Like the fluoroquinolones, fluconazole has fluorine in its molecular structure and is metabolically inert, so PET images of [^{18}F]fluconazole represent predominantly the distribution of the intact drug. The PK of [^{18}F]fluconazole has been examined in healthy adults using PET imaging for 2 h [17]. The drug was found to enter the brain in significant concentrations and to be relatively uniform (5.2–5.5 μg/g) throughout. In the periphery, concentrations of fluconazole were significant in all organs and ranged from 1.2 μg/g in bone to a high of 23 μg/g in spleen. The spleen was unusual in that a very high (40 μg/g) peak concentration occurred early after drug infusion and decreased to about half that value over the 2-h observation period. For all other organs, relatively constant levels were reached by 60 min.

The tissue levels of [^{18}F]fluconazole measured by PET were compared to the MIC of infectious organisms [17]. Because the majority of candidal species are inhibited by fluconazole concentrations >6 μg/g, the test dose of 5 mg/kg/d should be therapeutically effective according to the imaging assays. The low bone uptake of [^{18}F]fluconazole suggests that much greater doses (two to three times higher) would be needed to prevent therapeutic failure in the treatment of candidal osteomyelitis. Larger doses would also be needed for the treatment of less fluconazole-susceptible fungi, such as *Cryptococcus neoformans*, *Histoplasma capsulatum*, and *Coccidioides immitis*, which have MIC values ≥ 10 μg/g [17].

It is noteworthy that differences were seen in the PK of [^{18}F]fluconazole as determined by PET imaging of humans and rabbits [17]. Soft tissue accumulation of drug was lower in the rabbit compared to man, especially in the spleen and kidney. Also, bone accumulation of the antifungal was over twice as high in the rabbits compared to human subjects. These data underscore the disadvantage of using animal PK data as surrogate measurements for those of humans.

7.3.2 Oncology

Oncology is another medical specialty where PET measurement of drug PK is useful. A major reason for therapeutic failure in oncology is that insufficient concentrations of drug are achieved within the tumor. This adverse effect is commonly secondary to poor perfusion of the interior of tumors, accompanied by hypoxia and necrosis. Poor accumulation of drug at the target site results in suboptimal therapeutic effect and poor patient outcomes. Individual response rates to chemotherapy are difficult to predict because population response rates vary widely. This poor predictive information is problematic because not only is the patient possibly subjected to expensive and toxic therapy to which they do not respond, but it also delays initiation of therapy with the appropriate drugs to which the patient may respond.

This dilemma has led to an interest in the application of PET to identify those patients most likely to respond to specific chemotherapy drug regimens. PET imaging of positron-emitting chemotherapeutic drugs could thus provide valuable information on drug localization at the target site and expected therapeutic efficacy. Such image data promotes personalized treatment planning with these toxic drugs.

7.3.2.1 [^{18}F]Fluorouracil 5-Fluorouracil is a pyrimidine analog antimetabolite. It is cell cycle specific for the S phase and inhibits DNA and RNA synthesis. This drug has FDA indications for palliative care of breast cancer, carcinoma of the pancreas, colorectal cancer, and gastric cancer. 5- Fluorouracil is associated with significant toxicity, including serious adverse effects like myelosuppression, gastrointestinal ulceration, thrombophlebitis, and cerebellar syndrome. To assure effectiveness in treating carcinoma while minimizing toxicity, it is essential to establish optimum dosing so that therapeutic concentrations are delivered to tumor sites without unnecessary dosing of healthy tissue.

[^{18}F]Fluorouracil has been studied in several investigations using relatively large numbers of subjects [18–24]. One such study evaluated the radiolabeled chemotherapeutic drug in 50 colorectal cancer patients with 78 hepatic metastases [18]. In normal liver tissue, maximum liver activity following infusion of [^{18}F]fluorouracil was 11.3 SUV with a time to maximum of 30 min and a CL half-life of 22 min. Clearance from hepatic metastases was slower, with a half-life of 29.8 min. However, uptake in liver metastases was relatively low, with a concentration only one-third that of normal liver parenchyma at 120 min. There was a linear correlation found between early and late uptake in tumors, and it was concluded that high initial uptake into tumor cells is necessary for therapeutic response, which was indexed by late activity accumulation. Therapeutic failure can result either from poor initial uptake into the tumor cells or due to efflux of the drug from cancer cells. It was suggested that PET imaging of [^{18}F]fluorouracil tumor uptake kinetics be used to identify probable nonresponding patients prior to initiation of chemotherapy.

Perfusion and fluorouracil accumulation were evaluated using PET imaging of 36 patients with colorectal hepatic metastases [20]. [^{18}F]Fluorouracil was used to measure drug accumulation, whereas [^{15}O]water was used to assess perfusion. The tracers were administered intravenously and intra-arterially via the hepatic artery. For either route of administration, the correlation of perfusion with flurorouracil accumulation was only moderate ($r = 0.54$ and 0.61 for i.v. and i.a. injection, respectively). It was concluded that PET perfusion measurements predict fluorouracil uptake in tumors poorly and that imaging with [^{18}F]fluorouracil is a superior approach for measuring tumor accumulation of drug and for identification of patients that would benefit from intrahepatic versus i.v. chemotherapy.

A six-compartment PK model was developed to quantify the concentration of [^{18}F]fluorouracil in liver metastases [19]. Fourteen patients with 27 liver metastases from colorectal cancer were imaged dynamically with PET over a 2-h interval. Time-activity curves for tumor metastases, healthy liver, and the aorta were analyzed by the model. All measured time-activity curves could be described with only small interindividual differences in parameters. The most sensitive parameters for therapy monitoring were the rate constants for transport in (k_{in}) and out (k_{out}) of the metastases compartment. Trapping of fluorouracil, as defined by a high k_{in} and low k_{out}, corresponded to a high AUC for [^{18}F]fluorouracil. The model was used to evaluate the effect of a biomodulator, folinic acid, on the kinetics of the drug. This model may be used with PET-[^{18}F]fluorouracil image data to adjust the dose of fluorouracil on an individual basis to optimize treatment schedules.

7.3.2.2 [^{11}C]Docetaxel Docetaxel is an antimitotic drug that inhibits cell division by interfering with the mitotic and interphase functions of tumor cells. It is a chemotherapeutic drug indicated for treatment of breast, gastric, head and neck, and nonsmall cell lung cancers. Cardiovascular, dermatologic, gastrointestinal, hematologic, neurologic, and reproductive adverse effects commonly occur in over 50% of patients. Serious adverse effects are less prevalent but include edema, toxic epidermal necrosis, colitis, blood dyscrasias, hepatotoxicity, renal failure, and interstitial pneumonia. Careful adjustment of dose for optimum patient outcomes is thus necessary.

[^{11}C]Docetaxel has been evaluated using PET imaging of seven patients with metastatic solid tumors [25]. Gallbladder and liver demonstrated high uptake, with localization in the liver accounting for 47 ± 9% of the injected dose at 1 h. The high drug accumulation in liver warrants caution when docetaxel is used in patients with hepatic insufficiency. Accumulation of the labeled drug in the brain and lung was low. The labeled drug cleared rapidly from the blood compartment, and no radiolabeled metabolites were detected. The accumulation of [^{11}C]docetaxel in tumors was moderate and variable, with the tumor SUV ranging from 1 to 5.1. The low tumor accumulation of [^{11}C]docetaxel was attributed to the influence of the Pgp drug efflux pump.

7.3.2.3 [^{11}C]Temozolomide Temozolomide is an alkylating antineoplastic agent. It is indicated for treatment of glioblastoma and astrocytoma and is associated with many adverse effects. Common adverse effects are alopecia, nausea, vomiting, constipation, bone marrow depression, headache, seizure, and fatigue. Temozolomide is a prodrug that is converted to an active form by *in vivo* decarboxylation and ring opening to generate a reactive methyldiazonium ion that alkylates DNA.

[^{11}C]Temozolomide-PET has been used to evaluate the PK of this drug [26]. Paired studies of six patients with glioma were imaged twice, with the drug labeled in either the C-4 carboxyl position or the 3-*N*-methyl position. The dual label study was designed to test the hypothesized biological activation of the drug. There was a fivefold increase in exhaled $^{11}CO_2$ for the C-4-labeled drug compared to the C-3-labeled tracer. The AUC for [^{11}C]temozolomide was found to be higher in tumor relative to healthy brain tissue. These data demonstrate that metabolic conversion of temozolomide is an important aspect of the drug's PK and that tissue-directed activation may offer a therapeutic advantage in localizing the active alkylating species preferentially to the tumor site.

7.3.2.4 [^{11}C]Vinblastine Vinblastine is a cell cycle-specific agent that arrests tumor cells in the metaphase of mitosis. It is used in combination with other chemotherapeutic drugs to treat a variety of cancers, including testicular and breast cancer, Hodgkin's disease, and Kaposi's sarcoma. Adverse effects are numerous and affect most organ systems and include cardiovascular, dermatologic, metabolic, gastrointestinal, hematologic, hepatic, neurologic, psychiatric, renal, and respiratory effects. Optimum dosage regimens are thus essential to a successful use of this drug.

[^{11}C]Vinblastine was examined in two patients with renal carcinoma [27]. Liver accumulation predominated in both patients, which is expected based on the known hepatic metabolism of the drug. There was also focal accumulation of [^{11}C]vinblastine in bone metastases in one of the subjects, but there was rapid CL after 2 min in this nondrug responding patient.

7.3.3 Cardiovascular

7.3.3.1 [^{11}C]Verapamil Verapamil is a calcium channel blocker and class IV antiarrhythmic agent. It inhibits membrane transport of calcium in myocardial tissue, which produces negative inotropic and antiarrhythmic effects in the heart. It has therapeutic use in cardiac and cardiovascular disease, including treatment of angina, atrial fibrillation, supraventricular arrhythmia, and hypertension. However, verapamil is associated with severe adverse effects as well. These include bradyarrhythmia, atrioventricular block, congestive heart failure, pulmonary edema, hypotension, and shock.

The PK of the calcium channel blocker has been determined by PET imaging of humans with [^{11}C]verapamil [28]. Five cancer patients rather than healthy controls were used for these studies. The drug was found to accumulate in the lungs, with 43% of the injected dose in this organ at 1 h post injection. Localization to heart and tumor was relatively modest, with 1.3% and 0.9% accumulation, respectively, at 1 h post injection. The drug half-life in the lungs, heart, and tumor was 46.2 min, 73.8 min, and 23.7 min, respectively. It was concluded that the lungs are a major organ for accumulation of the drug and hypothesized that transport of the drug by the drug efflux pump Pgp is the primary cause for low tracer localization. Pgp is implicated in multidrug resistance (MDR) and affects the PK of other pharmacologically active agents.

7.3.3.2 [^{11}C]Epinephrine Epinephrine is a sympathomimetic catecholamine with activity at both alpha- and beta-adrenoceptors. Its alpha-agonist effect induces vasoconstriction, whereas its beta-agonist effect causes relaxation of bronchial smooth muscle. Epinephrine has FDA indications for treatment of serious conditions, including cardiac arrest, anaphylaxis, asthma, and hypersensitivity reaction. It is also associated with adverse effects that include angina, cardiac dysrhythmias, cerebral hemorrhage, and pulmonary edema. Proper dosing of epinephrine to minimize these unwanted effects is clearly essential.

[^{11}C]Epinephrine kinetics in the heart have been evaluated by PET [29]. Seven healthy control and 10 heart transplant patients were imaged. Metabolism of [^{11}C]epinephrine was rapid, with about 20% of the intact drug remaining at 20 min after injection. Retention of the drug in the myocardium was much higher than in the blood; the CL half-time from the heart is about 10 h. Myocardial CL was dramatically increased in transplant recipients; the CL half-life in these patients was ~4.0 h. Because [^{11}C]epinephrine accumulates in vesicular storage sites of adrenergic neurons, the difference in cardiac PK was attributed to cardiac denervation in the transplanted myocardium.

7.3.3.3 [^{11}C]Phenylephrine Phenylephrine is a sympathomimetic agent with properties similar to some of those of epinephrine. It is a powerful alpha-receptor agonist, with little activity at beta-receptors of the heart. It causes constriction of vascular beds and is used to treat hypotension associated with shock, hypersensitivity reactions, and drug-induced hypotension. It is also used topically as a decongestant and for inducing mydriasis in ophthalmology. Such adverse effects as hypertension, myocardial infarction, pulmonary edema, and pulmonary embolism have been reported with the use of phenylephrine.

PET studies of phenylephrine PK have been performed using [^{11}C]phenylephrine and [^{11}C]α-α-dideuterophenylephrine ([^{11}C]D2-phenylephrine) [30]. The rationale for this study was that because phenylephrine is transported into sympathetic nerves by the norepinephrine transporter (NET) and stored in vesicles, the drug is also metabolized by MAO. The reason for assessing the PK of [^{11}C]D2-phenylephrine was to identify the importance of MAO degradation on the overall PK of phenylephrine in the heart, since substitution of a deuterium at the α-carbon of an MAO substrate slows the rate of enzymatic metabolism. Paired dynamic PET studies of [^{11}C]phenylephrine and [^{11}C]D2-phenylephrine were performed in six healthy volunteers imaged for 60 min for each tracer. Initial myocardial uptake was similar for the versions of the drug, but [^{11}C]D2-phenylephrine cleared from the left ventricle 2.6 times more slowly than the nondeuterated tracer (CL half-times 155 ± 53 vs. 55 ± 10 min, respectively). The retention of activity in the myocardium was higher for the deuterated drug, but myocardial retention of [^{11}C]phenylephrine was also substantial. It was concluded that although MAO influenced the myocardial CL, the rate-limiting step in the elimination of the drug was the rate at which phenylephrine leaks from neuronal storage vesicles in the heart.

In an earlier study involving PET imaging of [^{11}C]phenylephrine in 14 healthy volunteers, it was found that myocardial localization is related to the number of neuronal NET and that blocking NET with oral desipramine results in dramatically reduced myocardial uptake [31]. It was further shown that correction of the blood data for metabolism is necessary for accurate measurements of neuronal retention of [^{11}C]phenylephrine in the heart.

7.4 DRUG DELIVERY SYSTEMS

The effect of the route of administration on the PK of drugs has also been evaluated by PET. PET can quantify the three-dimensional distribution of drugs as a function of time, which makes it ideally suited for such an application. PET radiopharmaceuticals are usually administered by i.v. injection for direct delivery

TABLE 7.3 PET Imaging of Drug Delivery Systems in Humans

Labeled drug	Route of administration	Therapeutic effect
[^{11}C]Vinpocetine [38]	Oral	Neuroprotectant
[^{18}F]Lomefloxacin [20]	Oral	Antibiotic
[^{11}C]Triamcinolone [55]	Nasal	Antiinflammatory
[^{11}C]Triamcinolone [56, 57]	Inhalation	Antiinflammatory
[^{11}C]Nicotine [64]	Inhalation	Recreational

into the blood. This route of drug administration is preferred for imaging applications since there is no loss of radioactive signal while drug absorption occurs, and measurement of drug distribution, metabolism, and elimination can commence immediately. When absorption of the drug from a drug delivery system is the key factor to evaluate, however, there is no reason why PET methodology cannot also be applied for this purpose. PET facilitates the noninvasive measurement of the tissue deposition of a drug from a delivery system, as well as the physiological fate of the drug after deposition. Table 7.3 lists examples of drug delivery systems that have been evaluated with the use of PET techniques.

7.4.1 [^{11}C]Vinpocetine

Vinpocetine is a neuroprotective drug used for treatment of cerebrovascular disease and Alzheimer's disease. The distribution of [^{11}C]vinpocetine in the whole body and within the brain has been examined with PET imaging of human subjects [32, 33]. Intravenously administered [^{11}C]vinpocetine was used for cerebral distribution studies in humans, and it was found that the drug localized in a heterogeneous pattern within the brain [32]. The brain tissues with the highest [^{11}C]vinpocetine concentrations correspond to those cerebral regions in which drug-induced increases in metabolism and perfusion occur.

Clinically, it is preferable to administer the drug orally rather than via i.v. injection due to convenience as well as patient safety concerns. Positron-emitting [^{11}C]vinpocetine was used with PET imaging to assess the PK of the drug after tablet ingestion [33]. Following administration of the oral formulation, the drug is rapidly absorbed into the bloodstream from the gastrointestinal tract and enters the brain within 10 min. It was evident from these studies that [^{11}C]vinpocetine and/or its metabolites are renally eliminated into the urine.

7.4.2 [^{18}F]Lomefloxacin

[^{18}F]Lomefloxacin-PET was used to image the PK of the drug following oral administration [16]. The reader is referred to the earlier discussion of antimicrobial drugs for details.

7.4.3 [^{11}C]Triamcinolone

Triamcinolone is a corticosteroid used to treat a variety of inflammatory conditions and is administered to patients via oral, parenteral, topical, ophthalmic, inhalation, or nasal routes. The drug is approved for the treatment of asthma and allergic rhinitis, for which it is given via inhalation or nasal spray. The efficacy of these drug delivery systems is dependent on achieving adequate drug concentration at the inflamed tissues.

PET imaging was used to evaluate the PK of triamcinolone for both routes of administration. Using [^{11}C]triamcinolone as a tracer, drug delivery was evaluated for a metered dose inhaler. PET imaging demonstrated that 10–15% of the inhaled drug was deposited in target airway regions and that the use of a spacer device significantly increased delivery of triamcinolone to target tissues [34]. The intrapulmonary kinetics of [^{11}C]triamcinolone were also studied over time, and it was seen that the drug distribution pattern changed from localization primarily in conducting airways early after inhalation to predominantly acinar airways after 47 min [35].

Regarding the nasal route of administration, PET has also been used with [^{11}C] triamcinolone to noninvasively examine drug PK. It was noted that following nasal inhalation, triamcinolone rapidly deposited onto target tissue, with penetration into sinuses and moderate redistribution [36]. PET is recommended as a unique tool for comparison of aerosol drug delivery devices that deliver drugs to pulmonary tissue via oral inhalation or nasal spray [37].

7.4.4 [^{11}C]Nicotine

When smoked, tobacco is a toxin rather than a drug and has no therapeutic benefit. However, nicotine is a popular recreational drug, and it is appropriate to discuss the PET studies that have been performed. [^{11}C]Nicotine in gaseous form has been studied by PET [38]. It was noted in these studies that 45% of the dose was delivered to the oral cavity and 10% to the esophagus, suggesting that a large portion of the inhaled drug is transferred to the stomach. Only 5% of the inhaled [^{11}C]nicotine was deposited into the lung tissue. These PK results demonstrate how systemic effects may arise from smoking tobacco products.

7.5 POSITRON-EMITTING DRUGS AS PET BIOMARKERS

It is apparent from the earlier discussion that in many cases, more sophisticated information than simple tissue distribution can be obtained from PET imaging of radiolabeled drugs. More detailed intelligence regarding the mechanism of action or the metabolism of radioactive drugs can be determined, which potentially means that the radiolabeled drug may be employed as a biomarker for noninvasive evaluation of its molecular target. Such an application is very valuable in nuclear medicine procedures, since the radiolabeled drug can be employed as a PET biomarker for noninvasive measurement of changes in the respective biological target. Such applications might include changes in the target during aging, disease, or therapeutic interventions.

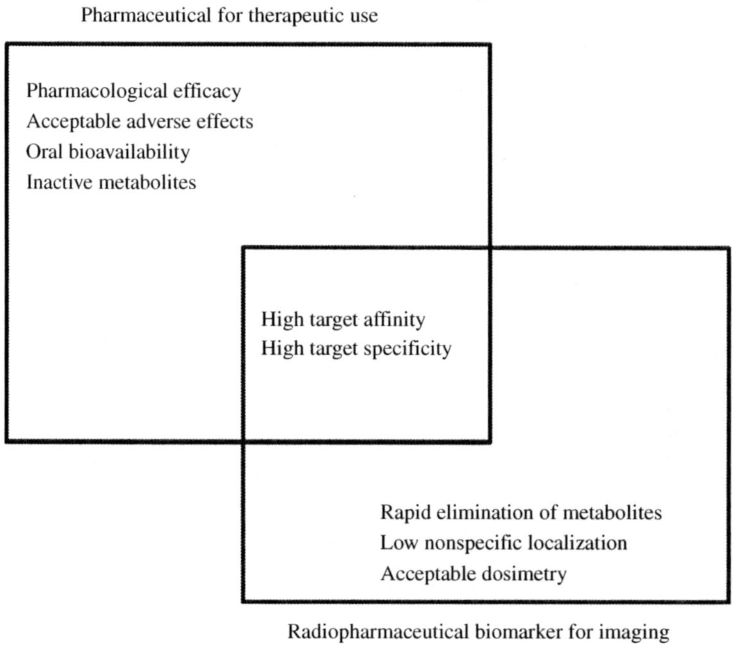

FIGURE 7.3 Characteristics of pharmaceuticals for therapeutic use and of imaging biomarkers. Note the overlapping characteristics that are common to both.

Before such applications can be realized, certain prerequisites must be met by the radiolabeled drug before it can be used as a radiopharmaceutical. Figure 7.3 demonstrates the idealized characteristics of a pharmaceutical for therapeutic use, those for an imaging biomarker, as well as common characteristics that overlap both. Characteristics that drugs and imaging biomarkers have in common are high affinity for binding to the target and specificity for particular targets. These characteristics foster potent, selective pharmacological action for a drug and generate images with high contrast and specificity in the case of a positron-emitting biomarker.

Despite these similarities, drugs and radiopharmaceuticals have very different requirements in order to be used in practice. A successful drug must demonstrate pharmacological efficacy and have acceptable adverse effects. In addition, oral bioavailability is preferred for ease of administration, and inactive metabolites improve the margin of safety for the patient. Depending on the pharmacological effect, rapid or slow CL of the drug from the blood compartment may have therapeutic benefit.

By contrast, for a radiopharmaceutical, oral bioavailability is not an advantageous characteristic, since they are used in a given patient only once or twice. Moreover, parenteral administration is preferred due to faster distribution of the tracer to the target *in vivo*. Low nonspecific binding by a biomarker is essential for image contrast and visualization of the target compared to the adjoining tissues. Note that for nonradioactive drugs, localization to nontarget sites is inconsequential, unless adverse

TABLE 7.4 Labeled Drugs as PET Biomarkers for Transporters

Labeled drug	Imaging target	Pharmacological effect	Therapeutic use [13, 14]
[^{11}C]Verapamil	Pgp	Ca channel antagonist	Arrhythmia; hypertension; angina
[^{11}C]Epinephrine	NET	α-/β-adrenergic agonist	Anaphylaxis; cardiac arrest; nasal; decongestant
[^{11}C]Phenylephrine	NET	α-/β-adrenergic agonist	Hypotension/shock; nasal congestion, hemorrhoids; ophthalmic procedures
[^{11}C]Cocaine	DAT	Local anesthetic	Topical anesthesia for procedures
[^{11}C]Methylphenidate	DAT	CNS stimulant	ADHD; narcolepsy
[^{11}C]Amphetamine	DAT	Anorexiant; stimulant	ADHD; exogenous obesity
[^{11}C]Tetrabenazine	VMAT	Monoamine-depleting agent	Huntington's disease

effects are induced in these tissues. Rapid elimination of radioactive metabolites (or no metabolism at all) derived from biomarkers is also an advantage, because it decreases nontarget-specific distribution of radioactivity and promotes a high signal-to-noise ratio. Finally, radiation dosimetry and associated limitations are critical aspects to consider for radiopharmaceuticals but play no role whatever in the use of traditional drugs.

Based on these divergent requirements, it is therefore not surprising that most clinically used drugs, when prepared in a radioactive form, do not necessarily achieve the requirements for a radiopharmaceutical. Referring to Figure 7.3, this would imply that the drug has characteristics listed for a drug, as well as those for a radiopharmaceutical. Such an occurrence would be unexpected and fortuitous.

There are, nonetheless, situations where positron-emitting drugs have demonstrated utility as imaging biomarkers. These PET biomarkers created from drugs are listed in Tables 7.4, 7.5, and 7.6, which are categorized according to type of imaging target. There are biomarkers listed that target transporters (Table 7.4), neurotransmitter receptors (Table 7.5), and enzymes (Table 7.6).

In some circumstances, the clinical indications (Tables 7.1 and 7.2) correspond to the imaging biomarker application, when the mechanism of action of the drug is taken into account. Thus, the psychiatric drugs listed in Table 7.2 could be predicted to be useful for imaging the targets in Table 7.5, because these are the same pharmacological targets that are used for pharmacological effects. In other cases, the imaging applications take totally unexpected directions from what might be predicted from the therapeutic use of the drug. An example of this is [^{11}C]verapamil, which is used clinically a cardiovascular drug to treat arrhythmias and hypertension (Table 7.1) but is used as a PET biomarker to study the Pgp drug efflux pump (Table 7.4).

TABLE 7.5 Labeled Drugs as PET Biomarkers for Receptors

Labeled drug	Receptor	Pharmacological effect	Therapeutic useC [13, 14]
[^{11}C]Flumazenil	BZD	BZD antagonist	Reversal of sedation; BZD toxicity
[^{11}C]Carfentanil	μ-Opioid	Opiate agonist	Animal tranquilizer
[^{11}C]Diprenorphine	Opioid	Opiate antagonist	Reversal of animal tranquilization
[^{11}C]Nicotine	nACh	Smoking cessation aid	Tobacco cessation
[^{11}C]Benztropine	mACh	Anticholinergic	Parkinson's disease; dystonia
[^{11}C]Mirtazapine	α_2	α_2-Adrenergic antagonist	Depression; PTSD
[^{11}C]Doxepin	H_1	Histamine H_1 antagonist	Depression; anxiety; pruritus
[^{11}C]Pyrilamine	H_1	Histamine H_1 antagonist	Cold remedy

TABLE 7.6 Labeled Drugs as PET Biomarkers for Enzymes

Labeled drug	Imaging target	Pharmacological effect	Therapeutic use [13, 14]
[^{11}C]Clorgyline	MAO-A	MAO-A inhibitor	Depression
[^{11}C]Deprenyl	MAO-B	MAO-B inhibitor	Parkinson's disease; depression
[^{11}C]Physostigmine	AChE	AChE inhibitor	Atropine reversal
[^{11}C]Donepezil	AChE	AChE inhibitor	Alzheimer's disease
[^{11}C]Rolipram	PDE4	PDE4 inhibitor	Depression

7.5.1 Drug Biomarkers for Transporters

There are several positron-emitting drugs that serve as biomarkers for physiological transport systems. These are listed in Table 7.4 and include PET probes for such transporters as Pgp, NET, dopamine (DAT), and vesicular monoamine (VMAT). Note that some of the biomarkers listed ([^{11}C]verapamil, [^{11}C]epinephrine, [^{11}C]phenylephrine) have been discussed with regard to the PK of cardiovascular drugs. Also worthy of mention is that many of these drugs are biomarkers for the same target as for therapeutic use (cocaine, amphetamine, and methylphenidate for DAT; epinephrine and phenylephrine for NET). Although any of these biomarkers can be used for evaluation of the respective target, generally speaking, one of these will have better imaging characteristics relative to the others.

7.5.1.1 [^{11}C]Verapamil Pgp-mediated efflux is responsible for MDR in tissues and is an important consideration in the design of therapeutic regimens. Noninvasive measurements of Pgp function in various tissues are thus very valuable in estimating whether therapeutic concentrations of drug will be attained in target organs. In the course of the PK studies discussed earlier, it was noted that tumor uptake of [^{11}C]verapamil was low, which was attributed to Pgp-mediated efflux [28]. Subsequent research led to the development of methodology for analysis of Pgp with [^{11}C]verapamil-PET in tumors [39] and at the blood–brain barrier [40].

Drugs such as cyclosporine [40–42], clarithromycin [43], and tariquidar [44, 45] have been evaluated for modulation of Pgp efflux activity in human subjects using [^{11}C]verapamil as a PET biomarker. The methodology has also been applied with this positron-emitting drug to examine changes in Pgp efflux activity in various conditions, including normal aging [46, 47], genetic predisposition [48, 49], epilepsy [50, 51], Parkinson's disease [52, 53], and schizophrenia [54].

7.5.1.2 [^{11}C]Epinephrine As described earlier, [^{11}C]epinephrine has been used as a PET biomarker for noninvasive measurement of sympathetic innervation of the heart [29]. In addition, the labeled drug has been used to evaluate sympathetic reinnervation in heart transplant recipients, together with exercise performance and treatment with the beta (β)-adrenergic antagonist drug propranolol [55].

7.5.1.3 [^{11}C]Phenylephrine Although phenylephrine is a biomarker for the NET, it is also a substrate for MAO. For this reason, [^{11}C]phenylephrine has been used with PET to examine the MAO-sensitive component of myocardial sympathetic neuronal innervations in humans [30, 31].

7.5.1.4 [^{11}C]Cocaine [^{11}C]Cocaine was the first, and is the most extensively studied, of the biomarkers that target the DAT. Cocaine binding sites were initially mapped in the human brain with [^{11}C]cocaine-PET [56, 57], and tracer kinetic modeling was developed for quantification of biomarker binding to cerebral DAT on a regional basis [58, 59]. Besides basal ganglia, which has the highest brain concentration of DAT, extrastriatal DAT can also be assessed by PET imaging of [^{11}C]cocaine [60].

Subsequent to these validation studies, [^{11}C]cocaine-PET was used to evaluate changes in DAT in human subjects with cocaine abuse [61, 62], alcohol intoxication [63], or HIV infection [64]. Methods were developed for using the biomarker to measure DAT occupancy in the brain [65], and these were applied for noninvasive evaluation of the pharmacological effect of the CNS stimulants cocaine [66] and methylphenidate [67, 68].

7.5.1.5 [^{11}C]Methylphenidate [^{11}C]Methylphenidate also binds to DAT in vivo [69, 70], and extrastriatal tracer kinetics for the biomarker have been compared to those of [^{11}C]cocaine [60]. PET imaging of the positron-emitting drug has been used to evaluate changes in human DAT during the aging process [71, 72], genetic predisposition to Parkinson's disease [73], as well as in patients with Parkinson's disease [74–77] and Tourette syndrome [78]. [^{11}C]Methylphenidate-PET has also been used to evaluate the pharmacological action of oral [79, 80] and i.v. [81] methylphenidate.

7.5.1.6 [^{11}C]Tetrabenazine Tetrabenazine is a unique drug biomarker that binds to the VMAT. VMAT differs from DAT in that it is an intracellular transport protein and unlike DAT is not located on the neuronal cell wall. [^{11}C]Tetrabenazine has been validated as a biomarker for VMAT in the human brain, and tracer kinetic modeling has been developed for dynamic PET imaging with high reproducibility

[82–84]. Extrastriatal VMAT density has been shown to be low, so that cortical areas of the brain can be used as a reference region [85].

The great value of [^{11}C]tetrabenazine as a drug biomarker is demonstrated by the large number of pathological conditions that have been studied with this labeled drug. Changes in VMAT with the normal human aging process have been investigated with this tracer [71, 86, 87]. [^{11}C]Tetrabenazine-PET has been used for examination of Parkinson's disease [88], genetically at-risk subjects [89], dystonia [90], Huntington's disease [91], familial paroxysmal choreoathetosis [92], multiple system atrophy [93, 94], Tourette syndrome [78, 95], and schizophrenia [96]. The biomarker can differentiate dementia with Lewy bodies with Parkinson's disease from Alzheimer's disease [97], and it has been shown that detailed kinetic analysis of [^{11}C]tetrabenazine alone provides more information than dual-tracer studies of the biomarker combined with metabolic measurements using [^{18}F]FDG [98]. Besides these PET studies of cerebral pathology, the positron-emitting drug is also useful for noninvasive measurements of VMAT in the pancreas [99]. This peripheral organ can be evaluated for changes in the transporter that are associated with diabetes pathology.

[^{11}C]Tetrabenazine-PET can also be used as a noninvasive tool for evaluation of pharmacological activity of drugs. Amphetamine [100], levodopa [101], and methamphetamine [102] pharmacology has been examined in human subjects using PET imaging of [^{11}C]tetrabenazine.

7.5.2 Drug Biomarkers for Receptors

Receptor binding sites are the target for numerous drugs. Some of the drugs that bind to these sites have suitable kinetics to be useful as biomarkers for PET imaging. The positron-emitting drugs that have such imaging properties are shown in Table 7.5. These PET probes are agonists/antagonists for the benzodiazepine (BZD), opioid, nicotinic acetylcholinergic (nACh), muscarinic acetylcholinergic (mACh), histamine-1 (H_1), and alpha-2 (α_2)-adrenergic receptors. Note that in some cases, there is more than one biomarker for the same target.

7.5.2.1 [^{11}C]Flumazenil Flumazenil is a BZD receptor antagonist that is used clinically for reversal of sedation or to treat BZD toxicity. BZD receptors are widespread throughout the brain and are associated with receptors for the inhibitory neurotransmitter gamma-aminobutyric acid (GABA).

[^{11}C]Flumazenil has been validated as a biomarker for PET measurements of cerebral BZD receptors [103–106]. In initial tracer kinetic analyses, the Scatchard approach with unlabeled flumazenil was used to derive measurements of receptor density B_{max} and dissociation rate constants K_d noninvasively with PET [103, 106]. Subsequent quantification efforts have involved several approaches including multicompartmental models, reference tissue methods, and parametric imaging [45, 107–125]. In tandem with these models, methods for analysis of radiometabolites in the plasma were developed for correction of input functions [126–129].

Because of the ubiquitous nature of GABA inhibition within the brain, deviations in BZD receptor binding occur with numerous pathological processes. Foremost among these is epilepsy, for which [^{11}C]flumazenil-PET has played an especially prominent role. Because GABA is homogeneously distributed throughout the brain and is an inhibitory neurotransmitter, epileptic foci appear as regions of decreased uptake when the cerebral distribution of the labeled drug is imaged. Virtually all types of epilepsy have been studied using [^{11}C]flumazenil-PET in adults [130–142] and sometimes in children as well [143–147]. The methodology has been compared to EEG, MRI, and metabolic measurements with [^{18}F]FDG-PET and found to be superior or complementary to these alternative methods [144, 147–158]. [^{11}C] Flumazenil-PET is useful for evaluation of intractable epilepsy and for pre- and post-surgical management of these patients [144, 146, 148, 159–162]. The PET biomarker has also been used to evaluate changes in BZD binding in subjects with genetic predisposition to epilepsy [163–166].

Another application for [^{11}C]flumazenil-PET is the evaluation of neuronal integrity. Because BZD receptors are only functional on living neurons, [^{11}C]flumazenil binding is a surrogate measurement for living tissue. Such PET measurements have been used for evaluation of infarct versus penumbra [167–174] associated with cardiovascular disease, ischemia, and stroke [175, 176]. The methodology has also been applied for examination of movement disorders [177–186], sclerosis of the brain [187–193], cerebral trauma [194–196], and the vegetative state of patients [197, 198]. [^{11}C]Flumazenil has been used with PET to investigate changes in cerebral BZD binding with visual disorders [199, 200], amputations [201], biochemical disturbances [202, 203], psychiatric conditions [204–207], and brain tumors [208].

[^{11}C]Flumazenil-PET has been used to noninvasively evaluate pharmacotherapy in human subjects for a variety of drugs. In these studies, changes in the PK of the biomarker were quantified to examine the interaction of the test drugs with the BZD receptor and estimate occupancy of the binding site. Benzodiazepines (flumazenil, midazolam) [103, 209, 210], the nonbenzodiazepine sedative zolpidem [211], anticonvulsants (vigabatrin, tiagabine) [212–214], ethanol [215–218] and drugs to treat alcoholism (disulfiram) [219], and anesthetics (propofol, sevoflurane, ketamine, xenon) [220–222] have all been evaluated for their pharmacological effect on BZD receptors in the living human brain.

7.5.2.2 [^{11}C]Carfentanil Carfentanil is a potent μ-opioid receptor agonist that is used in veterinary medicine as an animal tranquilizer. In the human, [^{11}C]carfentanil serves as a biomarker for PET measurement of cerebral opiate receptors using multicompartmental tracer kinetic modeling or reference tissue methods [223–225]. PET data can be corrected for partial volume effects using MRI [226–228], and high test–retest reproducibility has been demonstrated for PET measurements [229]. PET imaging of the biomarker binding to peripheral opioid receptors in cardiac tissue has been reported [230].

[^{11}C]Carfentanil-PET has been applied to the investigation of changes in cerebral opioid binding under various conditions. These include normal aging [231], pain [232, 233], epilepsy [234–236], cocaine abuse [237–239], alcoholism [240], and bulimia nervosa [241]. Besides these brain imaging studies, the biomarker has also been used for PET examination of opiate receptors located peripherally in lung carcinoma [242].

Beyond its utility for examination of pathology, [^{11}C]carfentanil-PET is also valuable for noninvasive study of human pharmacology. In control subjects, carfentanil [243], nalmefene [244], and traditional Chinese acupuncture [245] have been evaluated with the biomarker. The pharmacological action of naltrexone has been evaluated in alcoholics using [^{11}C]carfentanil-PET [246], and the technique has been used to study the effects of carfentanil [243] and buprenorphine [247] in substance abusers.

7.5.2.3 [^{11}C]Diprenorphine [^{11}C]Diprenorphine is a nonspecific opioid receptor antagonist that is used in veterinary medicine to reverse animal tranquilization. The labeled drug has been validated as a biomarker [248–250] and applied for noninvasive study of a number of disorders. Note that due to the nonspecific nature of receptor binding by this biomarker, its utility for pharmacological studies is limited. [^{11}C]Diprenorphine-PET has been used for study of human pain [251–257], epilepsy [234, 235, 258–261], movement disorders [262–265], and alcoholism [266]. The biomarker has also been applied for evaluation of pain therapies like acupuncture analgesia [267] and methadone administration [268].

7.5.2.4 [^{11}C]Nicotine Nicotine is a nACh receptor agonist used clinically as a smoking cessation aid. As discussed earlier, [^{11}C]nicotine has been used to evaluate the PK of gaseous nicotine [38], but as a biomarker it is administered intravenously. [^{11}C]Nicotine kinetics in the brain have been evaluated by PET imaging [269–271], and plasma radiometabolites generated from the biomarker have been measured [272].

[^{11}C]Nicotine-PET has been used to examine the brain of subjects with Alzheimer's disease [273, 274] and of tobacco smokers [275]. The biomarker has also been applied for evaluation of the effect of various drugs on nACh receptors in the brain. Tacrine [276–278], rivastigmine [279], galantamine [280], and nerve growth factor [281] have been studied using PET imaging of [^{11}C]nicotine in human subjects with Alzheimer's disease.

7.5.2.5 [^{11}C]Benztropine Benztropine is a mACh receptor antagonist that is used as an anticholinergic drug for the treatment of Parkinson's disease and dystonia. [^{11}C]Benztropine has been validated as a PET biomarker for mACh receptors [282]. The biomarker has been used in human subjects for noninvasive study of changes in brain mACh receptors in the normal aging process [283] and in Alzheimer's disease [277]. Human pharmacology experiments using [^{11}C]benztropine-PET have been used to study the effects of the Alzheimer's drug tacrine [277] and the sedative propofol [284].

7.5.2.6 [^{11}C]Mirtazapine Mirtazapine is an α2-adrenergic antagonist drug that is indicated for the treatment of depression and post-traumatic stress disorder (PTSD). The PK of [^{11}C]mirtazapine has been evaluated for peripheral organs [285] as well as within the brain [286]. It was used as a biomarker for cerebral α_2-adrenergic receptor binding in an investigation of nonresponder depressed patients [287].

[^{11}C]Mirtazapine-PET was also applied in a placebo-controlled PET study of receptor occupancy by unlabeled mirtazapine [288].

7.5.2.7 [^{11}C]Doxepin Doxepin is an H_1 receptor antagonist and tricyclic antidepressant used clinically to treat depression, anxiety, and pruritus. [^{11}C]Doxepin has been used with PET to visualize H_1 receptors in the brain [289], and tracer kinetic models and plasma metabolite analysis have been developed for quantification [290–292]. The biomarker has been used to evaluate changes in H_1 receptors in the human brain during normal aging [293], depression [294], and schizophrenia [295].

There has been extensive use of [^{11}C]doxepin-PET for evaluation of the human pharmacology of antihistamines. Drugs evaluated for H_1 receptor occupancy and CNS side effects are the first-generation (sedating) antihistamines chlorpheniramine [296–298], diphenhydramine [299, 300], and hydroxyzine [301] and the second-generation (nonsedating) antihistamines terfenadine [296], epinastine [297], ebastine [302], fexofenadine [303, 304], cetirizine [301, 303, 304], olopatadine [305, 306], ketotifen [305], bepotastine [299, 300], and loratadine [298].

7.5.2.8 [^{11}C]Pyrilamine Pyrilamine is an antihistamine (H_1 antagonist) commonly used in over-the-counter cold remedies for symptomatic relief of discharge associated with the common cold, rhinitis, and sinusitis. [^{11}C]Pyrilamine has been shown to be useful as a biomarker for PET measurement of H_1 receptor binding *in vivo* [307], and PET quantification methods have been developed [308]. Plasma metabolite analysis indicates that radiometabolites do not enter the brain and can be ignored in tracer kinetic models [308, 309]. Age-related decreases in H_1 receptor activity have been demonstrated in humans [293], but further application of the tracer as a biomarker has been largely supplanted by [^{11}C]doxepin, which has more favorable imaging characteristics.

7.5.3 Drug Biomarkers for Enzymes

Table 7.6 lists positron-emitting drugs that have been used as PET biomarkers for study of enzymes. These biomarkers target monoamine oxidase (MAO) (clorgyline, deprenyl), acetylcholinesterase (AChE) (physostigmine, donepezil), and phosphodiesterase (rolipram).

7.5.3.1 [^{11}C]Clorgyline MAO is an enzymatic target that has been extensively evaluated with PET biomarkers. It is a mitochondrial enzyme that deactivates several neurotransmitters in the CNS. MAO exists in isoforms A and B. MAO-A preferentially oxidizes serotonin and norepinephrine and is associated with depression and anxiety disorders. MAO-B preferentially oxidizes phenethylamine and is associated with neurodegenerative pathology like Parkinson's disease. Clorgyline selectively inhibits MAO-A, and L-deprenyl selectively inhibits MAO-B. These enzyme inhibitors, when labeled with carbon-11, provide a means for noninvasive study of changes in MAO activity caused by diet, disease, drugs, or genomics.

[^{11}C]Clorgyline and its deuterated analog [^{11}C]clorgyline-D2 have been used with PET to evaluate human MAO-A [310, 311]. For those organs or tissues in which binding of the tracer to MAO is the mechanism for localization, there is a robust isotope effect exhibited. In a PET study of the distribution of labeled clorgyline in peripheral organs, MAO activity in the lungs, kidney, thyroid, and heart could be demonstrated [310]. MAO-A activity in the liver could not be estimated, as there was no deuterium isotope effect for the tracer kinetics.

Within the brain, [^{11}C]clorgyline distributed in proportion to the distribution of MAO-A in an autopsied brain, and in a patient with the antidepressant MAO inhibitor phenelzine, there was a marked reduction in uptake [312]. The deuterium isotope effect was used to identify the MAO-A-specific and MAO-A-nonspecific uptake of tracer in the brain [313], and parametric imaging strategies have been developed [314].

[^{11}C]Clorgyline has been used for PET imaging studies of patients with schizophrenia [315]. It has also been used to noninvasively measure the effect of tobacco smoking on MAO-A activity in the brain [316], lungs [317], and peripheral organs [318]. The PET biomarker has also been used to evaluate whether the nontraditional drug ginkgo biloba extract affects MAO-A in the CNS [319].

7.5.3.2 L-[^{11}C]Deprenyl The positron-emitting biomarker has been used to measure MAO-B activity in the human brain [311, 312, 320, 321] as well as peripheral organs [322]. Studies have demonstrated that there is a deuterium isotope effect on the localization of L-[^{11}C]deprenyl [311, 321, 322].

L-[^{11}C]Deprenyl has been used to noninvasively evaluate changes in MAO-B activity in various pathological processes. These include the CNS conditions epilepsy [323–325], schizophrenia [315], amyotrophic lateral sclerosis (ALS) [326], and meningioma [327]. In the periphery, the PET biomarker has also found application in monitoring inflammation associated with rheumatoid arthritis [328].

The MAO-B biomarker has also been applied for PET evaluation of pharmacological or toxicological effects of substances. Reduction of MAO-B activity in the brain [329] and lungs [317] of chronic smokers has been demonstrated with L-[^{11}C]deprenyl-PET. Interestingly, nonsmokers smoking a single cigarette do not show this effect, suggesting that MAO-B inhibition by tobacco is a gradual and chronic condition [330]. Pharmacological effects of MAO-B inhibitor drugs have been noninvasively measured using L-[^{11}C]deprenyl in control subjects and patients with Parkinson's disease [331–333]. The tracer has also been used for PET evaluation of pharmacological effects of ginkgo biloba extract in human subjects [319].

7.5.3.3 [^{11}C]Donepezil Donepezil is an AChE inhibitor used for the treatment of dementia. [^{11}C]Donepezil binding in the brain was evaluated noninvasively in patients with Alzheimer's disease and elderly control subjects [334]. The biomarker distributed to brain regions rich in AChE such as striatum, thalamus, and cerebellum. Patients with Alzheimer's disease demonstrated reduced binding in proportion to disease severity, and the distribution volume of the labeled drug in hippocampus correlated with MMSE scores. Oral administration of donepezil (5 mg; 1 day before PET imaging) decreased the binding of [^{11}C]donepezil by over 60%. [^{11}C]Donepezil-PET is thus able to detect reduced tracer binding in early stages of disease and can be used to noninvasively measure AChE occupancy after orally administered donepezil.

Tracer kinetic models were developed for dynamic brain imaging of [^{11}C]donepezil using PET studies of control subjects [335]. The total distribution volume for the tracer using a one-compartment model, a two-compartment model, and Logan graphical analysis was compared. The rank order of the mean distribution volumes for 10 brain regions was in agreement with results from an independent postmortem evaluation, and correction for radiometabolites was not found to be essential for quantification.

7.5.3.4 [^{11}C]Physostigmine Like donepezil, physostigmine is also an AChE inhibitor, but in clinical practice, it is used to treat glaucoma or to reverse anticholinergic drug toxicity rather than treat dementia. [^{11}C]Physostigmine was evaluated by PET studies of control subjects [336]. The tracer is rapidly taken up by the brain and clears from the central compartment with a half-life of 20–35 min. Clearance rates were heterogeneous within the brain, and after 30 min, the cerebral distribution reflected local tissue concentrations of AChE. These results suggest that [^{11}C]physostigmine-PET may be a useful methodology for longitudinal measurement of changes in brain AChE with neurodegenerative disease.

7.5.3.5 [^{11}C]Rolipram Rolipram is an inhibitor of phosphodiesterase-4 (PDE4) and is under investigation as antidepressant. Whole-body PET imaging of [^{11}C] rolipram was performed in healthy adult human subjects, and time-activity curves were generated for major organs [337]. Imaging studies of the brain provide evidence that the tracer is suitable for measurement of PDE4 in the living brain and that a two-compartment model can be used to describe tracer kinetics [338].

7.6 CONCLUSIONS

The above information demonstrates the vast utility of PET for noninvasive assessment of PK information for small molecular-weight drugs. For those pharmaceuticals that can be labeled with positron-emitting nuclides without altering their molecular structure, PET provides a unique means for measurement of drug concentrations in deep organs as a function of time. Such information is particularly valuable in situations where tissue concentrations do not correspond to blood levels, such as infection, tumors, and centrally active drugs.

In addition to their use for PET measurement of drug PK, many positron-emitting drugs have properties that can be applied as biomarkers for different biological targets. This fortuitous circumstance occurs when the properties of the drug coincide with those of a radiopharmaceutical. These drug-based biomarkers can be used with PET for noninvasive evaluation of changes in transporters, receptors, or enzymes during pathological processes and therapeutic interventions. Such biomarkers can be used for surrogate measurements of drugs that cannot be synthesized with a positron-emitting nuclide, yet act at the same target as the labeled biomarker.

REFERENCES

1. Rowland M, Tozer TN. *Clinical Pharmacokinetics and Pharmacodynamics: Concepts and Applications*. 4th ed. Philadelphia: Wolters Kluwer Health/Lippincott William and Wilkins; 2011.
2. Phelps ME, Mazziotta JC, Schelbert HR. *Positron Emission Tomography and Autoradiography*. New York: Raven Press; 1986.

3. Stabin M. Nuclear medicine dosimetry. Phys Med Biol 2006;51:R187–R202.
4. Sgouros G, Frey E, Wahl R, et al. Three-dimensional imaging-based radiobiological dosimetry. Semin Nucl Med 2008;38:321–334.
5. Maziere B, Delforge J. Pharmacokinetic considerations in the PET and SPECT evaluation of CNS receptors. Q J Nucl Med 1997;41:119–126.
6. Logan J. Graphical analysis of PET data applied to reversible and irreversible tracers. Nucl Med Biol 2000;27:661–670.
7. Lammertsma AA. Radioligand studies: imaging and quantitative analysis. Eur Neuropsychopharmacol 2002;12:513–516.
8. Ichise M, Meyer JH, Yonekura Y. An introduction to PET and SPECT neuroreceptor quantification models. J Nucl Med 2001;42:755–763.
9. Heiss WD, Herholz K. Brain receptor imaging. J Nucl Med 2006;47:302–312.
10. Gunn RN, Gunn SR, Cunningham VJ. Positron emission tomography compartmental models. J Cereb Blood Flow Metab 2001;21:635–652.
11. Acton PD, Zhuang H, Alavi A. Quantification in PET. Radiol Clin North Am 2004;42:1055–1062, viii.
12. Price JC. Principles of tracer kinetic analysis. Neuroimaging Clin N Am 2003;13: 689–704.
13. Wollmer P, Pride NB, Rhodes CG, et al. Measurement of pulmonary erythromycin concentration in patients with lobar pneumonia by means of positron tomography. Lancet 1982;2:1361–1364.
14. Brunner M, Langer O, Dobrozemsky G, et al. [18F]Ciprofloxacin, a new positron emission tomography tracer for noninvasive assessment of the tissue distribution and pharmacokinetics of ciprofloxacin in humans. Antimicrob Agents Chemother 2004;48:3850–3857.
15. Fischman AJ, Babich JW, Bonab AA, et al. Pharmacokinetics of [18F]trovafloxacin in healthy human subjects studied with positron emission tomography. Antimicrob Agents Chemother 1998;42:2048–2054.
16. Tewson TJ, Yang D, Wong G, et al. The synthesis of fluorine-18 lomefloxacin and its preliminary use in human studies. Nucl Med Biol 1996;23:767–772.
17. Fischman AJ, Alpert NM, Livni E, et al. Pharmacokinetics of 18F-labeled fluconazole in healthy human subjects by positron emission tomography. Antimicrob Agents Chemother 1993;37:1270–1277.
18. Dimitrakopoulou A, Strauss LG, Clorius JH, et al. Studies with positron emission tomography after systemic administration of fluorine-18-uracil in patients with liver metastases from colorectal carcinoma. J Nucl Med 1993;34:1075–1081.
19. Kissel J, Brix G, Bellemann ME, et al. Pharmacokinetic analysis of 5-[18F]fluorouracil tissue concentrations measured with positron emission tomography in patients with liver metastases from colorectal adenocarcinoma. Cancer Res 1997;57:3415–3423.
20. Dimitrakopoulou-Strauss A, Strauss LG, Schlag P, et al. Intravenous and intra-arterial oxygen-15-labeled water and fluorine-18-labeled fluorouracil in patients with liver metastases from colorectal carcinoma. J Nucl Med 1998;39:465–473.
21. Dimitrakopoulou-Strauss A, Strauss LG, Schlag P, et al. Fluorine-18-fluorouracil to predict therapy response in liver metastases from colorectal carcinoma. J Nucl Med 1998;39:1197–1202.
22. Saleem A, Aboagye EO, Matthews JC, et al. Plasma pharmacokinetic evaluation of cytotoxic agents radiolabelled with positron emitting radioisotopes. Cancer Chemother Pharmacol 2008;61:865–873.
23. Harte RJ, Matthews JC, O'Reilly SM, et al. Sources of error in tissue and tumor measurements of 5-[18F]fluorouracil. J Nucl Med 1998;39:1370–1376.

24. Bading JR, Alauddin MM, Fissekis JD, et al. Blocking catabolism with eniluracil enhances PET studies of 5-[18F]fluorouracil pharmacokinetics. J Nucl Med 2000;41:1714–1724.
25. van der Veldt AA, Hendrikse NH, Smit EF, et al. Biodistribution and radiation dosimetry of 11C-labelled docetaxel in cancer patients. Eur J Nucl Med Mol Imaging 2010;37: 1950–1958.
26. Saleem A, Brown GD, Brady F, et al. Metabolic activation of temozolomide measured in vivo using positron emission tomography. Cancer Res 2003;63:2409–2415.
27. Solbach C, Patt M, Reimold M, et al. [11C]Vinblastine syntheses and preliminary imaging in cancer patients. J Pharm Pharm Sci 2007;10:266s–276s.
28. Hendrikse NH, de Vries EG, Franssen EJ, et al. In vivo measurement of [11C]verapamil kinetics in human tissues. Eur J Clin Pharmacol 2001;56:827–829.
29. Munch G, Nguyen NT, Nekolla S, et al. Evaluation of sympathetic nerve terminals with [(11)C]epinephrine and [(11)C]hydroxyephedrine and positron emission tomography. Circulation 2000;101:516–523.
30. Raffel DM, Corbett JR, del Rosario RB, et al. Sensitivity of [11C]phenylephrine kinetics to monoamine oxidase activity in normal human heart. J Nucl Med 1999;40:232–238.
31. Raffel DM, Corbett JR, del Rosario RB, et al. Clinical evaluation of carbon-11-phenylephrine: MAO-sensitive marker of cardiac sympathetic neurons. J Nucl Med 1996;37: 1923–1931.
32. Gulyas B, Halldin C, Sandell J, et al. PET studies on the brain uptake and regional distribution of [11C]vinpocetine in human subjects. Acta Neurol Scand 2002;106:325–332.
33. Gulyas B, Halldin C, Sovago J, et al. Drug distribution in man: a positron emission tomography study after oral administration of the labelled neuroprotective drug vinpocetine. Eur J Nucl Med Mol Imaging 2002;29:1031–1038.
34. Berridge MS, Lee Z, Heald DL. Pulmonary distribution and kinetics of inhaled [11C]triamcinolone acetonide. J Nucl Med 2000;41:1603–1611.
35. Lee Z, Berridge MS, Finlay WH, et al. Mapping PET-measured triamcinolone acetonide (TAA) aerosol distribution into deposition by airway generation. Int J Pharm 2000; 199:7–16.
36. Berridge MS, Heald DL, Muswick GJ, et al. Biodistribution and kinetics of nasal carbon-11-triamcinolone acetonide. J Nucl Med 1998;39:1972–1977.
37. Lee Z, Berridge MS. PET imaging-based evaluation of aerosol drugs and their delivery devices: nasal and pulmonary studies. IEEE Trans Med Imaging 2002;21:1324–1331.
38. Bergstrom M, Nordberg A, Lunell E, et al. Regional deposition of inhaled 11C-nicotine vapor in the human airway as visualized by positron emission tomography. Clin Pharmacol Ther 1995;57:309–317.
39. Hendrikse NH, Vaalburg W. Dynamics of multidrug resistance: P-glycoprotein analyses with positron emission tomography. Methods 2002;27:228–233.
40. Ikoma Y, Takano A, Ito H, et al. Quantitative analysis of 11C-verapamil transfer at the human blood-brain barrier for evaluation of P-glycoprotein function. J Nucl Med 2006;47:1531–1537.
41. Eyal S, Ke B, Muzi M, et al. Regional P-glycoprotein activity and inhibition at the human blood-brain barrier as imaged by positron emission tomography. Clin Pharmacol Ther 2010;87:579–585.
42. Muzi M, Mankoff DA, Link JM, et al. Imaging of cyclosporine inhibition of P-glycoprotein activity using 11C-verapamil in the brain: studies of healthy humans. J Nucl Med 2009;50:1267–1275.
43. Arakawa R, Ito H, Okumura M, et al. No inhibitory effect on P-glycoprotein function at blood-brain barrier by clinical dose of clarithromycin: a human PET study with [(11)C] verapamil. Ann Nucl Med 2010;24 (2):83–87.

44. Bauer M, Karch R, Neumann F, et al. Assessment of regional differences in tariquidar-induced P-glycoprotein modulation at the human blood-brain barrier. J Cereb Blood Flow Metab 2010;30:510–515.
45. Miederer I, Ziegler SI, Liedtke C, et al. Kinetic modelling of [11C]flumazenil using data-driven methods. Eur J Nucl Med Mol Imaging 2009;36:659–670.
46. Bartels AL, Kortekaas R, Bart J, et al. Blood-brain barrier P-glycoprotein function decreases in specific brain regions with aging: a possible role in progressive neurodegeneration. Neurobiol Aging 2009;30:1818–1824.
47. Bauer M, Karch R, Neumann F, et al. Age dependency of cerebral P-gp function measured with (R)-[11C]verapamil and PET. Eur J Clin Pharmacol 2009;65:941–946.
48. Brunner M, Langer O, Sunder-Plassmann R, et al. Influence of functional haplotypes in the drug transporter gene ABCB1 on central nervous system drug distribution in humans. Clin Pharmacol Ther 2005;78:182–190.
49. Takano A, Kusuhara H, Suhara T, et al. Evaluation of in vivo P-glycoprotein function at the blood-brain barrier among MDR1 gene polymorphisms by using 11C-verapamil. J Nucl Med 2006;47:1427–1433.
50. Langer O, Bauer M, Hammers A, et al. Pharmacoresistance in epilepsy: a pilot PET study with the P-glycoprotein substrate R-[(11)C]verapamil. Epilepsia 2007;48: 1774–1784.
51. Abrahim A, Luurtsema G, Bauer M, et al. Peripheral metabolism of (R)-[11C]verapamil in epilepsy patients. Eur J Nucl Med Mol Imaging 2008;35:116–123.
52. Bartels AL, Willemsen AT, Kortekaas R, et al. Decreased blood-brain barrier P-glycoprotein function in the progression of Parkinson's disease, PSP and MSA. J Neural Transm 2008;115:1001–1009.
53. Bartels AL, van Berckel BN, Lubberink M, et al. Blood-brain barrier P-glycoprotein function is not impaired in early Parkinson's disease. Parkinsonism Relat Disord 2008;14:505–508.
54. de Klerk OL, Willemsen AT, Bosker FJ, et al. Regional increase in P-glycoprotein function in the blood-brain barrier of patients with chronic schizophrenia: a PET study with [(11)C]verapamil as a probe for P-glycoprotein function. Psychiatry Res 2010;183:151–156.
55. Bengel FM, Ueberfuhr P, Karja J, et al. Sympathetic reinnervation, exercise performance and effects of beta-adrenergic blockade in cardiac transplant recipients. Eur Heart J 2004; 25:1726–1733.
56. Volkow ND, Fowler JS, Wolf AP, et al. Distribution and kinetics of carbon-11-cocaine in the human body measured with PET. J Nucl Med 1992;33:521–525.
57. Fowler JS, Volkow ND, Wolf AP, et al. Mapping cocaine binding sites in human and baboon brain in vivo. Synapse 1989;4:371–377.
58. Telang FW, Volkow ND, Levy A, et al. Distribution of tracer levels of cocaine in the human brain as assessed with averaged [11C]cocaine images. Synapse 1999;31: 290–296.
59. Logan J, Fowler JS, Volkow ND, et al. Graphical analysis of reversible radioligand binding from time-activity measurements applied to [N-11C-methyl]-(-)-cocaine PET studies in human subjects. J Cereb Blood Flow Metab 1990;10:740–747.
60. Wang GJ, Volkow ND, Fowler JS, et al. Comparison of two pet radioligands for imaging extrastriatal dopamine transporters in human brain. Life Sci 1995;57:PL187–PL191.
61. Wang GJ, Volkow ND, Fowler JS, et al. Cocaine abusers do not show loss of dopamine transporters with age. Life Sci 1997;61:1059–1065.
62. Logan J, Volkow ND, Fowler JS, et al. Concentration and occupancy of dopamine transporters in cocaine abusers with [11C]cocaine and PET. Synapse 1997;27:347–356.

63. Fowler JS, Volkow ND, Logan J, et al. Alcohol intoxication does not change [11C]cocaine pharmacokinetics in human brain and heart. Synapse 1992;12:228–235.

64. Chang L, Wang GJ, Volkow ND, et al. Decreased brain dopamine transporters are related to cognitive deficits in HIV patients with or without cocaine abuse. Neuroimage 2008;42: 869–878.

65. Fowler JS, Volkow ND, Wang GJ, et al. [(11)]Cocaine: PET studies of cocaine pharmacokinetics, dopamine transporter availability and dopamine transporter occupancy. Nucl Med Biol 2001;28:561–572.

66. Gatley SJ, Volkow ND, Gifford AN, et al. Model for estimating dopamine transporter occupancy and subsequent increases in synaptic dopamine using positron emission tomography and carbon-11-labeled cocaine. Biochem Pharmacol 1997;53:43–52.

67. Volkow ND, Wang GJ, Fowler JS, et al. Dopamine transporter occupancies in the human brain induced by therapeutic doses of oral methylphenidate. Am J Psychiatry 1998;155:1325–1331.

68. Volkow ND, Wang GJ, Fowler JS, et al. Blockade of striatal dopamine transporters by intravenous methylphenidate is not sufficient to induce self-reports of "high". J Pharmacol Exp Ther 1999;288:14–20.

69. Volkow ND, Ding YS, Fowler JS, et al. A new PET ligand for the dopamine transporter: studies in the human brain. J Nucl Med 1995;36:2162–2168.

70. Ding YS, Fowler JS, Volkow ND, et al. Chiral drugs: comparison of the pharmacokinetics of [11C]d-threo and L-threo-methylphenidate in the human and baboon brain. Psychopharmacology (Berl) 1997;131:71–78.

71. Troiano AR, Schulzer M, de la Fuente-Fernandez R, et al. Dopamine transporter PET in normal aging: dopamine transporter decline and its possible role in preservation of motor function. Synapse 2010;64:146–151.

72. Volkow ND, Ding YS, Fowler JS, et al. Dopamine transporters decrease with age. J Nucl Med 1996;37:554–559.

73. Adams JR, van Netten H, Schulzer M, et al. PET in LRRK2 mutations: comparison to sporadic Parkinson's disease and evidence for presymptomatic compensation. Brain 2005;128:2777–2785.

74. Lee CS, Samii A, Sossi V, et al. In vivo positron emission tomographic evidence for compensatory changes in presynaptic dopaminergic nerve terminals in Parkinson's disease. Ann Neurol 2000;47:493–503.

75. De La Fuente-Fernandez R, Lim AS, Sossi V, et al. Age and severity of nigrostriatal damage at onset of Parkinson's disease. Synapse 2003;47:152–158.

76. Breit S, Reimold M, Reischl G, et al. [(11)C]d-threo-methylphenidate PET in patients with Parkinson's disease and essential tremor. J Neural Transm 2006;113:187–193.

77. Troiano AR, de la Fuente-Fernandez R, Sossi V, et al. PET demonstrates reduced dopamine transporter expression in PD with dyskinesias. Neurology 2009;72:1211–1216.

78. Albin RL, Koeppe RA, Wernette K, et al. Striatal [11C]dihydrotetrabenazine and [11C] methylphenidate binding in Tourette syndrome. Neurology 2009;72:1390–1396.

79. Ding YS, Gatley SJ, Thanos PK, et al. Brain kinetics of methylphenidate (Ritalin) enantiomers after oral administration. Synapse 2004;53:168–175.

80. Volkow ND, Fowler JS, Wang G, et al. Mechanism of action of methylphenidate: insights from PET imaging studies. J Atten Disord 2002;6 Suppl 1:S31–S43.

81. Gatley SJ, Volkow ND, Gifford AN, et al. Dopamine-transporter occupancy after intravenous doses of cocaine and methylphenidate in mice and humans. Psychopharmacology (Berl) 1999;146:93–100.

82. Kilbourn MR, DaSilva JN, Frey KA, et al. In vivo imaging of vesicular monoamine transporters in human brain using [11C]tetrabenazine and positron emission tomography. J Neurochem 1993;60:2315–2318.

83. Koeppe RA, Frey KA, Vander Borght TM, et al. Kinetic evaluation of [11C]dihydrotetrabenazine by dynamic PET: measurement of vesicular monoamine transporter. J Cereb Blood Flow Metab 1996;16:1288–1299.

84. Chan GL, Holden JE, Stoessl AJ, et al. Reproducibility studies with 11C-DTBZ, a monoamine vesicular transporter inhibitor in healthy human subjects. J Nucl Med 1999;40:283–289.

85. Koeppe RA, Frey KA, Kuhl DE, et al. Assessment of extrastriatal vesicular monoamine transporter binding site density using stereoisomers of [11C]dihydrotetrabenazine. J Cereb Blood Flow Metab 1999;19:1376–1384.

86. Frey KA, Koeppe RA, Kilbourn MR, et al. Presynaptic monoaminergic vesicles in Parkinson's disease and normal aging. Ann Neurol 1996;40:873–884.

87. Bohnen NI, Albin RL, Koeppe RA, et al. Positron emission tomography of monoaminergic vesicular binding in aging and Parkinson disease. J Cereb Blood Flow Metab 2006;26:1198–1212.

88. de la Fuente-Fernandez R, Sossi V, McCormick S, et al. Visualizing vesicular dopamine dynamics in Parkinson's disease. Synapse 2009;63:713–716.

89. Hasegawa K, Stoessl AJ, Yokoyama T, et al. Familial parkinsonism: study of original Sagamihara PARK8 (I2020T) kindred with variable clinicopathologic outcomes. Parkinsonism Relat Disord 2009;15:300–306.

90. De La Fuente-Fernandez R, Furtado S, Guttman M, et al. VMAT2 binding is elevated in dopa-responsive dystonia: visualizing empty vesicles by PET. Synapse 2003;49: 20–28.

91. Bohnen NI, Koeppe RA, Meyer P, et al. Decreased striatal monoaminergic terminals in Huntington disease. Neurology 2000;54:1753–1759.

92. Bohnen NI, Albin RL, Frey KA, et al. (+)-alpha-[11C]Dihydrotetrabenazine PET imaging in familial paroxysmal dystonic choreoathetosis. Neurology 1999;52:1067–1069.

93. Gilman S, Chervin RD, Koeppe RA, et al. Obstructive sleep apnea is related to a thalamic cholinergic deficit in MSA. Neurology 2003;61:35–39.

94. Gilman S, Koeppe RA, Chervin RD, et al. REM sleep behavior disorder is related to striatal monoaminergic deficit in MSA. Neurology 2003;61:29–34.

95. Albin RL, Koeppe RA, Bohnen NI, et al. Increased ventral striatal monoaminergic innervation in Tourette syndrome. Neurology 2003;61:310–315.

96. Taylor SF, Koeppe RA, Tandon R, et al. In vivo measurement of the vesicular monoamine transporter in schizophrenia. Neuropsychopharmacology 2000;23:667–675.

97. Koeppe RA, Gilman S, Junck L, et al. Differentiating Alzheimer's disease from dementia with Lewy bodies and Parkinson's disease with (+)-[11C]dihydrotetrabenazine positron emission tomography. Alzheimers Dement 2008;4 (1 Suppl):S67–S76.

98. Koeppe RA, Gilman S, Joshi A, et al. 11C-DTBZ and 18F-FDG PET measures in differentiating dementias. J Nucl Med 2005;46:936–944.

99. Goland R, Freeby M, Parsey R, et al. 11C-dihydrotetrabenazine PET of the pancreas in subjects with long-standing type 1 diabetes and in healthy controls. J Nucl Med 2009;50:382–389.

100. Boileau I, Houle S, Rusjan PM, et al. Influence of a low dose of amphetamine on vesicular monoamine transporter binding: a PET (+)[11C]DTBZ study in humans. Synapse 2010;64:417–420.

101. Kumar A, Mann S, Sossi V, et al. [11C]DTBZ-PET correlates of levodopa responses in asymmetric Parkinson's disease. Brain 2003;126:2648–2655.

102. Boileau I, Rusjan P, Houle S, et al. Increased vesicular monoamine transporter binding during early abstinence in human methamphetamine users: is VMAT2 a stable dopamine neuron biomarker? J Neurosci 2008;28:9850–9856.

103. Pappata S, Samson Y, Chavoix C, et al. Regional specific binding of [11C]RO 15 1788 to central type benzodiazepine receptors in human brain: quantitative evaluation by PET. J Cereb Blood Flow Metab 1988;8:304–313.

104. Persson A, Ehrin E, Eriksson L, et al. Imaging of [11C]-labelled Ro 15-1788 binding to benzodiazepine receptors in the human brain by positron emission tomography. J Psychiatr Res 1985;19:609–622.

105. d'Argy R, Gillberg PG, Stalnacke CG, et al. In vivo and in vitro receptor autoradiography of the human brain using an 11C-labelled benzodiazepine analogue. Neurosci Lett 1988;85:304–310.

106. Abadie P, Baron JC, Bisserbe JC, et al. Central benzodiazepine receptors in human brain: estimation of regional Bmax and KD values with positron emission tomography. Eur J Pharmacol 1992;213:107–115.

107. Blomqvist G, Pauli S, Farde L, et al. Maps of receptor binding parameters in the human brain—a kinetic analysis of PET measurements. Eur J Nucl Med 1990;16: 257–265.

108. Iyo M, Itoh T, Yamasaki T, et al. Quantitative in vivo analysis of benzodiazepine binding sites in the human brain using positron emission tomography. Neuropharmacology 1991;30:207–215.

109. Lassen NA, Bartenstein PA, Lammertsma AA, et al. Benzodiazepine receptor quantification in vivo in humans using [11C]flumazenil and PET: application of the steady-state principle. J Cereb Blood Flow Metab 1995;15:152–165.

110. Holthoff VA, Koeppe RA, Frey KA, et al. Differentiation of radioligand delivery and binding in the brain: validation of a two-compartment model for [11C]flumazenil. J Cereb Blood Flow Metab 1991;11:745–752.

111. Delforge J, Pappata S, Millet P, et al. Quantification of benzodiazepine receptors in human brain using PET, [11C]flumazenil, and a single-experiment protocol. J Cereb Blood Flow Metab 1995;15:284–300.

112. Millet P, Delforge J, Mauguiere F, et al. Parameter and index images of benzodiazepine receptor concentration in the brain. J Nucl Med 1995;36:1462–1471.

113. Delforge J, Spelle L, Bendriem B, et al. Quantitation of benzodiazepine receptors in human brain using the partial saturation method. J Nucl Med 1996;37:5–11.

114. Pauli S, Sedvall G. Three-dimensional visualization and quantification of the benzodiazepine receptor population within a living human brain using PET and MRI. Eur Arch Psychiatry Clin Neurosci 1997;247:61–70.

115. Delforge J, Spelle L, Bendriem B, et al. Parametric images of benzodiazepine receptor concentration using a partial-saturation injection. J Cereb Blood Flow Metab 1997; 17:343–355.

116. Richardson MP, Friston KJ, Sisodiya SM, et al. Cortical grey matter and benzodiazepine receptors in malformations of cortical development. A voxel-based comparison of structural and functional imaging data. Brain 1997;120(Pt 11):1961–1973.

117. Millet P, Ibanez V, Delforge J, et al. Wavelet analysis of dynamic PET data: application to the parametric imaging of benzodiazepine receptor concentration. Neuroimage 2000; 11:458–472.

118. Mishina M, Senda M, Kimura Y, et al. Intrasubject correlation between static scan and distribution volume images for [11C]flumazenil PET. Ann Nucl Med 2000;14: 193–198.
119. Millet P, Graf C, Buck A, et al. Evaluation of the reference tissue models for PET and SPECT benzodiazepine binding parameters. Neuroimage 2002;17:928–942.
120. Sanabria-Bohorquez SM, Maes A, Dupont P, et al. Image-derived input function for [11C]flumazenil kinetic analysis in human brain. Mol Imaging Biol 2003;5:72–78.
121. Tsukamoto M, Katoh C, Shiga T, et al. Use of a standardized uptake value for parametric in vivo imaging of benzodiazepine receptor distribution on [11C]flumazenil brain PET. Eur J Nucl Med Mol Imaging 2004;31:846–851.
122. Mourik JE, Lubberink M, Klumpers UM, et al. Partial volume corrected image derived input functions for dynamic PET brain studies: methodology and validation for [11C] flumazenil. Neuroimage 2008;39:1041–1050.
123. Salmi E, Aalto S, Hirvonen J, et al. Measurement of GABAA receptor binding in vivo with [11C]flumazenil: a test-retest study in healthy subjects. Neuroimage 2008;41: 260–269.
124. Hammers A, Panagoda P, Heckemann RA, et al. [11C]Flumazenil PET in temporal lobe epilepsy: do we need an arterial input function or kinetic modeling? J Cereb Blood Flow Metab 2008;28:207–216.
125. Niimura K, Muzik O, Chugani DC, et al. [11C]flumazenil PET: activity images versus parametric images for the detection of neocortical epileptic foci. J Nucl Med 1999;40: 1985–1991.
126. Debruyne D, Abadie P, Barre L, et al. Plasma pharmacokinetics and metabolism of the benzodiazepine antagonist [11C] Ro 15-1788 (flumazenil) in baboon and human during positron emission tomography studies. Eur J Drug Metab Pharmacokinet 1991; 16: 141–152.
127. Ishiwata K, Itou T, Ohyama M, et al. Metabolite analysis of [11C]flumazenil in human plasma: assessment as the standardized value for quantitative PET studies. Ann Nucl Med 1998;12:55–59.
128. Laven M, Appel L, Moulder R, et al. Determination of flumazenil in human plasma by liquid chromatography-electrospray ionisation tandem mass spectrometry. J Chromatogr B Analyt Technol Biomed Life Sci 2004;808:221–227.
129. van Rij CM, Huitema AD, Swart EL, et al. Population plasma pharmacokinetics of 11C-flumazenil at tracer concentrations. Br J Clin Pharmacol 2005;60:477–485.
130. Savic I, Widen L, Thorell JO, et al. Cortical benzodiazepine receptor binding in patients with generalized and partial epilepsy. Epilepsia 1990;31:724–730.
131. Savic I, Pauli S, Thorell JO, et al. In vivo demonstration of altered benzodiazepine receptor density in patients with generalised epilepsy. J Neurol Neurosurg Psychiatry 1994;57:797–804.
132. Prevett MC, Lammertsma AA, Brooks DJ, et al. Benzodiazepine-GABAA receptor binding during absence seizures. Epilepsia 1995;36:592–599.
133. Savic I, Thorell JO, Roland P. [11C]flumazenil positron emission tomography visualizes frontal epileptogenic regions. Epilepsia 1995;36:1225–1232.
134. Burdette DE, Sakurai SY, Henry TR, et al. Temporal lobe central benzodiazepine binding in unilateral mesial temporal lobe epilepsy. Neurology 1995;45:934–941.
135. Prevett MC, Lammertsma AA, Brooks DJ, et al. Benzodiazepine-GABAA receptors in idiopathic generalized epilepsy measured with [11C]flumazenil and positron emission tomography. Epilepsia 1995;36:113–121.

136. Richardson MP, Koepp MJ, Brooks DJ, et al. Benzodiazepine receptors in focal epilepsy with cortical dysgenesis: an 11C-flumazenil PET study. Ann Neurol 1996;40: 188–198.
137. Savic I, Svanborg E, Thorell JO. Cortical benzodiazepine receptor changes are related to frequency of partial seizures: a positron emission tomography study. Epilepsia 1996;37:236–244.
138. Richardson MP, Koepp MJ, Brooks DJ, et al. 11C-flumazenil PET in neocortical epilepsy. Neurology 1998;51:485–492.
139. Juhasz C, Nagy F, Watson C, et al. Glucose and [11C]flumazenil positron emission tomography abnormalities of thalamic nuclei in temporal lobe epilepsy. Neurology 1999;53:2037–2045.
140. Hammers A, Koepp MJ, Labbe C, et al. Neocortical abnormalities of [11C]-flumazenil PET in mesial temporal lobe epilepsy. Neurology 2001;56:897–906.
141. Szelies B, Sobesky J, Pawlik G, et al. Impaired benzodiazepine receptor binding in perilesional cortex of patients with symptomatic epilepsies studied by [(11)C]-flumazenil PET. Eur J Neurol 2002;9:137–142.
142. Bouvard S, Costes N, Bonnefoi F, et al. Seizure-related short-term plasticity of benzodiazepine receptors in partial epilepsy: a [11C]flumazenil-PET study. Brain 2005;128: 1330–1343.
143. Nagy F, Chugani DC, Juhasz C, et al. Altered in vitro and in vivo flumazenil binding in human epileptogenic neocortex. J Cereb Blood Flow Metab 1999;19:939–947.
144. Muzik O, da Silva EA, Juhasz C, et al. Intracranial EEG versus flumazenil and glucose PET in children with extratemporal lobe epilepsy. Neurology 2000;54:171–179.
145. Chugani DC, Muzik O, Juhasz C, et al. Postnatal maturation of human GABAA receptors measured with positron emission tomography. Ann Neurol 2001;49:618–626.
146. Csaba J. Positron emission tomography in presurgical localization of epileptic foci. Ideggyogy Sz 2003;56:249–254.
147. Juhasz C, Chugani DC, Muzik O, et al. Electroclinical correlates of flumazenil and fluorodeoxyglucose PET abnormalities in lesional epilepsy. Neurology 2000;55:825–835.
148. Juhasz C, Chugani DC, Muzik O, et al. Relationship of flumazenil and glucose PET abnormalities to neocortical epilepsy surgery outcome. Neurology 2001;56: 1650–1658.
149. Savic I, Ingvar M, Stone-Elander S. Comparison of [11C]flumazenil and [18F]FDG as PET markers of epileptic foci. J Neurol Neurosurg Psychiatry 1993;56:615–621.
150. Szelies B, Weber-Luxenburger G, Pawlik G, et al. MRI-guided flumazenil- and FDG-PET in temporal lobe epilepsy. Neuroimage 1996;3:109–118.
151. Koepp MJ, Richardson MP, Labbe C, et al. 11C-flumazenil PET, volumetric MRI, and quantitative pathology in mesial temporal lobe epilepsy. Neurology 1997;49: 764–773.
152. Ryvlin P, Bouvard S, Le Bars D, et al. Clinical utility of flumazenil-PET versus [18F] fluorodeoxyglucose-PET and MRI in refractory partial epilepsy. A prospective study in 100 patients. Brain 1998;121(Pt 11):2067–2081.
153. Juhasz C, Nagy F, Muzik O, et al. [11C]Flumazenil PET in patients with epilepsy with dual pathology. Epilepsia 1999;40:566–574.
154. Szelies B, Weber-Luxenburger G, Mielke R, et al. Interictal hippocampal benzodiazepine receptors in temporal lobe epilepsy: comparison with coregistered hippocampal metabolism and volumetry. Eur J Neurol 2000;7:393–400.

155. Koepp MJ, Hammers A, Labbe C, et al. 11C-flumazenil PET in patients with refractory temporal lobe epilepsy and normal MRI. Neurology 2000;54:332–339.

156. Lamusuo S, Pitkanen A, Jutila L, et al. [11 C]Flumazenil binding in the medial temporal lobe in patients with temporal lobe epilepsy: correlation with hippocampal MR volumetry, T2 relaxometry, and neuropathology. Neurology 2000;54:2252–2260.

157. Hammers A, Koepp MJ, Hurlemann R, et al. Abnormalities of grey and white matter [11C]flumazenil binding in temporal lobe epilepsy with normal MRI. Brain 2002;125: 2257–2271.

158. Hammers A, Koepp MJ, Richardson MP, et al. Grey and white matter flumazenil binding in neocortical epilepsy with normal MRI. A PET study of 44 patients. Brain 2003;126: 1300–1318.

159. Savic I, Blomqvist G, Halldin C, et al. Regional increases in [11C]flumazenil binding after epilepsy surgery. Acta Neurol Scand 1998;97:279–286.

160. Padma MV, Simkins R, White P, et al. Clinical utility of 11C-flumazenil positron emission tomography in intractable temporal lobe epilepsy. Neurol India 2004;52: 457–462.

161. Ohta Y, Nariai T, Ishii K, et al. Voxel- and ROI-based statistical analyses of PET parameters for guidance in the surgical treatment of intractable mesial temporal lobe epilepsy. Ann Nucl Med 2008;22:495–503.

162. Juhasz C, Asano E, Shah A, et al. Focal decreases of cortical GABAA receptor binding remote from the primary seizure focus: what do they indicate? Epilepsia 2009;50: 240–250.

163. Fedi M, Berkovic SF, Marini C, et al. A GABAA receptor mutation causing generalized epilepsy reduces benzodiazepine receptor binding. Neuroimage 2006;32: 995–1000.

164. Turner MR, Osei-Lah AD, Hammers A, et al. Abnormal cortical excitability in sporadic but not homozygous D90A SOD1 ALS. J Neurol Neurosurg Psychiatry 2005;76:1279–1285.

165. Turner MR, Hammers A, Al-Chalabi A, et al. Distinct cerebral lesions in sporadic and 'D90A' SOD1 ALS: studies with [11C]flumazenil PET. Brain 2005;128:1323–1329.

166. Holopainen IE, Metsahonkala EL, Kokkonen H, et al. Decreased binding of [11C]flumazenil in Angelman syndrome patients with GABA(A) receptor beta3 subunit deletions. Ann Neurol 2001;49:110–113.

167. Dohmen C, Bosche B, Graf R, et al. Prediction of malignant course in MCA infarction by PET and microdialysis. Stroke 2003;34:2152–2158.

168. Kuroda S, Shiga T, Ishikawa T, et al. Reduced blood flow and preserved vasoreactivity characterize oxygen hypometabolism due to incomplete infarction in occlusive carotid artery diseases. J Nucl Med 2004;45:943–949.

169. Heiss WD, Sobesky J, Smekal U, et al. Probability of cortical infarction predicted by flumazenil binding and diffusion-weighted imaging signal intensity: a comparative positron emission tomography/magnetic resonance imaging study in early ischemic stroke. Stroke 2004;35:1892–1898.

170. Heiss WD, Sobesky J, Hesselmann V. Identifying thresholds for penumbra and irreversible tissue damage. Stroke 2004;35:2671–2674.

171. Yamauchi H, Kudoh T, Kishibe Y, et al. Selective neuronal damage and borderzone infarction in carotid artery occlusive disease: a 11C-flumazenil PET study. J Nucl Med 2005;46:1973–1979.

172. Kuroda S, Shiga T, Houkin K, et al. Cerebral oxygen metabolism and neuronal integrity in patients with impaired vasoreactivity attributable to occlusive carotid artery disease. Stroke 2006;37:393–398.

173. Guadagno JV, Jones PS, Aigbirhio FI, et al. Selective neuronal loss in rescued penumbra relates to initial hypoperfusion. Brain 2008;131:2666–2678.

174. Heiss WD, Sobesky J. Comparison of PET and DW/PW-MRI in acute ischemic stroke. Keio J Med 2008;57:125–131.

175. Heiss WD, Kracht L, Grond M, et al. Early [(11)C]Flumazenil/H(2)O positron emission tomography predicts irreversible ischemic cortical damage in stroke patients receiving acute thrombolytic therapy. Stroke 2000;31:366–369.

176. Heiss WD, Kracht LW, Thiel A, et al. Penumbral probability thresholds of cortical flumazenil binding and blood flow predicting tissue outcome in patients with cerebral ischaemia. Brain 2001;124:20–29.

177. Chavoix C, Samson Y, Pappata S, et al. Positron emission tomography study of brain benzodiazepine receptors in Friedreich's ataxia. Can J Neurol Sci 1990;17: 404–409.

178. Holthoff VA, Koeppe RA, Frey KA, et al. Positron emission tomography measures of benzodiazepine receptors in Huntington's disease. Ann Neurol 1993;34:76–81.

179. Mielke R, Hilker R, Weber-Luxenburger G, et al. Early-onset cerebellar ataxia (EOCA) with retained reflexes: reduced cerebellar benzodiazepine-receptor binding, progressive metabolic and cognitive impairment. Mov Disord 1998;13:739–745.

180. Kunig G, Leenders KL, Sanchez-Pernaute R, et al. Benzodiazepine receptor binding in Huntington's disease: [11C]flumazenil uptake measured using positron emission tomography. Ann Neurol 2000;47:644–648.

181. Foster NL, Minoshima S, Johanns J, et al. PET measures of benzodiazepine receptors in progressive supranuclear palsy. Neurology 2000;54:1768–1773.

182. Pradat PF, Dupel-Pottier C, Lacomblez L, et al. Case report of pallido-pyramidal disease with supplementary motor area involvement. Mov Disord 2001;16:762–764.

183. Perani D, Garibotto V, Moresco RM, et al. PET evidence of central GABAergic changes in stiff-person syndrome. Mov Disord 2007;22:1030–1033.

184. Ihara M, Tomimoto H, Ishizu K, et al. Association of vascular parkinsonism with impaired neuronal integrity in the striatum. J Neural Transm 2007;114:577–584.

185. Galldiks N, Thiel A, Haense C, et al. 11C-flumazenil positron emission tomography demonstrates reduction of both global and local cerebral benzodiazepine receptor binding in a patient with Stiff Person Syndrome. J Neurol 2008;255:1361–1364.

186. Boecker H, Weindl A, Brooks DJ, et al. GABAergic dysfunction in essential tremor: an 11C-flumazenil PET study. J Nucl Med 2010;51:1030–1035.

187. Koepp MJ, Richardson MP, Brooks DJ, et al. Cerebral benzodiazepine receptors in hippocampal sclerosis. An objective in vivo analysis. Brain 1996;119 (Pt 5): 1677–1687.

188. Koepp MJ, Labbe C, Richardson MP, et al. Regional hippocampal [11C]flumazenil PET in temporal lobe epilepsy with unilateral and bilateral hippocampal sclerosis. Brain 1997;120 (Pt 10):1865–1876.

189. Koepp MJ, Hand KS, Labbe C, et al. In vivo [11C]flumazenil-PET correlates with ex vivo [3H]flumazenil autoradiography in hippocampal sclerosis. Ann Neurol 1998;43 :618–626.

190. Lloyd CM, Richardson MP, Brooks DJ, et al. Extramotor involvement in ALS: PET studies with the GABA(A) ligand [(11)C]flumazenil. Brain 2000;123 (Pt 11): 2289–2296.

191. Hammers A, Koepp MJ, Brooks DJ, et al. Periventricular white matter flumazenil binding and postoperative outcome in hippocampal sclerosis. Epilepsia 2005;46: 944–948.

192. Turner MR, Hammers A, Al-Chalabi A, et al. Cortical involvement in four cases of primary lateral sclerosis using [(11)C]-flumazenil PET. J Neurol 2007;254:1033–1036.

193. Wicks P, Turner MR, Abrahams S, et al. Neuronal loss associated with cognitive performance in amyotrophic lateral sclerosis: an (11C)-flumazenil PET study. Amyotroph Lateral Scler 2008;9:43–49.

194. Kawai N, Maeda Y, Kudomi N, et al. Focal neuronal damage in patients with neuropsychological impairment after diffuse traumatic brain injury: evaluation using ^{11}C-flumazenil positron emission tomography with statistical image analysis. J Neurotrauma 2010;27 (12):2131–2138.

195. Niimura K, Chugani DC, Muzik O, et al. Cerebellar reorganization following cortical injury in humans: effects of lesion size and age. Neurology 1999;52:792–797.

196. Shiga T, Ikoma K, Katoh C, et al. Loss of neuronal integrity: a cause of hypometabolism in patients with traumatic brain injury without MRI abnormality in the chronic stage. Eur J Nucl Med Mol Imaging 2006;33:817–822.

197. Rudolf J, Sobesky J, Ghaemi M, et al. The correlation between cerebral glucose metabolism and benzodiazepine receptor density in the acute vegetative state. Eur J Neurol 2002;9:671–677.

198. Beuthien-Baumann B, Holthoff VA, Rudolf J. Functional imaging of vegetative state applying single photon emission tomography and positron emission tomography. Neuropsychol Rehabil 2005;15:276–282.

199. Mishina M, Senda M, Kiyosawa M, et al. Benzodiazepine receptor distribution and cerebral blood flow in early blindness—a PET study. Keio J Med 2000;49 Suppl 1: A114–A116.

200. Suzuki Y, Horie C, Kiyosawa M, et al. Measurement of the 11C-flumazenil binding in the visual cortex predicts the prognosis of hemianopia. J Neurol Sci 2008;268:102–107.

201. Capaday C, Richardson MP, Rothwell JC, et al. Long-term changes of GABAergic function in the sensorimotor cortex of amputees. A combined magnetic stimulation and 11C-flumazenil PET study. Exp Brain Res 2000;133:552–556.

202. Macdonald GA, Frey KA, Agranoff BW, et al. Cerebral benzodiazepine receptor binding in vivo in patients with recurrent hepatic encephalopathy. Hepatology 1997;26: 277–282.

203. Pearl PL, Gibson KM, Quezado Z, et al. Decreased GABA-A binding on FMZ-PET in succinic semialdehyde dehydrogenase deficiency. Neurology 2009;73:423–429.

204. Malizia AL, Cunningham VJ, Bell CJ, et al. Decreased brain GABA(A)-benzodiazepine receptor binding in panic disorder: preliminary results from a quantitative PET study. Arch Gen Psychiatry 1998;55:715–720.

205. Abadie P, Boulenger JP, Benali K, et al. Relationships between trait and state anxiety and the central benzodiazepine receptor: a PET study. Eur J Neurosci 1999;11: 1470–1478.

206. Geuze E, van Berckel BN, Lammertsma AA, et al. Reduced GABAA benzodiazepine receptor binding in veterans with post-traumatic stress disorder. Mol Psychiatry 2008;13 (1):74–83,3.

207. Klumpers UM, Veltman DJ, Drent ML, et al. Reduced parahippocampal and lateral temporal GABAA-[11C]flumazenil binding in major depression: preliminary results. Eur J Nucl Med Mol Imaging37:565–574.

208. Richardson MP, Hammers A, Brooks DJ, et al. Benzodiazepine-GABA(A) receptor binding is very low in dysembryoplastic neuroepithelial tumor: a PET study. Epilepsia 2001;42:1327–1334.

209. Shinotoh H, Iyo M, Yamada T, et al. Detection of benzodiazepine receptor occupancy in the human brain by positron emission tomography. Psychopharmacology (Berl) 1989;99:202–207.

210. Malizia AL, Gunn RN, Wilson SJ, et al. Benzodiazepine site pharmacokinetic/pharmacodynamic quantification in man: direct measurement of drug occupancy and effects on the human brain in vivo. Neuropharmacology 1996;35:1483–1491.

211. Abadie P, Rioux P, Scatton B, et al. Central benzodiazepine receptor occupancy by zolpidem in the human brain as assessed by positron emission tomography. Eur J Pharmacol 1996;295:35–44.

212. Weber OM, Verhagen A, Duc CO, et al. Effects of vigabatrin intake on brain GABA activity as monitored by spectrally edited magnetic resonance spectroscopy and positron emission tomography. Magn Reson Imaging 1999;17:417–425.

213. Frankle WG, Cho RY, Narendran R, et al. Tiagabine increases [11C]flumazenil binding in cortical brain regions in healthy control subjects. Neuropsychopharmacology 2009;34:624–633.

214. Juhasz C, Muzik O, Chugani DC, et al. Prolonged vigabatrin treatment modifies developmental changes of GABA(A)-receptor binding in young children with epilepsy. Epilepsia 2001;42:1320–1326.

215. Pauli S, Liljequist S, Farde L, et al. PET analysis of alcohol interaction with the brain disposition of [11C]flumazenil. Psychopharmacology (Berl) 1992;107:180–185.

216. Litton JE, Neiman J, Pauli S, et al. PET analysis of [11C]flumazenil binding to benzodiazepine receptors in chronic alcohol-dependent men and healthy controls. Psychiatry Res 1993;50:1–13.

217. Farde L, Pauli S, Litton JE, et al. PET-determination of benzodiazepine receptor binding in studies on alcoholism. EXS 1994;71:143–153.

218. Lingford-Hughes AR, Wilson SJ, Cunningham VJ, et al. GABA-benzodiazepine receptor function in alcohol dependence: a combined 11C-flumazenil PET and pharmacodynamic study. Psychopharmacology (Berl) 2005;180:595–606.

219. Gilman S, Adams KM, Johnson-Greene D, et al. Effects of disulfiram on positron emission tomography and neuropsychological studies in severe chronic alcoholism. Alcohol Clin Exp Res 1996;20:1456–1461.

220. Salmi E, Kaisti KK, Metsahonkala L, et al. Sevoflurane and propofol increase 11C-flumazenil binding to gamma-aminobutyric acid A receptors in humans. Anesth Analg 2004;99:1420–1426.

221. Salmi E, Langsjo JW, Aalto S, et al. Subanesthetic ketamine does not affect 11C-flumazenil binding in humans. Anesth Analg 2005;101:722–725.

222. Salmi E, Laitio RM, Aalto S, et al. Xenon does not affect gamma-aminobutyric acid type A receptor binding in humans. Anesth Analg 2008;106:129–134.

223. Frost JJ, Douglass KH, Mayberg HS, et al. Multicompartmental analysis of [11C]-carfentanil binding to opiate receptors in humans measured by positron emission tomography. J Cereb Blood Flow Metab 1989;9:398–409.

224. Endres CJ, Bencherif B, Hilton J, et al. Quantification of brain mu-opioid receptors with [11C]carfentanil: reference-tissue methods. Nucl Med Biol 2003;30:177–186.

225. Joshi A, Fessler JA, Koeppe RA. Improving PET receptor binding estimates from Logan plots using principal component analysis. J Cereb Blood Flow Metab 2008;28: 852–865.

226. Meltzer CC, Leal JP, Mayberg HS, et al. Correction of PET data for partial volume effects in human cerebral cortex by MR imaging. J Comput Assist Tomogr 1990;14:561–570.

227. Meltzer CC, Zubieta JK, Links JM, et al. MR-based correction of brain PET measurements for heterogeneous gray matter radioactivity distribution. J Cereb Blood Flow Metab 1996;16:650–658.

228. Bencherif B, Stumpf MJ, Links JM, et al. Application of MRI-based partial-volume correction to the analysis of PET images of mu-opioid receptors using statistical parametric mapping. J Nucl Med 2004;45:402–408.

229. Hirvonen J, Aalto S, Hagelberg N, et al. Measurement of central mu-opioid receptor binding in vivo with PET and [11C]carfentanil: a test-retest study in healthy subjects. Eur J Nucl Med Mol Imaging 2009;36:275–286.

230. Villemagne PS, Dannals RF, Ravert HT, et al. PET imaging of human cardiac opioid receptors. Eur J Nucl Med Mol Imaging 2002;29:1385–1388.

231. Zubieta JK, Dannals RF, Frost JJ. Gender and age influences on human brain mu-opioid receptor binding measured by PET. Am J Psychiatry 1999;156:842–848.

232. Bencherif B, Fuchs PN, Sheth R, et al. Pain activation of human supraspinal opioid pathways as demonstrated by [11C]-carfentanil and positron emission tomography (PET). Pain 2002;99:589–598.

233. Scott DJ, Stohler CS, Koeppe RA, et al. Time-course of change in [11C]carfentanil and [11C]raclopride binding potential after a nonpharmacological challenge. Synapse 2007;61:707–714.

234. Mayberg HS, Sadzot B, Meltzer CC, et al. Quantification of mu and non-mu opiate receptors in temporal lobe epilepsy using positron emission tomography. Ann Neurol 1991;30:3–11.

235. Bartenstein PA, Prevett MC, Duncan JS, et al. Quantification of opiate receptors in two patients with mesiobasal temporal lobe epilepsy, before and after selective amygdalohippocampectomy, using positron emission tomography. Epilepsy Res 1994;18:119–125.

236. Madar I, Lesser RP, Krauss G, et al. Imaging of delta- and mu-opioid receptors in temporal lobe epilepsy by positron emission tomography. Ann Neurol 1997;41: 358–367.

237. Zubieta JK, Gorelick DA, Stauffer R, et al. Increased mu opioid receptor binding detected by PET in cocaine-dependent men is associated with cocaine craving. Nat Med 1996;2:1225–1229.

238. Gorelick DA, Kim YK, Bencherif B, et al. Imaging brain mu-opioid receptors in abstinent cocaine users: time course and relation to cocaine craving. Biol Psychiatry 2005;57:1573–1582.

239. Gorelick DA, Kim YK, Bencherif B, et al. Brain mu-opioid receptor binding: relationship to relapse to cocaine use after monitored abstinence. Psychopharmacology (Berl) 2008;200:475–486.

240. Heinz A, Reimold M, Wrase J, et al. Correlation of stable elevations in striatal mu-opioid receptor availability in detoxified alcoholic patients with alcohol craving: a positron emission tomography study using carbon 11-labeled carfentanil. Arch Gen Psychiatry 2005;62:57–64.

241. Bencherif B, Guarda AS, Colantuoni C, et al. Regional mu-opioid receptor binding in insular cortex is decreased in bulimia nervosa and correlates inversely with fasting behavior. J Nucl Med 2005;46:1349–1351.

242. Madar I, Bencherif B, Lever J, et al. Imaging delta- and mu-opioid receptors by PET in lung carcinoma patients. J Nucl Med 2007;48:207–213.

243. Minkowski CP, Epstein D, Frost JJ, et al. Differential response to IV carfentanil in chronic cocaine users and healthy controls. Addict Biol 2012;17 (1):149–155.

244. Ingman K, Hagelberg N, Aalto S, et al. Prolonged central mu-opioid receptor occupancy after single and repeated nalmefene dosing. Neuropsychopharmacology 2005;30: 2245–2253.

245. Harris RE, Zubieta JK, Scott DJ, et al. Traditional Chinese acupuncture and placebo (sham) acupuncture are differentiated by their effects on mu-opioid receptors (MORs). Neuroimage 2009;47:1077–1085.

246. Weerts EM, Kim YK, Wand GS, et al. Differences in delta- and mu-opioid receptor blockade measured by positron emission tomography in naltrexone-treated recently abstinent alcohol-dependent subjects. Neuropsychopharmacology 2008;33: 653–665.

247. Greenwald MK, Johanson CE, Moody DE, et al. Effects of buprenorphine maintenance dose on mu-opioid receptor availability, plasma concentrations, and antagonist blockade in heroin-dependent volunteers. Neuropsychopharmacology 2003;28: 2000–2009.

248. Jones AK, Cunningham VJ, Ha-Kawa SK, et al. Quantitation of [11C]diprenorphine cerebral kinetics in man acquired by PET using presaturation, pulse-chase and tracer-only protocols. J Neurosci Methods 1994;51:123–134.

249. Schadrack J, Willoch F, Platzer S, et al. Opioid receptors in the human cerebellum: evidence from [11C]diprenorphine PET, mRNA expression and autoradiography. Neuroreport 1999;10:619–624.

250. Hammers A, Asselin MC, Turkheimer FE, et al. Balancing bias, reliability, noise properties and the need for parametric maps in quantitative ligand PET: [(11)C]diprenorphine test-retest data. Neuroimage 2007;38:82–94.

251. Jones AK, Cunningham VJ, Ha-Kawa S, et al. Changes in central opioid receptor binding in relation to inflammation and pain in patients with rheumatoid arthritis. Br J Rheumatol 1994;33:909–916.

252. Willoch F, Tolle TR, Wester HJ, et al. Central pain after pontine infarction is associated with changes in opioid receptor binding: a PET study with 11C-diprenorphine. Am J Neuroradiol 1999;20:686–690.

253. Jones AK, Watabe H, Cunningham VJ, et al. Cerebral decreases in opioid receptor binding in patients with central neuropathic pain measured by [11C]diprenorphine binding and PET. Eur J Pain 2004;8:479–485.

254. Willoch F, Schindler F, Wester HJ, et al. Central poststroke pain and reduced opioid receptor binding within pain processing circuitries: a [11C]diprenorphine PET study. Pain 2004;108:213–220.

255. Sprenger T, Willoch F, Miederer M, et al. Opioidergic changes in the pineal gland and hypothalamus in cluster headache: a ligand PET study. Neurology 2006;66: 1108–1110.

256. Maarrawi J, Peyron R, Mertens P, et al. Motor cortex stimulation for pain control induces changes in the endogenous opioid system. Neurology 2007;69:827–834.

257. Maarrawi J, Peyron R, Mertens P, et al. Differential brain opioid receptor availability in central and peripheral neuropathic pain. Pain 2007;127:183–194.

258. Prevett MC, Cunningham VJ, Brooks DJ, et al. Opiate receptors in idiopathic generalised epilepsy measured with [11C]diprenorphine and positron emission tomography. Epilepsy Res 1994;19:71–77.

259. Koepp MJ, Hansen ML, Pressler RM, et al. Comparison of EEG, MRI and PET in reading epilepsy: a case report. Epilepsy Res 1998;29:251–257.

260. Koepp MJ, Richardson MP, Brooks DJ, et al. Focal cortical release of endogenous opioids during reading-induced seizures. Lancet 1998;352:952–955.

261. Hammers A, Asselin MC, Hinz R, et al. Upregulation of opioid receptor binding following spontaneous epileptic seizures. Brain 2007;130:1009–1016.

262. Burn DJ, Rinne JO, Quinn NP, et al. Striatal opioid receptor binding in Parkinson's disease, striatonigral degeneration and Steele-Richardson-Olszewski syndrome, A [11C] diprenorphine PET study. Brain 1995;118 (Pt 4):951–958.

263. Piccini P, Weeks RA, Brooks DJ. Alterations in opioid receptor binding in Parkinson's disease patients with levodopa-induced dyskinesias. Ann Neurol 1997;42: 720–726.

264. Whone AL, Von Spiczak S, Edwards M, et al. Opioid binding in DYT1 primary torsion dystonia: an 11C-diprenorphine PET study. Mov Disord 2004;19:1498–1503.

265. von Spiczak S, Whone AL, Hammers A, et al. The role of opioids in restless legs syndrome: an [11C]diprenorphine PET study. Brain 2005;128:906–917.

266. Williams TM, Davies SJ, Taylor LG, et al. Brain opioid receptor binding in early abstinence from alcohol dependence and relationship to craving: an [11C]diprenorphine PET study. Eur Neuropsychopharmacol 2009;19:740–748.

267. Dougherty DD, Kong J, Webb M, et al. A combined [11C]diprenorphine PET study and fMRI study of acupuncture analgesia. Behav Brain Res 2008;193:63–68.

268. Melichar JK, Hume SP, Williams TM, et al. Using [11C]diprenorphine to image opioid receptor occupancy by methadone in opioid addiction: clinical and preclinical studies. J Pharmacol Exp Ther 2005;312:309–315.

269. Halldin C, Nagren K, Swahn CG, et al. (S)- and (R)-[11C]nicotine and the metabolite (R/S)-[11C]cotinine. Preparation, metabolite studies and in vivo distribution in the human brain using PET. Int J Rad Appl Instrum B 1992;19:871–880.

270. Yokoi F, Komiyama T, Ito T, et al. Application of carbon-11 labelled nicotine in the measurement of human cerebral blood flow and other physiological parameters. Eur J Nucl Med 1993;20:46–52.

271. Muzic RF, Jr., Berridge MS, Friedland RP, et al. PET quantification of specific binding of carbon-11-nicotine in human brain. J Nucl Med 1998;39:2048–2054.

272. Cumming P, Yokoi F, Chen A, et al. Pharmacokinetics of radiotracers in human plasma during positron emission tomography. Synapse 1999;34:124–134.

273. Nordberg A, Lundqvist H, Hartvig P, et al. Kinetic analysis of regional (S)(-)11C-nicotine binding in normal and Alzheimer brains--in vivo assessment using positron emission tomography. Alzheimer Dis Assoc Disord 1995;9:21–27.

274. Kadir A, Almkvist O, Wall A, et al. PET imaging of cortical 11C-nicotine binding correlates with the cognitive function of attention in Alzheimer's disease. Psychopharmacology (Berl) 2006;188:509–520.

275. Rose JE, Mukhin AG, Lokitz SJ, et al. Kinetics of brain nicotine accumulation in dependent and nondependent smokers assessed with PET and cigarettes containing 11C-nicotine. Proc Natl Acad Sci USA 2010;107:5190–5195.

276. Nordberg A. Effect of long-term treatment with tacrine (THA) in Alzheimer's disease as visualized by PET. Acta Neurol Scand Suppl 1993;149:62–65.

277. Nordberg A, Lundqvist H, Hartvig P, et al. Imaging of nicotinic and muscarinic receptors in Alzheimer's disease: effect of tacrine treatment. Dement Geriatr Cogn Disord 1997;8:78–84.

278. Nordberg A, Lilja A, Lundqvist H, et al. Tacrine restores cholinergic nicotinic receptors and glucose metabolism in Alzheimer patients as visualized by positron emission tomography. Neurobiol Aging 1992;13:747–758.

279. Kadir A, Darreh-Shori T, Almkvist O, et al. Changes in brain 11C-nicotine binding sites in patients with mild Alzheimer's disease following rivastigmine treatment as assessed by PET. Psychopharmacology (Berl) 2007;191:1005–1014.

280. Kadir A, Darreh-Shori T, Almkvist O, et al. PET imaging of the in vivo brain acetylcholinesterase activity and nicotine binding in galantamine-treated patients with AD. Neurobiol Aging 2008;29:1204–1217.

281. Olson L, Nordberg A, von Holst H, et al. Nerve growth factor affects 11C-nicotine binding, blood flow, EEG, and verbal episodic memory in an Alzheimer patient (case report). J Neural Transm Park Dis Dement Sect 1992;4:79–95.

282. Dewey SL, MacGregor RR, Brodie JD, et al. Mapping muscarinic receptors in human and baboon brain using [N-11C-methyl]-benztropine. Synapse 1990;5: 213–223.

283. Dewey SL, Volkow ND, Logan J, et al. Age-related decreases in muscarinic cholinergic receptor binding in the human brain measured with positron emission tomography (PET). J Neurosci Res 1990;27:569–575.

284. Xie G, Gunn RN, Dagher A, et al. PET quantification of muscarinic cholinergic receptors with [N-11C-methyl]-benztropine and application to studies of propofol-induced unconsciousness in healthy human volunteers. Synapse 2004;51:91–101.

285. Marthi K, Hansen SB, Jakobsen S, et al. Biodistribution and radiation dosimetry of [N-methyl-11C]mirtazapine, an antidepressant affecting adrenoceptors. Appl Radiat Isot 2003;59:175–179.

286. Marthi K, Jakobsen S, Bender D, et al. [N-methyl-11C]Mirtazapine for positron emission tomography neuroimaging of antidepressant actions in humans. Psychopharmacology (Berl) 2004;174:260–265.

287. Smith DF, Stork BS, Wegener G, et al. [11C]Mirtazapine binding in depressed antidepressant nonresponders studied by PET neuroimaging. Psychopharmacology (Berl) 2009;206:133–140.

288. Smith DF, Stork BS, Wegener G, et al. Receptor occupancy of mirtazapine determined by PET in healthy volunteers. Psychopharmacology (Berl) 2007;195:131–138.

289. Yanai K, Watanabe T, Yokoyama H, et al. Histamine H1 receptors in human brain visualized in vivo by [11C]doxepin and positron emission tomography. Neurosci Lett 1992;137:145–148.

290. Mochizuki H, Kimura Y, Ishii K, et al. Quantitative measurement of histamine H(1) receptors in human brains by PET and [11C]doxepin. Nucl Med Biol 2004;31: 165–171.

291. Mochizuki H, Kimura Y, Ishii K, et al. Simplified PET measurement for evaluating histamine H1 receptors in human brains using [11C]doxepin. Nucl Med Biol 2004;31: 1005–1011.

292. Ishiwata K, Yanai K, Iwata R, et al. Analysis of plasma metabolites during human PET-studies with three receptor ligands, [11C]YM-09151-2, [11C]doxepin and [11C] pyrilamine. Tohoku J Exp Med 1996;178:129–136.

293. Yanai K, Watanabe T, Meguro K, et al. Age-dependent decrease in histamine H1 receptor in human brains revealed by PET. Neuroreport 1992;3:433–436.

294. Kano M, Fukudo S, Tashiro A, et al. Decreased histamine H1 receptor binding in the brain of depressed patients. Eur J Neurosci 2004;20:803–810.

295. Iwabuchi K, Ito C, Tashiro M, et al. Histamine H1 receptors in schizophrenic patients measured by positron emission tomography. Eur Neuropsychopharmacol 2005;15: 185–191.

296. Yanai K, Ryu JH, Watanabe T, et al. Histamine H1 receptor occupancy in human brains after single oral doses of histamine H1 antagonists measured by positron emission tomography. Br J Pharmacol 1995;116:1649–1655.

297. Yanai K, Ryu JH, Watanabe T, et al. Positron emission tomographic study of central histamine H1-receptor occupancy in human subjects treated with epinastine, a second-generation antihistamine. Methods Find Exp Clin Pharmacol 1995;17 Suppl C: 64–69.

298. Kubo N, Senda M, Ohsumi Y, et al. Brain histamine H1 receptor occupancy of loratadine measured by positron emission topography: comparison of H1 receptor occupancy and proportional impairment ratio. Hum Psychopharmacol 2011;26 (2):133–139.

299. Tashiro M, Duan X, Kato M, et al. Brain histamine H1 receptor occupancy of orally administered antihistamines, bepotastine and diphenhydramine, measured by PET with 11C-doxepin. Br J Clin Pharmacol 2008;65:811–821.

300. Zhang D, Tashiro M, Shibuya K, et al. Next-day residual sedative effect after night-time administration of an over-the-counter antihistamine sleep aid, diphenhydramine, measured by positron emission tomography. J Clin Psychopharmacol 2010;30: 694–701.

301. Tashiro M, Kato M, Miyake M, et al. Dose dependency of brain histamine H(1) receptor occupancy following oral administration of cetirizine hydrochloride measured using PET with [11C]doxepin. Hum Psychopharmacol 2009;24:540–548.

302. Tagawa M, Kano M, Okamura N, et al. Neuroimaging of histamine H1-receptor occupancy in human brain by positron emission tomography (PET): a comparative study of ebastine, a second-generation antihistamine, and (+)-chlorpheniramine, a classical antihistamine. Br J Clin Pharmacol 2001;52:501–509.

303. Tashiro M, Mochizuki H, Iwabuchi K, et al. Roles of histamine in regulation of arousal and cognition: functional neuroimaging of histamine H1 receptors in human brain. Life Sci 2002;72:409–414.

304. Tashiro M, Sakurada Y, Iwabuchi K, et al. Central effects of fexofenadine and cetirizine: measurement of psychomotor performance, subjective sleepiness, and brain histamine H1-receptor occupancy using 11C-doxepin positron emission tomography. J Clin Pharmacol 2004;44:890–900.

305. Tashiro M, Mochizuki H, Sakurada Y, et al. Brain histamine H receptor occupancy of orally administered antihistamines measured by positron emission tomography with (11) C-doxepin in a placebo-controlled crossover study design in healthy subjects: a comparison of olopatadine and ketotifen. Br J Clin Pharmacol 2006;61:16–26.

306. Senda M, Kubo N, Adachi K, et al. Cerebral histamine H1 receptor binding potential measured with PET under a test dose of olopatadine, an antihistamine, is reduced after repeated administration of olopatadine. J Nucl Med 2009;50:887–892.

307. Yanai K, Watanabe T, Yokoyama H, et al. Mapping of histamine H1 receptors in the human brain using [11C]pyrilamine and positron emission tomography. J Neurochem 1992;59:128–136.

308. Kim SE, Szabo Z, Seki C, et al. Effect of tracer metabolism on PET measurement of [11C]pyrilamine binding to histamine H1 receptors. Ann Nucl Med 1999;13:101–107.

309. Szabo Z, Ravert HT, Gozukara I, et al. Noncompartmental and compartmental modeling of the kinetics of carbon-11 labeled pyrilamine in the human brain. Synapse 1993;15:263–275.

310. Fowler JS, Logan J, Wang GJ, et al. Monoamine oxidase A imaging in peripheral organs in healthy human subjects. Synapse 2003;49:178–187.

311. Fowler JS, Logan J, Wang GJ, et al. Comparison of the binding of the irreversible monoamine oxidase tracers, [(11)C]clorgyline and [(11)C]l-deprenyl in brain and peripheral organs in humans. Nucl Med Biol 2004;31:313–319.

312. Fowler JS, MacGregor RR, Wolf AP, et al. Mapping human brain monoamine oxidase A and B with 11C-labeled suicide inactivators and PET. Science 1987;235:481–485.

313. Fowler JS, Logan J, Ding YS, et al. Non-MAO A binding of clorgyline in white matter in human brain. J Neurochem 2001;79:1039–1046.

314. Logan J, Fowler JS, Ding YS, et al. Strategy for the formation of parametric images under conditions of low injected radioactivity applied to PET studies with the irreversible monoamine oxidase A tracers [11C]clorgyline and deuterium-substituted [11C]clorgyline. J Cereb Blood Flow Metab 2002;22:1367–1376.

315. Fowler CJ, Carlsson A, Winblad B. Monoamine oxidase-A and -B activities in the brain stem of schizophrenics and non-schizophrenic psychotics. J Neural Transm 1981;52: 23–32.

316. Fowler JS, Volkow ND, Wang GJ, et al. Brain monoamine oxidase A inhibition in cigarette smokers. Proc Natl Acad Sci USA 1996;93:14065–14069.

317. Logan J, Fowler JS. Evidence for reduced arterial plasma input, prolonged lung retention and reduced lung monoamine oxidase in smokers. Nucl Med Biol 2005;32:521–529.

318. Fowler JS, Logan J, Wang GJ, et al. Comparison of monoamine oxidase a in peripheral organs in nonsmokers and smokers. J Nucl Med 2005;46:1414–1420.

319. Fowler JS, Wang GJ, Volkow ND, et al. Evidence that gingko biloba extract does not inhibit MAO A and B in living human brain. Life Sci 2000;66:PL141–PL146.

320. Lammertsma AA, Bench CJ, Price GW, et al. Measurement of cerebral monoamine oxidase B activity using L-[11C]deprenyl and dynamic positron emission tomography. J Cereb Blood Flow Metab 1991;11:545–556.

321. Logan J, Fowler JS, Volkow ND, et al. Reproducibility of repeated measures of deuterium substituted [11C]L-deprenyl ([11C]L-deprenyl-D2) binding in the human brain. Nucl Med Biol 2000;27:43–49.

322. Fowler JS, Logan J, Wang GJ, et al. PET imaging of monoamine oxidase B in peripheral organs in humans. J Nucl Med 2002;43:1331–1338.

323. Kumlien E, Bergstrom M, Lilja A, et al. Positron emission tomography with [11C]deuterium-deprenyl in temporal lobe epilepsy. Epilepsia 1995;36:712–721.

324. Bergstrom M, Kumlien E, Lilja A, et al. Temporal lobe epilepsy visualized with PET with 11C-L-deuterium-deprenyl—analysis of kinetic data. Acta Neurol Scand 1998;98: 224–231.

325. Kumlien E, Nilsson A, Hagberg G, et al. PET with 11C-deuterium-deprenyl and 18F-FDG in focal epilepsy. Acta Neurol Scand 2001;103:360–366.

326. Johansson A, Engler H, Blomquist G, et al. Evidence for astrocytosis in ALS demonstrated by [11C](L)-deprenyl-D2 PET. J Neurol Sci 2007;255:17–22.

327. Bergstrom M, Muhr C, Jossan S, et al. Differentiation of pituitary adenoma and meningioma: visualization with positron emission tomography and [11C]-L-deprenyl. Neurosurgery 1992; 30:855–861.

328. Danfors T, Bergstrom M, Feltelius N, et al. Positron emission tomography with 11C-D-deprenyl in patients with rheumatoid arthritis. Evaluation of knee joint inflammation before and after intra-articular glucocorticoid treatment. Scand J Rheumatol 1997;26:43–48.

329. Fowler JS, Volkow ND, Wang GJ, et al. Neuropharmacological actions of cigarette smoke: brain monoamine oxidase B (MAO B) inhibition. J Addict Dis. 1998;17:23–34.

330. Fowler JS, Wang GJ, Volkow ND, et al. Smoking a single cigarette does not produce a measurable reduction in brain MAO B in non-smokers. Nicotine Tob Res 1999;1: 325–329.

331. Bench CJ, Price GW, Lammertsma AA, et al. Measurement of human cerebral monoamine oxidase type B (MAO-B) activity with positron emission tomography (PET): a dose ranging study with the reversible inhibitor Ro 19-6327. Eur J Clin Pharmacol 1991;40:169–173.

332. Fowler JS, Volkow ND, Logan J, et al. Monoamine oxidase B (MAO B) inhibitor therapy in Parkinson's disease: the degree and reversibility of human brain MAO B inhibition by Ro 19 6327. Neurology 1993;43:1984–1992.

333. Freedman NM, Mishani E, Krausz Y, et al. In vivo measurement of brain monoamine oxidase B occupancy by rasagiline, using (11)C-l-deprenyl and PET. J Nucl Med 2005;46:1618–1624.

334. Okamura N, Funaki Y, Tashiro M, et al. In vivo visualization of donepezil binding in the brain of patients with Alzheimer's disease. Br J Clin Pharmacol 2008;65:472–479.

335. Hiraoka K, Okamura N, Funaki Y, et al. Quantitative analysis of donepezil binding to acetylcholinesterase using positron emission tomography and [5-(11)C-methoxy]donepezil. Neuroimage 2009;46:616–623.

336. Pappata S, Tavitian B, Traykov L, et al. In vivo imaging of human cerebral acetylcholinesterase. J Neurochem 1996;67:876–879.

337. Sprague DR, Fujita M, Ryu YH, et al. Whole-body biodistribution and radiation dosimetry in monkeys and humans of the phosphodiesterase 4 radioligand [11C](R)-rolipram: comparison of two-dimensional planar, bisected and quadrisected image analyses. Nucl Med Biol 2008;35:493–500.

338. DaSilva JN, Lourenco CM, Meyer JH, et al. Imaging cAMP-specific phosphodiesterase-4 in human brain with R-[11C]rolipram and positron emission tomography. Eur J Nucl Med Mol Imaging 2002;29:1680–1683.

CHAPTER 8

IMAGING PEPTIDES, PROTEINS, AND ANTIBODY DELIVERY

FREDERIK L. GIESEL AND CLEMENS KRATOCHWIL

8.1 ANTIBODIES, PROTEINS, AND PEPTIDES

Antibodies offer the promise of highly selective binding and recognition of a variety of molecular targets such as receptors, adhesion molecules, and other cell surface molecules. Today, some antibodies such as trastuzumab (anti-HER2/neu) and bevacizumab (anti-VEGF) are already used in clinical applications. The idea is that the therapeutic efficacy of antibodies with specific binding but insufficient cytotoxic effect may be improved by labeling with therapeutic nuclides (i.e., beta-emitters), so-called radioimmunotherapy. Evaluating the tumor specificity of an identified target in comparison to the unspecific retention in normal organs, molecular imaging with labeled antibodies seems a promising tool to acquire dosimetric information for radioimmunoconjugates and biodistribution of cold antibodies.

However, a major drawback of antibodies as imaging probes is the prolonged retention in plasma and the fact that the target structures are addressed only slowly. For diagnostic imaging, examinations that last longer than one day constrain clinical application. In radioimmunotherapy, a long residence in blood causes high perfusion-dependent radiation to off-targeted organs such as bone marrow. Free glomeruale filtration and therefore fast plasma clearance is given for a molecular weight of less than 60 kDa. Therefore, instead of full antibodies, small molecules with specific target properties are preferable. However, the way from full antibodies to Fab fragments to single-chain variable fragments (scFv) or even short proteins or peptides is challenging. In comparison to antibodies, peptides are superior by faster plasma clearance and rapid tumor targeting. However, the major problem with peptide probes is their potentially fast degradation by plasma peptidases. This limitation can be overcome by replacing L-amino acids with D-amino acids at the ends of linear peptides or by a suitable quaternary structure, for example, a cyclic configuration.

Drug Delivery Applications of Noninvasive Imaging: Validation from Biodistribution to Sites of Action, First Edition. Edited by Chun Li and Mei Tian.

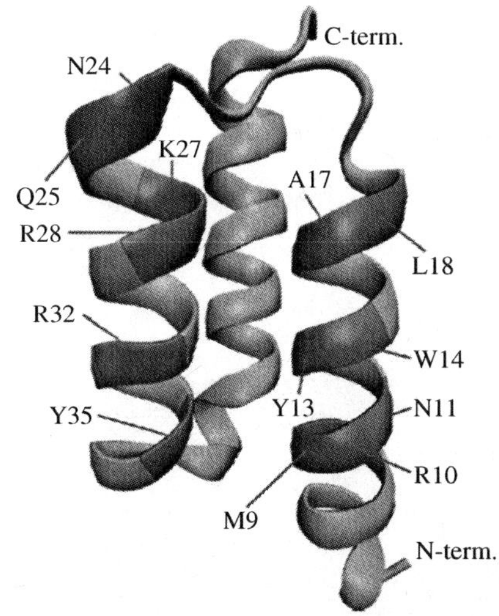

FIGURE 8.1 Three-dimensional structure of the affibody ZHer2:342. The structure was automatically modeled at Swiss Model Workspace 26 and visualized with visual molecular dynamics (VMD). Randomizable amino acids are painted in dark grey.

However, moving from antibody to *in vivo* stable protein has not been successfully realized until today. Another approach is the evaluation of antibody mimetics, which are a heterogeneous group of various artificial peptides or proteins with a molar mass of about 3–20 kDa stable toward plasma enzymes. A HER2/neu-specific affibody (Fig. 8.1) is currently under clinical development as an imaging agent for breast cancer [1].

Instead of starting with a complete antibodies and reducing the size of these molecules, another approach is to start directly with specific binding peptides such as peptide hormones. Several tumors overexpress receptors for these hormones and are therefore potential targets for targeted radiopeptide imaging or even therapy. Some promising probes are already identified and have been optimized in regard to their *in vivo* kinetics. Modifications include improved *in vivo* stability, higher receptor affinity, and accelerated plasma clearance by replacing amino acids that are not involved in the peptide/receptor interaction with more hydrophil alternatives. When linked to a chelator, these attributes are not allowed to change. While some of these probes are already sufficient for imaging, the ratio between tumor uptake and exposure of normal tissue needs to be further optimized for therapeutic application. Until today only radiolabeled somatostatin analogues such as DOTATOC or DOTATATE have reached clinical acceptance as PET diagnostics and agents for peptide receptor radiation therapy (PRRT).

8.1.1 Dynamic Analysis of Peptide Tracer Uptake Using DOTATOC

High accumulation of DOTATOC is normal for the pituitary gland, salivary gland, liver parenchyma, and spleen, but also in the kidney and (during the renal excretion) in the urinary bladder. Pathological somatostatin receptor subtype 2 (SSR2)

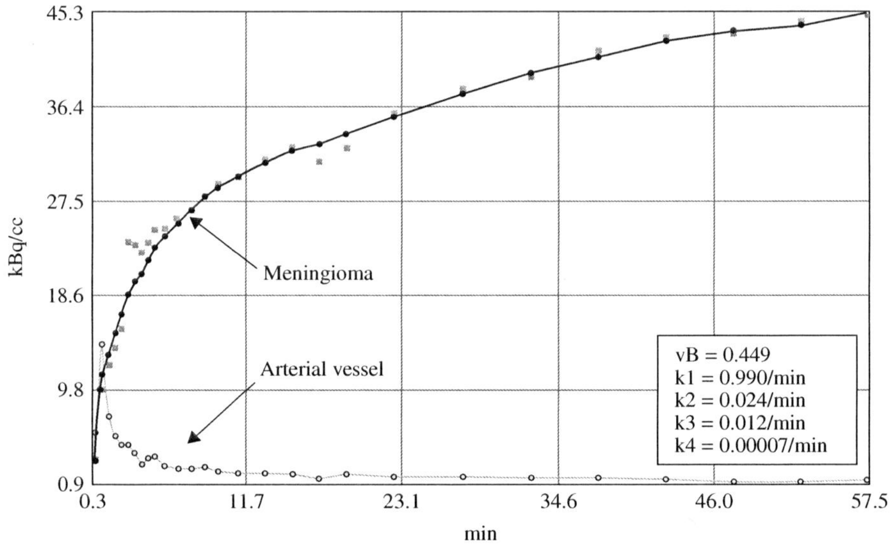

FIGURE 8.2 Fitted time-activity curves are shown for volumes of interest covering meningioma and an arterial vessel. Corresponding kinetic parameters of meningioma were calculated with PMod software.

overexpression is known in meningioma, carcinoids, and other neuroendocrine cancers of the lung and gastrointestinal tract. Curve fitting of the arterial input function and the time-activity curve of a tumor allows for compartment modeling if the tracer is stable *in vivo*. The two-tissue compartment model for DOTATOC was validated by Henze et al. [2] (Fig. 8.2).

The advantage of dynamic imaging/compartment modeling is the *in vivo* determination of flow rates; for DOTATOC, k1 represents receptor binding, k2 the displacement from the receptor, k3 the cellular internalization, and k4 the externalization (dimensions 1/min) [2]. In comparison to FDG, a remarkably higher k1 is found for the peptide tracer DOTATOC that implicates a high first-pass effect. This theoretical model predicted DOTATOC as a promising candidate for locoregional tumor addressing. A recent study performed by our group evaluated intra-arterial (i.a.) application of DOTATOC in patients with neuroendocrine liver metastases. One hour after the injection of ^{68}Ga-DOTATOC into the hepatic artery, the tumor uptake was 3.5 times higher compared to antecubital venous administration (Fig. 8.3).

8.1.2 Locoregional Radiopeptide Therapy

The first approach of locoregional therapy using DOTATOC was described for patients with gliomas in 1999 [3]. The radiopharmakon was injected intratumoral and presented a superior uptake of DOTATOC compared to the i.v. administration. However, the SSR2 expression in gliomas is only moderate. A better vector for addressing gliomas is substance P, another peptide tracer. Recently, successful targeted alpha-radionuclide therapy was demonstrated using this peptide for intratumoral

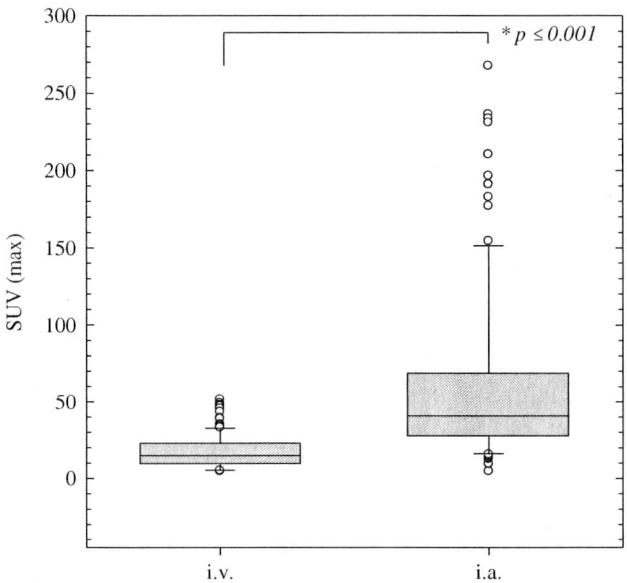

FIGURE 8.3 Comparison of maximal SUVs after i.v. and i.a. application, respectively, of ^{68}Ga-DOTATOC in 122 liver metastases of 11 patients suffering from metastasized gastroenteropancreatic cancer (paired *t*-test, $p < 0.001$).

injection [4]. In contrast to gliomas, which do not develop extracranial metastases, intratumoral injection is not suitable in neuroendocrine tumors, which commonly present with disseminated liver metastases. This promotes the use of DOTATOC tagged with a therapeutic radionuclide, such as Y-90 or Lu-177, as an i.a. therapy. However, some other preconditions are also essential to guarantee a relevant benefit of the i.a. approach.

1. A high first-pass effect in tumor. This has been validated in the mentioned i.v./i.a. comparison by Kratochwil et al.[5].
2. Nonretention in normal tissue of the target region. This has been validated in the same investigation—the uptake in normal liver was not enhanced after the injection of DOTATOC into the hepatic artery.
3. Less obviously, a fast plasma clearance, so the agent is not re-exposed to the target for several times. This has been validated by [6] who found a plasma half-life of DOTATOC of about 3 min.
4. Slow washout from the targeted tumor. In an investigation using ^{111}In-DOTATOC, we found a benefit of the arterial approach with an increased tumor uptake persisting until 72 h p.i. (Fig. 8.3 and Fig. 8.4). This should be sufficient to sustain a relevant increase of tumor dose using Y-90 (half-life 64 h); however, for radionuclides with longer half-life, the benefit might be less pronounced.
5. Nonsaturation of the target structure. Currently there is limited data available for this subject [7]. However, it might be possible to oversaturate the SSR using an arterial approach making high specific activity pivotal.

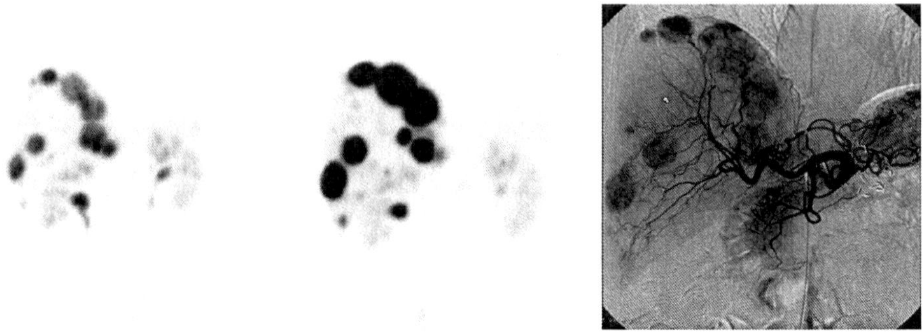

FIGURE 8.4 Comparison of intraindividual ^{68}Ga-DOTATOC PET/CT in a patient with multiple liver metastases, presented as a maximum intensity projection after i.v. versus i.a. application. Average SUV (max) presents a 3.2-fold higher value (122.9 vs. 38.5) after i.a. ^{68}Ga-DOTATOC infusion. Digital subtraction angiography (right image) illustrates the nearly exclusive arterial perfusion of neuroendocrine liver.

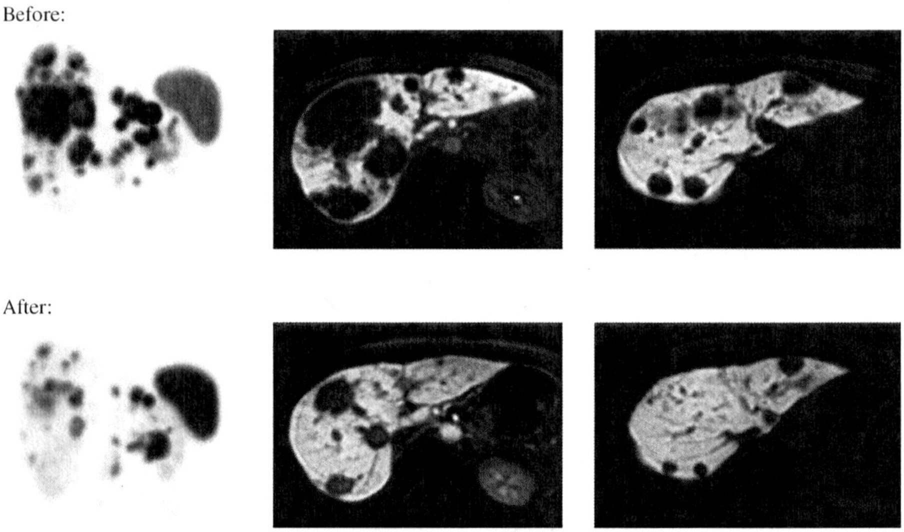

FIGURE 8.5 Prior (upper row) and after locoregional DOTATOC therapy (lower row) of a patient with liver metastases of a NET. In the DOTATOC-PET/CT (left image) and also in the late-enhanced MRI, the large tumor burden declined in bothreceptor DOTATOC occupancy and also in the tumor morphology.

Since DOTATOC fulfills most of the required preconditions, regional treatment of neuroendocrine tumors has been introduced recently. First results in the treatment of liver metastases were promising (Fig. 8.5) [8]. In comparison with selective radioembolization (SIRT), these approaches presented less side effects due to the specific targeted character of the peptide. In contrast, the nontargeted approach using radiolabeled microspheres also harms normal liver parenchyma and, therefore, is associated with off-target side effects (e.g., radiation-induced liver disease (RILD) or the postembolization syndrome).

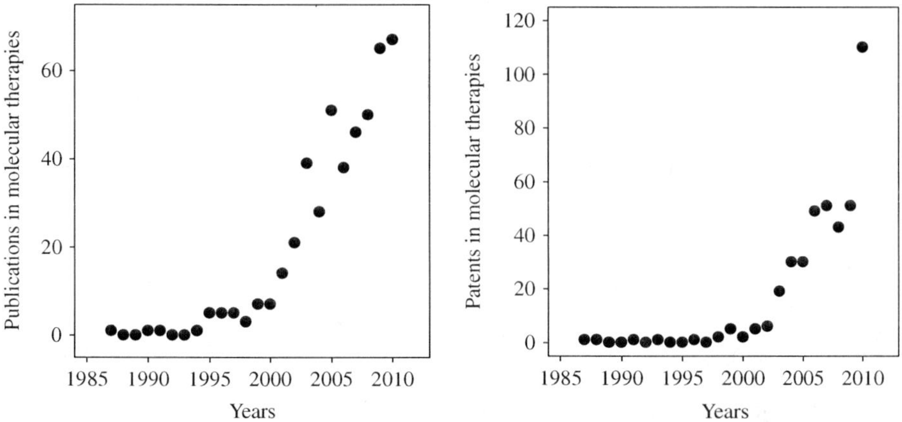

FIGURE 8.6 Increased interest in the field of molecular therapy can be seen both in the publication rate (left) and patent filing (right) in the last decade.

Since DOTATOC does not cause embolization, it can also be used outside the liver, for example, for downstaging irresectable primary tumors to resectable once.

Since the feasibility of tumor targeting using peptides was demonstrated by DOTATOC, radiopeptide as potential probes for diagnostics and therapy gained academic and commercial attention correlating with increased number of publications and patents in the field of molecular therapy (Fig. 8.6). Therefore, the impact of nuclear medicine in the diagnostics and therapy of various other diseases is predictable to increase.

REFERENCES

1. Wållberg H, Löfdahl PK, Tschapalda K, et al. Affinity recovery of eight HER2-binding affibody variants using an anti-idiotypic affibody molecule as capture ligand. Protein Expr Purif 2011;76 (1):127–135.
2. Henze M, Schuhmacher J, Hipp P, et al. PET imaging of somatostatin receptors using [68GA]DOTA-D-Phe1-Tyr3-octreotide: first results in patients with meningiomas. J Nucl Med 2001;42 (7):1053–1056.
3. Merlo A, Hausmann O, Wasner M, et al. Locoregional regulatory peptide receptor targeting with the diffusible somatostatin analogue 90Y-labeled DOTA0-D-Phe1-Tyr3-octreotide (DOTATOC): a pilot study in human gliomas. Clin Cancer Res 1999;5 (5):1025–1033.
4. Cordier D, Forrer F, Bruchertseifer F, et al. Targeted alpha-radionuclide therapy of functionally critically located gliomas with 213Bi-DOTA-[Thi8,Met(O2)11]-substance P: a pilot trial. Eur J Nucl Med Mol Imaging 2010;37 (7):1335–1344.
5. Kratochwil C, Giesel FL, López-Benítez R, et al. Intraindividual comparison of selective arterial versus venous 68Ga-DOTATOC PET/CT in patients with gastroenteropancreatic neuroendocrine tumors. Clin Cancer Res 2010;16 (10):2899–2905.
6. Henze M, Dimitrakopoulou-Strauss A, Milker-Zabel S, et al. Characterization of 68Ga-DOTA-D-Phe1-Tyr3-octreotide kinetics in patients with meningiomas. J Nucl Med 2005;46 (5):763–769.

7. Koukouraki S, Strauss LG, Georgoulias V, et al. Comparison of the pharmacokinetics of 68Ga-DOTATOC and [18F]FDG in patients with metastatic neuroendocrine tumours scheduled for 90Y-DOTATOC therapy. Eur J Nucl Med Mol Imaging 2006;33 (10):1115–1122.
8. Kratochwil C, López-Benítez R, Mier W, et al. Hepatic arterial infusion enhances DOTATOC radiopeptide therapy in patients with neuroendocrine liver metastases. Endocr Relat Cancer 2011;18 (5):595–602.
9. Nissen F, Kraft TE, Ruppert T, Eisenhut M, Haberkorn U, Mier W. Hot or not-the influence of elevated temperature and microwave irradiation on the solid phase synthesis of an affibody. Tetrahedron Letters. 2010;51:6216–6219.

CHAPTER 9

IN VIVO IMAGING AND DELIVERY OF siRNA

ZDRAVKA MEDAROVA AND ANNA MOORE

9.1 INTRODUCTION

Gene therapy is one of the fastest developing areas of experimental medicine. To date over 600 clinical trials involving gene therapy have been completed or are in progress (http://www.wiley.com/legacy/wileychi/genmed/clinical/). Different systems are utilized for both viral and nonviral gene delivery, the most popular being retroviral, adeno-associated, and adenoviral vectors, as well as naked DNA, DNA–liposome complexes, and DNA–microparticle complexes. All of these systems have been explored for therapy. The major hurdles to the application of gene therapy in a clinical setting are difficulties in achieving targeting specificity, regulated expression of recombinant genes to prevent genotoxicity, and control of vector immunogenicity [1]. Although target-specific treatments offer the promise of fewer systemic side effects, strategies such as antisense oligonucleotide treatments have been hampered by exacting target sequence requirements, low potency, and inadequate specificity [2].

Since 2001, when Elbashir et al. demonstrated functional RNA interference (RNAi) in mammalian cells [3], RNAi has emerged as one of the most promising platforms for gene therapeutic product development. Its broad applicability (virtually any gene can be silenced), superior efficiency (100–1000-fold compared to antisense oligonucleotides), and exquisite specificity (single nucleotide) could potentially be harnessed to develop a powerful novel treatment paradigm with global relevance to any disease amenable to manipulation at the level of gene expression.

Two basic methods for triggering RNAi have been adapted for use *in vivo*: delivery of small interfering RNAs (siRNAs) and delivery of plasmid and viral vectors that express a short-hairpin RNA (shRNA) that is subsequently processed into active siRNA. The use of siRNAs is more prominent than the use of shRNA expression vectors. For RNAi experiments in cultured cells, effective siRNAs to a particular target of interest are easier to obtain, and because they are small and only need to

Drug Delivery Applications of Noninvasive Imaging: Validation from Biodistribution to Sites of Action, First Edition. Edited by Chun Li and Mei Tian.

cross the cell membrane and not the nuclear membrane to be effective, they are easier to deliver. In contrast, shRNA expression vectors are time consuming to construct, particularly when one takes into account the time required to create and test several shRNA sequences to find an effective one. siRNAs are also easier to deliver than plasmid-based shRNA expression vectors, and they do not have the associated problems with insertional mutagenesis and immunogenicity that plague viral vectors (http://www.ambion.com/techlib/tn/131/5.html).

The most significant obstacle to the advancement of siRNA-based therapies is ensuring access of functional siRNA to the tissue of interest. The recent past has witnessed numerous reports of *in vivo* gene silencing using siRNAs. The feasibility of systemic *in vivo* siRNA delivery in mammals was first demonstrated using hydrodynamic tail vein injections in mice. In this procedure, unmodified siRNAs are rapidly injected into the tail vein in a large volume of aqueous solution, resulting in localization within hepatocytes [4]. Although not clinically relevant, this procedure does permit gene function and drug target validation studies. In support of this technique, Song and colleagues found that hydrodynamic injection (HDI) of a Fas siRNA resulted in silencing of Fas in mouse hepatocytes for a period of 10 days. This treatment protected hepatocytes from Fas antibody and concanavalin A-stimulated apoptosis and protected mice from fulminant hepatitis [4]. Similarly, hydrodynamic tail vein injection of a caspase 8 siRNA protected mice against acute liver failure induced by Fas antibody or expression of Fas ligand [5]. More recently, *in vivo* siRNA delivery has been accomplished using lipid-based agents [6–10], as well as various carriers, such as atelocollagen [11, 12], a protamine-antibody fusion protein [13], a transferrin-targeted cyclodextrin-containing polycation [14], and polyethyleneimine (PEI) [15–17]. In a seminal study, low volume, normal pressure intravenous (i.v.) delivery of a modified siRNA targeting apolipoprotein B in mice resulted in gene silencing in the liver and jejunum. The siRNA was conjugated with cholesterol to provide targeted delivery and included backbone and sugar modifications to enhance serum stability [18]. One of the most promising *in vivo* strategies involves encapsulation of siRNAs into stable nucleic acid lipid particles (SNALP) [19]. In 2006, it was reported that a single i.v. SNALP-siRNA injection into nonhuman primates resulted in >90% silencing of the target apolipoprotein gene [20]. Efficient silencing has also been reported following local administration of siRNAs [21–25]. The clinical relevance of this technology is underscored by the fact that several clinical trials are ongoing involving siRNA-mediated therapy. A notable example is the application of a siRNA to VEGF, developed by Acuity Pharmaceuticals and designed to be delivered via intravitreal injection as a treatment for wet age-related macular degeneration.

9.2 RNAi: A NOVEL THERAPEUTIC APPROACH FOR CANCER

The examples illustrating the value of siRNA as a cancer therapeutic tool are numerous. Here we will focus on the seminal early studies that demonstrate the feasibility of the approaches used today. With respect to cancer therapy, systemic delivery of siRNAs will likely be required to target most tumor types. Highly specific posttranscriptional suppression of gene expression by siRNA has been achieved *in vivo* using application of naked siRNA in mouse models of human cancer [26–30]. In addition, several groups have investigated the use of oligofectamine- [31], polyethyleneimine- [15, 16], liposome- [8, 10, 32], atelocollagen- [11, 12], and nanoparticle-based siRNA delivery complexes [13, 14, 17].

An important issue that has plagued tumor chemotherapy ever since its conception is chemoresistance. RNAi, with its ease of induction and superior knockdown efficiency, is a natural candidate for the genetic modification of tumor cells in order to enhance chemosensitivity. Indeed, the list of *in vitro* studies combining siRNA and chemotherapy is long. It includes both the validation of already established targets associated with tumor cell survival, proliferation, and metastasis [33–37] and the identification of new targets [38–41]. In addition, a large number of publications demonstrate the feasibility of directly reversing multidrug resistance by *in vitro* RNAi [42, 43]. *In vivo* reports attesting to the potential of RNAi as an "adjuvant" to chemotherapy include studies combining RNAi with newer experimental chemotherapy strategies [44–46] and with well-established chemotherapeutic agents [47–49]. The vast majority of these studies, however, employ tumor vector-transcribed shRNA in order to mediate silencing. All of the aforementioned *in vivo* investigations serve only as proof-of-principle validation studies, since they utilize animal tumor models derived from tumor cells stably transfected with shRNA expressing vectors or involve *in vivo* administration of the viral vector. Despite its immense value as a research tool, this experimental system is highly artificial without direct clinical relevance, since it involves transformation at the genetic level.

Alternatively, as described in the previous sections, direct administration of synthetic siRNA appears to be associated both with enhanced *in vivo* functionality and clear clinical relevance, since siRNAs can be administered as a drug. To date there are very few reports of coadministration of synthetic siRNA and chemotherapy for cancer treatment, due to the novelty of the field of siRNA administration to tumors. Both unmodified siRNAs [27, 29, 50] and siRNAs complexed with atelocollagen [51] have been used in combination with standard chemotherapy to mediate significant improvement in tumor outcome. In the atelocollagen study, for instance, average tumor weights in the treated group were as much as fivefold lower than in the nontreated group.

A common gene target whose silencing has been investigated extensively as a chemosensitizing tool is survivin. Survivin is a member of the inhibitor of apoptosis protein (IAP) family, which is highly expressed in most cancers and associated with chemotherapy resistance making antisurvivin therapy an attractive cancer treatment strategy [52]. *In vitro*, Tirro et al. [33] suppressed survivin in thyroid cancer cells and demonstrated significantly increased cytotoxicity of cisplatin and doxorubicin. A similar role for survivin and other members of the IAP family was demonstrated with relevance to pancreatic adenocarcinoma when RNAi was combined with 5-fluorouracil (5-FU) [53]; breast cancer in combination with taxanes [54], doxorubicin, and etoposide [55]; and small cell lung cancer in combination with cisplatin [56], to name a few. *In vivo*, Jiang et al. [47] utilized a lentiviral expression system to deliver survivin shRNA to oral squamous cell carcinoma cells and demonstrated a 2.1-fold increase in the sensitivity to adriamycin, translating into an inhibition of the development and growth of established tumor xenografts. All of these studies taken together underscore the value of RNAi in cancer therapy.

9.3 RNAi: A NOVEL THERAPEUTIC APPROACH FOR DIABETES

Various delivery approaches have been explored for the treatment of diabetes. A notable example involves the suppression of Fas (CD95), which is considered a key player in diabetes pathogenesis by mediating interactions between beta cells and cytotoxic T cells, resulting in apoptotic cell death. The effect of Fas-silencing siRNA (Fas siRNA) on type 1 diabetes development was evaluated in a cyclophosphamide (CY)-accelerated diabetes animal model after i.v. administration using a polymeric carrier, PEI. The systemic nonviral delivery of Fas siRNA showed significant delay in diabetes incidence [57].

Small interfering RNAs delivery to the liver using a lipid nanoparticle system was used to suppress the transcription factor, transducer of regulated cAMP responsive element-binding protein (CREB) activity 2 (TORC2), which is a major regulator of hepatic gluconeogenesis and is increased in hyperglycemic rodent models. This resulted in highly efficient, sustained (1–3 weeks) knockdown of TORC2 in the liver and a marked improvement of hyperglycemia in a variety of murine models. These results demonstrated the potential of the siRNA approach for the treatment of hyperglycemia and type 2 diabetes [58].

Other important examples include suppression of plasma cell membrane glycoprotein-1 or ectonucleotide pyrophosphatase/phosphodiesterase (PC-1/ENPP1), using an adenoviral vector expressing shRNA, for the improvement of glycemic control [59] and knockdown of liver ccaat/enhancer-binding protein alpha by adenovirus-transduced siRNA for the improvement of hepatic gluconeogenesis and lipid homeostasis in murine models of diabetes.

Small interfering RNAs has also been employed in the treatment of diabetic complications. For example, diabetic nephropathy is a complex and poorly understood disease process with limited treatment options. Recently, a developmental protein and one of the bone morphogenetic protein antagonists, Gremlin, has emerged as a novel modulator of diabetic nephropathy. Zhang et al. constructed a gremlin siRNA plasmid and examined the effect of Gremlin inhibition on the progression of diabetic nephropathy in CD-1 mice following uninephrectomy and STZ treatment. Inhibition of Gremlin by injection of the plasmid alleviated proteinuria and renal collagen IV accumulation 12 weeks after the STZ injection and inhibited renal cell proliferation and apoptosis [60].

Topical siRNA application has been used to treat a specific complication of diabetes, that is, wound healing. Specifically, p53, which is a master cell cycle regulator and is upregulated in diabetic wounds, was silenced by topical siRNA application in a wound bed. This resulted in improved diabetic wound healing and augmentation of vasculogenic mediators [61].

Other examples include target knockdown in retinal and choroidal endothelial cells by intravitreal and hydrodynamic plasmid siRNA delivery for the treatment of ocular complications from diabetes [62], cholesterol-tagged siRNA delivery for the treatment of diabetic nephropathy [63], and the application of adeno-associated virus for the suppression of a gene implicated in nonalcoholic fatty liver disease—a serious health problem linked to obesity and type 2 diabetes [64].

These examples illustrate the potential of RNAi as a therapeutic modality in diabetes. However, the field is still in its infancy and the relevant research is mostly restricted to proof-of-principle investigation.

9.4 RNA INTERFERENCE: A NOVEL THERAPEUTIC APPROACH FOR NEUROLOGICAL DISORDERS

Since its discovery, RNAi has become a mainstay in neurogenetic research. It has mostly involved local administration of siRNA to the brain for the study of basic biology [65–68]. Here we will focus on examples of systemic siRNA delivery for brain targeting.

The blood–brain barrier (BBB) is essentially impenetrable to any RNA moiety used in a potential RNAi therapy. Therefore, in applying RNAi to the brain, using systemic delivery, specific methods need to be designed to overcome the BBB. Three strategies that have been used toward this goal involve:

1. Coupling of siRNA to ligands, where the ligand first targets the BBB, followed by "homing in" on the target tissue once internalized into the CNS (reviewed in reference 69).
2. Altering the permeability of the BBB.
3. Coupling the siRNA to cell penetrating peptides.

As a notable example of the first strategy, short peptides derived from rabies virus glycoprotein (RVG), which binds to the acetylcholine receptor, were utilized for the delivery of siRNA across the BBB to target neuronal cells. dsRNA was coupled to RVG and delivered systemically [70]. Another investigation utilized coated nanoparticles for the delivery of encapsulated siRNA or siRNA-generating expression vectors through the BBB. These nanoparticles, by mimicking low-density lipoproteins (LDL), interact with the LDL receptors on the endothelium resulting in their uptake across the BBB [71]. More recently, a new siRNA delivery system was described. It incorporates a small peptide that binds siRNA and acetylcholine receptors (AchRs), acting as a molecular messenger for delivery to neurons, and cationic liposomes that protect siRNA–peptide complexes from serum degradation. These liposome–siRNA–peptide complexes were shown to cross the BBB and significantly downregulate the target gene in neuronal cells [72].

An example of the second technique is the mediation of increased permeability of the BBB capillary endothelium by first targeting cell adhesion molecules. siRNA against claudin-5 was used for that purpose to enhance the permeability of the BBB to larger molecules [73].

The third strategy is relatively novel and still being tested *in vitro*. In one study, two cell penetrating peptides—penetratin and transportan—were coupled to siRNA duplexes for delivery into mammalian cells [74]. A second study utilized Tat peptide to deliver siRNA against enhanced green fluorescent protein (EGFP) or cyclin-dependent kinase 9 into cells [75]. Our group described the generation and *in vitro* characterization of a probe, consisting of siRNA complexed with fluorescently labeled myristoylated polyarginine peptides (MPAP). This structure mediated successful delivery of siRNA to bEnd.3 brain endothelial cells and primary cortical neurons and astrocytes [76].

All of the examples given earlier utilize different strategies to address the major prerequisites for the *in vivo* applicability of siRNAs, comprising efficient transfer across the cell membrane, selective delivery into the desired tissue, adequate protection from degradation en route to the target tissue, and protection of the siRNA from rapid excretion. Surprisingly, rapid excretion has proven to be more of a problem than *in vivo* stability [77]. Although chemical stabilization is readily achieved via siRNA modification, it does not appear to be necessary in most cases, as excretion appears to occur prior to degradation. The use of nanoparticles or lipid complexes currently shows more promise than chemical modification to address the pharmacokinetics and tissue distribution issues endemic to *in vivo* siRNA delivery (http://www.ambion.com/techlib/tn/131/5.html). Importantly, in order to achieve therapeutic applicability, these requirements need to be satisfied in a clinically acceptable context defined by conventional minimally invasive administration. Furthermore, the development of clinically relevant imaging paradigms for the noninvasive assessment of siRNA delivery to tissues of interest is of paramount importance both for the conception and optimization of experimental treatment strategies.

9.5 IMAGING OF siRNA DELIVERY AND ITS EFFECTS

The field of *in vivo* RNAi imaging is still in its early stages of development. Studies that describe the imaging of RNAi include fluorescence or bioluminescence reporter imaging, nuclear imaging, and magnetic resonance imaging (MRI).

9.5.1 Fluorescence/Bioluminescence Imaging

The majority of the early proof-of-principle studies on *in vivo* RNAi utilized suppression of luciferase transgene expression by either synthetic siRNAs or shRNAs transcribed *in vivo*, achieving impressive 80–90% silencing efficiencies in the liver [78]. In the same study, the authors fused luciferase RNA with an endogenous gene and demonstrated, by measuring bioluminescence signal, that siRNAs against the nontransgene functioned with comparable efficiency, confirming that siRNA can be used *in vivo* as a therapeutic tool and establishing a method for the *in vivo* quantitative imaging of the silencing effect. In a similar study, Lewis et al. [79] co-injected a plasmid expressing luciferase with antiluciferase synthetic siRNA and observed significant 80–90% silencing, as measured by bioluminescence imaging (BLI) in a variety of organs, including the liver, spleen, lung, kidney, and pancreas. This investigation, for the first time, utilized *in vivo* imaging to identify, on a whole-body scale, the scope of siRNA biological activity, in a way representing a functional biodistribution characterization of siRNA function. More recently, BLI of luciferase expression was used to demonstrate the dose and time dependence of vector-based *in vivo* RNAi [80]. BLI has also been employed for the development of optimized methods for lipid-based siRNA delivery, resulting in nearly 80% inhibition with as little as 0.2 pmol siRNA [7], as well as for the establishment of the utility of atelocollagen as an efficient siRNA carrier to bone metastatic tumors [12]. In a comprehensive imaging study [46], bioluminescence was applied for the noninvasive assessment of p-glycoprotein silencing. P-glycoprotein, the product of the MDR1 gene, is responsible for multidrug resistance in a variety of cancers. Interestingly, coelenterazine is a known p-glycoprotein transport substrate, and also a Renilla luciferase fluorophore, suitable for BLI. Therefore, BLI of tumor xenografts stably transfected with Renilla luciferase revealed information about p-glycoprotein downregulation by lentivirally delivered shRNA, since in tumors in which MDR1 was silenced, p-glycoprotein-mediated coelenterazine efflux from cells was reduced, resulting in enhanced luminescence.

More recently, in a comprehensive new study [81], BLI was applied together with a mathematical model of siRNA delivery and function, in order to define, in a more general context, the effects of target-specific and treatment-specific parameters on siRNA-mediated gene silencing. This particular study demonstrated that, whether in rapidly dividing subcutaneous tumors, the silencing effect may persist for ~10 days; in nondividing hepatocytes, it may last as long as 3–4 weeks. More importantly, this study represents the origin of the current belief that siRNA dilution due to cell division, and not intracellular siRNA half-life, governs the duration of gene silencing. As a result of this study, now it is possible to apply this mathematical model in treatment design and to predict the dosing schedule required to maintain persistent silencing of target proteins with different half-lives in rapidly dividing or nondividing cells.

The same group used live-animal BLI to directly compare the kinetics of luciferase knockdown by unmodified and nuclease-stabilized siRNAs. In mice, a fourfold increase in maximum luciferase knockdown was observed after HDI of nuclease-stabilized siRNAs relative to unmodified siRNAs, yet the overall kinetics of the recovery after knockdown were nearly identical. Based on these findings, the authors hypothesized that the stabilization advantages of nuclease-stabilized siRNAs originate primarily from effects prior to and during internalization before the siRNAs can interact with the intracellular RNAi machinery [82].

Finally, the same authors studied the effect of tumor-specific targeting on the biodistribution, and the efficacy of siRNA nanoparticles as measured by multimodality BLI/PET/CT *in vivo* imaging [83]. Positron emission tomography (PET) was used to assess the biodistribution of nanoparticles formed with cyclodextrin-containing polycations and siRNA. BLI was used to quantify the *in vivo* function of the siRNA. BLI of mice bearing luciferase-expressing Neuro2A s.c. tumors before and after PET imaging enabled correlation of functional efficacy with biodistribution data. Although both nontargeted and transferrin-targeted siRNA nanoparticles exhibited similar biodistribution and tumor localization by PET, transferrin-targeted siRNA nanoparticles reduced tumor luciferase activity by ≈ 50% relative to nontargeted siRNA nanoparticles 1 day after injection (Fig. 9.1). The authors concluded that the primary advantage of targeted nanoparticles was associated with processes involved in cellular uptake in tumor cells rather than overall tumor localization.

Bioluminescence imaging has also been used to evaluate different lipid nanoparticle-based siRNA delivery formulations for targeting the liver. These include lipid nanoparticles consisting of cationic lipids, polyethylene glycol (PEG), and cholesterol [84]. This study is valuable because it illustrates the utility of imaging in guiding the design of siRNA delivery.

Numerous other interesting studies, which we cannot discuss in detail here, have explored BLI using luciferase as a reporter for the quantitative evaluation of RNAi efficiency [85–89]. As a result, this imaging modality has become a standard for the investigation of important issues, such as the time course of RNAi, the "biodistribution"

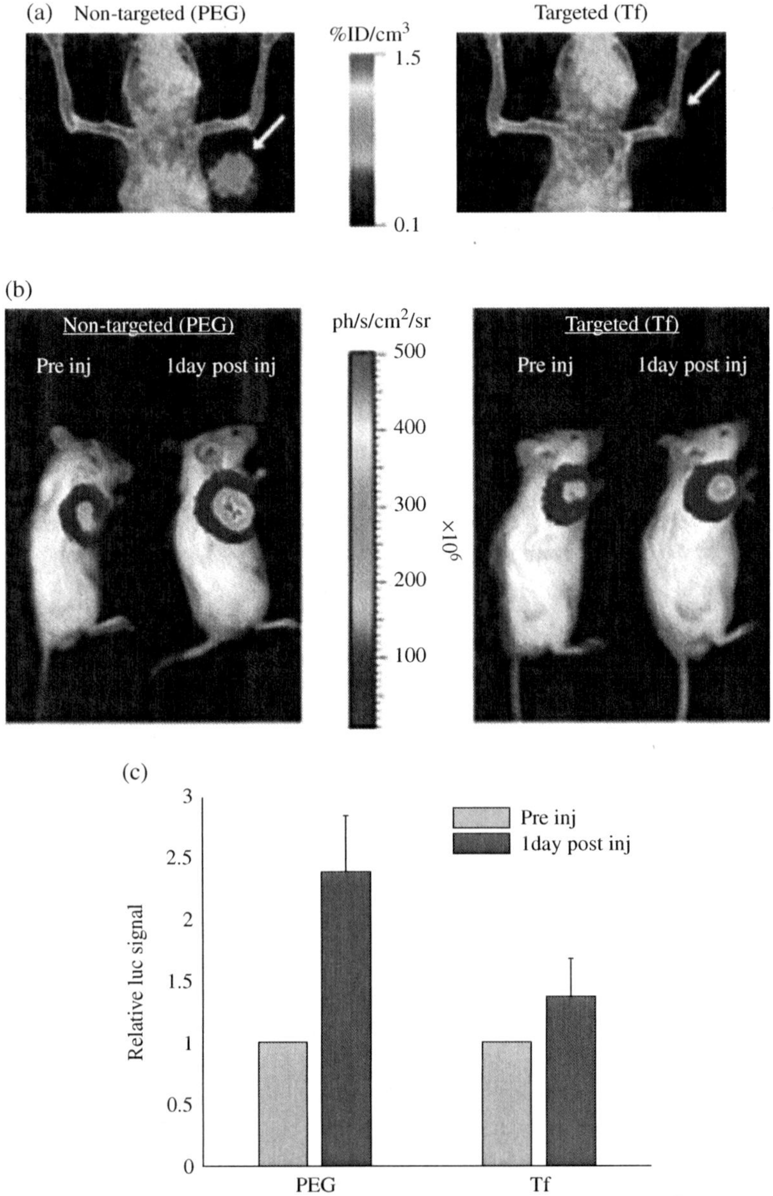

FIGURE 9.1 Multimodality *in vivo* imaging of siRNA nanoparticle delivery and function, using micro-PET/CT and BLI. (a) Fused micro-PET/CT images showing tumor-associated (arrow) activity 1 day after injection of targeted (Tf) and nontargeted (PEG) nanoparticles containing 64Cu-DOTA-siRNA. (b) BLI of the same mice shown in A before injection and 1 day after injection. (c) Relative change in luciferase expression 1 day after injection of Tf-targeted (Tf, n = 7) and nontargeted (PEG, n = 4) nanoparticles containing 64Cu-DOTA-siRNA for simultaneous PET imaging. Error bars indicate SE. (Reprinted from Bartlett et al., 2007, with permission from The National Academy of Sciences of the USA.) For color details, please see color plate section.

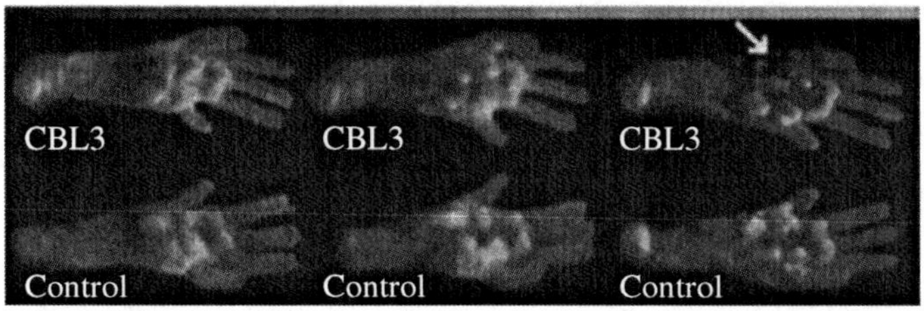

FIGURE 9.2 CBL3 Accell siRNA-loaded PADs inhibit hMGFP expression in mouse footpad skin. Transgenic CBL/hMGFP mouse footpads were treated every 2 days with three (3 × 5) PADs loaded with either CBL3 or nonspecific control Accell siRNA for 12 days. Enlarged view of mouse three treated paws, color-mapped to a binned black–blue–green color scheme to facilitate comparison of low-intensity regions. Arrow shows area of signal reduction. Color bar at top shows look-up table assignments corresponding to pixel bit values in increments of 10 (Reprinted from Gonzalez-Gonzalez et al., 2010 with permission from Nature Publishing Group).

of the silencing process, the optimization of delivery strategies, and the exquisite sensitivity of this knockdown method.

Similar questions have been addressed using GFP reporters [46, 90–92]. For example, the utility of lentiviral vectors for the *in vivo* delivery of shRNAs to the brain was demonstrated using an EGFP reporter [90]. However, this method has been restricted to *ex vivo* analysis with fluorescence microscopy or flow cytometry. *In vivo* GFP fluorescence imaging of transgenic mice in which a lentiviral vector carrying shRNA also incorporated EGFP as a reporter gene permitted the evaluation of shRNA delivery [93]. Importantly, the effect (94%) persisted into adulthood, due to integration of the vector into the host genome. However, this is a very artificial system, requiring transgenesis, and only applicable to a research scenario.

More recently, a reporter mouse model (transgenic click beetle luciferase/humanized monster GFP) was developed to enable the study of siRNA delivery to skin; in this transgenic mouse, GFP reporter gene expression is confined to the epidermis. Intradermal injection of siRNAs targeting the reporter gene resulted in marked reduction of GFP expression in the localized treatment areas as measured by intravital imaging using a dual-axes confocal fluorescence microscope [94] (Fig. 9.2).

Besides these studies, investigation of the delivery of siRNA to target tissues is largely based on *ex vivo* experiments [8–10, 17, 19, 95]. These studies provided answers to important new questions, such as the subcellular distribution of siRNAs delivered as part of various formulations and their bioavailability. For example, Schiffelers et al. [17] were one of the first to demonstrate by fluorescence microscopy that siRNAs complexed with a tumor-targeted carrier accumulated preferentially in tumor tissue and, due to their longer half-life, were present in tissue at much later time points than aqueous siRNA. This finding was significant because it established the feasibility of tumor-targeted siRNA delivery and provided evidence of the prominent dependence of siRNA functionality on its blood half-life.

An alternative technology involves the coupling of siRNA to quantum dots. One study utilized cationic liposomes to codeliver green quantum dots and siRNA targeting the lamin a/c gene (*Lmna*) into murine fibroblasts, followed by flow cytometry. The authors found that in cotransfected cells gene silencing correlated directly with intracellular fluorescence and resulted in about 90% knockdown in highly fluorescent cells [96]. More recently, the same group reported on a new system for siRNA delivery to tumor cells that consisted of a PEGylated quantum dot core as a scaffold with siRNA (to EGFP) and tumor-homing peptides attached to it. This system afforded intracellular localization of the quantum dot particle followed by endosomal escape mediated by the addition of cationic liposomes. A rather modest knockdown in EGFP fluorescence (about 29%) was achieved using this method [97]. In another study, quantum dots were encapsulated into chitosan nanoparticles labeled with antibody targeting the HER2 receptors to deliver HER2/neu siRNA. Targeted delivery of HER2 siRNA to HER2-overexpressing SKBR3 breast cancer cells was shown to be specific and confirmed by quantum dot microscopy [98]. While quantum dots remain an attractive tool for *in vitro* and animal testing, where fluorescence is the most accessible and common imaging modality, concerns over their cytotoxicity and the limited tissue penetration of light should be taken into account.

Besides bioluminescence and fluorescence, other novel optical imaging approaches have been employed in relationship to siRNA delivery and imaging. siRNA-encapsulating polyplexes were covalently coated with small gold nanoparticles (Au NPs) via acid-cleavable linkages in order to explore the possibility of achieving combined stimuli-responsive multimodal optical imaging and stimuli-enhanced gene silencing. In a mildly acidic tumor environment, Au NPs are dissociated from the siRNA-carrying polyplexes, generating various optical signal changes such as diminished scattering intensity, increased variance of Doppler frequency, and blue-shifted UV absorbance (stimuli-responsive imaging) [99]. Another nanotechnology approach utilized gold nanorod-DARPP-32 siRNA complexes (nanoplexes) that target a dopaminergic signaling pathway in the brain. The shift in the localized longitudinal plasmon resonance peak of gold nanorods (GNRs) was used to show their interaction with siRNA. Plasmonic enhanced dark field imaging was used to visualize the uptake of these nanoplexes in dopaminergic neurons *in vitro* [100].

9.5.2 Nuclear Imaging

There are a few examples of the application of nuclear imaging for monitoring siRNA delivery. The earliest study utilized SPECT imaging to visualize the cellular delivery and animal distribution of siRNA. The siRNA in this study was chemically modified with an S-acetyl N-hydroxysuccinimide (NHS) hydrazino nicotinamide.

The delivery of ^{99m}Tc-labeled RNAs *in vivo* was tested in nude mice bearing ACHN human kidney cancer xenografts. Both sense and antisense ^{99m}Tc-labeled RNAs accumulated in the tumors at the same rate following i.v. injection. The highest tumoral accumulation was observed at 4 h after injection, with radioactivity visualized in almost all tissues. While this study produced encouraging results regarding the absolute tumoral accumulation of radiolabeled RNAs, it is not clear whether the delivery of functional RNA was achieved and whether any therapeutic effect was observed. Nevertheless, this study paved the way for further investigations on the *in vivo* imaging of siRNA delivery [101].

Subsequently, a group from Germany optimized the labeling and purification of siRNA and applied siRNA or siRNA after complexation with PEI for *in vivo* real-time gamma camera and SPECT imaging [102]. The same group described the physico-biological characterization of PEI- and PEG-PEI polyplexes containing partially 2′-OMe modified 25/27mer dicer substrate siRNAs (DsiRNAs) and used three-dimensional SPECT imaging to assess their *in vivo* behavior regarding biodistribution and systemic bioavailability after pulmonary application [103]. In a more general context, the same group elegantly demonstrated the application of SPECT and fluorescence fluctuation spectroscopy to evaluate the pharmacokinetics and biodistribution of each component of siRNA polyplexes formed with 25-kDa PEI or PEGylated PEIs. Both pharmacokinetics and biodistribution of polymer-complexed siRNA were dominated by the polymer, which remained stably associated with the siRNA during passage through the liver [104] (Fig. 9.3).

With relevance to PET, in an early *in vivo* study, different methods for siRNAs labeling with ^{18}F were designed. The bioavailability of the siRNAs was then assessed by comparative dynamic and quantitative PET. After i.v. injections of [^{18}F]siRNAs in rodents, total radioactivity was rapidly eliminated by the kidneys and the liver [105]. More recently, the pharmacokinetics of naked ^{18}F-labeled siRNA was compared to that of cationic liposome/[^{18}F]-labeled siRNA complexes. The naked siRNA was cleared quite rapidly from the bloodstream and excreted from the kidneys. In contrast, the liposome complex tended to accumulate in the lungs [106].

Multimodal positron emission tomography/computed tomography (PET/CT) hybrid imaging has also been used to gain insights into siRNA delivery [107]. In this particular study, a synthetic siRNA delivery system targeted to the liver was labeled with ^{64}Cu to confirm that the siRNA delivery system was successfully targeted to the liver and that incorporation of the siRNA into the delivery system protected the siRNA from renal filtration long enough so that the siRNA could be delivered to the liver.

9.5.3 Magnetic Resonance Imaging

Our group has focused on the combined MRI and delivery of siRNA to target tissues [108]. In our earliest studies, we utilized dextran-coated superparamagnetic nanoparticles (MN), which, after modification with the Cy5.5 near-infrared optical dye (NIRF) and a membrane translocation peptide (MPAP), served as a core module for attaching synthetic siRNA molecules (MN-NIRF-siRNA). In this study, a NIRF dye was utilized for correlative optical imaging and microscopy. MPAP ensured translocation of the nanoparticle across the cell membrane into the cytoplasm in order to mediate RNAi. In our feasibility studies, we utilized siRNA targeting a model gene (GFP) expressed in tumors. In this model system, we used RFP-expressing tumors as controls. We expected that the probe, which is not tumor targeted, would accumulate in both tumors. However, we expected to observe silencing in the GFP-expressing tumor only. In order to monitor the delivery of the probe to GFP- and RFP-expressing tumors, we performed *in vivo* MRI of tumor-bearing mice. Superparamagnetic iron oxide nanoparticles are characterized by their strong T2 magnetic susceptibility effects. Their presence in tissue is reflected by marked shortening of T2 relaxation times resulting in a loss of signal (darkening) on MR

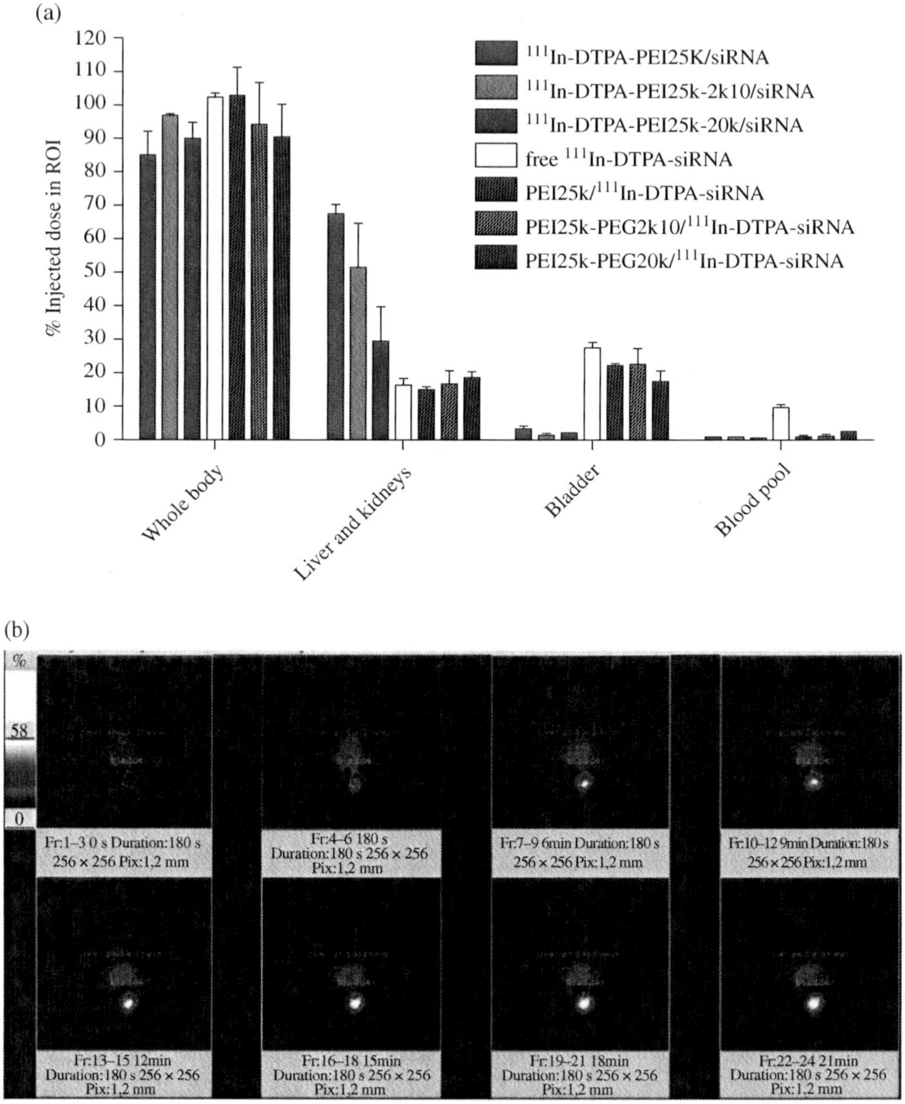

FIGURE 9.3 (a) Scintillation analysis of regions of interest 2 h after injection for evaluation of biodistribution. (b) Planar gamma camera perfusion movie with regions of interest for real-time quantification after injection of 35 μg siRNA, corresponding to 2.57 nmol and 0.255 MBq 111In, complexed at an N/P of 6 with 25-kDa bPEI (Reprinted from Merkel et al., 2009 with permission from Elsevier).

images. In our experiments, tumors appeared characteristically bright on precontrast T2 images. Following injection of the probe, there was a decrease in the T2 relaxation times of both tumors, as a result of MN-NIRF-siGFP accumulation (Fig. 9.4a). The accumulation of MN-NIRF-siGFP in tumors was also confirmed by *in vivo* NIRF imaging (Fig. 9.4b). *In vivo* optical imaging in the GFP and RFP

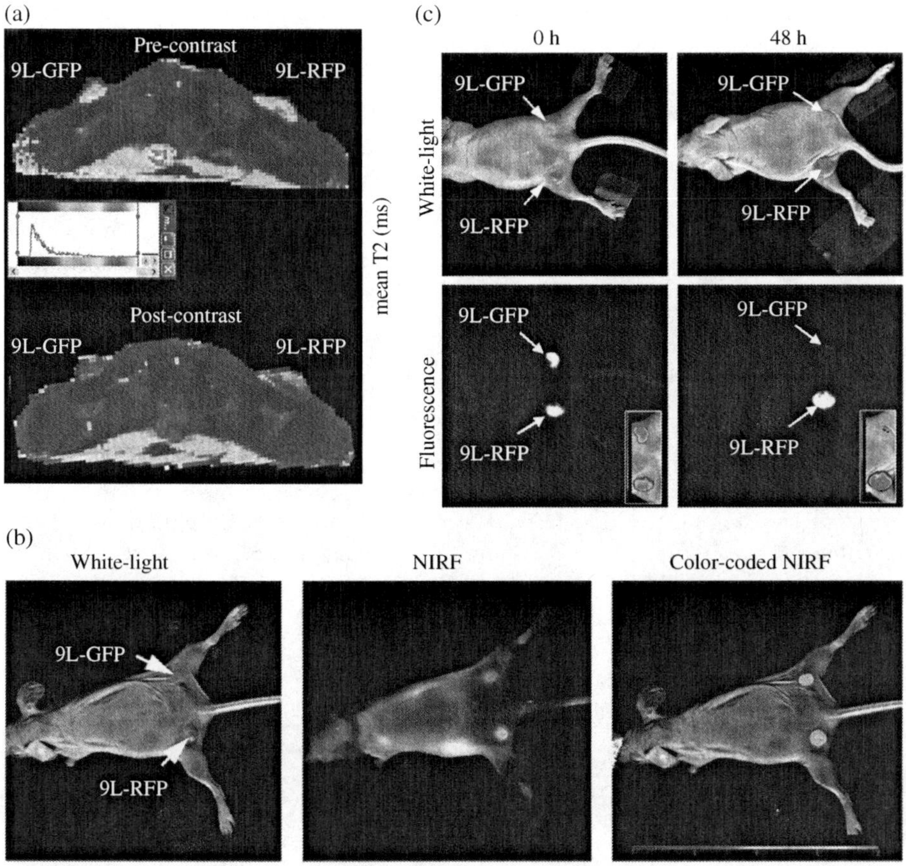

FIGURE 9.4 *In vivo* imaging of MN-NIRF-siGFP delivery to tumors. (a) *In vivo* MRI was performed on mice bearing bilateral 9L-GFP and 9L-RFP tumors before and 24 h after MN-NIRF-siGFP administration. Whereas on precontrast T2 images, tumors appeared characteristically bright, following injection of the probe, there was a significant decrease in T2 relaxivity associated with the tumors ($p = 0.0012$ for 9L-GFP, $p = 0.0049$ for 9L-RFP). Note that T2 relaxation times of muscle tissue remained unchanged. (b) *In vivo* NIRF optical imaging of the same animals was performed immediately after each MRI session. A high-intensity NIRF signal associated with the tumors confirmed the delivery of the MN-NIRF-siGFP probe to these tissues. (c) *In vivo* fluorescence optical imaging of animals bearing bilateral 9L-GFP and 9L-RFP tumors 48 h after i.v. probe injection. Forty-eight h after probe administration, 9L-GFP-associated fluorescence diminished dramatically ($p = 0.0083$), whereas 9L-RFP fluorescence remained unchanged compared to corresponding fluorescence intensities prior to injection of the probe (0 h) (Reprinted from Medarova et al., 2007 with permission from Nature Publishing Group). For color details, please see color plate section.

channels demonstrated marked differences in fluorescence 48 h after probe administration (Fig. 9.4c). Specific silencing of GFP was confirmed by real-time quantitative RT-PCR. GFP transcript levels in tumors treated with MN-NIRF-siGFP were 85 ± 2% lower than in saline-treated controls ($p < 0.001$) and 97 ± 1%, lower than in mice treated with mismatch control ($p < 0.001$). Importantly, no induction of type I interferon and inflammatory cytokines was noticed after treatment [108]. Similar nanoparticles were used for targeting the therapeutic antiapoptotic gene *birc5* (Survivin) in tumors (MN-NIRF-siSurvivin). The delivery of the nanoparticles to tumors was monitored by MR and NIRF imaging. This delivery resulted in 97 ± 2% lower survivin transcript levels than in controls treated with the parental MN nanoparticle ($p<0.01$) and in 83 ± 2% lower levels than in mice treated with mismatch control ($p<0.01$) [108].

To ascertain the therapeutic effect of targeting survivin, we synthesized and tested a multifunctional probe for the cell-specific delivery of antisurvivin siRNA to tumors [109]. The goal was to shift the overall biodistribution of the agent to increase relative tumor cell uptake. We synthesized this modified probe, MN-EPPT-siSurvivin, conjugated to a peptide, EPPT, directed to the underglycosylated mucin-1 antigen (uMUC-1) on the surface of tumor cells [110]. We reasoned that silencing of *birc5* could mediate a therapeutic effect by inducing necrotic/apoptotic tumor cell death. We tested the therapeutic potential of the probe by hematoxylin and eosin (H&E) staining (necrosis) and TUNEL assay (apoptosis) of tumor sections derived from breast tumor-bearing animals treated with the probe. We observed the induction of considerable levels of both necrosis and apoptosis. There was a fivefold increase in the fraction of apoptotic nuclei in experimental tumors versus controls (Fig. 9.5a). This translated into a nearly twofold decrease in tumor growth rate (Fig. 9.5b). These data attest to the feasibility of utilizing MN-conjugated knockdown oligonucleotide probes (siRNA) as a tool for efficient *in vivo* downregulation of target genes in tumors, as well as for the noninvasive semiquantitative imaging of the delivery process [109].

We have applied this technology not only to models of cancer but also in the context of diabetes. Since islet transplantation has recently emerged as the most promising clinical modality to achieve normoglycemia in diabetic patients [111], it is crucial to be able to monitor the distribution and fate of pancreatic islet grafts after transplantation. We first showed that transplanted pancreatic islets prelabeled with magnetic nanoparticles could be monitored longitudinally in live animals [112]. Since the success of transplantation depends on survival of the islets after the procedure, we suggested introducing a protection to islets prior to transplantation in the form of nanoparticles tagged to siRNA directed against pro-apoptotic components in islet cells. As proof of concept, we showed that siRNA conjugated to magnetic nanoparticles could accumulate in pancreatic islets at levels sufficient for detection by MRI *in vitro* and for silencing target genes (model GFP gene) [108]. Finally, we constructed a dual-purpose therapy/imaging nanoparticle probe to target the caspase-3 gene in pancreatic islets prior to transplantation. We reasoned that our "two-in-one" MN-siCaspase-3-imaging probe would silence the apoptotic-related gene, providing significant protection to the grafts from early loss after transplantation, and at the same time serve as an MRI label to assess the *in vivo* posttransplant fate of the grafts noninvasively. The results of this study indicated that human islets labeled with the probe showed superior survival after transplantation in mice compared with islets labeled with the parental probe, as demonstrated by *in vivo* MRI during the initial 2-week period when islet destruction is most prominent [113].

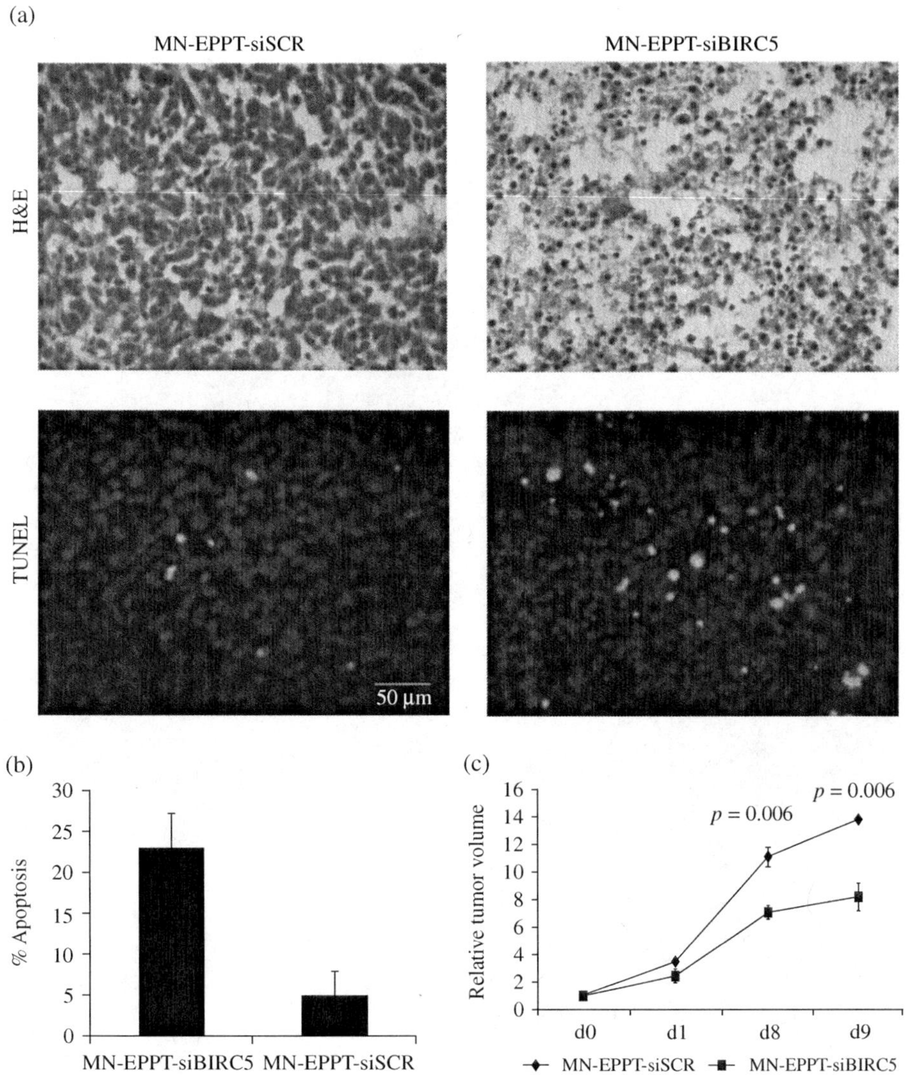

FIGURE 9.5 MN-EPPT-siBIRC5 therapeutic effects. (a) H&E (top) and TUNEL (bottom) staining of tumor tissue from mice treated with MN-EPPT-siBIRC5 or MN-EPPT-siSCR. In the H&E sections, there were extensive eosinophilic areas in the experimental but not control tumors, suggesting the induction of necrosis by the siBIRC5-bearing nanodrug. There was a visible increase in the fraction of apoptotic (green) nuclei in the TUNEL-stained tissues from experimental mice relative to controls, indicating the induction of apoptosis. The tissues were counterstained with DAPI (nuclei, blue). (b) Quantitative analysis of the TUNEL experiments. There was a fivefold induction of apoptosis in the tumor tissues of experimental animals relative to controls ($p = 0.003$). (c) Relative tumor volume measurements of MN-EPPT-siBIRC5—and MN-EPPT-siSCR—injected animals over the course of treatment. There was a significant decrease in tumor growth rate in the experimental animals relative to controls, apparent by 8 days after the beginning of treatment ($p < 0.01$; $n = 4$) (Reprinted from Medarova et al., 2010 with permission from AACR). For color details, please see color plate section.

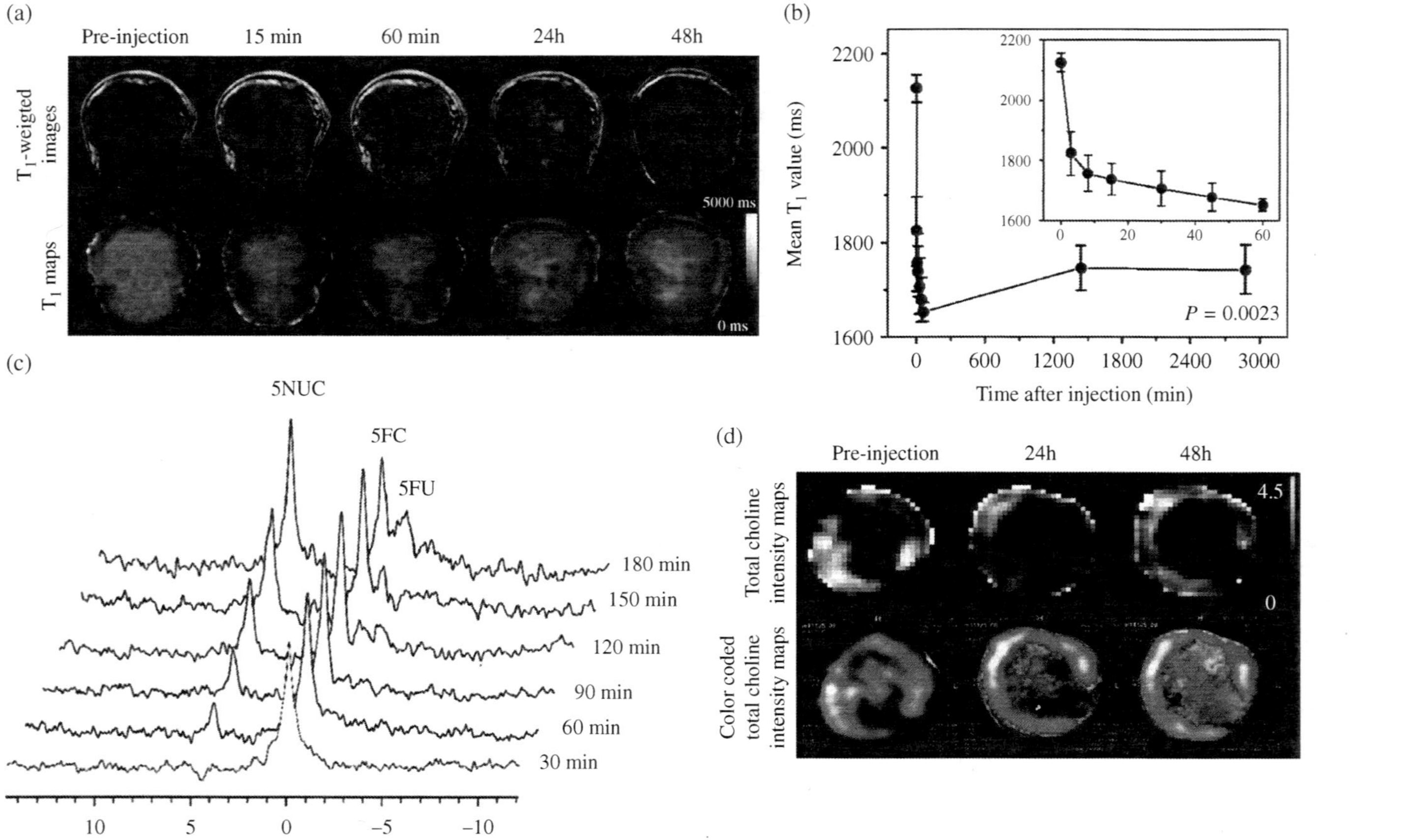

FIGURE 9.6 (a) Representative *in vivo* T1-weighted MR images (top panel) and quantitative T1 maps (bottom panel) of a tumor (400 mm^3) pre- and post-injection of nanoplex 11 (300 mg/kg, i.v.). (b) Time-dependent mean T1 values of tumors ($n = 4$) pre- and postinjection of nanoplex 11; a significant decrease of T1 ($p < 0.0023$) was observed up to 48 h. Inset panel shows the T1 variation within the first 60 min of injection. (c) *In vivo* 19F MRS demonstrated efficient conversion of prodrug 5-FC to 5-FU and its metabolites F-Nucl by nanoplex 11 localized in the tumor. 5-FC (200 mg/kg i.v. and 250 mg/kg i.p.) was injected at 24 h after nanoplex injection. (d) Representative *in vivo* tCho maps and color-coded tCho intensity maps overlaid on corresponding T1-weighted images of a tumor before and at 24 and 48 h after nanoplex injection (300 mg/kg, i.v.) (Reprinted from Li et al., 2010 with permission from ACS).

Valuable insight into the feasibility of monitoring siRNA delivery by MRI has also been obtained by a group from Johns Hopkins University. In their initial work, the authors used liposomes loaded with MR contrast agent as a useful means of delivering siRNA to tumors and imaging the delivery by MRI [114].

More recently, the same group developed a very novel technology for monitoring RNAi-mediated gene knockdown by MRI [115]. Lentivirus expressing shRNA against choline kinase (chk) was injected into breast tumor-bearing mice. Transduction efficiency was assessed optically by EGFP expression. Phosphatidylcholine (PC) levels were monitored noninvasively by ^{31}P MRS in tumors showing lower PC and phosphomonoester levels that were associated with reduced tumor growth and proliferation. In the past year, this technology was extended further to combine siRNA delivery and imaging with tumor therapy [116]. The authors synthesized a nanoplex carrying MRI reporters, siRNA, and a functional prodrug enzyme for delivery in breast tumors. siRNA targeting of choline kinase-α (chk-α), an enzyme significantly upregulated in aggressive breast cancer cells, was combined with the prodrug enzyme bacterial cytosine deaminase (bCD) that converts the nontoxic prodrug 5-fluorocytosine (5-FC) to cytotoxic 5-FU. *In vivo* MRI and optical imaging showed efficient intratumoral nanoplex delivery (Fig. 9.6a and b). siRNA-mediated downregulation of chk-α and the conversion of 5-FC to 5-FU by bCD were detected noninvasively with MR spectroscopic imaging (Fig. 9.6c and d). Combined siRNA and prodrug enzyme-activated treatment achieved higher growth delay than either treatment alone.

Other promising MRI-detectable siRNA delivery vehicles include nanovectors consisting of a superparamagnetic iron oxide nanoparticle core coated with PEG-grafted chitosan [117], PEI, and surface functionalized hollow manganese oxide nanoparticles [118]. These probes are still under development and have only undergone *in vitro* testing.

9.6 FUTURE OUTLOOK

The field of image-guided siRNA delivery is developing very rapidly due to the clear potential of RNAi as a therapeutic strategy. siRNA represents a unique treatment agent, since it can form the backbone for the modular design of a wide array of molecular drugs relevant to many diseases. In order to develop siRNA-based therapeutic agents, however, it is key to obtain comprehensive quantitative information about their bioavailability *in vivo*, the efficiency of the silencing effect mediated by them, and the therapeutic sequelae to this effect. As illustrated by the examples given in this chapter, noninvasive imaging can address all of these issues and bring about the transformation of many of the described technologies into standard clinical practice in the near future.

REFERENCES

1. Nabel G. Genetic, cellular and immune approach to disease therapy: past and future. Nat Med 2004;10:135–141.
2. Braasch DA, Corey DR. Novel antisense and peptide nucleic acid strategies for controlling gene expression. Biochemistry 2002;41:4503–4510.

3. Elbashir SM, Harborth J, Lendeckel W, et al. Duplexes of 21-nucleotide RNAs mediate RNA interference in cultured mammalian cells. Nature 2001;411:494–498.
4. Song E, Lee SK, Wang J, et al. RNA interference targeting Fas protects mice from fulminant hepatitis. Nat Med 2003;9:347–351.
5. Zender L, Hutker S, Liedtke C, et al. Caspase 8 small interfering RNA prevents acute liver failure in mice. Proc Natl Acad Sci USA 2003;100:7797–7802.
6. Layzer JM, McCaffrey AP, Tanner AK, et al. In vivo activity of nuclease-resistant siRNAs. RNA 2004;10:766–771.
7. Hassani Z, Lemkine GF, Erbacher P, et al. Lipid-mediated siRNA delivery down-regulates exogenous gene expression in the mouse brain at picomolar levels. J Gene Med 2005;7:198–207.
8. Landen CN, Jr., Chavez-Reyes A, Bucana C, et al. Therapeutic EphA2 gene targeting in vivo using neutral liposomal small interfering RNA delivery. Cancer Res 2005;65: 6910–6918.
9. Santel A, Aleku M, Keil O, et al. A novel siRNA-lipoplex technology for RNA interference in the mouse vascular endothelium. Gene Ther 2006;13:1222–1234.
10. Santel A, Aleku M, Keil O, et al. RNA interference in the mouse vascular endothelium by systemic administration of siRNA-lipoplexes for cancer therapy. Gene Ther 2006;13 (18):1360–1370.
11. Takei Y, Kadomatsu K, Yuzawa Y, et al. A small interfering RNA targeting vascular endothelial growth factor as cancer therapeutics. Cancer Res 2004;64:3365–3370.
12. Takeshita F, Minakuchi Y, Nagahara S, et al. Efficient delivery of small interfering RNA to bone-metastatic tumors by using atelocollagen in vivo. Proc Natl Acad Sci USA 2005;102 (34):12177–12182.
13. Song E, Zhu P, Lee SK, et al. Antibody mediated in vivo delivery of small interfering RNAs via cell-surface receptors. Nat Biotechnol 2005;23:709–717.
14. Hu-Lieskovan S, Heidel JD, Bartlett DW, et al. Sequence-specific knockdown of EWS-FLI1 by targeted, nonviral delivery of small interfering RNA inhibits tumor growth in a murine model of metastatic Ewing's sarcoma. Cancer Res 2005;65:8984–8992.
15. Grzelinski M, Urban-Klein B, Martens T, et al. RNA interference-mediated gene silencing of pleiotrophin through polyethylenimine-complexed small interfering RNAs in vivo exerts antitumoral effects in glioblastoma xenografts. Hum Gene Ther 2006;17 (7): 751–766.
16. Urban-Klein B, Werth S, Abuharbeid S,. RNAi-mediated gene-targeting through systemic application of polyethylenimine (PEI)-complexed siRNA in vivo. Gene Ther 2005;12: 461–466.
17. Schiffelers RM, Ansari A, Xu J, et al. Cancer siRNA therapy by tumor selective delivery with ligand-targeted sterically stabilized nanoparticle. Nucleic Acids Res 2004;32:e149.
18. Soutschek J, Akinc A, Bramlage B, et al. Therapeutic silencing of an endogenous gene by systemic administration of modified siRNAs. Nature 2004;432:173–178.
19. Morrissey DV, Lockridge JA, Shaw L, et al. Potent and persistent in vivo anti-HBV activity of chemically modified siRNAs. Nat Biotechnol 2005;23:1002–1007.
20. Zimmermann TS, Lee AC, Akinc A, et al. RNAi-mediated gene silencing in non-human primates. Nature 2006;441:111–114.
21. Dorn G, Patel S, Wotherspoon G, et al. siRNA relieves chronic neuropathic pain. Nucleic Acids Res 2004;32:e49.

22. Hamar P, Song E, Kokeny G, et al. Small interfering RNA targeting Fas protects mice against renal ischemia-reperfusion injury. Proc Natl Acad Sci USA 2004;101:14883–14888.

23. Bitko V, Musiyenko A, Shulyayeva O, et al. Inhibition of respiratory viruses by nasally administered siRNA. Nat Med 2005;11:50–55.

24. Reich SJ, Fosnot J, Kuroki A, et al. Small interfering RNA (siRNA) targeting VEGF effectively inhibits ocular neovascularization in a mouse model. Mol Vis 2003;9:210–216.

25. Li BJ, Tang Q, Cheng D, et al. Using siRNA in prophylactic and therapeutic regimens against SARS coronavirus in Rhesus macaque. Nat Med 2005;11:944–951.

26. Filleur S, Courtin A, Ait-Si-Ali S, et al. SiRNA-mediated inhibition of vascular endothelial growth factor severely limits tumor resistance to antiangiogenic thrombospondin-1 and slows tumor vascularization and growth. Cancer Res 2003;63:3919–3922.

27. Duxbury M, Ito H, Benoit E, et al. RNA interference targeting focal adhesion kinase enhances pancreatic adenocarcinoma gemcitabine chemosensitivity. Biochem Biophys Res Commun 2003;311:786–792.

28. Duxbury MS, Ito H, Zinner MJ, et al. EphA2: a determinant of malignant cellular behavior and a potential therapeutic target in pancreatic adenocarcinoma. Oncogene 2004;23: 1448–1456.

29. Duxbury M, Ito H, Zinner M, et al. RNA interference targeting the M2 subunit of ribonucleotide reductase enhances pancreatic adenocarcinoma chemosensitivity to gemcitabine. Oncogene 2004;23:1539–1548.

30. Duxbury MS, Matros E, Ito H, et al. Systemic siRNA-mediated gene silencing: a new approach to targeted therapy of cancer. Ann Surg 2004;240:667–674; discussion 75–76.

31. Verma UN, Surabhi RM, Schmaltieg A, et al. Small interfering RNAs directed against beta-catenin inhibit the in vitro and in vivo growth of colon cancer cells. Clin Cancer Res 2003;9:1291–1300.

32. Pal A, Ahmad A, Khan S, et al. Systemic delivery of RafsiRNA using cationic cardiolipin liposomes silences Raf-1 expression and inhibits tumor growth in xenograft model of human prostate cancer. Int J Oncol 2005;26:1087–1091.

33. Tirro E, Consoli ML, Massimino M, et al. Altered expression of c-IAP1, survivin, and Smac contributes to chemotherapy resistance in thyroid cancer cells. Cancer Res 2006;66:4263–4272.

34. Shi Y, Zou M, Collison K, et al. Ribonucleic acid interference targeting S100A4 (Mts1) suppresses tumor growth and metastasis of anaplastic thyroid carcinoma in a mouse model. J Clin Endocrinol Metab 2006;91:2373–2379.

35. Song X, Liu X, Chi W, et al. Hypoxia-induced resistance to cisplatin and doxorubicin in non-small cell lung cancer is inhibited by silencing of HIF-1alpha gene. Cancer Chemother Pharmacol 2006;58 (6):776–784.

36. Halder J, Landen CN, Jr., Lutgendorf SK, et al. Focal adhesion kinase silencing augments docetaxel-mediated apoptosis in ovarian cancer cells. Clin Cancer Res 2005;11: 8829–8836.

37. Nakamura M, Masutomi K, Kyo S, et al. Efficient inhibition of human telomerase reverse transcriptase expression by RNA interference sensitizes cancer cells to ionizing radiation and chemotherapy. Hum Gene Ther 2005;16:859–868.

38. Leong CT, Ong CK, Tay SK, et al. Silencing expression of UO-44 (CUZD1) using small interfering RNA sensitizes human ovarian cancer cells to cisplatin in vitro. Oncogene 2007;26 (6):870–80.

39. Olivero M, Ruggiero T, Saviozzi S, et al. Genes regulated by hepatocyte growth factor as targets to sensitize ovarian cancer cells to cisplatin. Mol Cancer Ther 2006;5: 1126–1135.

40. Cheung HW, Chun AC, Wang Q, et al. Inactivation of human MAD2B in nasopharyngeal carcinoma cells leads to chemosensitization to DNA-damaging agents. Cancer Res 2006;66:4357–4367.

41. Paik J, Duncan T, Lindahl T, et al. Sensitization of human carcinoma cells to alkylating agents by small interfering RNA suppression of 3-alkyladenine-DNA glycosylase. Cancer Res 2005;65:10472–10477.

42. Chen XP, Wang Q, Guan J, et al. Reversing multidrug resistance by RNA interference through the suppression of MDR1 gene in human hepatoma cells. World J Gastroenterol 2006;12:3332–3337.

43. Sun YL, Zhou GY, Li KN, et al. Suppression of glucosylceramide synthase by RNA interference reverses multidrug resistance in human breast cancer cells. Neoplasma 2006;53: 1–8.

44. Rossi A, Ciafre S, Balsamo M, et al. Targeting the heat shock factor 1 by RNA interference: a potent tool to enhance hyperthermochemotherapy efficacy in cervical cancer. Cancer Res 2006;66:7678–7685.

45. Li L, Lin X, Shoemaker AR, et al. Hypoxia-inducible factor-1 inhibition in combination with temozolomide treatment exhibits robust antitumor efficacy in vivo. Clin Cancer Res 2006;12:4747–4754.

46. Pichler A, Zelcer N, Prior JL, et al. In vivo RNA interference-mediated ablation of MDR1 P-glycoprotein. Clin Cancer Res 2005;11:4487–4494.

47. Jiang G, Li J, Zeng Z, et al. Lentivirus-mediated gene therapy by suppressing survivin in BALB/c nude mice bearing oral squamous cell carcinoma. Cancer Biol Ther 2006;5: 435–440.

48. Duxbury MS, Ito H, Benoit E, et al. Retrovirally mediated RNA interference targeting the M2 subunit of ribonucleotide reductase: a novel therapeutic strategy in pancreatic cancer. Surgery 2004;136:261–269.

49. Hu X, Su F, Qin L, et al. Stable RNA interference of ErbB-2 gene synergistic with epirubicin suppresses breast cancer growth in vitro and in vivo. Biochem Biophys Res Commun 2006;346:778–785.

50. Guo J, Verma UN, Gaynor RB, et al. Enhanced chemosensitivity to irinotecan by RNA interference-mediated down-regulation of the nuclear factor-kappaB p65 subunit. Clin Cancer Res 2004;10:3333–3341.

51. Takei Y, Kadomatsu K, Goto T, et al. Combinational antitumor effect of siRNA against midkine and paclitaxel on growth of human prostate cancer xenografts. Cancer 2006;107 (4):864–873.

52. Fukuda S, Pelus LM. Survivin, a cancer target with an emerging role in normal adult tissues. Mol Cancer Ther 2006;5:1087–1098.

53. Li Y, Jian Z, Xia K, et al. XIAP is related to the chemoresistance and inhibited its expression by RNA interference sensitize pancreatic carcinoma cells to chemotherapeutics. Pancreas 2006;32:288–296.

54. McManus DC, Lefebvre CA, Cherton-Horvat G, et al. Loss of XIAP protein expression by RNAi and antisense approaches sensitizes cancer cells to functionally diverse chemotherapeutics. Oncogene 2004;23:8105–8117.

55. Lima RT, Martins LM, Guimaraes JE, et al. Specific downregulation of bcl-2 and xIAP by RNAi enhances the effects of chemotherapeutic agents in MCF-7 human breast cancer cells. Cancer Gene Ther 2004;11:309–316.

56. Belyanskaya LL, Hopkins-Donaldson S, Kurtz S, et al. Cisplatin activates Akt in small cell lung cancer cells and attenuates apoptosis by survivin upregulation. Int J Cancer 2005;117:755–763.

57. Jeong JH, Kim SH, Lee M, et al. Non-viral systemic delivery of Fas siRNA suppresses cyclophosphamide-induced diabetes in NOD mice. J Control Release 2010;143:88–94.

58. Saberi M, Bjelica D, Schenk S, et al. Novel liver-specific TORC2 siRNA corrects hyperglycemia in rodent models of type 2 diabetes. Am J Physiol Endocrinol Metab 2009;297(5):E1137–E1146.

59. Zhou HH, Chin CN, Wu M, et al. Suppression of PC-1/ENPP-1 expression improves insulin sensitivity in vitro and in vivo. Eur J Pharmacol 2009;616:346–352.

60. Zhang Q, Shi Y, Wada J, et al. In vivo delivery of Gremlin siRNA plasmid reveals therapeutic potential against diabetic nephropathy by recovering bone morphogenetic protein-7. PLoS One 2010;5:e11709.

61. Nguyen PD, Tutela JP, Thanik VD, et al. Improved diabetic wound healing through topical silencing of p53 is associated with augmented vasculogenic mediators. Wound Repair Regen 2010;18:553–559.

62. Hirano Y, Sakurai E, Matsubara A, et al. Suppression of ICAM-1 in retinal and choroidal endothelial cells by plasmid small-interfering RNAs in vivo. Invest Ophthalmol Vis Sci 2010;51:508–515.

63. Yuan H, Lanting L, Xu ZG, et al. Effects of cholesterol-tagged small interfering RNAs targeting 12/15-lipoxygenase on parameters of diabetic nephropathy in a mouse model of type 1 diabetes. Am J Physiol Renal Physiol 2008;295:F605–F617.

64. Doege H, Grimm D, Falcon A, et al. Silencing of hepatic fatty acid transporter protein 5 in vivo reverses diet-induced non-alcoholic fatty liver disease and improves hyperglycemia. J Biol Chem 2008;283:22186–22192.

65. Ortiz CM, Ito T, Hashimoto Y, et al. Effects of small interfering RNAs targeting fascin on human esophageal squamous cell carcinoma cell lines. Diagn Pathol 2010;5:41.

66. McCormack AL, Mak SK, Henderson JM, et al. Alpha-synuclein suppression by targeted small interfering RNA in the primate substantia nigra. PLoS One 2010;5:e12122.

67. Gorbatyuk OS, Li S, Nash K, et al. In vivo RNAi-mediated alpha-synuclein silencing induces nigrostriatal degeneration. Mol Ther 2010;18:1450–1457.

68. Bevers MB, Ingleton LP, Che D, et al. RNAi targeting micro-calpain increases neuron survival and preserves hippocampal function after global brain ischemia. Exp Neurol 2010;224:170–177.

69. Mathupala SP. Delivery of small-interfering RNA (siRNA) to the brain. Expert Opin Ther Pat 2009;19:137–140.

70. Kumar P, Wu H, McBride JL, et al. Transvascular delivery of small interfering RNA to the central nervous system. Nature 2007;448:39–43.

71. Lesniak MS. Novel advances in drug delivery to brain cancer. Technol Cancer Res Treat 2005;4:417–428.

72. Pulford B, Reim N, Bell A, et al. Liposome-siRNA-peptide complexes cross the blood-brain barrier and significantly decrease PrP on neuronal cells and PrP in infected cell cultures. PLoS One 2010;5:e11085.

73. Campbell M, Kiang AS, Kenna PF, et al. RNAi-mediated reversible opening of the blood-brain barrier. J Gene Med 2008;10:930–947.

74. Muratovska A, Eccles MR. Conjugate for efficient delivery of short interfering RNA (siRNA) into mammalian cells. FEBS Lett 2004;558:63–68.

75. Chiu YL, Ali A, Chu CY, et al. Visualizing a correlation between siRNA localization, cellular uptake, and RNAi in living cells. Chem Biol 2004;11:1165–1175.
76. Ifediba MA, Medarova Z, Ng SW, et al. siRNA delivery to CNS cells using a membrane translocation peptide. Bioconjug Chem 2010;21:803–806.
77. Lu PY, Xie F, Woodle MC. In vivo application of RNA interference: from functional genomics to therapeutics. Adv Genet 2005;54:117–142.
78. McCaffrey AP, Meuse L, Pham TT, et al. RNA interference in adult mice. Nature 2002;418:38–39.
79. Lewis DL, Hagstrom JE, Loomis AG, et al. Efficient delivery of siRNA for inhibition of gene expression in postnatal mice. Nat Genet 2002;32:107–108.
80. Kobayashi N, Matsui Y, Kawase A, et al. Vector-based in vivo RNA interference: dose- and time-dependent suppression of transgene expression. J Pharmacol Exp Ther 2004; 308:688–693.
81. Bartlett DW, Davis ME. Insights into the kinetics of siRNA-mediated gene silencing from live-cell and live-animal bioluminescent imaging. Nucleic Acids Res 2006;34: 322–333.
82. Bartlett DW, Davis ME. Effect of siRNA nuclease stability on the in vitro and in vivo kinetics of siRNA-mediated gene silencing. Biotechnol Bioeng 2007;97:909–921.
83. Bartlett DW, Su H, Hildebrandt IJ, et al. Impact of tumor-specific targeting on the biodistribution and efficacy of siRNA nanoparticles measured by multimodality in vivo imaging. Proc Natl Acad Sci USA 2007;104:15549–15554.
84. Tao W, Davide JP, Cai M, et al. Noninvasive imaging of lipid nanoparticle-mediated systemic delivery of small-interfering RNA to the liver. Mol Ther 2010;18:1657–1666.
85. Rousseau J, Escriou V, Perrot P, et al. Advantages of bioluminescence imaging to follow siRNA or chemotherapeutic treatments in osteosarcoma preclinical models. Cancer Gene Ther 2010;17:387–397.
86. Takahashi Y, Nishikawa M, Kobayashi N, et al. Gene silencing in primary and metastatic tumors by small interfering RNA delivery in mice: quantitative analysis using melanoma cells expressing firefly and sea pansy luciferases. J Control Release 2005;105: 332–343.
87. Takabatake Y, Isaka Y, Mizui M, et al. Exploring RNA interference as a therapeutic strategy for renal disease. Gene Ther 2005;12:965–973.
88. Zhang Y, Boado RJ, Pardridge WM. In vivo knockdown of gene expression in brain cancer with intravenous RNAi in adult rats. J Gene Med 2003;5:1039–1045.
89. Minakuchi Y, Takeshita F, Kosaka N, et al. Atelocollagen-mediated synthetic small interfering RNA delivery for effective gene silencing in vitro and in vivo. Nucleic Acids Res 2004;32:e109.
90. Lewis J, Achilefu S, Garbow J, et al. Small animal imaging: current technology and perspectives for oncological imaging. Eur J Cancer 2002;38:2173–2188.
91. Sorensen DR, Leirdal M, Sioud M. Gene silencing by systemic delivery of synthetic siRNAs in adult mice. J Mol Biol 2003;327:761–766.
92. Xia H, Mao Q, Paulson HL, et al. siRNA-mediated gene silencing in vitro and in vivo. Nat Biotechnol 2002;20:1006–1010.
93. Rubinson DA, Dillon CP, Kwiatkowski AV, et al. A lentivirus-based system to functionally silence genes in primary mammalian cells, stem cells and transgenic mice by RNA interference. Nat Genet 2003;33:401–406.
94. Gonzalez-Gonzalez E, Ra H, Hickerson RP, Wang Q, et al. siRNA silencing of keratinocyte-specific GFP expression in a transgenic mouse skin model. Gene Ther 2009;16:963–972.

95. Sioud M, Sorensen DR. Cationic liposome-mediated delivery of siRNAs in adult mice. Biochem Biophys Res Commun 2003;312:1220–1225.

96. Chen AA, Derfus AM, Khetani SR, et al. Quantum dots to monitor RNAi delivery and improve gene silencing. Nucleic Acids Res 2005;33:e190.

97. Derfus AM, Chen AA, Min DH, et al. Targeted quantum dot conjugates for siRNA delivery. Bioconjug Chem 2007;18:1391–1396.

98. Tan WB, Jiang S, Zhang Y. Quantum-dot based nanoparticles for targeted silencing of HER2/neu gene via RNA interference. Biomaterials 2007;28:1565–1571.

99. Shim MS, Kim CS, Ahn YC, et al. Combined multimodal optical imaging and targeted gene silencing using stimuli-transforming nanotheragnostics. J Am Chem Soc 2010;132: 8316–8324.

100. Bonoiu AC, Mahajan SD, Ding H, et al. Nanotechnology approach for drug addiction therapy: gene silencing using delivery of gold nanorod-siRNA nanoplex in dopaminergic neurons. Proc Natl Acad Sci USA 2009;106:5546–5550.

101. Liu N, Ding H, Vanderheyden JL, et al. Radiolabeling small RNA with technetium-99m for visualizing cellular delivery and mouse biodistribution. Nucl Med Biol 2007;34: 399–404.

102. Merkel OM, Librizzi D, Pfestroff A, et al. In vivo SPECT and real-time gamma camera imaging of biodistribution and pharmacokinetics of siRNA delivery using an optimized radiolabeling and purification procedure. Bioconjug Chem 2009;20:174–182.

103. Merkel OM, Beyerle A, Librizzi D, et al. Nonviral siRNA delivery to the lung: investigation of PEG-PEI polyplexes and their in vivo performance. Mol Pharm 2009;6: 1246–1260.

104. Merkel OM, Librizzi D, Pfestroff A, et al. Stability of siRNA polyplexes from poly(ethylenimine) and poly(ethylenimine)-g-poly(ethylene glycol) under in vivo conditions: effects on pharmacokinetics and biodistribution measured by Fluorescence fluctuation spectroscopy and single photon emission computed tomography (SPECT) imaging. J Control Release 2009;138:148–159.

105. Viel T, Boisgard R, Kuhnast B, et al. Molecular imaging study on in vivo distribution and pharmacokinetics of modified small interfering RNAs (siRNAs). Oligonucleotides 2008; 18:201–212.

106. Hatanaka K, Asai T, Koide H, et al. Development of double-stranded siRNA labeling method using positron emitter and its in vivo trafficking analyzed by positron emission tomography. Bioconjug Chem 2010;21:756–763.

107. Mudd SR, Trubetskoy VS, Blokhin AV, et al. Hybrid PET/CT for noninvasive pharmacokinetic evaluation of dynamic PolyConjugates, a synthetic siRNA delivery system. Bioconjug Chem 2010;21:1183–1189.

108. Medarova Z, Pham W, Farrar C, et al. In vivo imaging of siRNA delivery and silencing in tumors. Nat Med 2007;13:372–377.

109. Medarova Z, Kumar M, Ng SW, et al. Multifunctional magnetic nanocarriers for image-tagged SiRNA delivery to intact pancreatic islets. Transplantation 2008;86:1170–1177.

110. Moore A, Medarova Z, Potthast A, et al. In vivo targeting of underglycosylated MUC-1 tumor antigen using a multi-modal imaging probe. Cancer Res 2004;64:1821–1827.

111. Fiorina P, Shapiro A, Ricordi C, et al. The clinical impact of islet transplantation. Am J Transplant 2008;8:1990–1997.

112. Evgenov N, Medarova Z, Dai G, et al. In vivo imaging of islet transplantation. Nat Med 2006;12:144–1448.

113. Wang P, Yigit M, Medarova Z, et al. Combined small interfering RNA therapy and in vivo magnetic resonance imaging in islet transplantation. Diabetes 2011;60:565–571.

114. Mikhaylova M, Stasinopoulos I, Kato Y, et al. Imaging of cationic multifunctional liposome-mediated delivery of COX-2 siRNA. Cancer Gene Ther 2009;16:217–226.

115. Krishnamachary B, Glunde K, Wildes F, et al. Noninvasive detection of lentiviral-mediated choline kinase targeting in a human breast cancer xenograft. Cancer Res 2009;69: 3464–3471.

116. Li C, Penet MF, Wildes F, et al. Nanoplex delivery of siRNA and prodrug enzyme for multimodality image-guided molecular pathway targeted cancer therapy. ACS Nano 2010;4:6707–6716.

117. Veiseh O, Kievit FM, Fang C, et al. Chlorotoxin bound magnetic nanovector tailored for cancer cell targeting, imaging, and siRNA delivery. Biomaterials 2010;31:8032–8042.

118. Bae KH, Lee K, Kim C, et al. Surface functionalized hollow manganese oxide nanoparticles for cancer targeted siRNA delivery and magnetic resonance imaging. Biomaterials 2010;32:176–184.

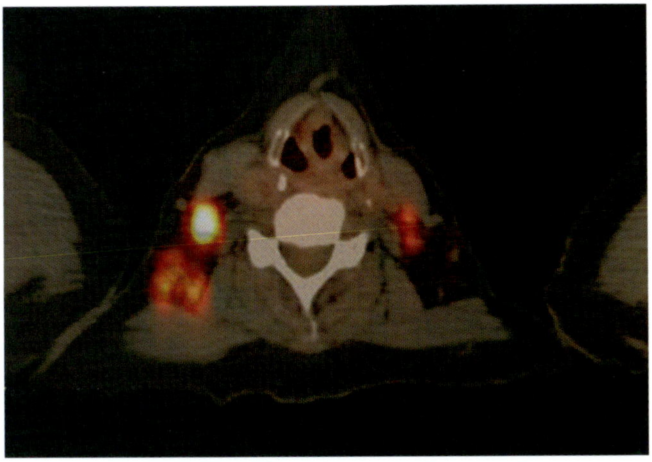

FIGURE 1.1 ^{18}F-FDG PET/CT. Axial view of the neck demonstrates increased uptake (orange) of ^{18}F-FDG in lymph nodes signifying involvement by lymphoma.

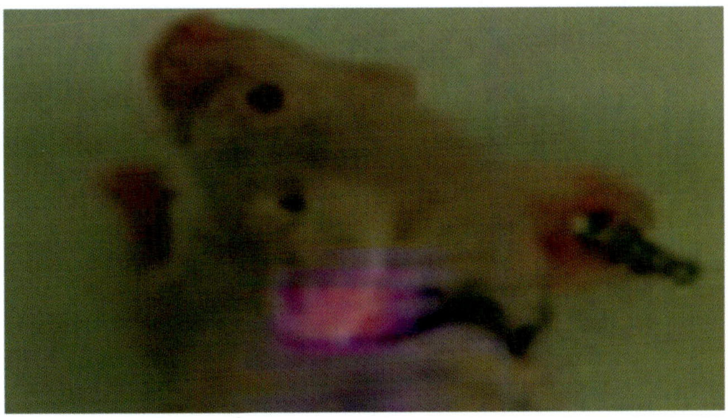

FIGURE 1.2 SPECT imaging of SSTR2-based gene expression. Coronal view of the lungs demonstrates imaging of gene expression after *in vivo* transfer of a somatostatin receptor-based reporter. The reporter was made to express in a human lung tumor on the left side and bound its ligand, ^{111}In-octreotide, as demonstrated by the increased uptake (pink).

Drug Delivery Applications of Noninvasive Imaging: Validation from Biodistribution to Sites of Action, First Edition. Edited by Chun Li and Mei Tian.

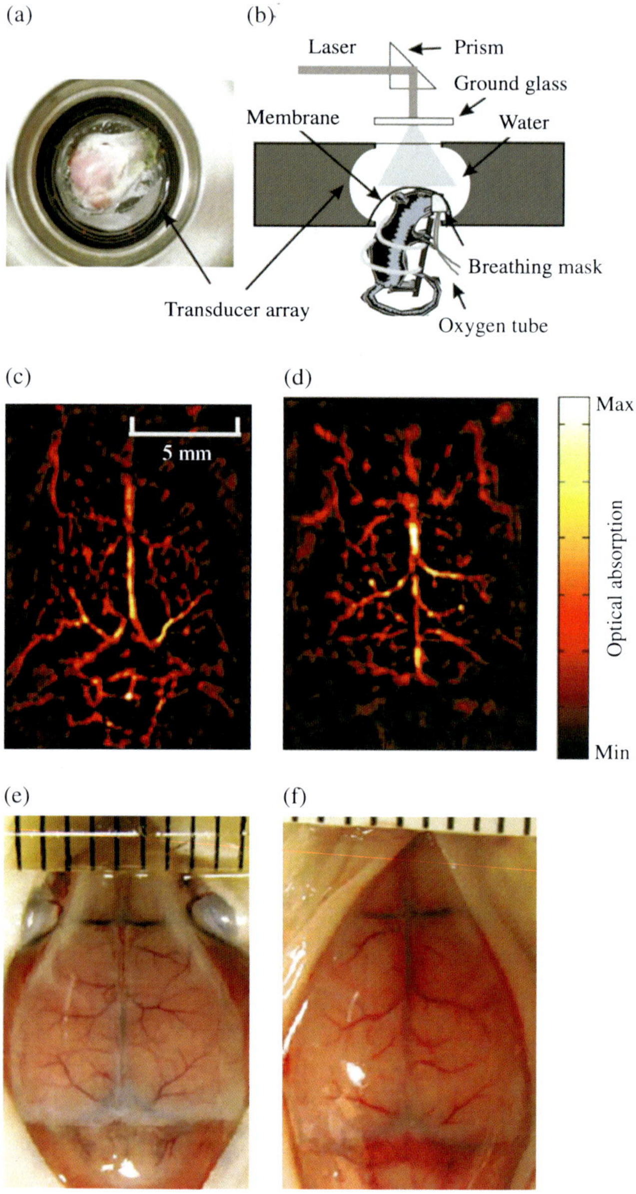

FIGURE 3.1 Full-ring ultrasonic array PACT system and brain cortex imaging. (a) Top view of the full-ring ultrasonic array and the animal head. (b) Side view of the system and the position of the animal. (c, d) *In vivo* photoacoustic cerebral cortex images acquired with scalp and skull intact. (e, f) Open-skull photographs of the cerebral cortexes corresponding to images in (c, d) (Reprinted with permission from reference [40]).

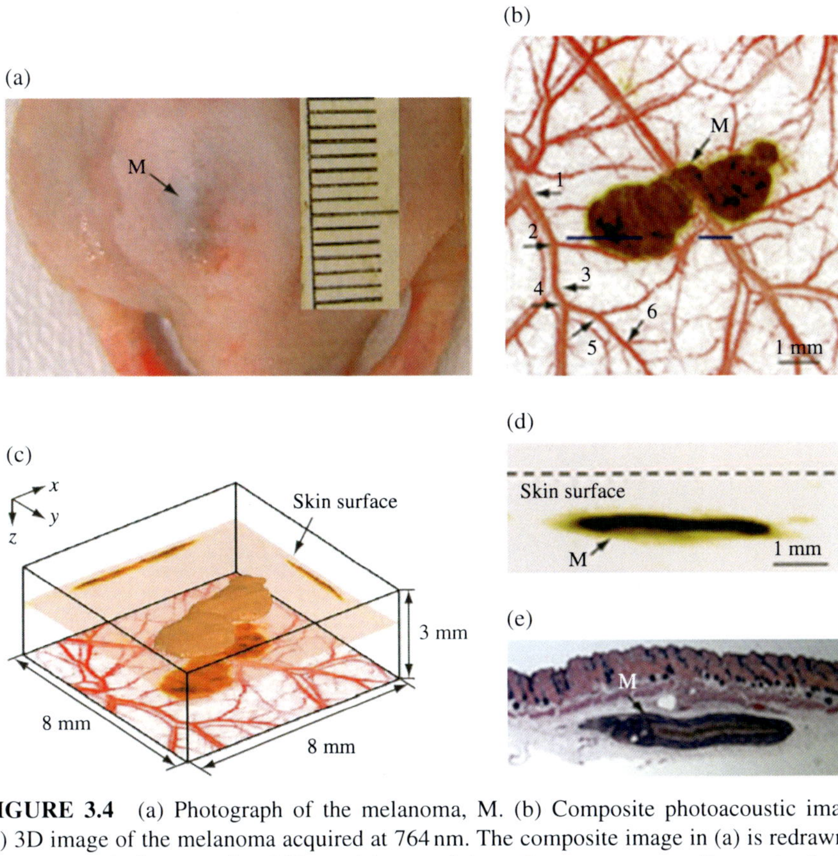

FIGURE 3.4 (a) Photograph of the melanoma, M. (b) Composite photoacoustic image. (c) 3D image of the melanoma acquired at 764 nm. The composite image in (a) is redrawn at the bottom. (d) Cross-sectional (B-scan) image of the melanoma corresponding to the dashed line in (b). (e) Hematoxylin and eosin (HE)-stained section corresponding to the region in (d) (Reprinted with permission from reference [5]).

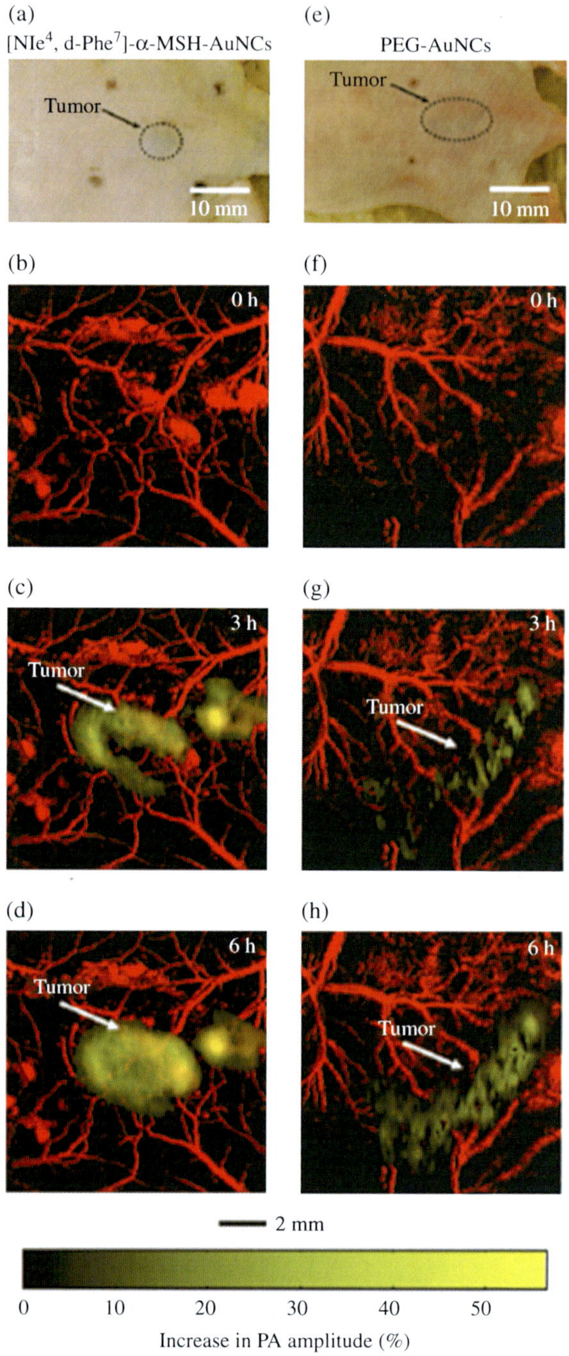

FIGURE 3.11 Noninvasive *in vivo* PAT of B16 melanomas using [Nle^4, d-Phe^7]-α-MSH- and PEG-gold nanocages. (a, e) Photographs of nude mice transplanted with B16 melanomas before nanocage injection. PA images of the B16 melanomas after intravenous injection of 100-μL 10 nM [Nle^4, d-Phe^7]-α-MSH- (b–d) and (f–h) PEG-AuNCs through the tail vein. The background vasculature images were obtained using PAM at 570 nm, and the melanoma images were obtained using PAM at 778 nm (Reprinted with permission from reference [87]).

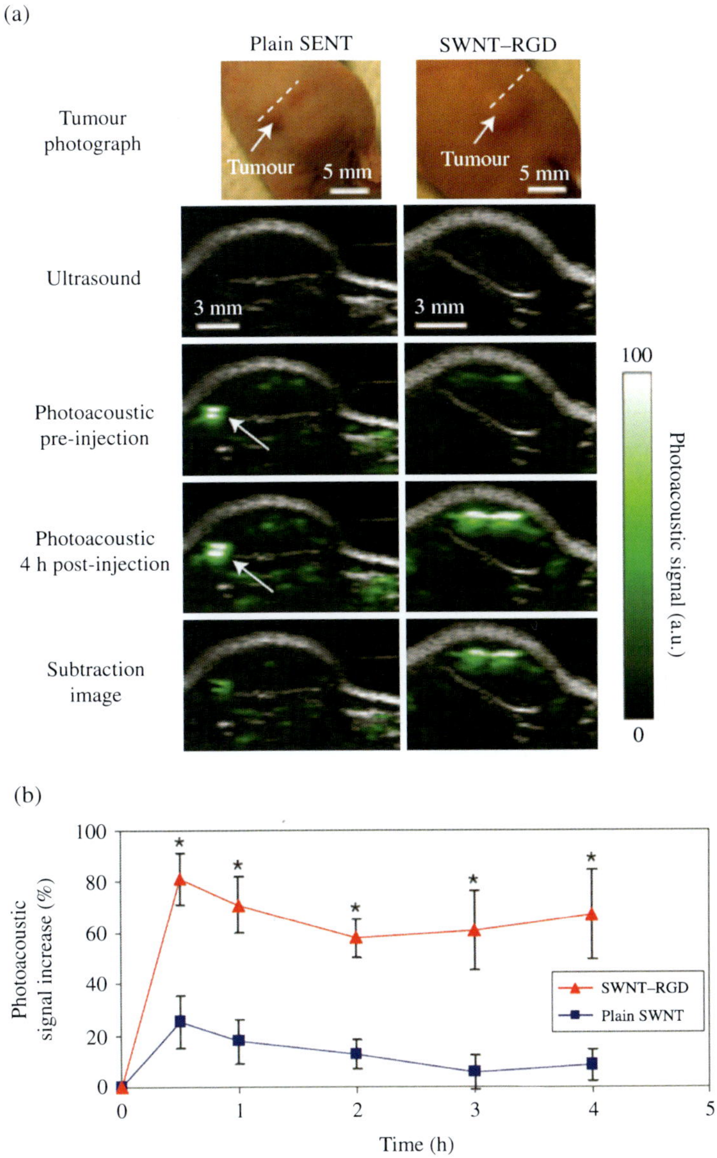

FIGURE 3.13 (a) Ultrasonic (grey) and photoacoustic (green) cross-sectional (along the dotted line) images of the tumor, before and 4 h after the injection of carbon nanotubes. (b) Mice injected with SWNT-RGDs showed a significantly higher photoacoustic signal than mice injected with plain SWNTs. The error bars represent standard error ($n=4$) ($*p<0.05$) (Reprinted with permission from reference [98]).

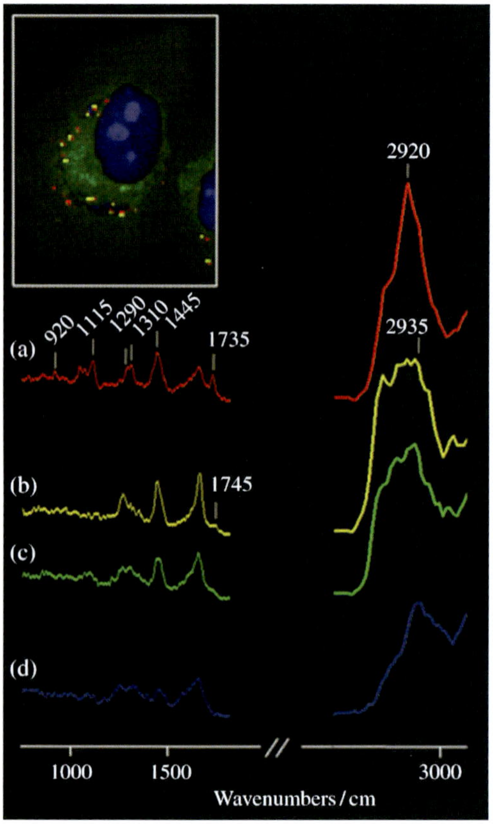

FIGURE 4.3 *Inset*: Pseudo-color Raman image (VCA analysis) of a HeLa cell incubated with PEO-modified PCL nanoparticle-containing medium for 7 h. Red regions: PCL nanoparticle aggregates, yellow regions: lipid/phospholipid inclusions, green: membrane-rich organelles such as ER, Golgi apparatus, and mitochondria; blue: proteins of the cell body. Traces (a–d) represent the Raman spectra of the corresponding regions.

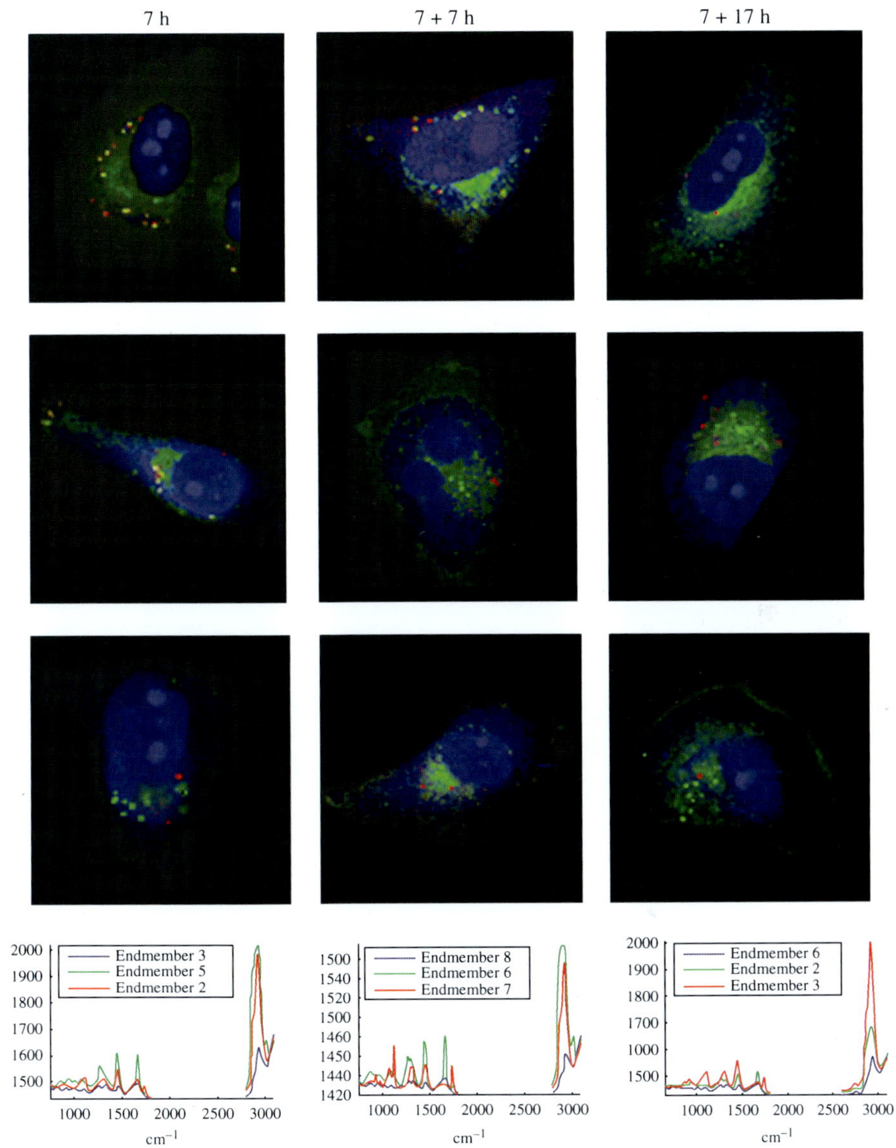

FIGURE 4.4 Time course of PEO-modified PCL nanoparticle degradation patterns obtained from Raman imaging/VCA. Red regions represent PCL aggregates, green regions high lipid/phospholipid content, and blue the proteins of the cell body. Top row: images of cells after 7-h incubation with PCL-containing medium. Middle row: images of cells incubated with PCL-containing medium for 7 h followed by incubation with fresh medium for another 7 h. Bottom row: images of cells incubated in PCL-containing medium for 7 h followed by incubation for another 17 h in fresh medium. EM spectra for the three incubation times are shown on left, corresponding to cells of the last row.

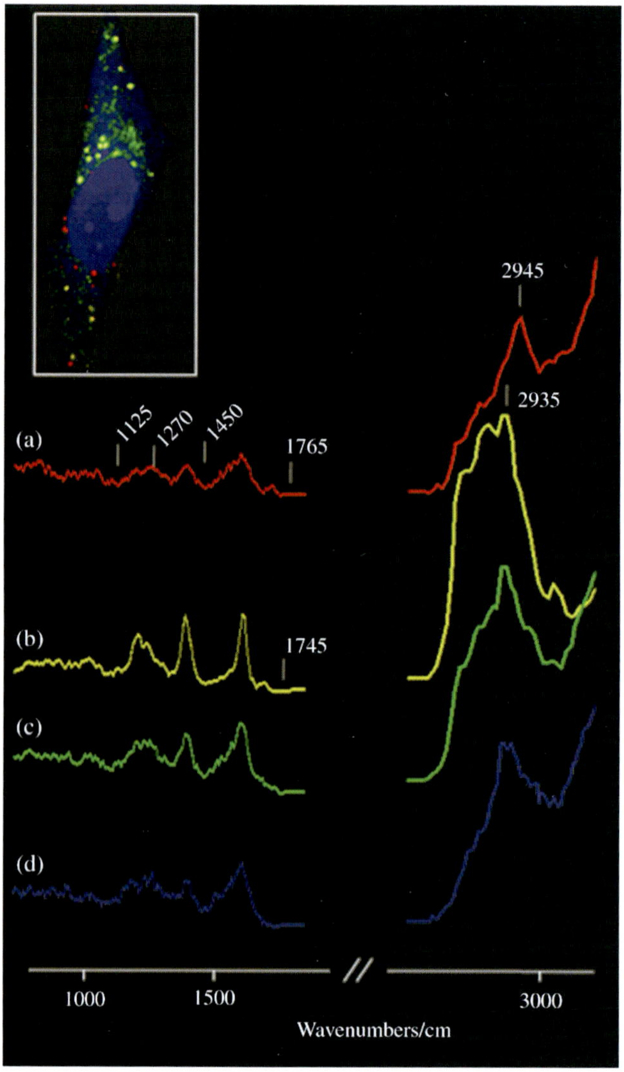

FIGURE 4.5 *Inset*: Pseudo-color Raman image (VCA analysis) of a HeLa cell incubated with PEO-modified PLGA-containing medium for 2 h. Red regions: PLGA aggregates, yellow regions: lipid/phospholipid inclusions, green regions: membrane-rich organelles such as ER, Golgi apparatus, and mitochondria; blue regions: proteins of cell body. Traces (a–d) represent the Raman spectra of the corresponding regions.

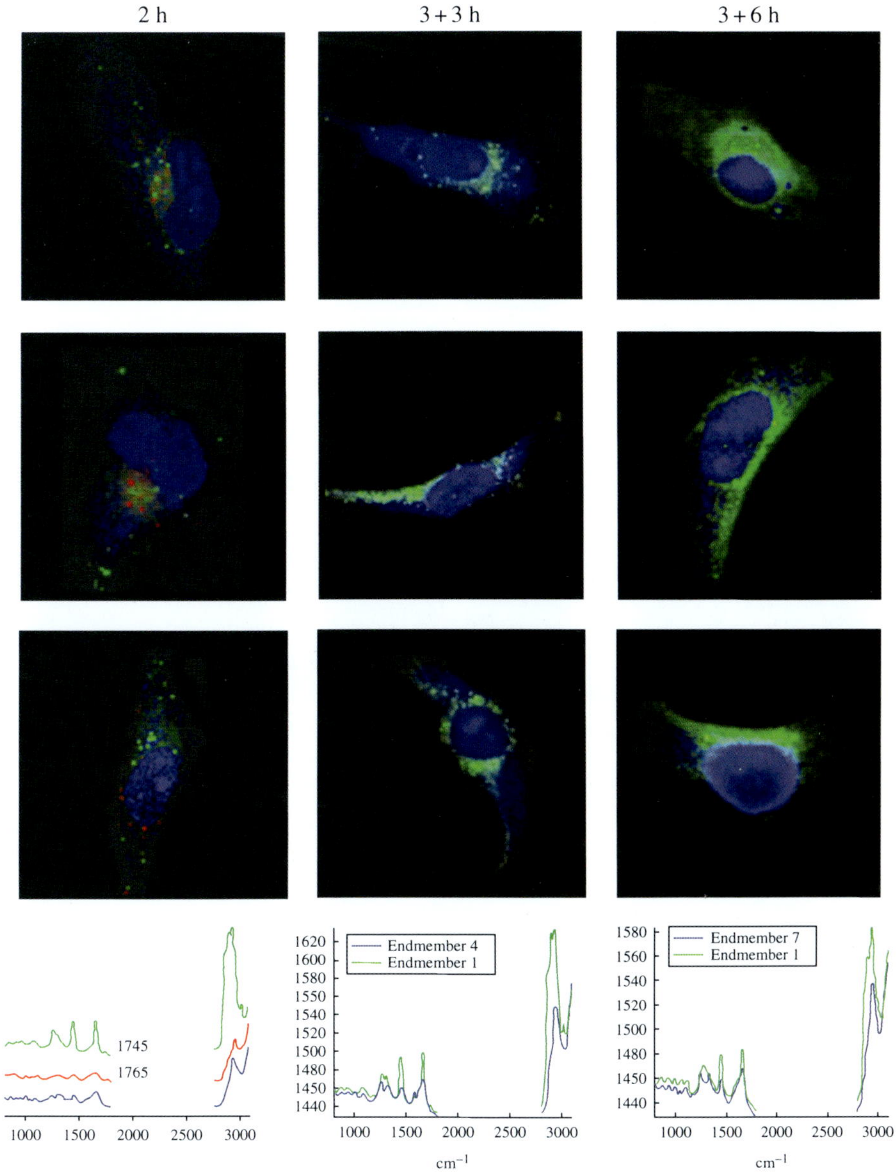

FIGURE 4.6 Time course of PEO-modified PLGA nanoparticle degradation patterns obtained from Raman imaging/VCA. Red regions represent PLGA aggregates, green regions high lipid/phospholipid content, and blue the proteins of the cell body. Top row: images of cells after 2-h incubation with PLGA-containing medium. Middle row: images of cells incubated with PLGA-containing medium for 3 h followed by incubation with fresh medium for another 3 h. Bottom row: images of cells incubated in PLGA-containing medium for 3 h followed by incubation for another 6 h in fresh medium. EM spectra for the three incubation times are shown on left, corresponding to cells of the last row.

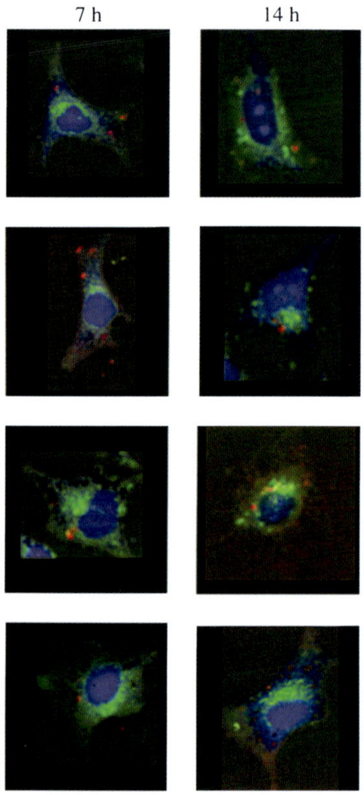

FIGURE 4.8 Time-course experiment of ceramide-loaded nanoparticle distribution within a HeLa cell culture. Top row: 7-h incubation period of HeLa cells. Bottom row: 14-h incubation period of HeLa cells. All cell data analyzed via VCA shows cell body: blue intensity, membrane-rich organelles: green intensity, internalized nanoparticle: red intensity.

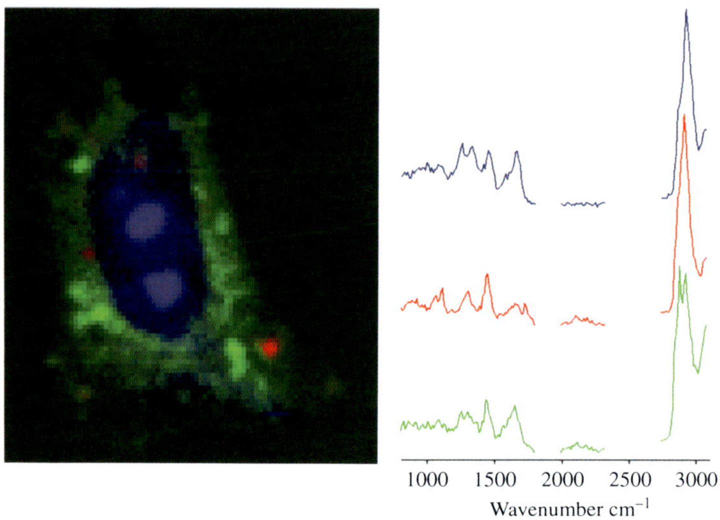

FIGURE 4.9 HeLa cell analyzed via VCA of a 14-h incubation period. Blue region denotes highest protein content; green abundance denotes membrane-rich regions, while the red clusters delineate nanoparticle aggregations. Spectra on the right of the image are of the three overlaid abundances rendered by the VCA.

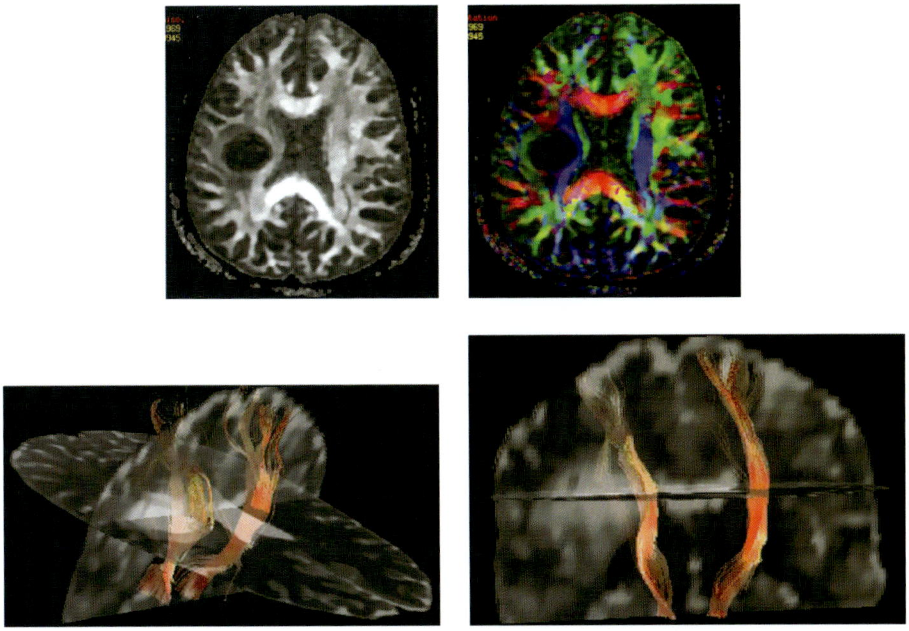

FIGURE 5.16 Top left: FA index image in which the pixel intensity is proportional to the degree of diffusion anisotropy providing a means of imaging white matter tracts. Top right: A color FA map in which white matter tracts running left to right are red, those running anterior–posterior are green, and those running superior–inferior are blue. Middle and bottom: Fiber-tracking images obtained by placing seed pixels in the primary motor cortex in a patient with a high-grade glioma causing a medial shift of the corticospinal white matter tract.

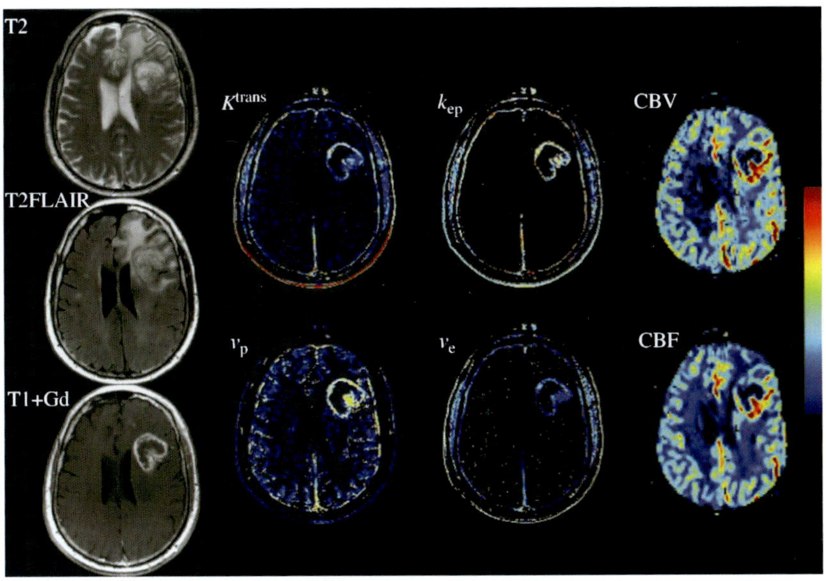

FIGURE 5.19 Various anatomic images and corresponding functional parameter maps in a patient with pathologically proven glioblastoma. The images on the left are standard anatomic images (T2W, T2-FLAIR, and T1W post-Gd contrast agent images). The functional parametric maps obtained from a DCE-MRI acquisition (K^{trans}, k_{ep}, v_p, and v_e) and obtained from a DSC-MRI acquisition (rCBV and rCBF) are shown.

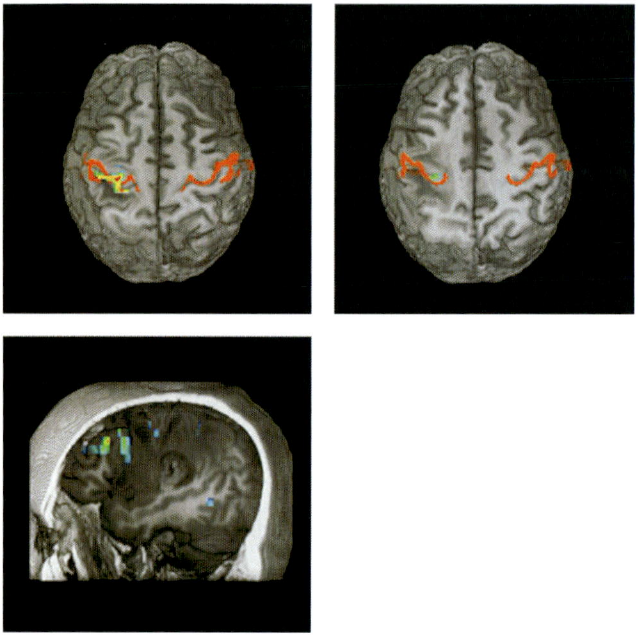

FIGURE 5.20 BOLD fMRI activation maps obtained by cross-correlation of the known stimulus pattern (a simple left-hand finger-tapping task) with the signal intensity time course on a pixel-by-pixel basis in a patient with a neoplasm (two different sections locations, top left and top right, are shown). The yellow/blue/green pixels represent the measured BOLD fMRI response, while the orange pixels represent the Brodmann areas 1, 2, and 3 based on a standard stereotactic atlas. The fMRI BOLD response obtained during an expressive speech task (silently producing as many words as possible beginning with a specified letter in 20-s intervals) is shown on the bottom left.

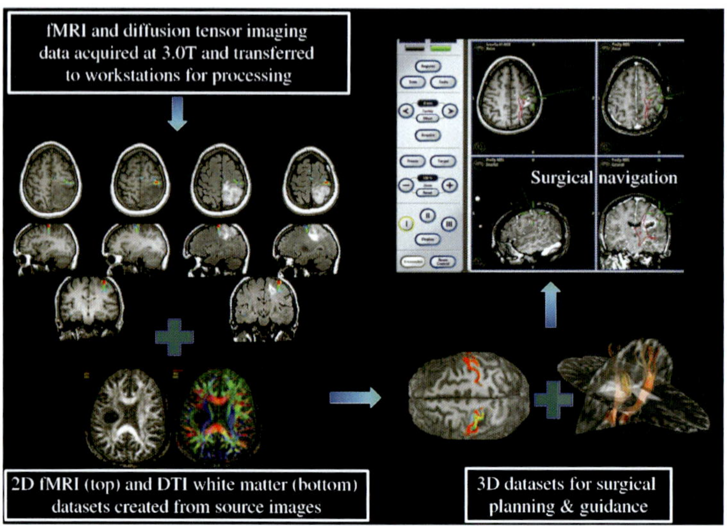

FIGURE 5.21 Combination of BOLD fMRI and diffusion tensor imaging (FA and cortico-spinal fiber-tracking maps) results and transfer to image-guided surgery navigation system. While this application is for identification of regions of motor "eloquent cortex" and associated white matter tracts, any of the parameter maps discussed, for example, K^{trans} and rCBV, can be similarly used for targeting of optimal biopsy targets or for targeting local delivery of drug.

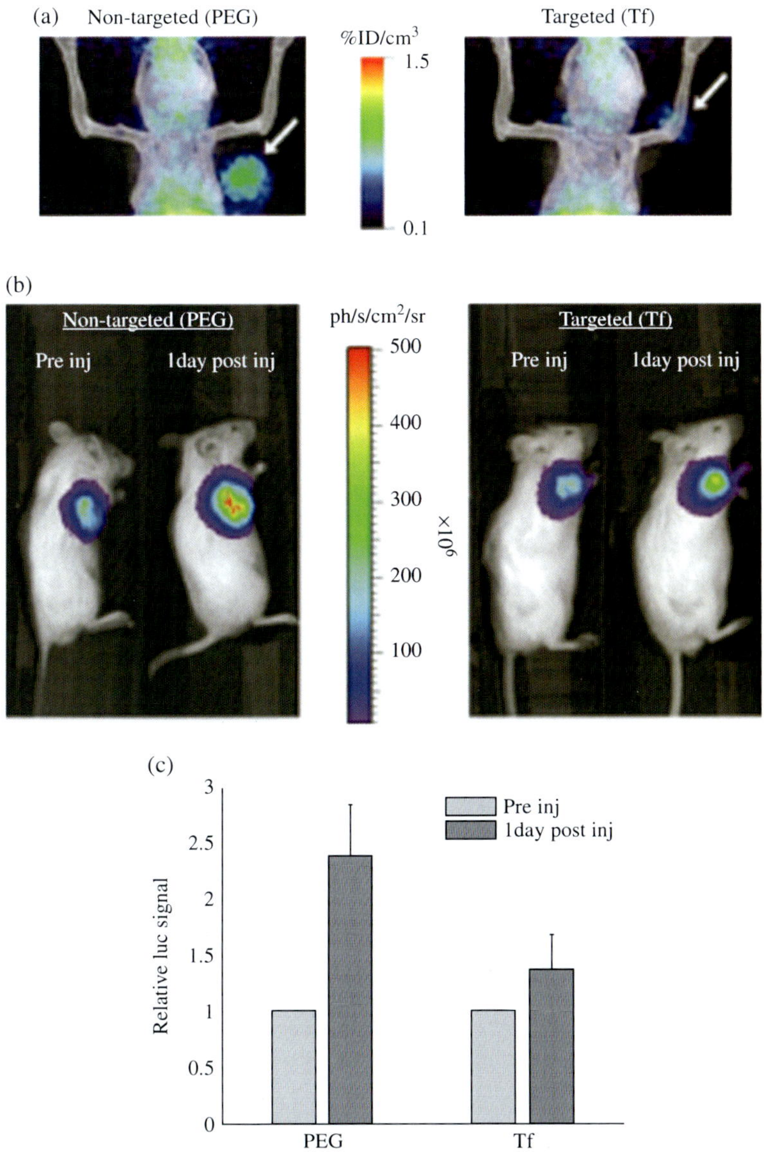

FIGURE 9.1 Multimodality *in vivo* imaging of siRNA nanoparticle delivery and function, using micro-PET/CT and BLI. (a) Fused micro-PET/CT images showing tumor-associated (arrow) activity 1 day after injection of targeted (Tf) and nontargeted (PEG) nanoparticles containing 64Cu-DOTA-siRNA. (b) BLI of the same mice shown in A before injection and 1 day after injection. (c) Relative change in luciferase expression 1 day after injection of Tf-targeted (Tf, $n=7$) and nontargeted (PEG, $n=4$) nanoparticles containing 64Cu-DOTA-siRNA for simultaneous PET imaging. Error bars indicate SE. (Reprinted from Bartlett et al., 2007, with permission from The National Academy of Sciences of the USA.)

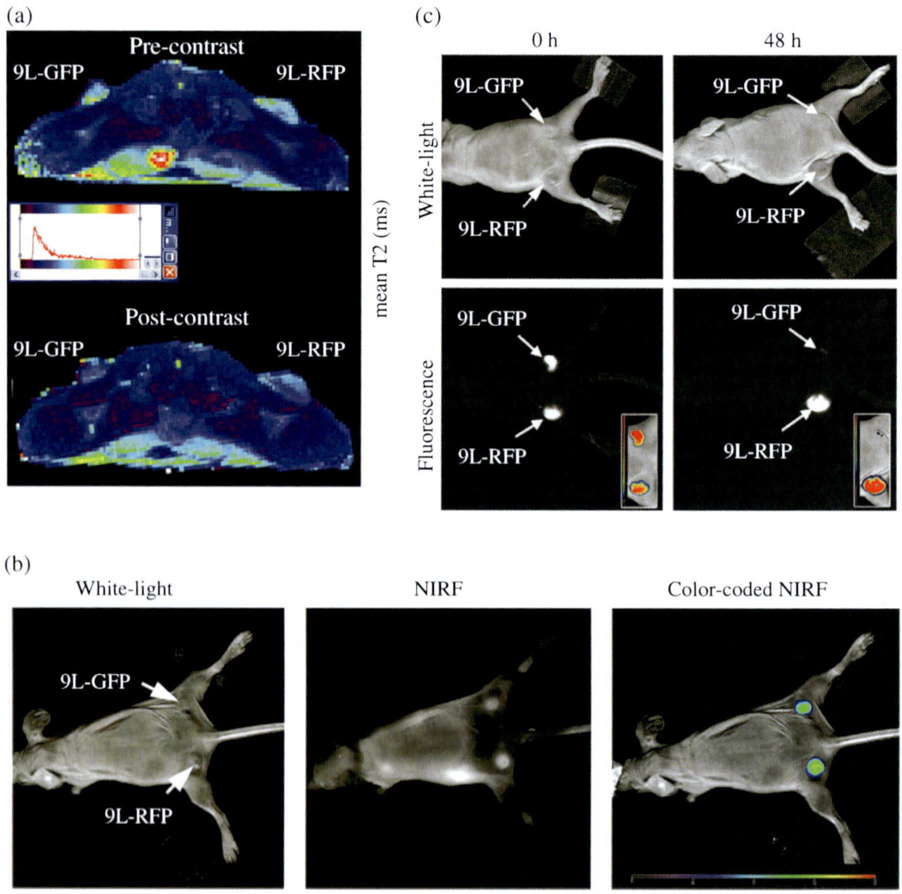

FIGURE 9.4 *In vivo* imaging of MN-NIRF-siGFP delivery to tumors. (a) *In vivo* MRI was performed on mice bearing bilateral 9L-GFP and 9L-RFP tumors before and 24h after MN-NIRF-siGFP administration. Whereas on precontrast T2 images, tumors appeared characteristically bright, following injection of the probe, there was a significant decrease in T2 relaxivity associated with the tumors ($p=0.0012$ for 9L-GFP, $p=0.0049$ for 9L-RFP). Note that T2 relaxation times of muscle tissue remained unchanged. (b) *In vivo* NIRF optical imaging of the same animals was performed immediately after each MRI session. A high-intensity NIRF signal associated with the tumors confirmed the delivery of the MN-NIRF-siGFP probe to these tissues. (c) *In vivo* fluorescence optical imaging of animals bearing bilateral 9L-GFP and 9L-RFP tumors 48h after i.v. probe injection. Forty-eight h after probe administration, 9L-GFP-associated fluorescence diminished dramatically ($p=0.0083$), whereas 9L-RFP fluorescence remained unchanged compared to corresponding fluorescence intensities prior to injection of the probe (0h) (Reprinted from Medarova et al., 2007 with permission from Nature Publishing Group).

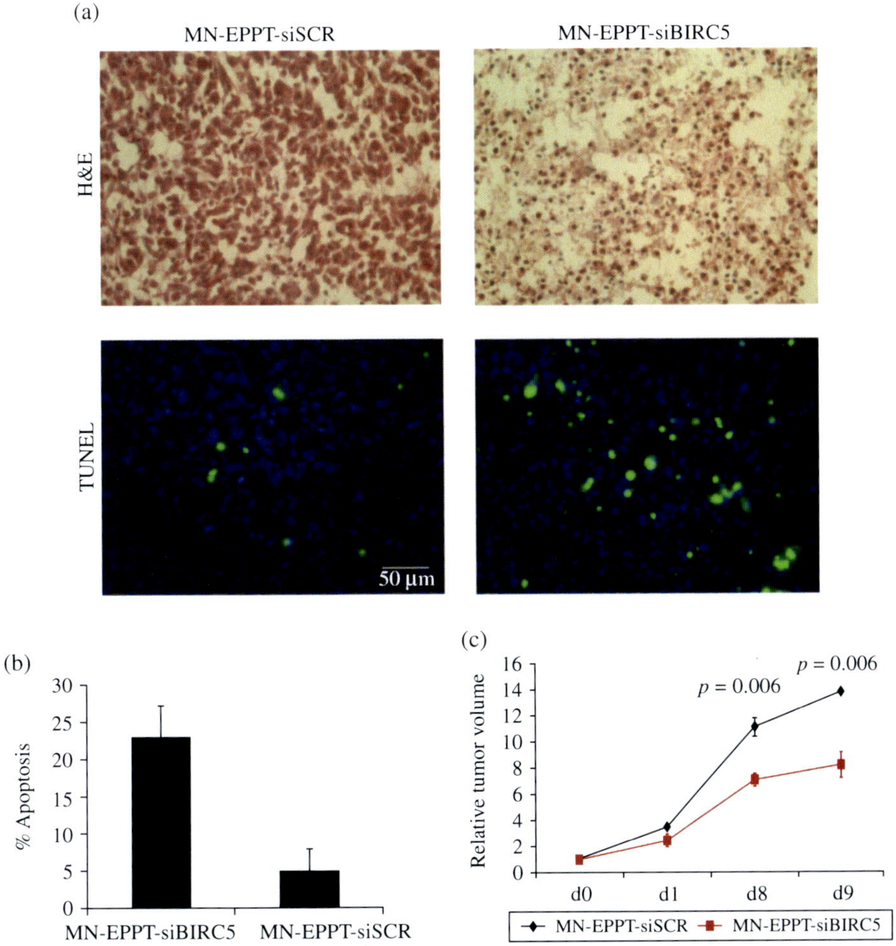

FIGURE 9.5 MN-EPPT-siBIRC5 therapeutic effects. (a) H&E (top) and TUNEL (bottom) staining of tumor tissue from mice treated with MN-EPPT-siBIRC5 or MN-EPPT-siSCR. In the H&E sections, there were extensive eosinophilic areas in the experimental but not control tumors, suggesting the induction of necrosis by the siBIRC5-bearing nanodrug. There was a visible increase in the fraction of apoptotic (green) nuclei in the TUNEL-stained tissues from experimental mice relative to controls, indicating the induction of apoptosis. The tissues were counterstained with DAPI (nuclei, blue). (b) Quantitative analysis of the TUNEL experiments. There was a fivefold induction of apoptosis in the tumor tissues of experimental animals relative to controls ($p = 0.003$). (c) Relative tumor volume measurements of MN-EPPT-siBIRC5—and MN-EPPT-siSCR—injected animals over the course of treatment. There was a significant decrease in tumor growth rate in the experimental animals relative to controls, apparent by 8 days after the beginning of treatment ($p < 0.01$; $n = 4$) (Reprinted from Medarova et al., 2010 with permission from AACR).

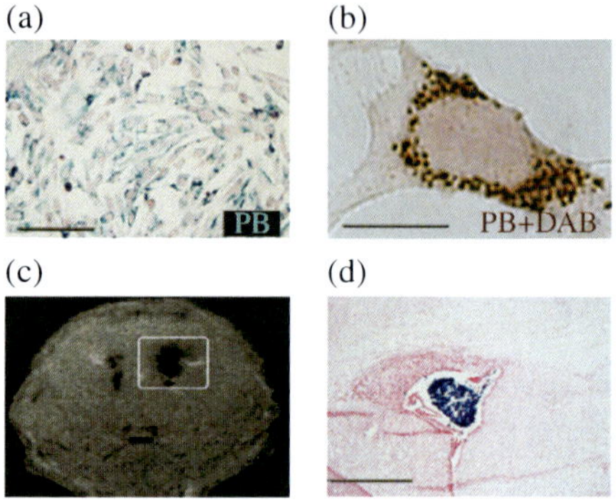

FIGURE 10.3 Direct labeling of cells with SPIO. (a) C17.2 cells (a neural stem cell line) are labeled with Feridex®, as visualized by Prussian Blue staining and (b) diaminobenzidine-enhanced Prussian Blue staining. Nearly all cells are labeled, with an endosomal cytoplasmic distribution of Feridex® particles. (c) Magnetically labeled C17.2 cells induced a strong hypointense signal on the MR images after injection into the lateral ventricles. (d) Prussian Blue staining of the corresponding tissue slice shows a close overlap with the MR signal. Scale bars: *a*, *d* = 100 μm; *b* = 10 μm. Reprinted from [14] with permission.

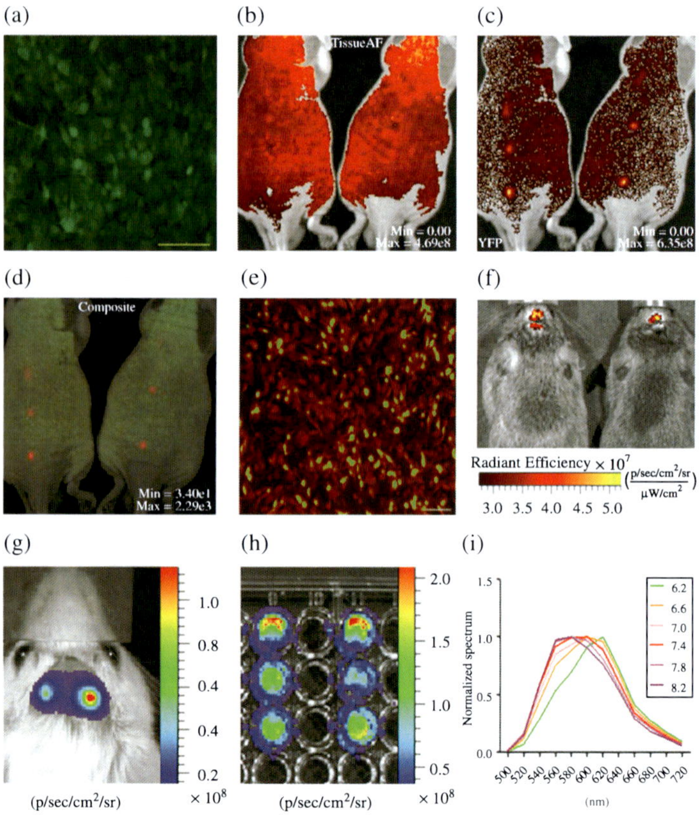

FIGURE 10.4 Fluorescent and luminescent reporter genes. C17.2 cells stably expressing (a) GFP, (e) tdTomato, or (h) firefly luciferase were transplanted into mice and imaged using an IVIS 200 optical system. (b, c, d) Spectral unmixing result of GFP for better visualization of cells that were injected subcutaneously into nude mice. (b) Tissue autofluorescence; (c) Fluorescence from GFP. (d) Unmixing result with red dots showing positive signals. (f) tdTomato expressing C17.2 cells were transplanted into the striatum of rag2 mice, which were tracked by fluorescent imaging. Intensity of luminescence from firefly luciferase expressing C17.2 cells could be quantified *in vitro* (h) or *in vivo* (g). (i) Shift of the emission spectrum of firefly luciferase in response to extracellular pH change. Adapted from [52]. Scale bars = 100 μm.

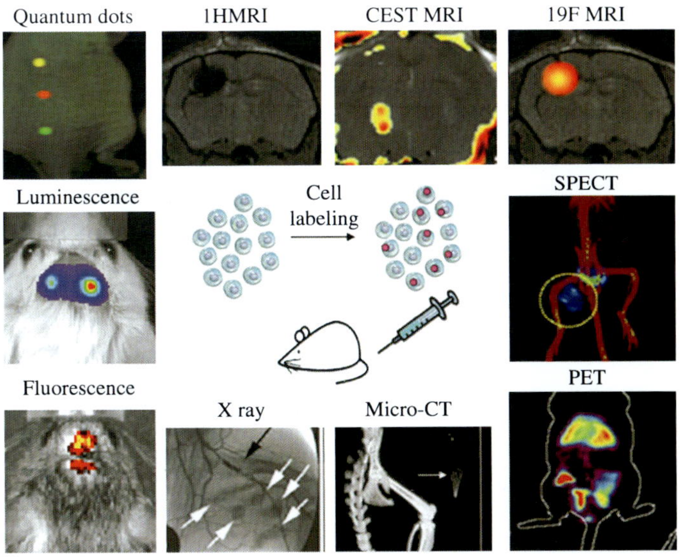

FIGURE 10.7 Imaging modalities for *in vivo* cell tracking. Labeled cells can be noninvasively tracked by more than 10 imaging modalities. Fluorescence, luminescence, and QDs use visible and near-infrared light emitted from labeled cells, which are extremely valuable for small animal studies. X-ray and micro-CT rely on X-ray (ionizing radiation). PET and SPECT (including scintigraphy) detect annihilation photon and gamma rays, respectively, from radioactive nuclides (ionizing radiation). MRI, including proton-based, ^{19}fluorine-based, and CEST MRI, use radiowaves for imaging and represents the most noninvasive imaging modality for cells and hosts. Images for QDs and ^{19}F MRI are simulated drawings. ^{1}H MR image reproduced from [100], CEST MR image reproduced from [58], SPECT-CT image reproduced from [59], PET image adapted from http://www.cibm.ch/page65843-en.html,and X-ray picture reproduced from [69] (all with permission). Micro-CT image by courtesy of Dr. Dian Arifin.

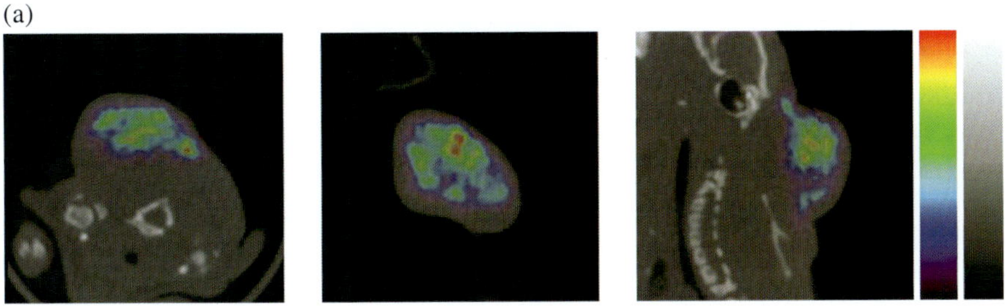

FIGURE 11.3 Intratumoral delivery of liposomes in a head and neck tumor model. (a) Transaxial, coronal, and sagittal views of fused pinhole SPECT/CT images acquired at 20 h after intratumoral CED injection of ^{99m}Tc-liposomes. The gray and color scale bars show the pixel values from 0 to 100% for the CT and SPECT images, respectively.

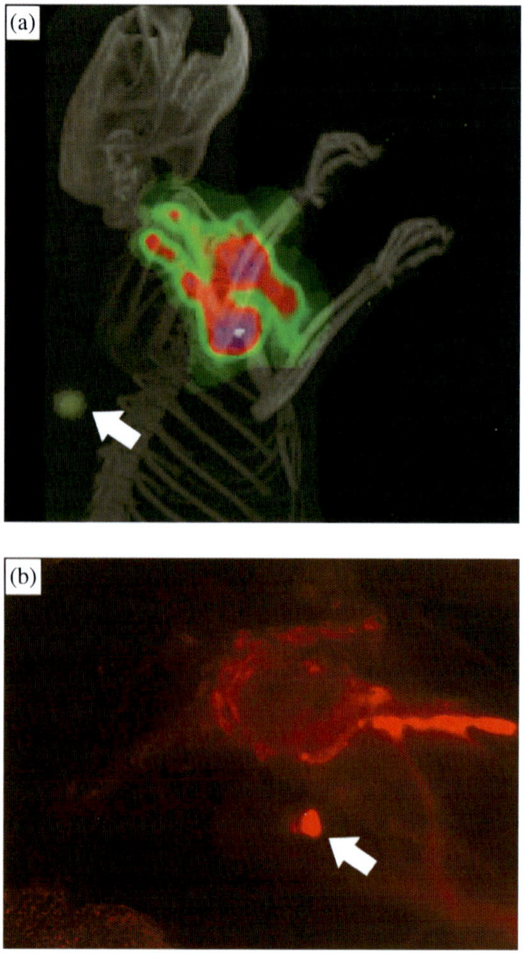

FIGURE 11.4 Use of imaging for monitoring liposome distribution in breast cancer post-lumpectomy brachytherapy model. (a) Fused SPECT/CT projection image of rat acquired 20 h after intracavitary injection of dual-labeled ^{99m}Tc-rhodamine fluorescent liposomes. Most of the ^{99m}Tc-radioactivity was retained locally within surgical cavity with some accumulation in axillary lymph node (white arrow). (b) Fluorescent stereomicroscope image acquired 44 h after intracavitary injection of dual-labeled ^{99m}Tc-rhodamine fluorescent liposomes. This image shows the distribution of the ^{99m}Tc-rhodamine fluorescent liposomes lining the surgical cavity and lymphatic drainage to the lymph node (white arrow).

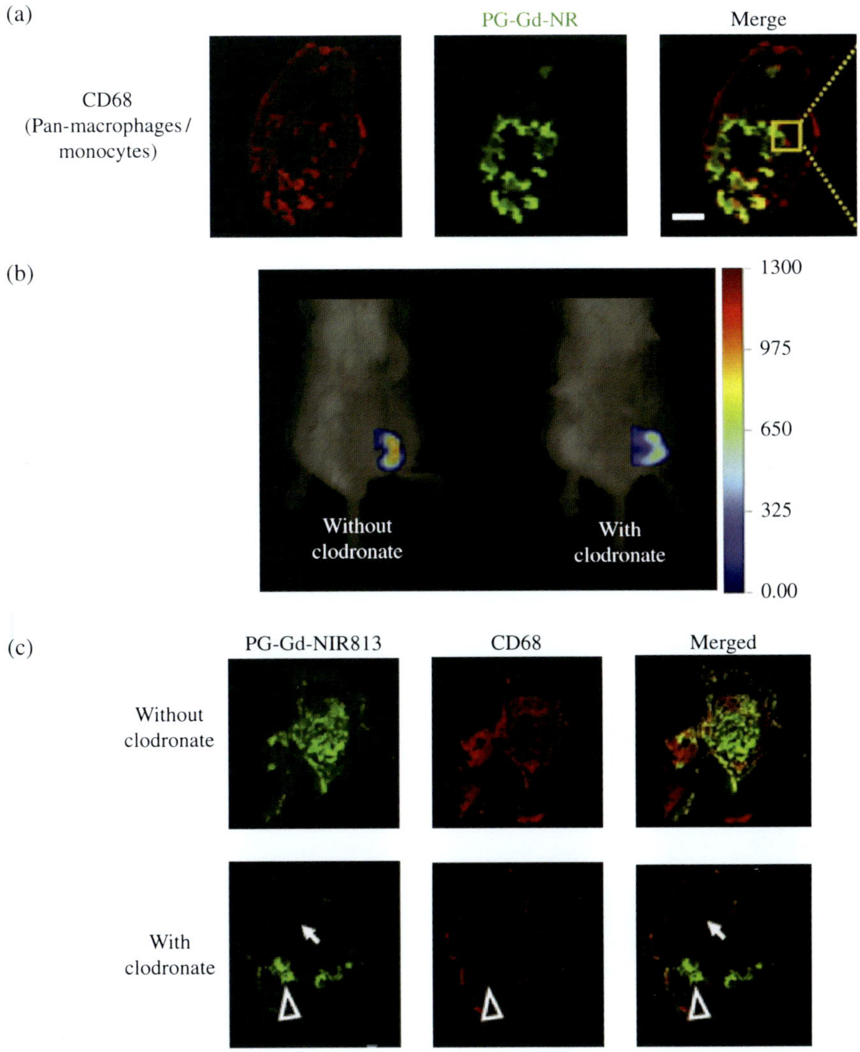

FIGURE 12.7 Noninvasive NIFR imaging of M2 macrophages using PG-DTPA-Gd-NIR813. (a) Co-localization of CD68 and PG-DTPA-Gd-NIR813 in C6 tumors 48 h after intravenous injection. (b) Fluorescence intensities of PG-Gd-NIR813 in A20 tumors in mice injected with clodronate, a macrophage-depleting agent, compared with saline control. (c) Fluorescence micrographs of C6 tumors in nude rats injected with PG-DTPA-Gd-NIR813 with or without macrophage depletion. (Reprinted with permission from Biomaterials, 2010;31[25]: 6567–6573 [72].)

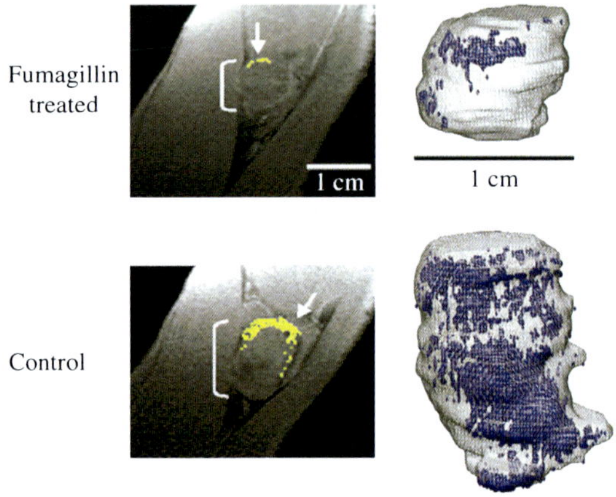

FIGURE 13.2 Diminished $\alpha_v\beta_3$-integrin contrast enhancement in T_1-weighted, fat-suppressed, 3D gradient-echo MR single slice images (250 × 250 μm, 500 μm slices, TR/TE = 40/5.6 ms, 65° flip angle, 1.5 T) in rabbits administered $\alpha_v\beta_3$-targeted fumagillin nanoparticles (top) versus those given $\alpha_v\beta_3$-targeted nanoparticles without drug (bottom). **Left**: Enhancing pixels, color coded in yellow (arrows), demonstrate sparse areas of angiogenesis in fumagillin-treated animal (top). **Right**: 3D neovascular maps of example V × 2 tumors on day 16 following $\alpha_v\beta_3$-targeted fumagillin nanoparticles (top) versus $\alpha_v\beta_3$-nanoparticles without drug (bottom). Note the asymmetric distribution of angiogenic signal (color coded in blue) over the tumor surface in both the control and treated animals. Neovessel dense islands and the interspersed fine network of angiogenic proliferation over the tumor surface are diminished in rabbits receiving the targeted fumagillin treatment. Reproduced with permission from [25].

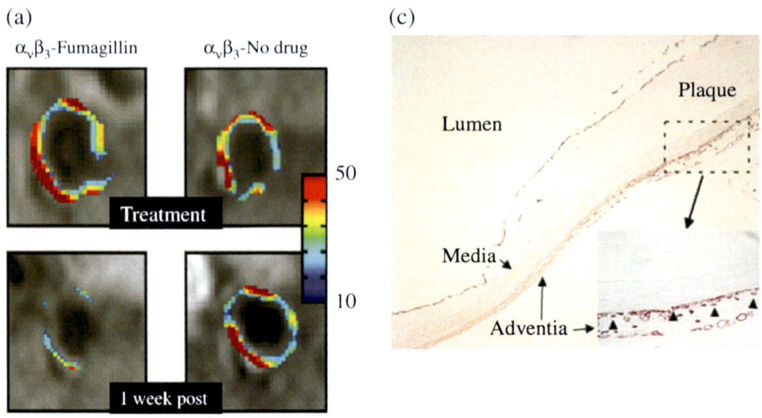

FIGURE 13.3 (a) MRI of abdominal aorta showing false-colored overlay of percent signal enhancement at the time of treatment (top) and one week post treatment (bottom). (c) PECAM-stained section (4 ×) of abdominal aorta from a hyperlipidemic rabbit displaying very early atherosclerotic changes in the adventitia, media, and plaque showing adventitial vasa vasorum expansion exclusively in region of plaque. Higher magnification inset (20 ×) shows that microvessels were predominantly located in the adventitia associated with thickening neo-intima. Arrowheads illustrate the type of PECAM microvessels counted within each section to assess fumagillin antiangiogenic effects.

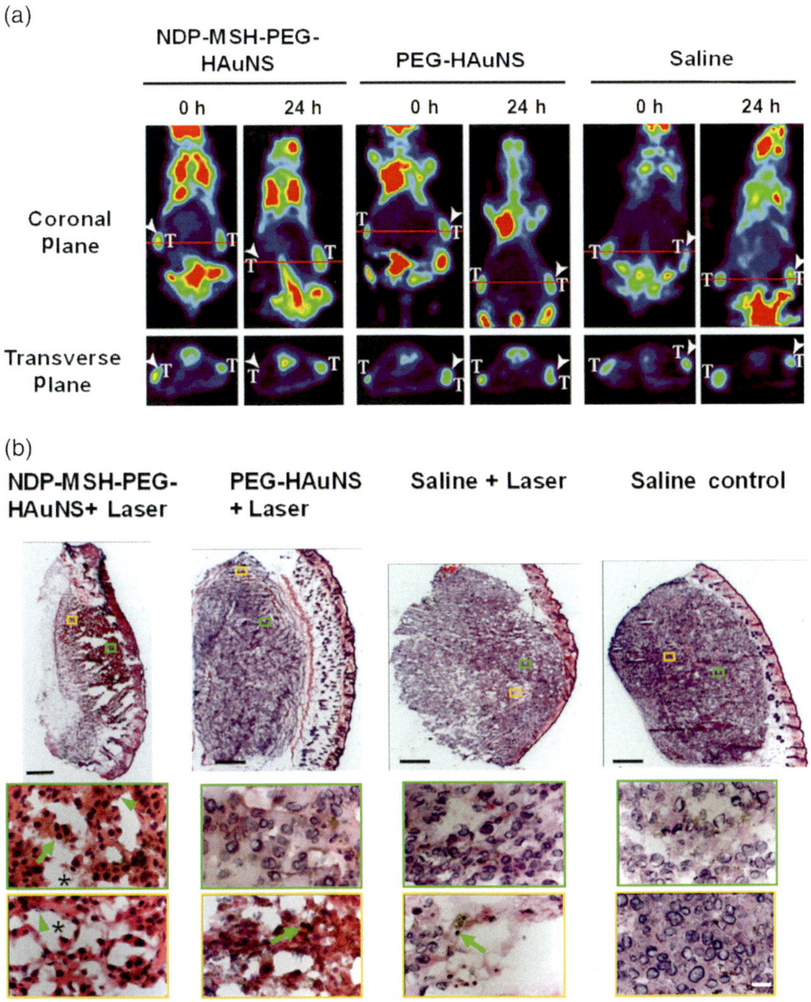

FIGURE 14.4 *In vivo* PTA-induced selective destruction of B16/F10 melanoma in nude mice. (a) [^{18}F]FDG-PET shows significantly reduced tumor metabolic activity after PTA only in mice pretreated with NDP-MSH-PEG-HAuNS. [^{18}F]FDG-PET was conducted before (0 h) and 24 h after NIR laser irradiation, which was commenced 4 h after intravenous injection of NDP-MSH-PEG-HAuNS, PEG-HAuNS, or saline control. For each mouse, tumors on one flank were randomly selected for laser treatment (0.5 W/cm^2 at 808 nm for 1 min). T: tumor; Arrowheads: tumors irradiated with NIR light. (b) Histological assessment of tumor necrosis. The whole tumor from each mouse 24 h after NIR irradiation was removed, fixed, and stained with hematoxylin–eosin (bar = 500 μm). Representative images show that in the group treated with NDP-MSH-PEG-HAuNS plus laser, most of tumor cells were necrotic as indicated by pyknosis (arrows), karyolysis (arrowheads), cytoplasmic acidophilia, and degradation and corruption of the extracellular matrix of the tumor (asterisks). Only a small fraction of necrotic zone was found in the tumors treated with PEG-HAuNS plus laser, laser alone, and tumor without any treatment. Bar = 50 μm. From reference [83]. Copyright American Association for Cancer Research. Reprinted with permission.

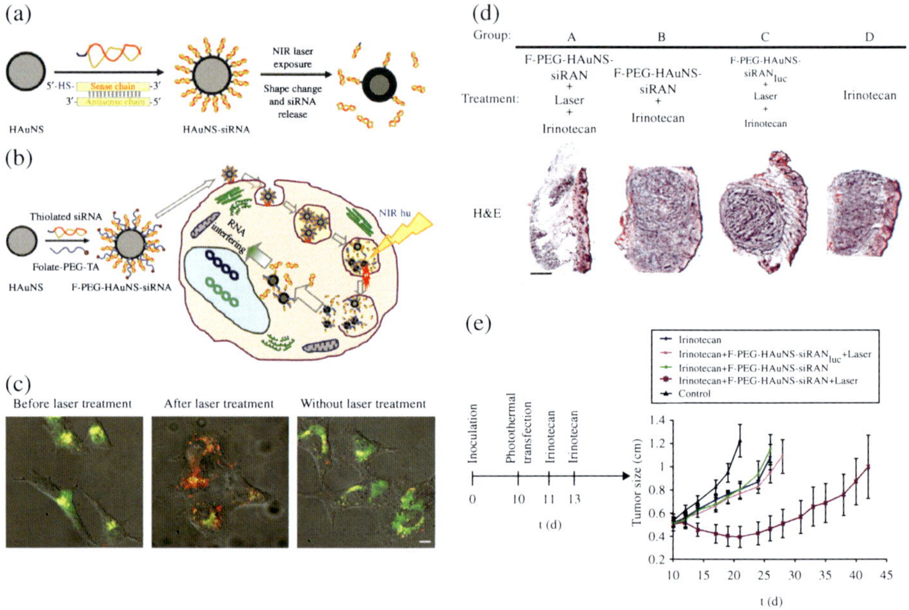

FIGURE 14.7 Photothermal transfection. (a) Scheme for the conjugation of HAuNS-siRNA and photothermal-induced release of siRNA. (b) Scheme for the synthesis of folate-modified HAuNS-siRNA conjugates (F-PEG-HAuNS-siRNA) and their proposed intracellular itinerary following NIR laser irradiation. TA, thioctic acid. (c) Photothermal-induced endolysosomal escape of Dy547-labeled siRNA. Green represents LysoTracker green-labeled endolysosomes; red represents Dy547-labeled siRNA. Bar, 10 μm. (d and e) Effect of p65 siRNA photothermal transfection combined with irinotecan on nude mice bearing HeLa cancer xenografts. (d) Representative micrographs of tumors stained with hematoxylin–eosin. Bar, 1 mm. (e) Tumor size versus time curve. Control, tumor-bearing mice that did not receive any treatment. From reference [33]. Copyright American Association for Cancer Research. Reprinted with permission.

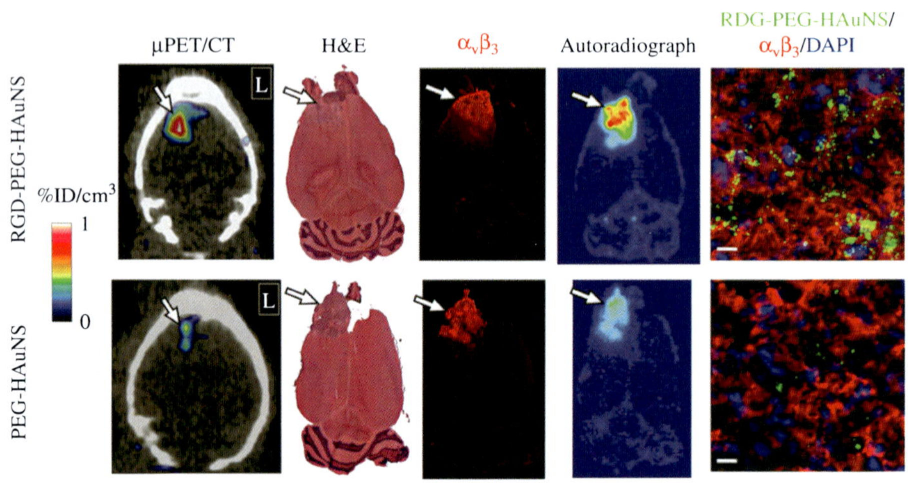

FIGURE 14.9 Targeted delivery of ^{64}Cu-labeled c(KRGDf)-PEG-HAuNS to U87 human glioma in mouse brains. Shown are micro-PET/CT images (obtained 24 h after intravenous injection of nanoparticles), photographs of mouse brains stained with hematoxylin–eosin and $\alpha_v\beta_3$ and autoradiographs of mouse brains, and immunofluorescence micrographs of tumor tissue at high magnification. Red, $\alpha_v\beta_3$; green, scattering signal of HAuNS under dark field; blue, DAPI-stained cell nuclei; arrows, tumors; bar, 10 μm. From reference [96]. Copyright American Association for Cancer Research. Reprinted with permission.

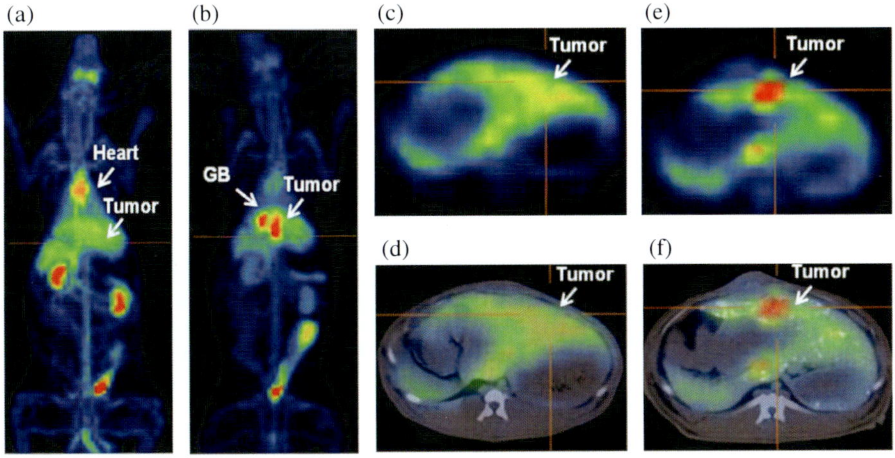

FIGURE 17.1 PET imaging of ^{64}Cu-labeled nanoparticles in liver tumor rabbits at 60 min after injection. Whole-body (a), transaxial (c), and PET/CT fused (d) images after IV injection of ^{64}Cu-HAuNS; whole-body (b), transaxial (e) and PET/CT fused (f) images after IA injection of ^{64}Cu-HAuNS-lipiodol. GB: gallbladder.

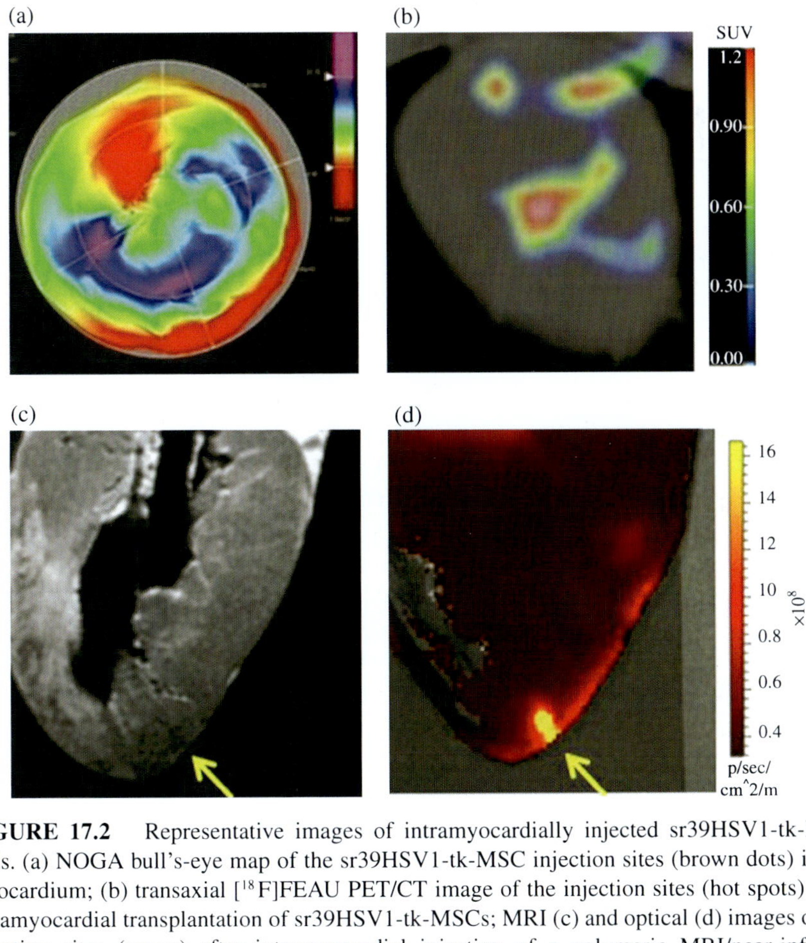

FIGURE 17.2 Representative images of intramyocardially injected sr39HSV1-tk-MSC cells. (a) NOGA bull's-eye map of the sr39HSV1-tk-MSC injection sites (brown dots) in the myocardium; (b) transaxial [^{18}F]FEAU PET/CT image of the injection sites (hot spots) after intramyocardial transplantation of sr39HSV1-tk-MSCs; MRI (c) and optical (d) images of the injection sites (arrow) after intramyocardial injection of a polymeric MRI/near-infrared fluorescent contrast agent PG-Gd-NIR813.

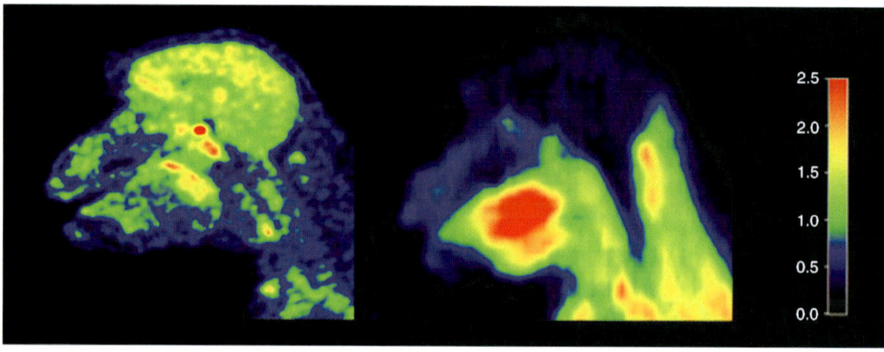

FIGURE 20.2 PET studies using two ^{11}C-labeled drug candidates injected as intravenous microdoses in cynomolgus monkeys. Left: Drug candidate targeting the central nervous system. Right: Drug candidate targeting peripheral tissues. Average PET images from 9 to 93 min after intravenous injection.

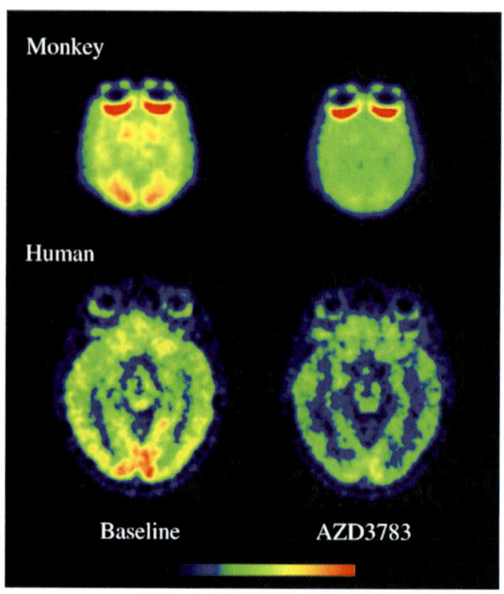

FIGURE 20.3 PET images of [^{11}C]AZ10419369 brain disposition in a cynomolgus monkey (upper panel) and a human subject (lower panel) at baseline (left) and after administration of the 5-HT_{1B} receptor antagonist AZD3783 (right).

CHAPTER 10

IMAGING CELL THERAPY

YAJIE LIANG AND JEFF W. M. BULTE

10.1 INTRODUCTION

Cell therapy can be defined as the transfer (infusion or transplantation) of living cells into a patient for treatment of a specific disease or condition. Historically, cell therapy is an old discipline. Written in 1552 B.C., the oldest preserved medical document, the Ebers Papyrus, recommended medicinal preparations of animal organs to improve human vitality. Animal organs for therapeutic use were also mentioned many times by Aristotle in the Materia Medica. Philippus Aureolus Paracelsus (1493–1541) was a German-Swiss physician and alchemist, who wrote in his *Der Grossen Wundartzney* (*Great Surgery Book*) in 1536 that "the heart heals the heart, lung heals the lung, spleen heals the spleen; like cures like." Paracelsus and other sixteenth-century physicians believed that the best way to treat illness was to use living tissue to rebuild and revitalize ailing or aging tissue. Blood transfusion can be viewed as one of the oldest forms of the practice of cell therapy. A fully documented blood transfusion was performed in 1667, when Dr. Jean-Baptiste Denis, physician to King Louis XIV of France, published his experience of transfusing blood from a lamb (due to its presumed soothing qualities) to a 15-year-old boy, who recovered (Fig. 10.1a). The first recorded attempt at nonblood cellular therapy occurred in 1912 when German physicians attempted to treat children with hypothyroidism with thyroid cells. In general, however, the development of modern cell therapy is credited to the Swiss physician Paul Niehans (1882–1971) (Fig. 10.1b), for his contributions to the advancement of the technique of organ extraction and cell therapy. However, Paul Niehans became "the father of cell therapy" quite by chance. In 1931, Dr. Niehans was summoned to an emergency operation where he was requested to perform a transplant for an elderly woman whose parathyroid glands were accidentally removed during a thyroid surgery. The patient was in critical condition and Paul Niehans did not have time to implant the whole gland. So he decided to dice a steer's parathyroid gland into fine pieces, mix the pieces in a saline solution, and inject them into the dying patient. Then amazingly, the patient's condition quickly stabilized and continued to improve.

Drug Delivery Applications of Noninvasive Imaging: Validation from Biodistribution to Sites of Action, First Edition. Edited by Chun Li and Mei Tian.

FIGURE 10.1 History of cell therapy. (a) Cell therapy in the early days: illustration of blood transfusion from lamb to man, 1705 (Credits: Wellcome Library, London). (b) Dr. Paul Niehans (1882–1971), arguably "the father of cell therapy" (http://www.biotherapycenter.com/xe/ch_ biomolecular_therapy).

In fact, she went on to live another 30 years. The success of this therapy led Paul Niehans to abandon the surgical transplantation of the entire gland and opt instead for cell implants through injection. This experience paved the way to future progress in the field of cellular therapy. Paul Niehans went on to apply his discoveries in cellular therapy to treat over 50,000 patients, some of whom included Winston Churchill, Charles De Gaulle, and Dwight Eisenhower. Due to its phenomenal success, cellular therapy has achieved global prominence as an effective regenerative technique. But, at that early stage of cell therapy, it involved the administration of heterologous biological material, which very often led to life-threatening, and even fatal, complications [1, 2]. In addition, people knew very little about the science of the therapeutic effect in the fortunate patients. Thus, this early stage should be viewed as the empirical stage of modern cell therapy.

At present, the most common form of clinical cell therapy (other than blood transfusion) is bone marrow transplantation. The field of cell therapy is growing, with the promise to become an approved treatment for a variety of conditions. In principle, cell therapy can be categorized into two aspects according to different purposes: renewal of stable tissues, such as stem cell transplantation into postmyocardial infarction tissues, and regulation of the immune system, such as administering cytotoxic immune cells or dendritic cell vaccination for cancer therapy. In light of our growing knowledge about inducible pluripotent stem cells (iPS cells) and their capacity to differentiate into all sorts of cells in the body, stem cell therapy would be a most promising treatment for conditions involving damaged tissues throughout the entire human body. As such, cell-based therapies may gain future importance in defeating the most prevalent disorders of our time, including cancer, cardiac infarction, stroke, and neurodegenerative diseases, among others. Currently, over 17,000 ongoing clinical trials involve some form of cell therapy (http://www.clinicaltrials.gov).

Although these treatments are promising and tremendous efforts are expended to make them work, translation from preclinical studies to humans has been difficult and largely disappointing. Consider stem cell therapy for cardiac diseases, for example. A well-publicized study in 2006 found no benefit from intracoronary autologous bone marrow cell (BMC) infusion in patients with ST-segment elevation myocardial infarction [3]. However, another larger, randomized, multicenter study indicated that intracoronary infusion of autologous bone marrow-derived mononuclear cells provided only short-term improvement in left ventricular function, compared to a placebo [4].

The conflicting clinical data underscores the need for suitable techniques that can monitor cell-based treatment trials. The number of cells to be delivered, the optimal stem cell types, the most suitable route for cell delivery, and engraftment and survival of transplanted cells are all factors that need to be addressed. Cellular imaging techniques are currently under investigation for their ability to identify, localize, and monitor transplanted cells *in vivo* longitudinally after implantation. Whole-body imaging will serve to both elucidate the biodistribution of injected cells and help reveal undesired dissemination in nontarget organs. In this context, noninvasive localization of transplanted cells and monitoring of their migration can facilitate basic research for understanding the underlying mechanisms of action and finally improve clinical translation.

10.2 CELL LABELING

Cells must be labeled to provide contrast for certain imaging modalities. Admittedly, cell labeling with contrast agents to obtain optimal *in vivo* imaging conditions remains a daunting task. Hence, an ideal contrast agent is one that can produce detectable contrast with a sufficient signal-to-noise ratio. Generally speaking, labeling methods can be divided into three categories: direct labeling, reporter gene labeling, and microencapsulation. Figure 10.2 is a schematic summary of the three cell labeling techniques.

10.2.1 Direct Labeling

One of the most commonly used strategies for the labeling of stem cells for imaging in living subjects is that of direct labeling. It is accomplished by introducing a marker into the cells or onto the cell surface before implantation. Imaging is then performed of the previously introduced marker as a surrogate for the number of transplanted cells. The simplest method is direct incubation of cells with the markers. Most currently available contrast agents for *in vivo* imaging techniques require a direct labeling technique.

10.2.1.1 Direct Labeling with Fluorescent Probes Direct cell labeling with fluorescent probes (fluorescent dyes and nanoparticles) forms a historical basis from which many other direct labeling techniques were developed. Typically, a cytoplasm/cell membrane is labeled with a fluorescent probe (e.g., DiI, DiO, DiD, or CM-DiI).

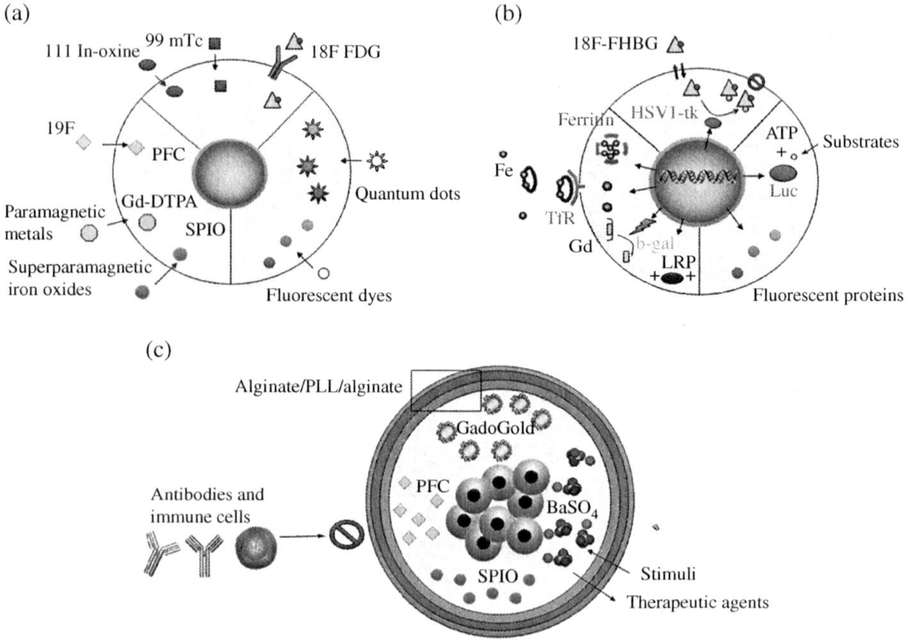

FIGURE 10.2 Summary of cell labeling methods. (a) *Direct labeling*. Labeling agents for optical imaging (fluorescent dyes and QDs), MRI (PFC, Gd-DTPA, and SPIO), and radionuclide imaging (^{111}In oxine, ^{99m}Tc, and ^{18}F-FDG) are used to directly label cells *ex vivo*. (b) *Reporter gene labeling*. Fluorescent proteins and luciferases are commonly used reporter genes for optical imaging. HSV1-tk is a reporter gene for PET or SPECT, in which phosphorylated probes, such as ^{18}F-FHBG or other acycloguanosines, result in intracellular trapping of the probes and accumulation of signal. Overexpression of TfR or ferritin increases intracellular iron levels, reducing T2-weighted MRI signal from cells expressing them. β-galactosidase cleaves off a galactose group that hinders the inner sphere relaxation enhancement effect of the paramagnetic chelated gadolinium (Gd), leading to inner sphere relaxation enhancement and increased contrast. The artificial LRP gene codes for a lysine-rich peptide that provides high CEST contrast. (c) *Microcapsule labeling*. Contrast agents (GadoGold, BaSO4, SPIO, or PFC) were coencapsulated with cells using alginate–poly-L-lysine or alginate–protamine sulfate microcapsules, which block the entrance of antibodies or immune cells, but allow for the free flow of metabolites and secreted molecules including insulin.

Abbreviations: BaSO4, barium sulfate; b-gal = β-galactosidase; ^{18}F-FDG = ^{18}F-fluorodeoxy glucose; ^{18}F-FHBG = 9-[4-[^{18}F]fluoro-3- (hydroxymethyl)butyl] guanine; Gd = gadolinium; GadoGold = gold-gadolinium dithiolated diethylenetriamine pentaacetic acid; Gd-DTPA = gadolinium-diethylenetriamine pentaacetic acid; HSV1-tk = herpes simplex virus type 1 thymidine kinase; LRP = lysine-rich protein; Luc = luciferase; PFC = perfluorocarbon; SPIO = superparamagnetic iron oxide; ^{99m}Tc = ^{99m}Tc-HMPAO (^{99m}technetium -hexamethylpropyleneamine oxime); TfR = transferrin receptor.

Due to their stability and nontoxic properties, CellTracker™ fluorescent probes have gained popularity for cell labeling. They are transferred to daughter cells after cell division and are not transferred to adjacent cells in a population. These probes contain a chloromethyl or bromomethyl group that reacts with thiols and are transformed into a cell-impermeable, fluorescent dye–thioether adduct. Fluorescent probe labeling allows direct visualization of stem cells by optical imaging at high spatial resolution [5]. As such, it is often combined with labels for other imaging modalities, such as magnetic resonance imaging (MRI) contrast agents, to validate the noninvasive imaging results by postmortem fluorescent histological studies. Fluorescent semiconductor nanocrystals (known as quantum dots, QDs) have been recently adopted for *in vitro* and *in vivo* stem cell labeling and bioimaging [6, 7]. QDs are multifunctional, inorganic, fluorescent, and semiconductor nanoparticles. Compared to organic fluorescent dyes, QDs have unique optical advantages, such as a broader band excitation and narrower band emission, with a high quantum yield, exceptional photostability, and resistance to chemical and metabolic degradation [8]. Both mesenchymal stem cells (MSCs) and embryonic stem cells (ESCs) have been effectively labeled with QDs using passive incubation or peptide-mediated uptake [9, 10]. However, the long-term effects of QDs on stem cell differentiation and proliferation remain unknown. Moreover, the nonspecific binding of QDs to multiple cells and the clearance mechanism from the body of the highly toxic cadmium base of QDs may hamper translation to the clinic.

10.2.1.2 Magnetic Labeling with Contrast Agents The contrast agents that directly label cells for MRI can be classified into different types according to their different mechanisms. Contrast agents may modulate the longitudinal (T1) and transverse (T2) relaxation time of surrounding water and lipids, influence the local magnetic field homogeneity (T2* relaxation time), or utilize different nuclei that are not present in the host (e.g., ^{19}F-based contrast agents). Based on those mechanisms, areas of contrast agent accumulation can be distinguished against background signal. One can generally distinguish between (i) superparamagnetic materials, (ii) paramagnetic compounds containing lanthanide chelates or other paramagnetic ions like Mn^{2+}, and (iii) contrast agents that contain MR-detectable nuclei other than hydrogen.

Prelabeling with superparamagnetic iron oxide (SPIO) contrast agents is currently the most widely used method. Two compounds within the SPIO family were initially approved for clinical use as liver agent and used off-label for cell labeling: one is Feridex® in the United States (labeled as Endorem® in Europe), and the other is Resovist®, which was available in Europe and Japan. Both compounds are no longer available. Due to their safety in clinical diagnostic imaging, SPIOs have been widely used for cell tracking. Stem cells can be directly labeled with SPIOs with no or minimal side effects on cell viability, proliferation, and differentiation. While SPIOs are not internalized by nonphagocytic cells without further modification, simple methods to induce internalization and uptake have been developed and tested in a variety of stem cells. One of the most commonly used methods is a method where transfection agents (TAs), such as poly-L-lysine or protamine sulfate, are complexed with SPIOs to provide the formation of SPIO oligomers with a highly positive surface charge, which induces macropinocytosis (endocytosis) of the SPIO-TA complexes [11]. Concentrations of 2–20 pg iron/cell can be easily achieved after 24–48h incubation *in vitro* [12]. Because of particle clustering of SPIOs within endosomes, T2* effects are more pronounced, and thus, the signal is hypointense [13]. Figure 10.3 is a demonstration of SPIO staining [14]. Magnetoelectroporation (MEP) is another method of SPIO cellular labeling. MEP uses small pulsed voltages to induce intracellular uptake of SPIOs [15, 16], without the need for TAs, which may aid in more rapid clinical translation. In addition, millions of cells can be labeled

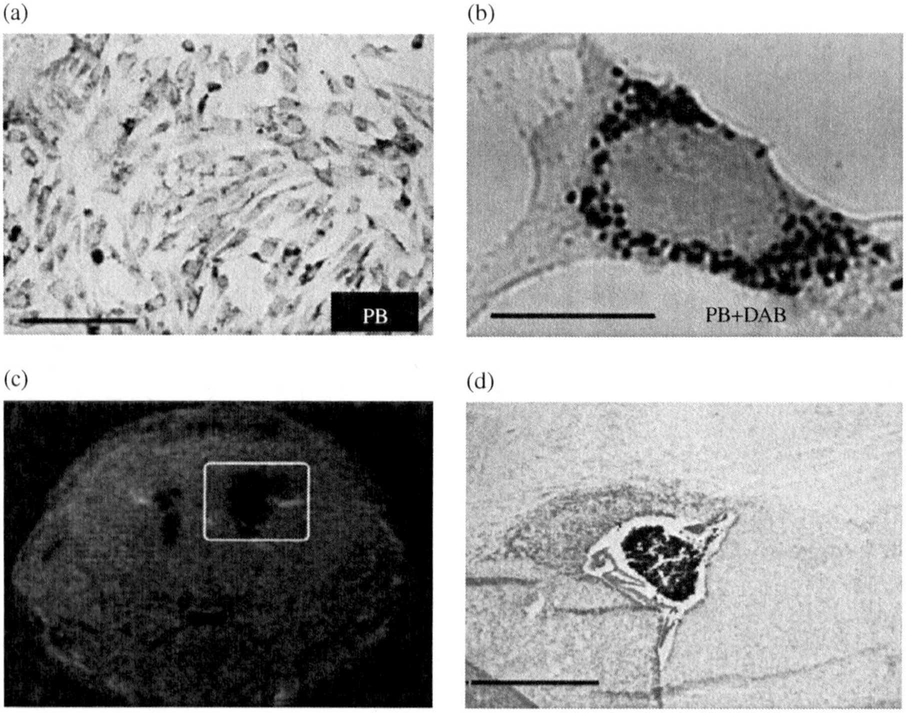

FIGURE 10.3 Direct labeling of cells with SPIO. (a) C17.2 cells (a neural stem cell line) are labeled with Feridex®, as visualized by Prussian Blue staining and (b) diaminobenzidine-enhanced Prussian Blue staining. Nearly all cells are labeled, with an endosomal cytoplasmic distribution of Feridex® particles. (c) Magnetically labeled C17.2 cells induced a strong hypointense signal on the MR images after injection into the lateral ventricles. (d) Prussian Blue staining of the corresponding tissue slice shows a close overlap with the MR signal. Scale bars: $a, d = 100\,\mu m$; $b = 10\,\mu m$. Reprinted from [14] with permission. For color details, please see color plate section.

in seconds using MEP, which may be important in certain cell lines that are altered by culturing *in vitro* due to adhesion to tissue culture plastics that lead to morphological changes. Both TA-mediated labeling and MEP can be used to label cells with a variety of contrast agents, including manganese oxide nanoparticles [17].

Cell labeling with paramagnetic metals, such as gadolinium (Gd) and manganese, enables the visualization of cells on T1-weighted MRI (hyperintense contrast). Because unchelated paramagnetic metals are toxic, clinical formulations, such as gadolinium-diethylene triamine pentaacetic acid (Gd-DTPA) and gadolinium tetraazacyclododecanetetraacetic acid (Gd-DOTA), are always chelated for biocompatibility. Because of the low longitudinal relaxivity (4 mmol/l/s) in Gd chelates, high intracellular accumulation is required for the utility of these contrast media as labeling agents. As a result of the nonparticulate nature of the Gd chelates, pinocytosis (used by cells for absorption of extracellular fluids), rather than phagocytosis, is a major mechanism for cellular uptake, which appears to be a very effective way to label stem cells by incubating cells with high concentrations of Gd chelates for a few hours [18]. In contrast to the membrane vesicular distribution of Gd chelates after pinocytosis, the electroporation procedure results in intracellular diffusion of Gd molecules and a greater longitudinal relaxation rate constant (R1) and higher MR signal intensity of labeled cells [19]. However, the potential for metal dechelation of paramagnetically labeled cells that fail to engraft and are removed from the body has hampered clinical translation of these agents.

In addition to the nuclei of ^{1}H, some other nuclei, including ^{19}F, ^{31}P, ^{23}Na, and ^{13}C, also have nuclear spins that can be detected by MRI. Because of their existence (background) in biological tissues and the lower gyromagnetic ratio of these elements compared to that of protons, ^{31}P, ^{23}Na, and ^{13}C experience a loss of sensitivity and are therefore not suitable for MRI. In contrast, the gyromagnetic ratio of ^{19}F nuclei is approximately 90% that of protons. ^{19}F atoms are naturally absent in the body, making them sensitive nuclei for MR detection of the labeled stem cells [20]. Anatomical localization can be achieved by conventional proton MRI. Fluorinated nanoparticles have been investigated for direct cell labeling for MRI detection [21, 22]. However, the requirement for specialized hardware and software to image nonproton species may limit a rapid and widespread adoption of these techniques.

10.2.1.3 Radionuclide Direct Labeling Similar to ^{19}F MRI, the lack of radioactive isotopes in the body results in a higher signal-to-noise ratio when compared to routine ^{1}H MRI for cell tracking. ^{111}Indium (^{111}In) oxyquinoline, ^{99m}technicium (^{99m}Tc)-hexamethylprophylene amine oxine (HMPAO), and ^{18}F fluoro-2-deoxy-D-glucose (^{18}F-FDG) are widely used in the clinic for single photoelectron computed tomography (SPECT) (^{111}In and ^{99m}Tc) or positron emission tomography (PET) (^{18}F-FDG). Because of their exquisite picomolar (10^{-11}–10^{-12} mol/l) sensitivity [23], PET and SPECT imaging modalities are able to detect a tracer quantity of radioisotopes for studying biological processes in living subjects.

Analogous to cell labeling for MRI and optical imaging, cells can be labeled by direct incubation with the radioactive agent [24, 25] or by receptor-mediated binding [26]. *Ex vivo* labeling with ^{111}In oxine is a well-established method for leucocyte or thrombocyte scintigraphy. Labeling efficiencies close to 100% are not uncommon, although cellular concentrations of the isotope depend on several factors, including cell type, incubation time, and concentrations. Radionuclides have different physical half-lives (e.g., ^{99m}Tc, 6 h; ^{111}In, 2.8 days; ^{18}F, 109 min; ^{64}Cu, 12 h). This determines the amount of time that cells can be monitored noninvasively after cell labeling. In general, radioisotopes with a relatively long decay half-life are used to track cells during a period of several hours or even days.

Among the many options, labeling cells with ^{111}In oxine has been used most extensively for stem cell labeling due to its favorable half-life (about 2.8 days) and prior FDA approval for labeling lymphocytes to study infection and inflammation [27]. Direct incubation of stem cells with ^{111}In oxine can be performed in less than 2 h, during which it passively diffuses into cells, dissociates, and becomes trapped in the cytoplasm [27]. In a canine reperfused myocardial infarction model, our group detected the focal and diffuse uptake of ^{111}In oxine- and SPIO-labeled MSCs in the infarcted myocardium using a clinical SPECT/CT scan within 24 h after intravenous (IV) injection, and those signals persisted up to seven days after stem cell delivery [25]. However, MRI, with a much lower sensitivity than SPECT, was unable to detect the same signal (SPIO-labeled MSCs) in the myocardium [25]. Depending on the radiotracer used and cell type, the minimum detection limits of direct radiotracer labeling vary from 2,900 to 25,000 cells [28].

10.2.1.4 Limitations of Direct Labeling Prerequisites for the clinical applicability of contrast agents for cell labeling are accuracy, safety, and nontoxicity for cells and tissues. But, for direct labeling, there are issues with all those aspects. First, there is the problem of leakage of contrast agent toward neighboring cells and potential transfer to nonlabeled cells, such as macrophages following cell death. This could lead to false-positive imaging and a misinterpretation of the mechanisms relating to stem–progenitor cell transfer and engraftment. Second, the use of SPIOs (at least at doses of 20 pg of SPIO/cell) may have an effect on the gross morphology and proliferation capacity of stem cells [29, 30]. In the case of radionuclide labeling, cell toxicity will likely vary depending on the radionuclide and dosage used [31, 32]. Third, cell division results in "dilution" of the signal over time. This aspect of direct labeling limits its use for long-term monitoring. In summary, direct labeling strategies appear to be a good imaging strategy for detection of cells shortly after transplantation (e.g., to ensure that cells were delivered to the intended organ or region of an organ), providing a good signal-to-noise ratio, but are less suited for long-term monitoring of stem cell viability.

10.2.2 Reporter Gene Labeling

Reporter gene imaging is a method used to characterize cellular and molecular processes *in vitro*. For cellular imaging, target cells are first genetically modified to express the reporter gene(s) before transplantation. The reporter gene DNA construct can be transferred into cells via a viral (e.g., lentivirus, retrovirus, adenovirus, or adeno-associated virus) or nonviral (e.g., plasmid) vector. Translation of the reporter gene mRNA produces a protein that provides contrast when it is detected by a charged-coupled device (CCD), PET, SPECT, or MRI. An advantage of reporter gene imaging is that the construct can be custom designed for the intended application by varying the promoter sequence to regulate the expression of the gene within the cell. Commonly used promoters include constitutive, inducible, or cell/tissue specific ones. Another advantage of reporter gene labeling is that, in contrast to direct labeling techniques, imaging signals are generated only from viable cells of interest. Finally, cell division should carry over a constant expression of the marker, without dilution as is the case for direct labeling methods.

10.2.2.1 Fluorescent Reporter Genes Fluorescence is generated by the excitation of a fluorophore. The heterologous expression of green fluorescent protein (GFP), cloned from bioluminescent jellyfish, is an excellent tool for labeling proliferating cells [33]. Site-directed mutagenesis and the isolation of fluorescent proteins from different organisms have led to multiwavelength applications that span excitation wavelengths from 350 to 630 nm and emission wavelengths from 450 to 650 nm. Due to the limited tissue penetration of green light, imaging in the intact organism using GFP is restricted to superficial tissues. In our experiments, C17.2 cells stably expressing GFP (Fig. 10.4a) were transplanted subcutaneously in nude mice, and signal could be detected by excitation at 500 nm, which could be enhanced by subtraction of autofluorescence from tissue (Fig. 10.4b–d). However, in most studies, GFP is used to analyze tissue postmortem. *In vivo* optical imaging of deep tissues in animals is most feasible between 650 and 900 nm because such wavelengths minimize the absorbance by hemoglobin, water, and lipids, as well as light scattering [34]. Thus,

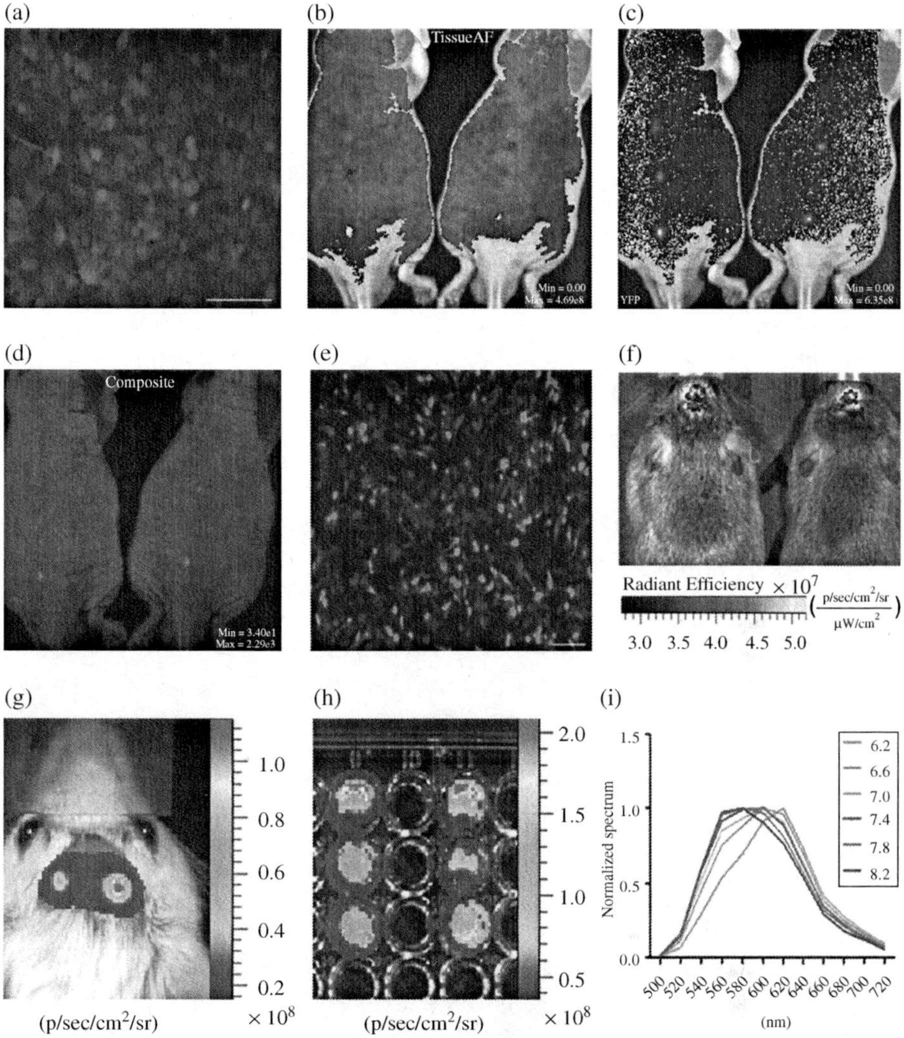

FIGURE 10.4 Fluorescent and luminescent reporter genes. C17.2 cells stably expressing (a) GFP, (e) tdTomato, or (h) firefly luciferase were transplanted into mice and imaged using an IVIS 200 optical system. (b, c, d) Spectral unmixing result of GFP for better visualization of cells that were injected subcutaneously into nude mice. (b) Tissue autofluorescence; (c) Fluorescence from GFP. (d) Unmixing result with red dots showing positive signals. (e, f) tdTomato expressing C17.2 cells were transplanted into the striatum of Rag2 mice, which were tracked by fluorescent imaging. Intensity of luminescence from firefly luciferase expressing C17.2 cells could be quantified *in vitro* (h) or *in vivo* (g). (i) Shift of the emission spectrum of firefly luciferase in response to extracellular pH change. Adapted from [52]. Scale bars = 100 µm. For color details, please see color plate section.

fluorescent proteins with a longer wavelength of excitation and emission would be preferable. Monomeric red fluorescent protein (mRFP), a monomeric mutant of DsRed from Discosoma coral, is such a candidate, with a longer wavelength emission than

GFP and, consequently, deeper tissue penetration [35]. The latest version of mRFP is mCherry, which matures more completely, is more tolerant of N-terminal fusions, and is over ten-fold more photostable than mRFP [35]. But, the version that is most suitable for *in vivo* imaging is tdTomato, which is a genetic fusion of two copies of the mCherry gene. Its tandem dimer structure plays an important role in the exceptional brightness of tdTomato [36]. TdTomato's emission wavelength (581 nm) and brightness make it ideal for live-animal imaging studies. In our experiments, we transduced C17.2 cells with this reporter gene and injected them into the striatum of immunodeficient mice. Cells stably expressing tdTomato gave a very high fluorescent signal in culture (Fig. 10.4e) and also much signal when they were transplanted (Fig. 10.4f). Recently, Roger Tsien's group reported a bacteriophytochrome from *Deinococcus radiodurans*, incorporating biliverdin as the chromophore, which was engineered into monomeric, infrared-fluorescent proteins (IFPs), with excitation and emission maxima of 684 and 708 nm, respectively [37]. Since their wavelengths penetrate the tissue well, those IFPs are suitable for whole-body imaging, although more engineering is needed to increase their brightness. One inherent problem for current optical imaging is the technical difficulty of tomographic detection of light signal, because of high photon absorbance and scattering in the visible range. Shallow penetration is a severe limitation for *in vivo* detection even in small animals, not to mention in humans. Further developments in optical imaging techniques, such as diffuse optical tomography [38] and optical coherence tomography [39], offer potential solutions for these current problems; however, they are primarily limited to preclinical research.

10.2.2.2 Luminescent Reporter Genes Luminescence in the form of light is generated from an enzymatic reaction catalyzed by, for example, firefly luciferase (Fluc); when cells expressing Fluc are supplied with D-luciferin, a substrate of the enzyme, cells will emit luminescence. Bioluminescence imaging (BLI) has been intensively used for *in vivo* tracking of transplanted stem cells, including hematopoietic [40], embryonic [41], mesenchymal [42], or neural stem cells [43] in rodents. The expression of luciferase reporter genes enables researchers to observe luminescence upon injection of its substrates and therefore provides the ability to monitor the survival, migration, or differentiation of tested stem cells over time, with high sensitivity at relative low cost. Although there are many versions of luciferase, such as luciferases from fireflies, *Renilla reniformis*, or click beetles, luciferases from fireflies are more suitable for *in vivo* imaging due to the broad spectrum, higher magnitude of yield, and stability. To date, two major directions of improvements have been pursued for optimization of luciferase genes for *in vivo* imaging. One is the optimization of expression in mammalian cells. Since the first report of the Fluc cDNA sequence [44], this sequence has undergone several important modifications to optimize its expression in mammalian cells, including optimized mammalian codon usage, removal of peroxisome targeting sites, cryptic regulatory sequence removal, and degradation signal addition [45]. The commercially available luc2 gene from Promega is the latest version of such an effort, which is now prevalently used for tracking of transplanted cells [46, 47]. Another direction of improvement is toward the red-shifted spectrum. Blood hemoglobin and myoglobin represent the main endogenous absorbers of BLI signal *in vivo*. The absorption of blue and green light is very strong, but it is considerably less at wavelengths above 600 nm [48]. As such, the development of red-shifted luciferase for BLI has been intensively pursued in recent years. For instance, amino acid substitutions are made so that wavelength of emission is shifted toward the red region of the visible spectrum [49–51]. In our experience, the luc2 reporter gene is very sensitive for tracking cells (Fig. 10.4 g) and may be the brightest among commercially available luciferases, although its emission spectrum would change in response to an extracellular pH change [52] (Fig. 10.4i). Because of limited tissue penetration, light scattering, and the fact that luciferase is a nonmammalian protein, this imaging modality is limited to small animal models (usually mice and rats). In addition, unlike PET/SPECT/MR imaging modalities, in which data are acquired in a tomographic fashion, luminescence detection relies on a continuous wave technique, leading to a diffusive image.

10.2.2.3 MRI Reporter Genes Over the past few years, significant efforts have been devoted to develop MRI reporter genes. Although less sensitive, several approaches for MRI-based imaging reporter genes have been tested in proof-of-principle experiments. According to the different mechanisms of imaging, MRI reporter genes can be divided into three classes: enzyme-based reporter genes, metalloprotein-based reporter genes, and chemical exchange saturation transfer (CEST) reporter genes. Although none of those MRI reporter genes has been used to track stem cells or therapeutic cells for treatment of disease, primarily due to the sensitivity and potential toxicity to cells, MRI reporter genes have become an active and growing area of scientific interest.

One of the earlier examples of the enzyme-based MRI reporter gene is the genetic expression of β-galactosidase, which can convert the substrate X-gal into a Prussian Blue-like product that can be readily detected by light microscopy. For MRI, Moats et al. [53] have developed a gadolinium-based substrate (EgadMe) that contains a galactose group that hinders the inner sphere relaxation enhancement effect of the paramagnetic chelated Gd. In the presence of lacZ-transfected cells expressing β-galactosidase, the galactose is cleaved off, leading to an increase in water molecules that diffuse toward the inner Gd atom [54]. This, in turn, results in inner sphere relaxation enhancement and increased contrast.

For the second class of MRI reporter genes, proteins involved in iron transport and metabolism, such as the transferrin receptor (TfR), tyrosinase, and ferritin, are produced to accumulate irons intracellularly, inducing changes in relaxivity (primarily T2* effects) for the detection of reporter protein activity. Iron, depending on its oxidation state and its structure, can act as a paramagnetic reactive metal. In the bloodstream, iron is bound to the plasma protein transferrin and enters cells via binding of the iron–transferrin complex to the TfR. The administration of transferrin linked to monocrystalline iron oxide nanocompounds to nude mice that had hTfR-transfected 9-L gliosarcoma tumors led to a significant difference in MRI signal 24 h after administration [55]. Although iron has a crucial role in body homeostasis, an excess of iron can cause damage to cells by catalyzing the Fenton reaction, which generates reactive oxygen species. Therefore, excess intracellular iron is stored in the ferrihydrite core of the ferritin protein. The first use of ferritin in the design of C6 glioma MRI reporter cells involved a construct that carried enhanced GFP and the heavy chain of ferritin, which exhibits ferroxidase activity, under tetracycline control. Inoculation of these tumor cells in nude mice revealed a significant decrease in T2 on tetracycline withdrawal [56].

Chemical exchange saturation transfer (CEST) MRI is a new type of MRI contrast, which relies on the chemical exchange of the protons of solutes with those present on CEST contrast agents [57]. A variety of organic and organometallic compounds have sufficient numbers of protons with suitable chemical exchange rates and specific MRI frequencies to be excited selectively and detected sensitively. CEST contrast agents have two major advantages. First, they are switchable; that is, the contrast is detectable only when a saturation pulse is applied at the specific frequency characteristic of the exchangeable protons of an agent. Otherwise, the contrast agent is MRI invisible. The second advantage is that different contrast agents with different excitation frequencies can be used for imaging more than one target simultaneously. This property might be exploited for imaging multiple cells (e.g., for studying neuronal circuits) or the expression of multiple genes (such as in genetic circuits). In an initial experiment by our group, a lysine-rich protein (LRP), containing a high density of amide protons, was cloned and expressed in 9-L rat glioma cells. The LRP amino acid sequence is similar to poly-L-lysine and provides MRI contrast at the amide proton frequency. LRP overexpression was detectable *in vitro* in cell extracts and *in vivo* in xenografts of LRP-expressing 9-L cells in the mouse brain (see Fig. 10.7 CEST MRI) [58].

10.2.2.4 SPECT/PET Reporter Gene Both SPECT and PET reporter gene imaging are based on the interaction between an exogenously administered probe (that contains the tracer) and the reporter gene product (e.g., enzyme, receptor), which results in the retention of the radionuclide substrate that can then be imaged noninvasively. There are predominantly three reporter gene systems (for PET or SPECT) that have been used for cell imaging. The system used most commonly is based on the production of an intracellular enzyme (e.g., herpes simplex virus type 1 thymidine kinase, HSV1-tk) that phosphorylates an exogenously administered substrate that is retained in the cell because of its negative charge. While normal cells (without the HSV1-tk) do carry the mammalian version of this enzyme, this thymidine kinase only minimally phosphorylates the radionuclide probes used in this system. On the other hand, in cells carrying the HSV1-tk, the exogenously administered probe undergoes significant phosphorylation and intracellular retention, leading to a robust signal-to-background ratio, enabling accurate monitoring of these cells. Figure 10.5 shows data on using HSV1-tk to image the formation of a teratoma after human ESCs (hESCs) were transduced with this reporter gene [59]. This strategy is very powerful because the enzyme can phosphorylate many molecules of the radionuclide substrate, increasing the signal retained in the cells of interest and improving the signal-to-background ratio. However, the probe has to cross the cell membrane, which may limit the interaction between substrate and enzyme, resulting in reduced signal. In addition, because HSV1-tk is a nonmammalian protein (of bacterial origin), it has the potential to trigger an immunological response, resulting in a reaction in the organism, with decreased overall signal. Attempts to circumvent this problem have included the use of a mammalian protein (dopamine receptor, see below), a mammalian mitochondrial thymidine kinase [60], or a destabilized HSV1-tk [61]. This imaging approach has been used extensively to monitor the viability of different stem cells after transplantation [59, 62].

A second reporter gene imaging strategy is based on the imaging of dopamine receptors using PET (dopamine 2-like receptor, D2R) [63]. In this case, the reporter

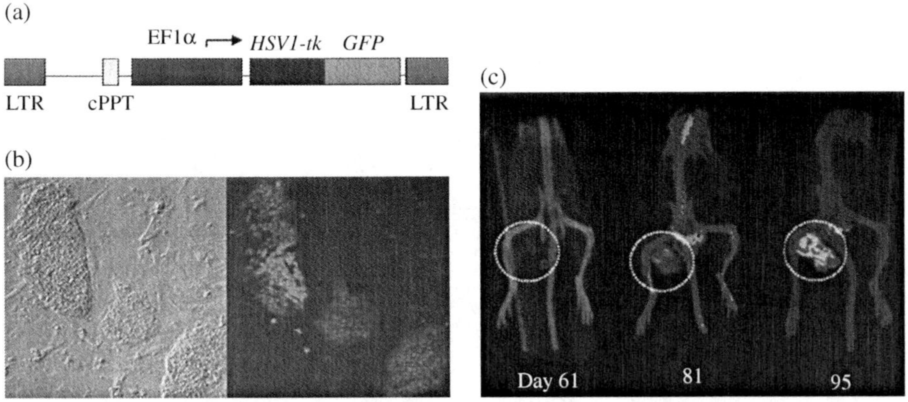

FIGURE 10.5 Noninvasive whole-body SPECT imaging of hESCs *in vivo*. (a) cEF.tk-GFP, the cDNA encoding the HSV1-tk-GFP fusion protein, was inserted into the lentivector, driven by a ubiquitous promoter from the human housekeeping gene, EF1a. (b) hESCs transduced with the tk-GFP reporter under a fluorescent microscope *in vitro*. (c) Longitudinal SPECT-CT imaging of a SCID mouse harboring a teratoma from hESCs transduced with the cEF.tk-GFP vector. The mouse was imaged at 61, 81, and 95 days after inoculation. The animal was treated with ~1 mCi of [^{125}I] FIAU at each time point and imaged at 24 h after injection. Reprinted from [59] with permission.

gene encodes for a cell membrane protein, which binds to an exogenously given probe, and the "bound probe" is then imaged noninvasively with PET. It is important to mention that the wild-type D2R has the potential to elicit a downstream biological response, which can be prevented if a mutant version of the dopamine receptor is used [64]. Notably, the mutant version of the D2R (mutation of Asp80 or Ser194) maintains the affinity for the PET probe (3-(2′-[^{18}F]-fluoroethyl)-spiperone) used. This strategy can be advantageous because the probe does not have to cross the cell membrane to interact with the reporter protein. However, this approach is limited by the amount of signal that can be produced, as one receptor interacts with only one molecule of the ligand. This monitoring strategy has been used to monitor cell biology or even the expression of a therapeutic gene expressed in cells transplanted to the myocardium [65]. A third approach consists of the encoding of the sodium–iodide symporter (NIS) [66], a thyroid transmembrane protein, which, under physiological conditions, transports iodine into the cells in exchange for sodium. One advantage of the NIS system is that it can be used for PET (with ^{124}I as the tracer) and SPECT imaging (using ^{123}I or ^{99}Tc-pertechnetate as the tracer).

A major advantage of PET and SPECT (when compared to BLI) resides in their potential for clinical use. Both PET and SPECT radionuclides are of relatively high energy (PET: 511 KeV, SPECT: 80–250 KeV) and do not undergo significant tissue attenuation. While SPECT (due to its relatively low energy) does have some tissue attenuation, it is still used extensively in clinical practice.

10.2.3 Microencapsulation

Initially, encapsulating cells with biomaterials was used to prevent an immune attack against transplanted islet cells. A semipermeable membrane was created to restrict large immunoglobulins and immune cells without affecting the free diffusion of insulin and nutrients [67]. Then, it was realized that contrast reagents could also be included in a microcapsule to allow for visualization of the cells. Our group has modified the alginate–poly-L-lysine–alginate microencapsulation method to include contrast agents that will enable monitoring of microencapsulated cell delivery [68, 69]. This resulted in enhanced imaging sensitivity by the addition of higher concentrations of the contrast agent in the capsule. The first MRI-visible microcapsules were developed for pancreatic islet cell delivery using SPIO (20% weight volume) [68]. Infusion of these "magnetocapsules" through the portal vein into the liver could be monitored in real time using MRI. Another advantage of microencapsulation is that it allows for multiple-modality imaging by the inclusion of contrast agents for other imaging modalities. For example, alginate microcapsules containing barium sulfate remain visible on X-ray fluoroscopy for several months. Formulation of the alginate microcapsules has been an area of intense interest with the addition of substances, such as perfluorocarbons (PFC), to increase oxygenation within the microcapsules [70]. A slight modification to this formulation has led to a microcapsule that was potentially visible by ^{19}F MRI, X-ray, CT, and ultrasound [71]. Our group recently optimized this strategy by using novel gold nanoparticles, functionalized with DTDTPA:gadolinium chelates (GadoGold). GadoGold particles were coencapsulated with pancreatic islet cells, using protamine sulfate as a clinical-grade alginate cross-linker. We found that labeled, single microcapsules could be readily visualized by (positive) contrast high-field MRI, micro-CT, and ultrasound imaging, both *in vitro* and *in vivo* [72] (Fig. 10.6). Those data demonstrate that cell encapsulation with GadoGold nanoparticles provides a means for simultaneous immunoprotection and trimodal, noninvasive monitoring of cell engraftment.

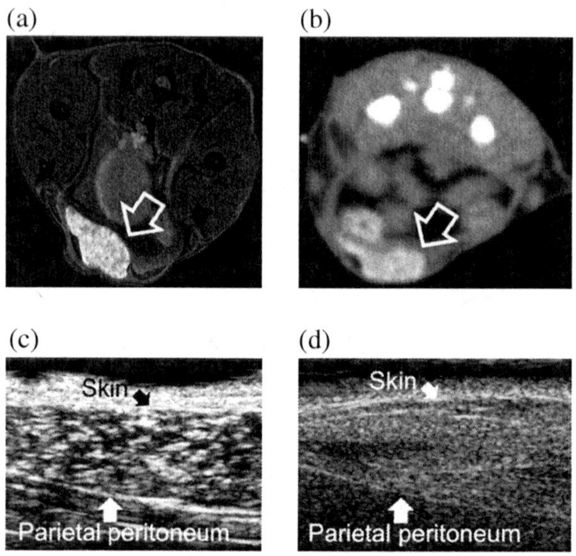

FIGURE 10.6 Multimodal imaging of microencapsulated cells. Positive contrast 9.4T MRI (a) and micro-CT images (b) of GadoGold microcapsules injected subcutaneously in the mouse abdomen (arrows). (c, d) 40-MHz ultrasound images of the fluid cavity between abdominal skin and parietal peritoneum of a mouse with (c) and without (d) injection of microcapsules. Reprinted from [72] with permission.

10.3 CHOOSING IMAGING MODALITIES

Many techniques for *in vivo* cellular imaging are now available for characterization of transplanted cells in the environment of a living organism. Figure 10.7 is a brief schematic summary of commonly used imaging modalities for cell tracking. All these imaging methods are based on different principles, and each provides distinctive, usually complementary information.

Generally speaking, an ideal imaging method of cell trafficking would ensure that cell viability and characteristics are truly depicted in real time. This imaging technology should allow quantification of cell numbers at any anatomical location and at any given time point. It should also permit cell tracking over a long period of time, both serially and noninvasively, in living subjects. In reality, although molecular imaging

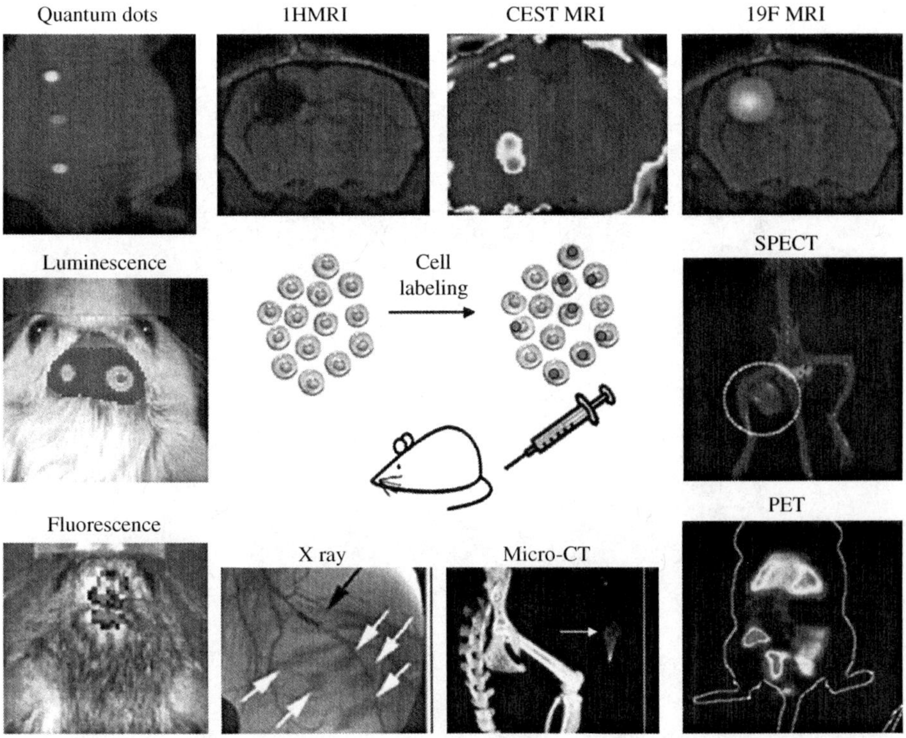

FIGURE 10.7 Imaging modalities for *in vivo* cell tracking. Labeled cells can be noninvasively tracked by more than 10 imaging modalities. Fluorescence, luminescence, and QDs use visible and near-infrared light emitted from labeled cells, which are extremely valuable for small animal studies. X-ray and micro-CT rely on X-ray (ionizing radiation). PET and SPECT (including scintigraphy) detect annihilation photon and gamma rays, respectively, from radioactive nuclides (ionizing radiation). MRI, including proton-based, ^{19}fluorine-based, and CEST MRI, use radiowaves for imaging and represents the most noninvasive imaging modality for cells and hosts. Images for QDs and ^{19}F MRI are simulated drawings. ^{1}H MR image reproduced from [100], CEST MR image reproduced from [58], SPECT-CT image reproduced from [59], PET image adapted from http://www.cibm.ch/page65843-en.html,and X-ray picture reproduced from [69] (all with permission). Micro-CT image by courtesy of Dr. Dian Arifin. For color details, please see color plate section.

TABLE 10.1 Available Noninvasive Cellular Imaging Modalities, with a Comparison for Clinical Use, Spatial Resolution, Sensitivity (Lowest Number of Cells That Can Be Detected), Anatomy (Providing Anatomical Images), Longevity (Duration of Signal), Reporter Gene (Availability of Reporter Gene), and Quantification (Whether It Allows Quantification of the Number of Cells)

	Clinical use	Resolution	Sensitivity (number of cells)	Anatomy	Longevity of labels	Reporter gene	Quantification
Fluorescence	No	+++*	N/A**	No	+++	Yes	No
QD	No	+	N/A	No	++++	No	No
BLI	No	+	10,000	No	+++	Yes	Yes
PET	Yes	+	50,000	No	+	Yes	Yes
γ-Irradiation	Yes	++	1,000	No	+	No	Yes
^{1}H MRI	Yes	++++	1	Yes	++	No	No
^{19}F MRI	No	++	7,500	No	++	No	Yes
CEST MRI	Yes	+++	25,000	Yes	+++	Yes	Yes
X-ray	Yes	++++	N/A	Yes	++	No	Yes
US	Yes	+	1	Yes	++	No	Yes

QD, quantum dots; BLI, bioluminescent imaging; PET, positron emission tomography; ^{1}H MRI, proton MRI; ^{19}F MRI, fluorine MRI; CEST MRI, chemical exchange saturation transfer MRI; US, ultrasound imaging.

*+,weak; ++, =moderate; +++, good; ++++, excellent.

**N/A, data not available.

modalities, like PET, SPECT, MRI, and optical imaging, are powerful tools for the evaluation of transplanted cell survival and function, no single imaging modality possesses all the desired qualities for optimal evaluation of cell therapy. Every method has its own advantages and disadvantages and can provide relevant and useful information. Table 10.1 shows the basic parameters and characteristics of each imaging modality.

Imaging modalities like radiolabeling have high sensitivity, but are limited to short-term follow-up because of radioactive decay. MRI has long-term imaging capability, but lacks the sensitivity of SPECT and may detect false signal in nonviable, apoptotic, or neighboring cells. BLI is convenient and widely available, but is not capable of deep tissue imaging. The reporter gene technique can provide long-term tracking and can directly detect cell viability, but it requires genetic manipulation of cells before transplantation.

There are two ways to address the problem of the imperfections in each imaging modality. One way is to carefully choose an imaging modality depending on the question that is being investigated. If the objective of the study is to image the delivery and short-term homing of stem cells in different organs, a direct labeling approach may fulfill that need. MRI provides the highest spatial resolution and almost real-time image guidance for cell delivery. If the objective of the study is to monitor the long-term viability of stem cells, reporter gene imaging is the only choice and can be achieved using optical imaging (bioluminescence, fluorescence) or PET/SPECT imaging. Although optical imaging is in general more sensitive, it does not provide much anatomical localization and is limited to small animals. PET and SPECT provide good sensitivity with better anatomical localization, and these methods have the potential to be translated to human applications.

Another solution is to use a multimodality approach, which is increasingly preferred to address the multiplicity of issues posed in this field. Combinations of two, three, or even more *in vivo* imaging modalities allow for the integration of the strengths of each modality and, at the same time, eliminate one or more weaknesses of an individual modality. One direction of multimodality imaging is to merge imaging instruments. PET/CT, SPECT/CT, and PET/MRI are examples of multimodality systems that have become widely adopted. Research is underway to couple optical imaging with MRI, PET, and/or SPECT to allow multimodal functional and molecular imaging [73]. Another direction is to develop multimodality contrast agents and reporter genes. The advancement in design of microencapsulation (such as GadoGold microcapsules as mentioned earlier) and the combination of reporter genes from different imaging modalities (like HSV1-tk plus luciferase and fluorescent proteins) [59] are in support of multimodality imaging systems.

10.4 APPLICATIONS FOR *IN VIVO* CELL IMAGING: STEM CELL THERAPY FOR CARDIAC DISEASES

Due to the numerous unresolved and controversial issues in stem cell-based therapy of ischemic heart disease, visualization of injected stem or progenitor cells is a much desired approach for better understanding of the fate of grafted cells and how it may affect the therapeutic outcome. The ideal imaging modality should provide integrated information related to the entire process of cell engraftment, survival, and functional outcome. Radionuclide technology and MRI are best suited to meet such an objective owing to their resolution and clinical applicability, with MRI having some advantages in terms of spatial resolution. The vision is that progress in imaging and therapy will go hand in hand to close existing gaps between basic science and clinical trials of cardiac cell therapy.

10.4.1 Radionuclide Approaches

The major advantages of using radionuclide probes for stem cell tracking include the inherently high signal-to-noise ratio, quantifiable levels of radionuclide probes, and the correlation with cell viability in combination with reporter genes. Stem cells can be labeled with radionuclide or PET agents by direct coincubation of cells with the probes, such as ^{111}In or ^{99m}Tc for SPECT and ^{18}F-FDG for PET imaging. ^{111}In-labeled endothelial progenitor cells have been found to accumulate in infarcted rat myocardium after intraventricular injection (4.7% in rats with myocardial infarction vs. 2.7% in sham-operated rats), but approximately 70% of the radioactivity was localized in the spleen and liver. For IV injection, the percentage of homing cells in the heart decreased to 2% compared to 1% in sham-operated rats [74], suggesting that only a small number of cells ultimately home to ischemic myocardium. Results from our group also revealed that injected ^{111}In-labeled MSCs could not be detected to localize into the myocardium after IV injection in pigs, due to early entrapment of MSCs in the lungs [75]. This observation was confirmed in rats: systemic IV delivery of ^{99m}Tc-HMPAO-labeled bone marrow-derived MSCs leads to entrapment of the donor cells in the lungs, while delivery by left ventricular cavity infusion resulted in drastically lower lung uptake, better uptake in the heart, and specifically higher uptake in infarcted compared with sham-operated hearts [24]. However, the entrapment of cells in the lung may not always be a problem for IV delivery of cells. Using a dog model of myocardial infarction, our group later demonstrated that ^{111}In-labeled MSCs redistributed from initial localization in the lungs to the liver and spleen within 24–48 h after IV injection. Although only a few percent of injected cells effectively localized in the myocardium, this enabled a clear SPECT-CT visualization of the homing of MSCs to infarcted myocardium (Fig. 10.8a) [25].

Although improvements in SPECT-CT imaging might enable the detection of labeled cells at higher resolution, PET still tends to be regarded as having higher temporal and spatial resolution compared to SPECT. The PET tracer ^{18}F-FDG has been used for labeling and imaging of BMCs in humans [76, 77].

Unselected autologous BMCs labeled with ^{18}F-FDG were delivered by either an intracoronary or an IV route in patients with infarcted myocardium. Using PET, only 1.3–2.6% of the cells were detected around the infarct border zone within 1–1.5 h after intracoronary injection, whereas only background activity was detected in the infarcted myocardium after IV injection. However, with immunomagnetically enriched CD34-positive cells from the same population, 14–39% of total injected radioactivity was detected in infarcted myocardium after intracoronary injection [76]. Kang and colleagues reported that, 2 h after intracoronary infusion of peripheral hematopoietic stem cells labeled with ^{18}F-FDG, 0.2–3.3% of injected activity accumulated at the infarcted myocardium in 17 patients with myocardial infarction, whereas high activity was found in the spleen, liver, bladder, and brain. At 20 h after the injection of cells, PET scans revealed only 1.5% of injected activity in the myocardium. However, no activity could be detected in the hearts of three patients who

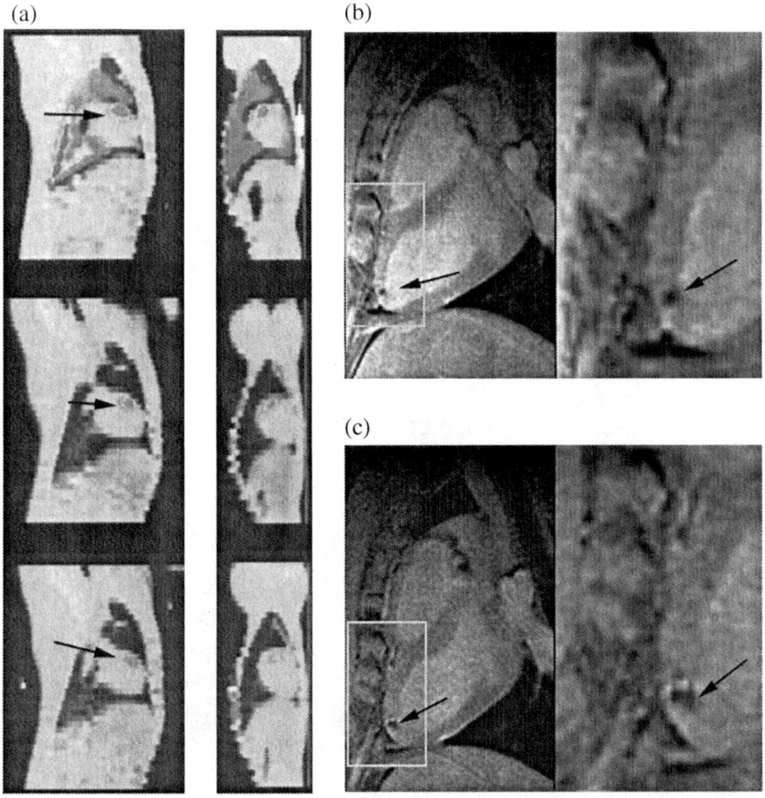

FIGURE 10.8 Radionuclide imaging and MRI of grafted stem cells homing to the heart. (a) Sagittal (left) and coronal (right) view of fused SPECT/CT images on days 1, 2, and 7 (from top to bottom) in an animal that demonstrated focal uptake of MSCs labeled with ^{111}In in the anterior midventricular region of the heart (arrows). (b,c) Long-axis MR images of a swine model of reperfused myocardial infarction showing hypointense lesions (arrows) derived from magnetically labeled MSCs at (b) 24 h or (c) 1 week post injection. Inserts demonstrate expansion of the lesion over 1 week. Reproduced from [25, 81] with permission.

received IV injection of ^{18}F-FDG-labeled cells [77]. These data support the translational potential of nuclear imaging to guide cell therapy approaches from preclinical to clinical applications. Future work should aim to prolong the half-life of radioisotopes in order to prevent loss of the imaging signal within a few days.

Reporter genes for PET/SPECT have been used to track cardiovascular cells. Imaging of genetically labeled cells was obtained using rat cardiomyoblasts, which were transfected *ex vivo* with adenovirus carrying the HSV1-tk and luciferase reporter genes. Cell-specific *in vivo* optical and micro-PET imaging was performed for up to 2 weeks after direct injection of cells into the myocardium of nude rats [62]. Imaging of cells with the use of transduced reporter genes has significant future potential. Reporter genes may be expressed under control of restrictive promoters so that they are only expressed in a defined group of cells. If the expression of a transduced reporter gene were to be dependent on the expression of certain genes that are active only in mature cardiac cells, reporter imaging would allow for identification of cell differentiation.

10.4.2 Stem Cell Tracking by MRI

Magnetic resonance imaging has a long history of applications for clinical diagnosis and is rapidly developing as a method for noninvasive imaging of cells during animal development and in laboratory models of disease. At present, many of the cellular imaging studies, including initial clinical applications, are performed with MRI. MRI has several advantages over other modalities. It is non-invasive and entirely clinically translatable. It carries no radiation, allows whole-body imaging, and has superior resolution, ranging from a near-cellular level (50 μm) in animals to up to 300 μm in clinical-grade scanners. Currently, most research groups conduct magnetic cell labeling using commercially available SPIO particles, as these are clinically translatable [78].

With regard to imaging of the heart, MRI provides excellent spatial and temporal resolution for imaging cardiac anatomy and function, allowing detailed delineation of cardiac and surrounding soft tissues. The rapid beating of the heart provides a substantial challenge to MR imaging in this context due to the sensitivity of this imaging modality to motion artifacts. As a result, both respiratory and cardiac gating is required to generate good quality images [79]. Initial animal model studies, using micron-sized particles [80] or nanoparticles of iron oxide [25, 81] to label stem cells, showed the potential for nontoxic labeling of hematopoietic bone marrow-derived stem cell and MSC populations. Several groups have attempted to visualize the incorporation of stem cells into cardiac tissue following (MR-guided) injection of SPIO-labeled MSCs. Our group [81], Hill et al. [80], and Garot et al. [82] have demonstrated that implantation of SPIO-labeled transplanted cells can be detected using clinical MR scanners (Fig. 10.8b and c). Injection of SPIO particles alone (without cells) into the area of damage in the heart only results in a transient (less than 12 h) signal attenuation on T2-weighted scans; unlabeled control cells do not affect the MR signal *in vivo*, suggesting that the long-term identification of SPIO-labeled cells is not a mere artifact of the surgical procedure or the incorporation of free iron oxide particles into host cells [83].

Although trials of cardiac stem cell therapy have used intracoronary or IV administration of cells, most SPIO-labeled stem cell MRI studies performed thus far have been limited to direct intramyocardial delivery of cells via percutaneous catheters. Systemic administration of MSCs through IV administration would be highly desirable as it is less invasive. However, the number of MSCs that home to injured areas (e.g., the heart or kidney) may be below the threshold of detection for *in vivo* MRI, as many cells become initially entrapped in the liver [84]. As such, cellular MRI, aside from imaging of migration or trafficking, is more suited to verify accurate cell delivery in the targeted organ (region). This is the case for MSC delivery in myocardial infarction, as well as MSC delivery to organs (liver, kidney) through their feeding arteries [29, 85]. Significant advances have been achieved in designing catheters that allow MRI-guided delivery and monitoring of transplanted cells, even in patients with implanted devices, such as pacemakers.

These studies have shown that it is possible to use MRI for stem cell tracking in the context of cardiovascular disease, and future clinical trials should benefit greatly from this new ability to precisely inject and track the biological fate of these cells.

10.5 APPLICATIONS FOR *IN VIVO* CELL IMAGING: MR TRACKING OF TRANSPLANTED STEM CELLS FOR TREATMENT OF STROKE

Recent advancements in stem cell research offer novel therapeutic strategies for treating a variety of neurological disorders. It has been shown repeatedly in animal models that neurological deficits can be diminished by the introduction of therapeutic cells [86, 87]. These observations in animal models provided the basis for the first clinical trials in Parkinson's disease [88] and stroke patients [89]. Successful treatment of CNS disorders is largely dependent on the targeted, uniform distribution of transplanted cells. Thus, a reliable method for the *in vivo* evaluation of global cell distribution in a dynamic manner is highly desirable. Our lab has developed methods of labeling neural stem cells from neurospheres for *in vivo* MR tracking [12, 90]. This method has allowed the labeling of mouse and human neural stem cells and MR detection in mice for up to 4 weeks following intracranial injection [91, 92]. Groups have shown that these labeled neurosphere cells will migrate in response to inflammatory cues present in experimental autoimmune encephalomyelitis (EAE, a model of multiple sclerosis) following intracerebral injection, thus providing a key role for MR cell tracking to identify the optimal time for administering cellular therapy, as well the delivery route and cell numbers [93].

Stroke is the most common cause of adult disability in the Western world. Yet, to date, therapeutic interventions have met with limited success. Cellular MRI in stroke allows monitoring of the engraftment of transplanted stem cells designed to improve behavioral impairment. In stroke, MRI has been used in rats to demonstrate that ESCs [94] or neural stem cells [95], grafted into the hemisphere contralateral to the ischemic lesion, migrate along white matter tracts and populate the border zone of the ischemic brain lesion. In the first study, investigators observed the migration of ESCs along the corpus callosum by means of double labeling cells for transplantation with GFP as a reporter gene and iron oxide particles (pretreated with a transfection agent) [94]. Modo et al. used a bimodal contrast agent, Gadolinium-RhodamIne Dextran (GRID), detectable both *in vivo* by MRI and, subsequently, at postmortem by fluorescent microscopy, as a tool to track migrating neural stem cells [95]. Both studies demonstrated the advantage of a bimodal contrast agent to visualize the migration of a large number of cells both *in vivo* and *ex vivo* to corroborate the presence of the contrast agent within transplanted cells. The Chopp group [96, 97] also used MRI to track neural stem cells (labeled with SPIO particles) after intracisternal implantation. Similarly, they reported the migration of transplanted subventricular zone cells toward the ischemic lesion in animals with stroke damage. From those experiments involving the use of MRI for cell tracking, it can be concluded that the stroke damage functions as an attractant to the neural stem cells, since, in intact brains, no significant migration (apart from a few scattered cells in the surrounding striatum) away from the injection tract was observed. The migration and differentiation of stem cells, therefore, appears to depend on the presence of "damaged" tissue.

Due to the invasive nature of intraparenchymal or intracisternal injection for the transplantation of stem cells, several other routes of cell administration were explored with the help of MRI. IV injection is the least invasive method, but may lead to low numbers of engrafted cells at the lesion and major trapping in the lung, liver, and spleen [84]. Intra-arterial (IA) injection, on the other hand, may be used to bypass the

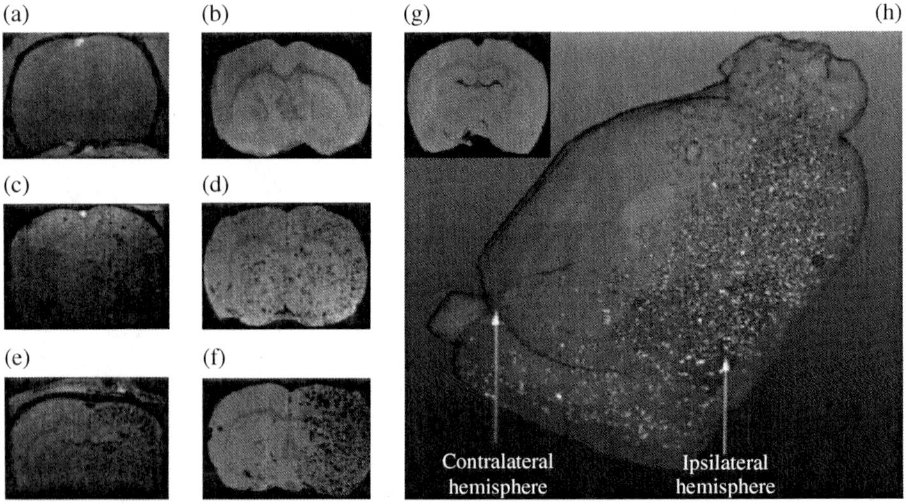

FIGURE 10.9 MR images of intracerebral cell engraftment after (a–f, h) IA and (**g**) IV injection. MSCs appear as hypointense spots on T2*-weighted images. (a) *in vivo* and (b) *ex vivo* MR images of an animal without cerebral engraftment of MSCs. (c) *In vivo* and (d) *ex vivo* MR images of an animal showing a moderate brain engraftment of MSCs with some cells entering the other hemisphere. (e) *In vivo* and (f) *ex vivo* MR images of an animal exhibiting a massive engraftment of MSCs, with a limited cell distribution within the right hemisphere perfused by the injected right carotid artery. (g) No engraftment could be detected after IV injection. (h) 3D reconstruction of cellular distribution in the brain shown in (e, f). Cells distributed widely throughout the right brain hemisphere, with a prevalence for the area directly perfused by the right internal carotid artery; the territory of the anterior and posterior cerebral arteries also exhibit significant engraftment. Few cells were detected in the contralateral (left) hemisphere, suggesting that cells did not circulate systemically before arriving at their final location. Reprinted from [98] with permission.

initial uptake by the systemic organs and deliver larger numbers of cells directly to the ischemic lesion, once vessels are reperfused. We compared the two routes of injection and have demonstrated, using dual-modality (laser Doppler flow signal measurements and MRI) monitoring, that IA delivery of stem cells offers the advantage of directly targeting the damaged tissue region (see Fig. 10.9) [98].

MRI has also been combined with optical imaging for non-invasive tracking of transplanted stem cells in a stroke model. Daadi et al. transduced ESC-derived neural stem cells with GFP and Fluc reporter genes, labeled them with SPIO, and grafted them intracerebrally in stroke-lesioned rats [99]. Cell survival and engraftment were tracked noninvasively with BLI and MRI, respectively, for 2 months after transplantation and confirmed histologically. Those data suggested that the combination of BLI and MRI modalities provides reliable real-time monitoring of cell fate. In our recent study, we also used a combination of MRI and BLI to monitor long-term (93 days) survival and distribution of neural stem cells labeled with SPIO and transfected with Fluc after intracerebral transplantation into immunodeficient or immunocompetent mice [100]. Unexpectedly, in mice that rejected the cells, the hypointense MR signal persisted throughout the entire time course, whereas in the nonrejecting mice, the contrast cleared at a faster rate (see Fig. 10.10). In immunocompetent,

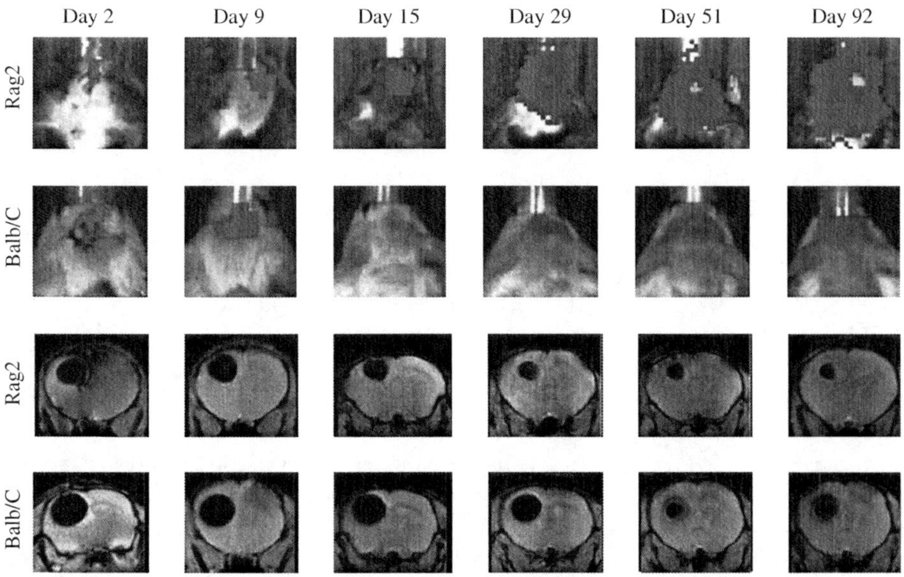

FIGURE 10.10 Monitoring the survival of transplanted C17.2 cells by BLI and MRI. C17.2 cells stably expressed firefly luciferase and were labeled with SPIO before transplantation. BLI (upper two rows) and MRI (lower two rows) of C17.2 cells transplanted into graft-accepting Rag2 and graft-rejecting Balb/c mice. Mice were imaged weekly from day 2 to day 92. In all mice, hypointensities were detectable within the corpus callosum the day after transplantation and persisted for up to 93 days, although they were rejected in Balb/C mice. BLI, but not MRI, accurately depicted the time course of survival of transplanted cells. Adapted from [100] with permission.

graft-rejecting Balb/c mice, infiltrating leukocytes and microglia were found surrounding dead cells and internalizing the SPIO clusters. These results indicate that live cell proliferation and associated label dilution may dominate contrast clearance, compared with cell death and subsequent transfer and retention of SPIO within phagocytes (microglia cells) and the brain interstitium. Thus, interpretation of signal changes during long-term MR cell tracking is complex and requires caution. In the same way, the Gambhir and Wu groups compared Fluc reporter gene and SPIO as cell markers for longitudinal monitoring of grafted cell survival (cardiomyoblast and ES cells, respectively) using BLI and MRI [30, 101]. Their results, along with our findings, indicate that BLI reporter gene imaging is the only accurate gauge of transplanted cell survival, compared to MRI of SPIO-labeled cells, while iron particle labeling is a better marker for high-resolution detection of cell location by MRI.

10.6 CONCLUSIONS AND PERSPECTIVES

Cellular therapy promises to revolutionize medicine by restoring tissue and organ function. Stem cells and progenitor cells are being intensively explored in regenerative medicine for therapeutic purposes. As with all major developments, new therapies must be introduced to allow optimization. For cell therapy, the key tool is *in vivo* imaging for real-time assessment of parameters such as cell localization, numbers, and viability. Such data are critical to modulate and tailor the therapy for each patient. In fact, the development of successful cell therapy is intricately coupled to the development of medical imaging technologies that can allow noninvasive, longitudinal, and quantitative monitoring of the transferred cells.

Over the last decade, we have seen a revolution in noninvasive stem cell imaging in the living subject. Each imaging modality has been evolving to become more capable of observing the lowest amount of cells and providing better signal-to-noise ratio and be less invasive, more versatile, and also compatible with other modalities, although each still has its own weakness inherent in the design of its detection system. Thus, it is likely that not one technique will answer all questions, but the use of a multimodality approach will be the most appropriate approach to address the myriad aspects of the *in vivo* cell fate. One issue, with regard to the development of a more advanced *in vivo* imaging modality, is the cost of expensive imaging scanners—often a multimillion dollar investment. Multimodal scanners are even more expensive. It would be ideal if the cheaper imaging techniques, such as ultrasound or fluorescence-based approaches, could be adapted to clinical cell tracking, although much development is needed for the successful adaptation of these techniques.

One big challenge for *in vivo* cell tracking is the translation of different imaging techniques into clinical protocols. The clinical implementation of multimodal cellular imaging, in particular, will be one of the major focuses in the coming years. For example, the clinical implementation of cell-tracking methods using MRI may be considerably informative to confirm the correct transplantation of high numbers of labeled cells, for example, of dendritic cells [78] or cytolytic T cells [102] in tumor immunotherapy, in Parkinson's disease, or in myocardial infarction. Undoubtedly, increasing the involvement of imaging in grafting procedures will greatly advance the field of stem cell therapy in general and help to improve cell transplantation to make it a more efficient and successful clinical procedure.

REFERENCES

1. Bohl JR, Goebel HH, Potsch L, et al. Cell therapy and its risks. Fortschr Med 1994;112:261–265.
2. Bohl J, Goebel HH, Potsch L, et al. Complications following cell therapy. Z Rechtsmed 1989;103:1–20.
3. Lunde K, Solheim S, Aakhus S, et al. Intracoronary injection of mononuclear bone marrow cells in acute myocardial infarction. N Engl J Med 2006;355:1199–1209.
4. Schachinger V, Erbs S, Elsasser A, et al. Intracoronary bone marrow-derived progenitor cells in acute myocardial infarction. N Engl J Med 2006;355:1210–1221.
5. Sutton EJ, Henning TD, Pichler BJ, et al. Cell tracking with optical imaging. Eur Radiol 2008;18:2021–2032.
6. Lin S, Xie X, Patel MR, et al. Quantum dot imaging for embryonic stem cells. BMC Biotechnol 2007;7:67.
7. Smith AM, Duan H, Mohs AM, et al. Bioconjugated quantum dots for in vivo molecular and cellular imaging. Adv Drug Deliv Rev 2008;60:1226–1240.
8. Medintz IL, Uyeda HT, Goldman ER, et al. Quantum dot bioconjugates for imaging, labelling and sensing. Nat Mater 2005;4:435–446.

9. Muller-Borer BJ, Collins MC, Gunst PR, et al. Quantum dot labeling of mesenchymal stem cells. J Nanobiotechnology 2007;5:9.
10. Slotkin JR, Chakrabarti L, Dai HN, et al. In vivo quantum dot labeling of mammalian stem and progenitor cells. Dev Dyn 2007;236:3393–3401.
11. Frank JA, Zywicke H, Jordan EK, et al. Magnetic intracellular labeling of mammalian cells by combining (FDA-approved) superparamagnetic iron oxide MR contrast agents and commonly used transfection agents. Acad.Radiol 2002;9 Suppl 2:S484–S487.
12. Frank JA, Miller BR, Arbab AS, et al. Clinically applicable labeling of mammalian and stem cells by combining superparamagnetic iron oxides and transfection agents. Radiology 2003;228:480–487.
13. Kraitchman DL, Bulte JW. Imaging of stem cells using MRI. Basic Res Cardiol 2008; 103:105–113.
14. Walczak P, Kedziorek DA, Gilad AA, et al. Applicability and limitations of MR tracking of neural stem cells with asymmetric cell division and rapid turnover: the case of the shiverer dysmyelinated mouse brain. Magn Reson Med 2007;58:261–269.
15. Walczak P, Kedziorek DA, Gilad AA, et al. Instant MR labeling of stem cells using magnetoelectroporation. Magn Reson Med 2005;54:769–774.
16. Walczak P, Ruiz-Cabello J, Kedziorek DA, et al. Magnetoelectroporation: improved labeling of neural stem cells and leukocytes for cellular magnetic resonance imaging using a single FDA-approved agent. Nanomedicine 2006;2:89–94.
17. Gilad AA, Walczak P, McMahon MT, et al. MR tracking of transplanted cells with "positive contrast" using manganese oxide nanoparticles. Magn Reson Med 2008; 60:1–7.
18. Aime S, Barge A, Cabella C, et al. Targeting cells with MR imaging probes based on paramagnetic Gd(III) chelates. Curr Pharm Biotechnol 2004;5:509–518.
19. Terreno E, Geninatti CS, Belfiore S, et al. Effect of the intracellular localization of a Gd-based imaging probe on the relaxation enhancement of water protons. Magn Reson Med 2006;55:491–497.
20. Srinivas M, Heerschap A, Ahrens ET, et al. (19)F MRI for quantitative in vivo cell tracking. Trends Biotechnol 2010;28:363–370.
21. Partlow KC, Chen J, Brant JA, et al. 19F magnetic resonance imaging for stem/progenitor cell tracking with multiple unique perfluorocarbon nanobeacons. FASEB J 2007;21: 1647–1654.
22. Ruiz-Cabello J, Walczak P, Kedziorek DA, et al. In vivo "hot spot" MR imaging of neural stem cells using fluorinated nanoparticles. Magn Reson Med 2008;60:1506–1511.
23. Massoud TF, Gambhir SS. Molecular imaging in living subjects: seeing fundamental biological processes in a new light. Genes Dev 2003;17:545–580.
24. Barbash IM, Chouraqui P, Baron J, et al. Systemic delivery of bone marrow-derived mesenchymal stem cells to the infarcted myocardium: feasibility, cell migration, and body distribution. Circulation 2003;108:863–868.
25. Kraitchman DL, Tatsumi M, Gilson WD, et al. Dynamic imaging of allogeneic mesenchymal stem cells trafficking to myocardial infarction. Circulation 2005;112: 1451–1461.
26. Simonova M, Shtanko O, Sergeyev N, et al. Engineering of technetium-99m-binding artificial receptors for imaging gene expression. J Gene Med 2003;5:1056–1066.
27. Thakur ML, Segal AW, Louis L, et al. Indium-111-labeled cellular blood components: mechanism of labeling and intracellular location in human neutrophils. J Nucl Med 1977;18:1022–1026.

28. Jin Y, Kong H, Stodilka RZ, et al. Determining the minimum number of detectable cardiac-transplanted 111In-tropolone-labelled bone-marrow-derived mesenchymal stem cells by SPECT. Phys Med Biol 2005;50:4445–4455.
29. Bos C, Delmas Y, Desmouliere A, et al. In vivo MR imaging of intravascularly injected magnetically labeled mesenchymal stem cells in rat kidney and liver. Radiology 2004;233:781–789.
30. Chen IY, Greve JM, Gheysens O, et al. Comparison of optical bioluminescence reporter gene and superparamagnetic iron oxide MR contrast agent as cell markers for noninvasive imaging of cardiac cell transplantation. Mol Imaging Biol 2009;11:178–187.
31. Carr HM, Smyth JV, Rooney OB, et al. Limitations of in-vitro labeling of endothelial cells with indium-111 oxine. Cell Transplant 1995;4:291–296.
32. Zanzonico P, Koehne G, Gallardo HF, et al. [131I]FIAU labeling of genetically transduced, tumor-reactive lymphocytes: cell-level dosimetry and dose-dependent toxicity. Eur J Nucl Med Mol Imaging 2006;33:988–997.
33. Chudakov DM, Matz MV, Lukyanov S, et al. Fluorescent proteins and their applications in imaging living cells and tissues. Physiol Rev 2010;90:1103–1163.
34. Weissleder R, Ntziachristos V. Shedding light onto live molecular targets. Nat Med 2003;9:123–128.
35. Campbell RE, Tour O, Palmer AE, et al. A monomeric red fluorescent protein. Proc Natl Acad Sci USA 2002;99:7877–7882.
36. Shaner NC, Campbell RE, Steinbach PA, et al. Improved monomeric red, orange and yellow fluorescent proteins derived from Discosoma sp. red fluorescent protein. Nat Biotechnol 2004;22:1567–1572.
37. Shu X, Royant A, Lin MZ, et al. Mammalian expression of infrared fluorescent proteins engineered from a bacterial phytochrome. Science 2009;324:804–807.
38. Hielscher AH. Optical tomographic imaging of small animals. Curr Opin Biotechnol 2005;16:79–88.
39. Fujimoto JG. Optical coherence tomography for ultrahigh resolution in vivo imaging. Nat Biotechnol 2003;21:1361–1367.
40. Lin Y, Molter J, Lee Z, et al. Bioluminescence imaging of hematopoietic stem cell repopulation in murine models. Methods Mol Biol 2008;430:295–306.
41. Wilson KD, Huang M, Wu JC. Bioluminescence reporter gene imaging of human embryonic stem cell survival, proliferation, and fate. Methods Mol Biol 2009;574: 87–103.
42. Love Z, Wang F, Dennis J, et al. Imaging of mesenchymal stem cell transplant by bioluminescence and PET. J Nucl Med 2007;48:2011–2020.
43. Sher F, van DG, Boddeke E, et al. Bioluminescence imaging of Olig2-neural stem cells reveals improved engraftment in a demyelination mouse model. Stem Cells 2009;27:1582–1591.
44. de Wet JR, Wood KV, Helinski DR, et al. Cloning of firefly luciferase cDNA and the expression of active luciferase in Escherichia coli. Proc Natl Acad Sci USA 1985;82: 7870–7873.
45. Sherf BA, Wood KV. Firefly luciferase engineered for improved genetic reporting. Promega Notes Mag 1994;49:14.
46. Kim DE, Schellingerhout D, Ishii K, et al. Imaging of stem cell recruitment to ischemic infarcts in a murine model. Stroke 2004;35:952–957.
47. Kim JB, Urban K, Cochran E, et al. Non-invasive detection of a small number of bioluminescent cancer cells in vivo. PLoS One 2010;5:e9364.
48. Rice BW, Cable MD, Nelson MB. In vivo imaging of light-emitting probes. J Biomed Opt 2001;6:432–440.

49. Branchini BR, Southworth TL, Khattak NF, et al. Red- and green-emitting firefly luciferase mutants for bioluminescent reporter applications. Anal Biochem 2005;345: 140–148.

50. Branchini BR, Ablamsky DM, Murtiashaw MH, et al. Thermostable red and green light-producing firefly luciferase mutants for bioluminescent reporter applications. Anal Biochem 2007;361:253–262.

51. Branchini BR, Ablamsky DM, Davis AL, et al. Red-emitting luciferases for bioluminescence reporter and imaging applications. Anal Biochem 2010;396:290–297.

52. Liang Y, Walczak P, Bulte JW. Comparison of red-shifted firefly luciferase Ppy RE9 and conventional Luc2 as bioluminescence imaging reporter genes for in vivo imaging of stem cells. J Biomed Opt 2012;17:016004.

53. Louie AY, Huber MM, Ahrens ET, et al. In vivo visualization of gene expression using magnetic resonance imaging. Nat Biotechnol 2000;18:321–325.

54. Kodibagkar VD, Yu J, Liu L, et al. Imaging beta-galactosidase activity using 19F chemical shift imaging of LacZ gene-reporter molecule 2-fluoro-4-nitrophenol-beta-D-galactopyranoside. Magn Reson Imaging 2006;24:959–962.

55. Weissleder R, Moore A, Mahmood U, et al. In vivo magnetic resonance imaging of transgene expression. Nat Med 2000;6:351–355.

56. Cohen B, Dafni H, Meir G, et al. Ferritin as an endogenous MRI reporter for noninvasive imaging of gene expression in C6 glioma tumors. Neoplasia 2005;7:109–117.

57. Sherry AD, Woods M. Chemical exchange saturation transfer contrast agents for magnetic resonance imaging. Annu Rev Biomed Eng 2008;10:391–411.

58. Gilad AA, McMahon MT, Walczak P, et al. Artificial reporter gene providing MRI contrast based on proton exchange. Nat Biotechnol 2007;25:217–219.

59. Pomper MG, Hammond H, Yu X, et al. Serial imaging of human embryonic stem-cell engraftment and teratoma formation in live mouse models. Cell Res 2009;19:370–379.

60. Ponomarev V, Doubrovin M, Shavrin A, et al. A human-derived reporter gene for noninvasive imaging in humans: mitochondrial thymidine kinase type 2. J Nucl Med 2007;48:819–826.

61. Hsieh CH, Chen FD, Wang HE, et al. Generation of destabilized herpes simplex virus type 1 thymidine kinase as transcription reporter for PET reporter systems in molecular genetic imaging. J Nucl Med 2008;49:142–150.

62. Wu JC, Chen IY, Sundaresan G, et al. Molecular imaging of cardiac cell transplantation in living animals using optical bioluminescence and positron emission tomography. Circulation 2003;108:1302–1305.

63. MacLaren DC, Gambhir SS, Satyamurthy N, et al. Repetitive, non-invasive imaging of the dopamine D2 receptor as a reporter gene in living animals. Gene Ther 1999;6:785–791.

64. Liang Q, Satyamurthy N, Barrio JR, et al. Noninvasive, quantitative imaging in living animals of a mutant dopamine D2 receptor reporter gene in which ligand binding is uncoupled from signal transduction. Gene Ther 2001;8:1490–1498.

65. Chen IY, Wu JC, Min JJ, et al. Micro-positron emission tomography imaging of cardiac gene expression in rats using bicistronic adenoviral vector-mediated gene delivery. Circulation 2004;109:1415–1420.

66. Terrovitis J, Kwok KF, Lautamaki R, et al. Ectopic expression of the sodium-iodide symporter enables imaging of transplanted cardiac stem cells in vivo by single-photon emission computed tomography or positron emission tomography. J Am Coll Cardiol 2008;52:1652–1660.

67. Lim F, Sun AM. Microencapsulated islets as bioartificial endocrine pancreas. Science 1980;210:908–910.

68. Barnett BP, Arepally A, Karmarkar PV, et al. Magnetic resonance-guided, real-time targeted delivery and imaging of magnetocapsules immunoprotecting pancreatic islet cells. Nat Med 2007;13:986–991.

69. Barnett BP, Kraitchman DL, Lauzon C, et al. Radiopaque alginate microcapsules for X-ray visualization and immunoprotection of cellular therapeutics. Mol Pharm 2006;3:531–538.

70. Khattak SF, Chin KS, Bhatia SR, et al. Enhancing oxygen tension and cellular function in alginate cell encapsulation devices through the use of perfluorocarbons. Biotechnol Bioeng 2007;96:156–166.

71. Barnett BP, Ruiz-Cabello J, Hota P, et al. Fluorocapsules for improved function, immunoprotection, and visualization of cellular therapeutics with MR, US, and CT imaging. Radiology 2011;258:182–191.

72. Arifin DR, Long CM, Gilad AA, et al. Trimodal gadolinium-gold microcapsules containing pancreatic islet cells restore normoglycemia in diabetic mice and can be tracked by using US, CT, and positive-contrast MR imaging. Radiology 2011;260:790–798.

73. Cherry SR. Multimodality in vivo imaging systems: twice the power or double the trouble? Annu Rev Biomed Eng 2006;8:35–62.

74. Aicher A, Brenner W, Zuhayra M, et al. Assessment of the tissue distribution of transplanted human endothelial progenitor cells by radioactive labeling. Circulation 2003;107: 2134–2139.

75. Chin BB, Nakamoto Y, Bulte JW, et al. 111In oxine labelled mesenchymal stem cell SPECT after intravenous administration in myocardial infarction. Nucl Med Commun 2003;24:1149–1154.

76. Hofmann M, Wollert KC, Meyer GP, et al. Monitoring of bone marrow cell homing into the infarcted human myocardium. Circulation 2005;111:2198–2202.

77. Kang WJ, Kang HJ, Kim HS, et al. Tissue distribution of 18F-FDG-labeled peripheral hematopoietic stem cells after intracoronary administration in patients with myocardial infarction. J Nucl Med 2006;47:1295–1301.

78. de VI, Lesterhuis WJ, Barentsz JO, et al. Magnetic resonance tracking of dendritic cells in melanoma patients for monitoring of cellular therapy. Nat Biotechnol 2005;23: 1407–1413.

79. Leergaard TB, Bjaalie JG, Devor A, et al. In vivo tracing of major rat brain pathways using manganese-enhanced magnetic resonance imaging and three-dimensional digital atlasing. Neuroimage 2003;20:1591–1600.

80. Hill JM, Dick AJ, Raman VK, et al. Serial cardiac magnetic resonance imaging of injected mesenchymal stem cells. Circulation 2003;108:1009–1014.

81. Kraitchman DL, Heldman AW, Atalar E, et al. In vivo magnetic resonance imaging of mesenchymal stem cells in myocardial infarction. Circulation 2003;107:2290–2293.

82. Garot J, Unterseeh T, Teiger E, et al. Magnetic resonance imaging of targeted catheter-based implantation of myogenic precursor cells into infarcted left ventricular myocardium. J Am Coll Cardiol 2003;41:1841–1846.

83. Himes N, Min JY, Lee R, et al. In vivo MRI of embryonic stem cells in a mouse model of myocardial infarction. Magn Reson Med 2004;52:1214–1219.

84. Hauger O, Frost EE, van HR, et al. MR evaluation of the glomerular homing of magnetically labeled mesenchymal stem cells in a rat model of nephropathy. Radiology 2006;238:200–210.

85. Mahmood U. Can MR imaging be used to track delivery of intravascularly administered stem cells? Radiology 2004;233:625–626.

86. Bjorklund A, Stenevi U. Reconstruction of the nigrostriatal dopamine pathway by intracerebral nigral transplants. Brain Res 1979;177:555–560.

87. Borlongan CV, Tajima Y, Trojanowski JQ, et al. Transplantation of cryopreserved human embryonal carcinoma-derived neurons (NT2N cells) promotes functional recovery in ischemic rats. Exp Neurol 1998;149:310–321.

88. Kordower JH, Freeman TB, Snow BJ, et al. Neuropathological evidence of graft survival and striatal reinnervation after the transplantation of fetal mesencephalic tissue in a patient with Parkinson's disease. N Engl J Med 1995;332:1118–1124.

89. Kondziolka D, Wechsler L, Goldstein S, et al. Transplantation of cultured human neuronal cells for patients with stroke. Neurology 2000;55:565–569.

90. Bulte JW, Douglas T, Witwer B, et al. Magnetodendrimers allow endosomal magnetic labeling and in vivo tracking of stem cells. Nat Biotechnol 19:1141–1147.

91. Neri M, Maderna C, Cavazzin C, et al. Efficient in vitro labeling of human neural precursor cells with superparamagnetic iron oxide particles: relevance for in vivo cell tracking. Stem Cells 2008;26:505–516.

92. Guzman R, Uchida N, Bliss TM, et al. Long-term monitoring of transplanted human neural stem cells in developmental and pathological contexts with MRI. Proc Natl Acad Sci USA 2007;104:10211–10216.

93. Bulte JW, Ben-Hur T, Miller BR, et al. MR microscopy of magnetically labeled neurospheres transplanted into the Lewis EAE rat brain. Magn Reson Med 2003;50:201–205.

94. Hoehn M, Kustermann E, Blunk J, et al. Monitoring of implanted stem cell migration in vivo: a highly resolved in vivo magnetic resonance imaging investigation of experimental stroke in rat. Proc Natl Acad Sci USA 2002;99:16267–16272.

95. Modo M, Mellodew K, Cash D, et al. Mapping transplanted stem cell migration after a stroke: a serial, in vivo magnetic resonance imaging study. Neuroimage 2004;21: 311–317.

96. Zheng ZG, Jiang Q, Zhang RL, et al. Magnetic resonance imaging and neurosphere therapy of stroke in rat. Ann Neurol 2003;53:259–263, 2003.

97. Jiang Q, Zhang ZG, Ding GL, et al. Investigation of neural progenitor cell induced angiogenesis after embolic stroke in rat using MRI. Neuroimage 2005;28:698–707.

98. Walczak P, Zhang J, Gilad AA, et al. Dual-modality monitoring of targeted intraarterial delivery of mesenchymal stem cells after transient ischemia. Stroke 2008;39: 1569–1574.

99. Daadi MM, Li Z, Arac A, et al. Molecular and magnetic resonance imaging of human embryonic stem cell-derived neural stem cell grafts in ischemic rat brain. Mol Ther 2009;17:1282–1291.

100. Berman SC, Galpoththawela C, Gilad AA, et al. Long-term MR cell tracking of neural stem cells grafted in immunocompetent versus immunodeficient mice reveals distinct differences in contrast between live and dead cells. Magn Reson Med 2010;65 (2):564–574.

101. Li Z, Suzuki Y, Huang M, et al. Comparison of reporter gene and iron particle labeling for tracking fate of human embryonic stem cells and differentiated endothelial cells in living subjects. Stem Cells 2008;26:864–873.

102. Yaghoubi SS, Jensen MC, Satyamurthy N, et al. Noninvasive detection of therapeutic cytolytic T cells with 18F-FHBG PET in a patient with glioma. Nat Clin Pract Oncol 2009;6:53–58.

CHAPTER 11

RADIOLABELED LIPOSOMES AS DRUG DELIVERY NANOTHERANOSTICS

WILLIAM T. PHILLIPS, ANDE BAO, KEITARO SOU, SHIHONG LI, AND BETH GOINS

11.1 INTRODUCTION

Liposomes are spontaneously forming lipid spheres that have proven to be valuable drug carriers [1, 2]. Liposomes in the nanoparticle size range of less than 250 nm have been shown to have drug carrier advantages that are not found in larger liposomes so that these smaller liposomes have been referred to as nanoliposomes [3–5]. Nanoliposomes have relatively long intravenous circulation times not found with larger liposome formulations [6], making them ideal carriers for therapeutic agents. A longer circulation time generally results in increased uptake by the targeted tissue because of the greater number of circulatory passes by the target [7]. Several clinically approved anticancer and antifungal products are liposomal formulations with size ranges that are smaller than 100 nm [8, 9]. Due to their longer circulation time, these liposomal products have a much higher uptake in their tumor or fungal infection targets as compared with the same free unencapsulated drug. Nevertheless, even with this improved drug delivery to targeted tissue, the majority of the intravenously administered nanoencapsulated drug is taken up by nontarget tissue, generally macrophages of the liver and spleen.

Liposomes have a myriad of possible compositions and modifications that make them extremely flexible drug carriers. In the last 30 years, there has been significant progress in liposome technology that has greatly increased their functionality as drug carriers including the development of (i) methods to produce homogeneously sized liposomes in the nanometer size range, (ii) methods to stably encapsulate sufficient quantities of therapeutic agents within the liposomes, (iii) manufacturing technology for large-scale liposome manufacture, (iv) modifications to prolong the circulation time of liposomes in the blood [10–12], (v) attachment of targeted ligands to the surface of liposomes to improve their *in vivo* distribution following intravenous administration [13–16], and (vi) effective methods of noninvasively tracking and

Drug Delivery Applications of Noninvasive Imaging: Validation from Biodistribution to Sites of Action, First Edition. Edited by Chun Li and Mei Tian.

quantifying the distribution of liposomes in the body [17–22]. This chapter will review the uses and advantages of nuclear imaging technology for monitoring the biodistribution of radiolabeled liposome formulations and highlight several examples of liposomes labeled with radionuclides that have diagnostic, monitoring, and therapeutic functions (theranostic) [23, 24].

11.2 NUCLEAR IMAGING FOR *IN VIVO* RADIOLABELED LIPOSOME IMAGING

Nuclear imaging, also known as gamma photon imaging, is a noninvasive imaging technology that involves the detection of gamma photons emitted from radioactive molecules. It is commonly used in clinical patients for the diagnosis of diseases following administration of a specific radionuclide-labeled molecule or liposome. This imaging modality has a very high sensitivity for detecting and quantitating radiolabeled molecules in the body with relatively high sensitivities that are not possible with standard X-ray, X-ray computed tomography (CT) imaging, or magnetic resonance imaging (MRI). Because it is possible to detect very small quantities of radioactive matter (<0.01–1 ng of actual physical matter) with nuclear imaging, it is ideally suited to track radiolabeled liposomes in the body.

There are two types of nuclear imaging, which are based on either (i) single gamma photon detection or (ii) dual-annihilation photon detection. Both types of nuclear imaging permit noninvasive quantification of the distribution of radiolabeled agents within the body. With single photon imaging, the images can be acquired as a two-dimensional planar projection image or as a tomographic three-dimensional image known as single photon emission computed tomography (SPECT) image. Commonly used single photon-emitting radionuclides for liposome imaging are technetium-99m (^{99m}Tc) with a 6-h half-life and indium-111 (^{111}In) with a half-life of 67 h. These radionuclides are readily available from local clinical radiopharmacies, and ^{99m}Tc is the most commonly used radionuclide in clinical nuclear medicine. The approaches of labeling liposomes with single photon radionuclides have been covered in a more detailed review of labeling technology in previous reviews by the authors [25]. Several detailed reviews have been recently written describing the methodology for radiolabeling liposomes [26–28].

One of the methods for radiolabeling liposomes with the single photon radionuclide, ^{99m}Tc, can also be used to radiolabel liposomes with the therapeutic beta particle-emitting radionuclides, rhenium-186 (^{186}Re) and rhenium-188 (^{188}Re) [29]. These beta particle-emitting radionuclides have excellent promise for treating solid cancers of a wide variety of tumor types [30, 31]. This therapy is very similar to the very successful use of iodine-131 (^{131}I) for the treatment of thyroid cancer. One significant advantage of both ^{186}Re and ^{188}Re therapeutic radionuclides compared with many other therapeutic radionuclides is that they emit a low ratio of photons in the ideal range of gamma camera imaging. For every 10 and 6.5 beta emissions, ^{186}Re and ^{188}Re radionuclides emit a gamma photon in the ideal energy range near that of ^{99m}Tc (140 keV). This low ratio of emitted photon to beta particles is an advantageous feature for use as a theranostic agent because it lowers the dose delivered to the whole body while still permitting biodistribution monitoring with either planar or SPECT imaging.

Having a limited number of gamma photons per beta particle lowers the dose to radiation handlers and to the patient themselves as 137–155-keV gamma photons have an average tissue penetration of ~4.6 cm. These advantages for ^{186}Re and ^{188}Re contrast significantly with the most commonly used therapeutic radionuclide, ^{131}I, which emits ~1 high-energy 364-keV gamma photon (average tissue penetration length of 6.3 cm) per every one beta particle emitted and therefore delivers significant radiation to the whole body of the patient as well as radiation workers administering doses of ^{131}I.

The second type of imaging, positron emission tomography (PET) imaging, is based on the detection of two annihilation photons that are emitted simultaneously at ~180° from each other. These dual photon emissions can be localized along a line by the simultaneous detection of the two 180° emitted photons by crystal scintillation detectors that are part of a ring of detectors surrounding the body. The most commonly used radionuclides for PET imaging have relatively short half-life such as oxygen-15 (2-min half-life), carbon-11 (20-min half-life), and fluorine-18 (110-min half-life). Although some studies have been performed with shorter-lived PET radionuclides and liposome imaging [32, 33], for most liposome imaging applications, these PET isotopes have too short of a half-life to track the liposome *in vivo*. PET radionuclides with longer half-lives, such as copper-64 (^{64}Cu) (12.7-h half-life) and iodine-124 (^{124}I) (4.2-day half-life), are more promising for tracking the distribution of long-circulating nanoparticles such as liposomes. Recently, several methods have been described for radiolabeling liposomes with ^{64}Cu [21, 34, 35].

Recent technological advances have resulted in great improvements in the resolution and sensitivity of SPECT and PET cameras used for nuclear imaging [36–38]. These include the development of imaging systems for small animal imaging. Nuclear cameras that combine CT with nuclear imaging have greatly increased the information that can be derived from the tracking of radiolabeled liposomes [39–41]. Instead of knowing the general location of the uptake of radiolabeled liposomes in the body, the superimposition of high anatomic resolution CT on the nuclear image permits determination of the specific organ in which the radiolabeled liposomes have accumulated.

One particular advantage of nuclear imaging for the study of liposomes is that it can be performed in almost all mammalian species, making nuclear imaging particularly valuable for development of therapeutic liposomes because findings in small animals have the potential to be directly translated to humans who can also be imaged by nuclear imaging.

11.3 VALUE OF NUCLEAR IMAGING OF LIPOSOMES

11.3.1 Theranostic Use

Nuclear imaging of radiolabeled liposomes provides many advantages for clinical monitoring of the distribution of liposome therapeutics in real time to predict therapeutic efficacy based on imaging of the distribution of the liposomes and to determine the specific dose required to achieve the required distribution at the target site. When the radiolabeled liposomes are carrying a chemotherapeutic agent, this type of theranostic monitoring can be called chemodosimetry [42]. Several studies have shown a very good correlation of liposome uptake in targeted tissue derived from nuclear imaging measurements with measured tissue drug levels following tissue extraction and analytical chemistry analysis [42, 43].

The ability to perform chemodosimetry is very new to the field of medical oncology even though dosimetry has always been considered to be an essential part of radiation oncology. Imaging of radiolabeled liposomes provides a tool for medical oncologists to actually verify proper delivery of liposome-encapsulated drugs in human patients. Future studies of drug chemodosimetry in tumors could provide a valuable tool to determine why therapy has been ineffective and to adjust the dose of liposome chemotherapy administered to a particular patient in order to limit toxicity. Our group has developed a method to radiolabel the common clinically approved anticancer agent, liposomal doxorubicin [17, 43]. This ability to track the distribution of liposomal drugs in clinical patients could be valuable in the future development of technologies aimed at increasing the uptake of liposome-encapsulated drugs at the site of the tumor. Technologies such as image-guided hyperthermia with ultrasound [42, 44, 45] or radiofrequency (RF) ablation [43, 46–48] have already shown promise as a means to increase drug delivery and effectiveness of these therapies for the treatment of solid tumors.

11.3.2 Liposome Formulation Development

Nuclear imaging is very valuable for liposome formulation development. It provides an ability to obtain an overall global perspective of the *in vivo* distribution of administered liposomes in the body over time, providing guidance for interpretation of data and development of future research directions. Imaging can also ensure the proper collection of tissue samples for quantitative counting and determination of liposome accumulation on a per gram basis. Imaging can provide rapid quantitative feedback to select the best liposome formulation from a series of liposomes with different surface modifications such as changes in the amount of the surface percentage of polyethylene glycol (PEG). For example, imaging has been used to determine the best particle size of PEG liposomes for increased hemoglobin encapsulation [49] and the optimal percentage of PEG to add to the surface of a liposome to allow for the specific uptake by an *in vivo* target [50]. In many cases, too small amount of PEG on the liposome surface can result in rapid clearance from the circulation and low accumulation at the specific target, whereas too much PEG can result in overlapping surface coverage with the specific targeted ligand so that there is also very little accumulation at the target. Imaging can provide a rapid means of determining the optimal amount of PEG that results in the best uptake by the target.

11.4 EXAMPLE OF USE OF IMAGING TO DEVELOP BONE MARROW-TARGETED LIPOSOME FORMULATION

Recently, using nuclear imaging, we serendipitously discovered that a particular surface-modified liposome formulation had very high uptake in rabbit bone marrow with over 60% of the injected dose of liposomes being taken up in the bone marrow in rabbits by 24 h compared with 1–2% of conventional liposomes, with a high fraction of this uptake occurring within 6 h of injection. These liposomes were surface

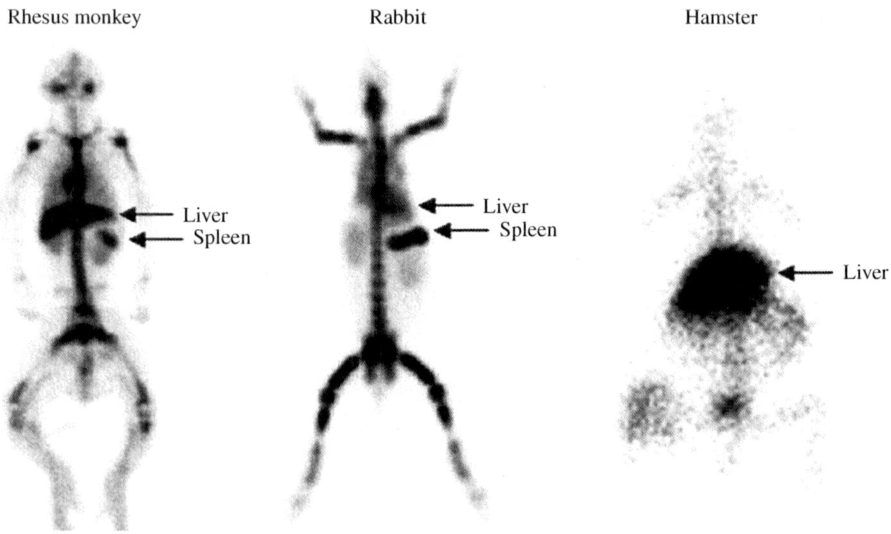

FIGURE 11.1 Nuclear images of rhesus monkey, rabbit, and hamster acquired after intravenous injection of ^{99m}Tc-bone marrow-targeted liposomes.

modified with an anionic amphiphile: L-glutamic acid, *N*-(3-carboxy-1-oxopropyl)-,1, 5-dihexadecyl ester (SA) (5% of total lipid) for use as a blood substitute [50]. The particular amphiphilic lipid was developed to increase the amount of hemoglobin that the liposome would encapsulate. This unexpected biodistribution might not have been discovered without the use of nuclear imaging that clearly demonstrated high bone marrow uptake (Fig. 11.1). In the initial studies in rabbits, we investigated a series of different PEG concentrations that were added to the surface of these liposomes to determine the ideal PEG concentration that resulted in the highest bone marrow uptake [50]. In these studies, a PEG concentration of 0.6 mol% was found to have the highest bone marrow uptake. The bone marrow uptake was significantly higher than the original formulation at the time of the serendipitous discovery. PEG concentrations greater than 0.6 mol% had lower uptake in the bone marrow and higher uptake in the spleen, while PEG concentrations of less than 0.6 mol% had lower uptake in the bone marrow and higher uptake in the liver [50]. Imaging provided for a convenient and quantitative assessment of bone marrow uptake that would have been difficult if not impossible to accurately assess with standard tissue sampling methods.

Imaging also provided a reliable and relatively easy method of assessing the uptake of bone marrow-targeted liposomes in other species. The formulation that had the highest bone marrow uptake in rabbits was also tested in several rodent species including rat, mouse, and hamster. To our surprise, the uptake in bone marrow was significantly lower in rat and mouse at ~8% of the total injected dose, much lower than the ~60% injected dose observed in the rabbit (Fig. 11.1). Interestingly, the hamster was found to have ~30% uptake in bone marrow. The reason for the difference in the uptake in the bone marrow was not obviously apparent; however, we have speculated that the uptake was related to rates of cell production in different organs of rodents and the size of the spleen [51]. For instance, the ratio of spleen size in the rat is much higher than in rabbits, hamsters, and primates. The spleen-to-body weight ratio is nearly three times greater in the rat as compared with the hamster and rhesus primate and nearly six times greater than in the rabbit.

With these very big differences between species, it was important to know whether a high bone marrow uptake with these liposomes would be found in species more closely related to humans. Fortunately, nuclear imaging made it relatively easy to determine bone marrow uptake noninvasively in primates. We therefore carried out studies in the rhesus monkey and found that the formulation with the highest uptake in rabbits had even higher uptake in the rhesus monkey at ~70% (Fig. 11.1). This high uptake shows that this liposome formulation is very likely to have high uptake in humans although a clinical investigation with nuclear imaging will need to be performed.

We are very interested in this liposome formulation for the delivery of therapeutic agents to bone marrow due to the importance of the bone marrow for erythrocyte [52, 53], platelet [54], immune cell production [55], and bone development [56]. Bone marrow, a hematopoietic organ, has generally been considered a challenging target for drug delivery. Surface-modified liposomes offer the possibility of delivering high quantities of liposome-encapsulated drugs to bone marrow via specific bone marrow phagocytosis. This bone marrow delivery system has potential as a drug delivery carrier for chemotherapy of hematopoietic malignancies such as myelocytic leukemia and multiple myeloma. Potential clinical uses of this bone marrow delivery system include the delivery of agents that protect the marrow from the toxic effects of chemotherapy and radiation and the delivery of agents to effectively and safely ablate bone marrow prior to bone marrow transplant. Also, the ability to specifically deliver stimulants of hematopoietic cell proliferation may open novel therapeutic systems for increasing hematopoiesis.

The significant differences in bone marrow uptake between the rats and rabbits/rhesus monkeys demonstrate the importance of studying liposome distributions in a variety of animal species, particular larger animals such as rabbits and monkeys. Quantitative imaging studies of the distribution of liposomes in larger animals can best be done with nuclear imaging.

11.5 USE OF IMAGING TO MONITOR DISTRIBUTION OF LIPOSOMAL DRUGS FOLLOWING IMAGE-GUIDED INTERVENTION

As previously discussed, imaging can be used to noninvasively monitor the *in vivo* uptake of drugs at a target site following an image-guided intervention. Previous research has shown that thermal tumor therapy with RF ablation greatly increases the deposition of liposomal drugs in the region of the tumor [47, 57]. Additionally, Goldberg and colleagues have shown that intravenously administrated liposomal doxorubicin increases the size of the zone of coagulation induced by RF ablation, demonstrated in animal tumor models by pathological assessment [57–60] and in human patients by CT follow-up [47].

Recently, our group has investigated nuclear imaging of radiolabeled liposomal doxorubicin (Doxil™) for noninvasively monitoring the effect of RF ablation on tumor uptake. In these studies, Doxil was labeled with the single photon agent, ^{99m}Tc, using a method previously described by our group [17]. Prior to this investigation, the only type of imaging that had been used to monitor drug delivery following RF ablation had been the autoradiography of excised tumors treated with combination RF ablation and intravenous tritiated liposomes [61]. Using nuclear imaging, the pharmacokinetics and biodistribution of the ^{99m}Tc-liposomal doxorubicin could be determined over multiple time points (Fig. 11.2a–c). The results show that doxorubicin delivery is greatly increased when RF ablation is combined with the ^{99m}Tc-liposomal doxorubicin just as it is when using nonradiolabeled liposomes (Fig. 11.2b–c). Image analysis quantification of the tumoral uptake of the radiolabeled

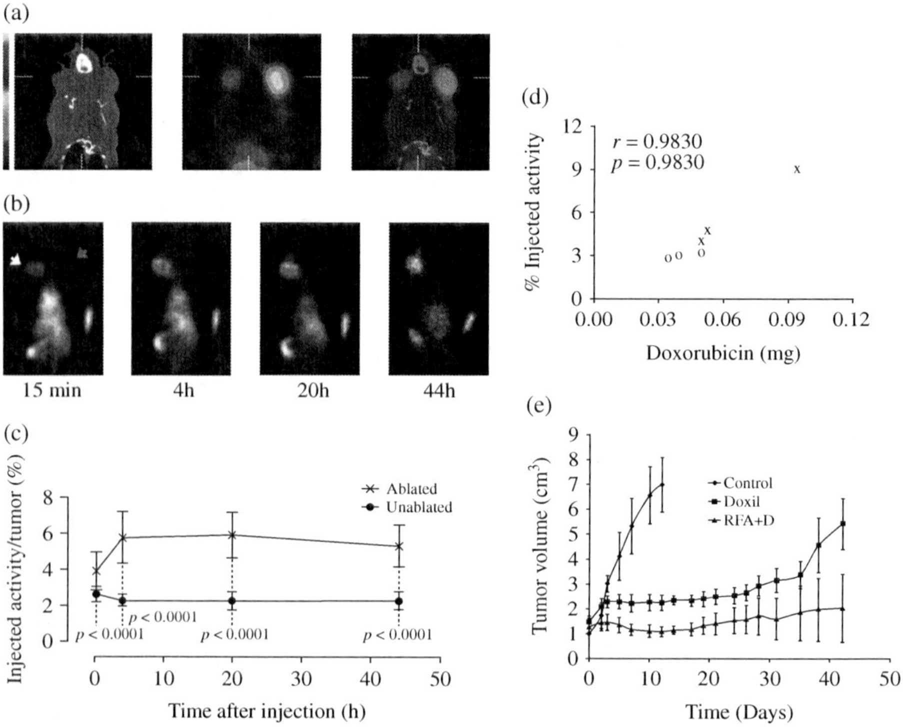

FIGURE 11.2 Use of imaging to monitor pharmacokinetics and distribution of liposomal doxorubicin in combination with RF ablation therapy. (a) Coronal view of CT, SPECT, and fused SPECT/CT images of rat with two head and necks tumors at base of neck acquired 4 h after intravenous injection of ^{99m}Tc-liposomal doxorubicin and RF ablation therapy. The tumor receiving RF ablation therapy (right) had significantly more accumulation of ^{99m}Tc-liposomal doxorubicin compared with unablated tumor (left). (b) Planar nuclear images of rat with two head and neck tumors acquired at various times after intravenous injection of ^{99m}Tc-liposomal doxorubicin and RF ablation. Ablated tumor (left, white arrow) had significantly higher uptake of ^{99m}Tc-liposomal doxorubicin compared with unablated tumor (right, gray arrow). (c) Injected ^{99m}Tc-activity in tumor over time determined from region-of-interest analysis of planar images. (d) Percentage of ^{99m}Tc-liposomal doxorubicin activity determined by planar images at 44 h post injection as a function of the quantity of doxorubicin (mg/g tissue) extracted from the ablated and unablated tumors. There was a significant positive correlation. (e) Tumor volume measured at various time points for rats with head and neck tumors following no therapy (control), intravenous injection of liposomal doxorubicin (Doxil), or intravenous injection of liposomal doxorubicin and RF ablation (RFA + D). Tumor growth was controlled for 14 days longer in liposomal doxorubicin/RF ablation group compared with liposomal doxorubicin alone therapy.

liposomal doxorubicin was found to correlate closely to actual doxorubicin drug concentration within tumors (Fig. 11.2d). These results were similar to previous results with noninvasive imaging of following thermal tumor therapy delivered by a hot water bath [42] as a model for ultrasound-based thermal therapy. These imaging studies demonstrated that the increase in drug delivery that occurs with combination therapy can be not only depicted but also quantified with noninvasive imaging. As previously discussed, this quantification could lead to a future application of clinically useful imaging-based chemodosimetry [42, 62]. Clinical image-based chemodosimetry may provide useful prognostic information as to the likely success of the tumor treatment and whether a repeat treatment will be needed.

Using the information gained in the ^{99m}Tc-liposomal doxorubicin/RF ablation imaging study, a separate group of rats with head and neck xenograft tumors were monitored for therapeutic efficacy of liposomal doxorubicin compared with liposomal doxorubicin/RF ablation therapy (Fig. 11.2e). Both liposomal doxorubicin and liposomal doxorubicin/RF ablation extended tumor growth control compared with no therapy group. The addition of RF ablation therapy with liposomal doxorubicin also extended tumor control by 7–10 days over liposomal doxorubicin alone. Imaging can play a vital role in liposome drug therapy studies by aiding in the design and optimization of the delivery methods.

11.6 USE OF IMAGING TO MONITOR THE DELIVERY OF NANOPARTICLE RADIONUCLIDE BRACHYTHERAPY

Brachytherapy (from the Greek word *brachys*, meaning "short distance") is a form of radiotherapy where a radiation source is placed inside or next to the area requiring treatment. In current clinical practice, brachytherapy is commonly used as a treatment for cervical [63], prostate [64], and breast cancer [65]. In the past, most brachytherapy agents have been low-energy gamma photon emitters that are locally applied as sealed sources such as within metal seeds. Until recently, beta-emitting radionuclides could only be administered as unsealed free molecules such as the application of orally administered free ^{131}I molecules for the treatment of thyroid cancer therapy. Before the research by our group and others, there existed no effective means of applying beta-emitting radionuclides directly into the region of the tumor as a brachytherapy agent. This inability to directly apply beta-emitting radionuclides into the tumor region for brachytherapy applications was due to the fact that any container used to seal the beta-emitting radionuclide would absorb and block the emitted beta particle from reaching the tumor. It was also not possible to apply small molecules directly into the tumor because their distribution could not be controlled and they were generally rapidly absorbed into the blood or lymphatics with little retention in the region of the tumor.

Recent research by Bao et al. in our group [29–31, 66] as well as other investigators [67, 68] has shown that locally applied liposomes carrying beta-emitting radionuclides can be an effective means of delivering radiation directly to local solid tumors. The liposome beta-emitting radionuclide carrier provided the ability to control the distribution within the tumor without blocking the effectiveness of the emitted beta particle. Using SPECT/CT imaging, intratumorally administered radionuclides encapsulated in a liposomal nanoparticle were shown to disperse through the tumor tissue (Fig. 11.3a). Liposomal nanoparticles were also shown to

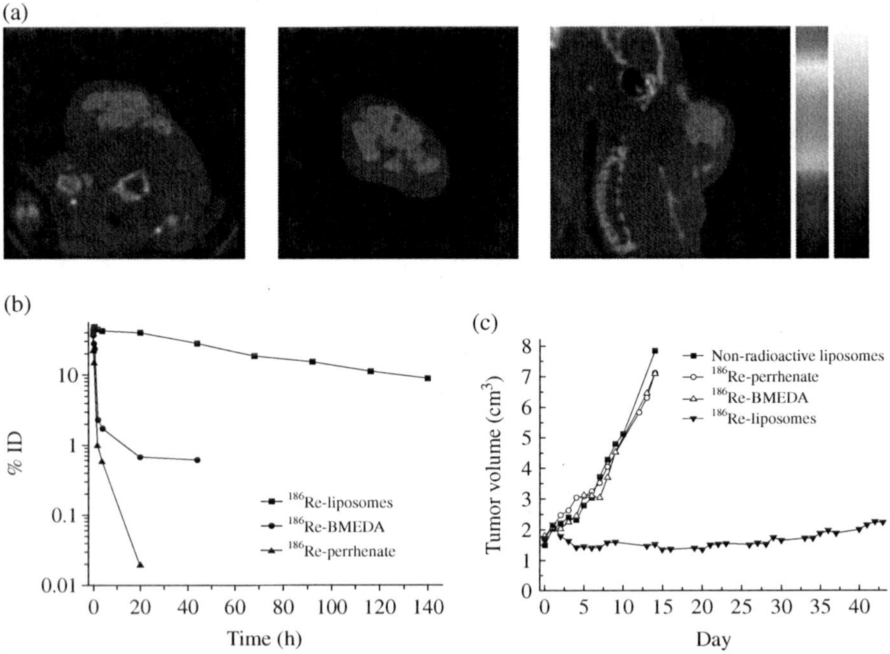

FIGURE 11.3 Intratumoral delivery of liposomes in a head and neck tumor model. (a) Transaxial, coronal, and sagittal views of fused pinhole SPECT/CT images acquired at 20 h after intratumoral CED injection of ^{99m}Tc-liposomes. The gray and color scale bars show the pixel values from 0 to 100% for the CT and SPECT images, respectively. For color details, please see color plate section. (b) Percent injected dose of ^{186}Re-radioactivity retained in head and neck tumors over time following intratumoral CED injection of ^{186}Re-liposomes, ^{186}Re-BMEDA, or ^{186}Re-perrhenate. Retention was significantly improved by encapsulation in liposomes. (c) Tumor growth in head and neck tumors after intratumoral CED injection of ^{186}Re-liposomes compared with control groups of nonradioactive liposomes, ^{186}Re-perrhenate or ^{186}Re-BMEDA. Encapsulation of ^{186}Re inside liposomes greatly improved tumor control.

enhance the retention of the beta particle-emitting radionuclides in the local region of the tumor (Fig. 11.3b) [30, 69].

For the local treatment of tumors, an area of great progress has been in the investigation of methods of infusing local therapies into tumors. Most of this research has been conducted for the treatment of brain tumors due to the fact that brain tumors grow and cause damage in the region that they originate and drug delivery to the brain following intravenous therapy is limited by the blood–brain barrier (BBB). Investigators in this area have developed a technique termed convection-enhanced delivery (CED) for infusion of brain tumors. CED is a method of infusing the tumors with the goal of achieving the largest dispersion from the needle tip throughout the tumor. CED, in this regard, is the administration of therapeutic agents directly into the targeted tissue through a needle using a slow infusion pump that provides constant pressure (convection) that greatly improves the dispersion of the therapeutic agent through the tissue [70–72].

Our group has used small animal SPECT imaging to monitor the intratumoral distribution of radiolabeled nanoliposomes in a 1.5-cm diameter head and neck tumor xenograft in nude rats following CED administration. Retention and microdistribution studies of liposomes encapsulating radionuclides using an animal SPECT/CT scanner have been performed by the authors' group [69, 73, 74]. Figure 11.3a clearly shows that the liposomes have spread throughout the tumor although their distribution is not perfectly homogeneous. It is anticipated that future research will lead to much progress in improving the homogeneity of liposomes infused by CED as well as their retention. Our studies have shown that ~50% of the CED-infused liposomes are retained at their site of infusion, while the other 50% returns to the blood circulation (Fig. 11.3b) [30, 69]. A significant advantage in the use of beta particle therapeutics is that their 2–4-mm average path length in tissue can greatly compensate for any inhomogeneities in the distribution of the liposomes so that the radiation dose delivered to the tumor can be reasonably homogeneous even though the actual distribution of the nanoparticles is less so. Another significant advantage of the use of beta particle-emitting nanoparticles is that unlike nanoparticles that carry drugs, the therapeutic agent does not have to move from the liposome and into the tumor cell in order for the therapy to be effective. Its 2–4-mm microfield of therapy, that covers 100–200 cell diameters, delivers the therapy to tumor cells while the radionuclide remains encapsulated in the liposome.

Using a brachytherapy approach, we and other groups have shown excellent promise for the use of liposomes as a carrier for image-guided treatment of solid tumors via direct intratumoral administration [69, 75–79]. For example, our group reported that ^{186}Re-liposomes injected into a head and neck tumor xenograft model were reduced to 26% of their original size at day 35 post therapy. In contrast, tumors treated with control liposomes containing no ^{186}Re increased to ~290% of original size by day 15 post administration at which time the rats were sacrificed (Fig. 11.3c) [31]. In addition, we reported that free unencapsulated ^{186}Re infused into a head and neck tumor was rapidly removed from the tumor and had no therapeutic effect, while the same amount of radioactivity encapsulated in a liposomal nanoparticle was well retained in the tumor and associated with a significant therapeutic effect [30]. All animals treated with ^{186}Re-liposomes survived out to 45 days, whereas animals treated with free ^{186}Re or a ^{186}Re-labeled small molecule (N,N-bis (2-mercaptoethyl) -N',N'-diethylethylenediamine (BMEDA)) had rapid tumor growth and required sacrifice at 15 days post treatment. Because of the much longer retention in the tumor, ^{186}Re-liposomes were calculated to have delivered an average dose of 526 Gy to the tumor, while the tumors treated with the free ^{186}Re or the ^{186}Re-labeled small molecule received doses of 3.3 and 13.4 Gy, respectively [30]. Other groups have also shown similar promising results with ^{188}Re-labeled solid lipid nanoparticles [67] and gold-198 (^{198}Au) nanoparticles [80].

11.7 USE OF IMAGING TO MONITOR LIPOSOME BRACHYTHERAPY TO BREAST CANCER POSTLUMPECTOMY CAVITY AND DRAINING LYMPH NODES

Adjuvant radiotherapy after breast-conserving surgery performed with either standard methods or accelerated partial breast irradiation techniques has been shown to lower rates of local reoccurrence compared with surgery alone [81, 82]. Our group has investigated the use of liposomes radiolabeled with the beta particle-emitting radionuclides of ^{186}Re and ^{188}Re and delivered locally to the surgical cavity as a

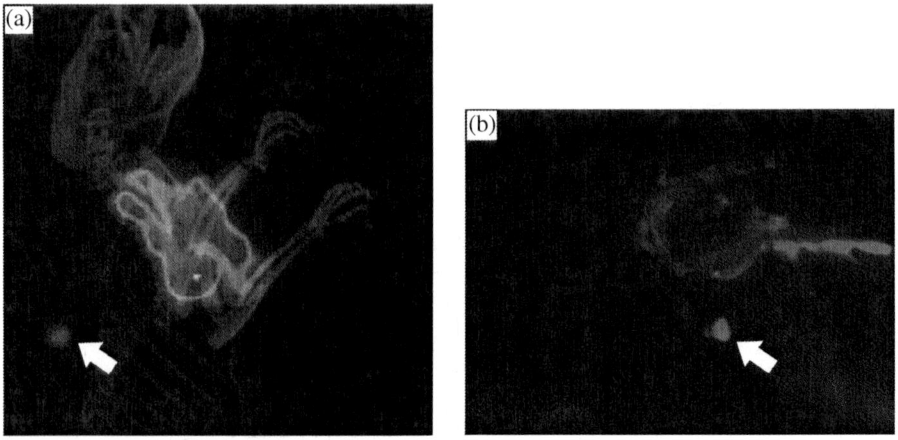

FIGURE 11.4 Use of imaging for monitoring liposome distribution in breast cancer post-lumpectomy brachytherapy model. (a) Fused SPECT/CT projection image of rat acquired 20 h after intracavitary injection of dual-labeled ^{99m}Tc-rhodamine fluorescent liposomes. Most of the ^{99m}Tc-radioactivity was retained locally within surgical cavity with some accumulation in axillary lymph node (white arrow). (b) Fluorescent stereomicroscope image acquired 44 h after intracavitary injection of dual-labeled ^{99m}Tc-rhodamine fluorescent liposomes. This image shows the distribution of the ^{99m}Tc-rhodamine fluorescent liposomes lining the surgical cavity and lymphatic drainage to the lymph node (white arrow). For color details, please see color plate section.

method to decrease the high radiation dose that is delivered to normal breast tissue by current accelerated irradiation techniques [83–85].

Imaging of the distribution of liposomes, dual labeled with ^{99m}Tc and rhodamine-lipid, after intracavitary injection in a positive surgical margin lumpectomy model has provided valuable information. Serving as a surrogate for the Re-liposomes, these imaging studies greatly aided understanding of the retention and clearance patterns of different liposome formulations after intracavitary injection [86]. Nuclear imaging of the ^{99m}Tc-rhodamine fluorescent liposomes (Fig. 11.4a) revealed that most of the liposomes were retained in the surgical cavity with some accumulation in regional lymph nodes draining the cavity. Despite the surface charge of the liposomes tested, 100-nm liposomes were better retained in the surgical cavity compared with 400-nm liposomes. In addition, 100-nm anionic liposomes had higher lymph node accumulation compared with the other formulations tested. Fluorescent images of the surgical cavity acquired at the time of sacrifice confirmed the nuclear image distribution but provided greater detail of the distribution pattern (Fig. 11.4b). However, attenuation and scattering of the fluorescence signal prevented quantitative noninvasive longitudinal monitoring of the liposomes, which is a major advantage of the nuclear imaging methods.

11.8 SUMMARY

Nuclear imaging of liposomal nanoparticles in combination with other anatomic based imaging modalities is providing important quantitative data that can be noninvasively obtained over time in the same individual. Imaging provides an excellent tool for the customization of therapy in individual patients following liposome administration giving these agents a dual-mode theranostic capability. Imaging can provide valuable guidance in the development of new liposomal carriers. It is also one of the best tools for discovering and exploiting the novel capabilities of liposomes and other nanoparticles so their unique advantages can improve the treatment of many different disease processes.

ACKNOWLEDGMENTS

The authors would like to thank Cristina Santoyo, Ricardo Perez III, Marcela Saenz, Dr. Anuradha Soundararajan, and Dr. Hayden W. Head for their technical assistance. The authors also appreciate Jonathan Sumner for his help in figure preparation. This research was partly supported by NIH grants 5 P30 CA054174-16 and R01 CA131039, Susan G. Komen for the Cure research grant BCTR0707169, and the Japan Society for the Promotion of Science (JSPS) Bilateral Joint Projects between Japan and the United States.

REFERENCES

1. Bangham AD. Liposomes: the Babraham connection. Chem Phys Lipids 1993;64: 275–285.
2. Gregoriadis G, Ryman BE. Liposomes as carriers of enzymes or drugs: a new approach to the treatment of storage diseases. Biochem J 1971;124:58P.
3. Chen LC, Chang CH, Yu CY, et al. Biodistribution, pharmacokinetics and imaging of (188)Re-BMEDA-labeled pegylated liposomes after intraperitoneal injection in a C26 colon carcinoma ascites mouse model. Nucl Med Biol 2007;34:415–423.
4. Drummond DC, Noble CO, Guo Z, et al. Development of a highly stable and targetable nanoliposomal formulation of topotecan. J Control Release 2010;141:13–21.
5. Mozafari MR, Johnson C, Hatziantoniou S, et al. Nanoliposomes and their applications in food nanotechnology. J Liposome Res 2008;18:309–327.
6. Awasthi VD, Garcia D, Goins BA, et al. Circulation and biodistribution profiles of long-circulating PEG-liposomes of various sizes in rabbits. Int J Pharm 2003;253:121–132.
7. Gabizon A, Catane R, Uziely B, et al. Prolonged circulation time and enhanced accumulation in malignant exudates of doxorubicin encapsulated in polyethylene-glycol coated liposomes. Cancer Research 1994;54:987–992.
8. Barratt G, Legrand P. Comparison of the efficacy and pharmacology of formulations of amphotericin B used in treatment of leishmaniasis. Curr Opin Infect Dis 2005;18:527–530.
9. Gabizon A, Shmeeda H, Barenholz Y. Pharmacokinetics of pegylated liposomal Doxorubicin: review of animal and human studies. Clin Pharmacokinet 2003;42:419–436.
10. Allen TM, Hansen C, Martin F, et al. Liposomes containing synthetic lipid derivative of poly(ethylene) show prolonged circulation half-lives in vivo. Biochimica et Biophysica Acta 1991;1066:29–36.
11. Phillips WT, Klipper RW, Awasthi VD, et al. Polyethylene glycol-modified liposome-encapsulated hemoglobin: a long circulating red cell substitute. J Pharmacol Exp Ther 1999;288:665–670.

12. Torchilin VP, Papisov MI. Why do polyethylene glycol-coated liposomes circulate so long? J Liposome Res 1994;4:725–739.
13. Allen TM, Sapra P, Moase E. Use of the post-insertion method for the formation of ligand-coupled liposomes. Cell Mol Biol Lett 2002;7:889–894.
14. Erdogan S, Roby A, Torchilin VP. Enhanced tumor visualization by gamma-scintigraphy with 111In-labeled polychelating-polymer-containing immunoliposomes. Mol Pharm 2006;3:525–530.
15. Gupta B, Torchilin VP. Monoclonal antibody 2C5-modified doxorubicin-loaded liposomes with significantly enhanced therapeutic activity against intracranial human brain U-87 MG tumor xenografts in nude mice. Cancer Immunol Immunother 2007;56:1215–1223.
16. Sapra P, Tyagi P, Allen TM. Ligand-targeted liposomes for cancer treatment. Curr Drug Deliv 2005;2:369–381.
17. Bao A, Goins B, Klipper R, et al. Direct 99mTc labeling of pegylated liposomal doxorubicin (Doxil) for pharmacokinetic and non-invasive imaging studies. J Pharmacol Exp Ther 2004;308:419–425.
18. Gabizon A, Huberty J, Straubinger RM, et al. An improved method for in vivo tracing and imaging of liposomes using a gallium 67-deferoxamine complex. J Liposome Res 1988;1:123–135.
19. Goins B, Klipper R, Rudolph AS, et al. Biodistribution and imaging studies of technetium-99m labeled liposomes in rats with focal infection. J Nucl Med 1993;34:2160–2168.
20. Laverman P, Dams ETM, Oyen WJG, et al. A novel method to label liposomes with Tc-99m via the hydrazino nicotinyl derivative. J Nucl Med 1998;39:7P.
21. Petersen AL, Binderup T, Rasmussen P, et al. 64Cu loaded liposomes as positron emission tomography imaging agents. Biomaterials 2011;32:2334–2341.
22. Phillips WT, Rudolph AS, Goins B, et al. A simple method for producing a technetium-99m-labeled liposome which is stable in vivo. Nucl Med Biol 1992;19:539–547.
23. Lucignani G. Nanoparticles for concurrent multimodality imaging and therapy: the dawn of new theragnostic synergies. Eur J Nucl Med Mol Imaging 2009;36:869–874.
24. Pene F, Courtine E, Cariou A, et al. Toward theragnostics. Crit Care Med 2009;37: S50–S58.
25. Goins B, Bao A, Phillips WT. Techniques for loading technetium-99m and rhenium-186/188 radionuclides into pre-formed liposomes for diagnostic imaging and radionuclide therapy. Methods Mol Biol 2010;606:469–491.
26. Goins BA, Phillips WT. Methods for tracking radiolabeled liposomes after injection in the body. In: Gregoriadis G, editor. *Liposome Technology*. Vol. 3, New York: Informa Healthcare; 2007. p 191–209.
27. Laverman P, Boerman OC, Storm G. Radiolabeling of liposomes for scintigraphic imaging. Methods Enzymol 2003;373:234–248.
28. Laverman P, Storm G, Phillips WT, et al. Radiolabeling of liposome for scintigraphic imaging. In: Gregoriadis G, editor. *Liposome Technology*. Vol. 3, New York: Informa Healthcare; 2007. p 169–185.
29. Bao A, Goins B, Klipper R, et al. 186Re-liposome labeling using 186Re-SNS/S complexes: in vitro stability, imaging, and biodistribution in rats. J Nucl Med 2003;44:1992–1999.
30. French JT, Goins B, Saenz M, et al. Interventional therapy of head and neck cancer with lipid nanoparticle-carried rhenium 186 radionuclide. J Vasc Interv Radiol 2010;21:1271–1279.
31. Wang SX, Bao A, Herrera SJ, et al. Intraoperative 186Re-liposome radionuclide therapy in a head and neck squamous cell carcinoma xenograft positive surgical margin model. Clin Cancer Res 2008;14:3975–3983.

32. Awasthi V, Yee SH, Jerabek P, et al. Cerebral oxygen delivery by liposome-encapsulated hemoglobin: a positron-emission tomographic evaluation in a rat model of hemorrhagic shock. J Appl Physiol 2007;103:28–38.
33. Urakami T, Akai S, Katayama Y, et al. Novel amphiphilic probes for [18F]-radiolabeling preformed liposomes and determination of liposomal trafficking by positron emission tomography. J Med Chem 2007;50:6454–6457.
34. Paoli EE, Kruse DE, Seo JW, et al. An optical and microPET assessment of thermally-sensitive liposome biodistribution in the Met-1 tumor model: importance of formulation. J Control Release 2010;143:13–22.
35. Seo JW, Zhang H, Kukis DL, et al. A novel method to label preformed liposomes with 64Cu for positron emission tomography (PET) imaging. Bioconjug Chem 2008;19:2577–2584.
36. O'Connor MK, Kemp BJ. Single-photon emission computed tomography/computed tomography: basic instrumentation and innovations. Semin Nucl Med 2006;36:258–266.
37. Townsend DW, Carney JP, Yap JT, et al. PET/CT today and tomorrow. J Nucl Med 2004;45 Suppl:4S–14S.
38. Zaidi H. Recent developments and future trends in nuclear medicine instrumentation. Z Med Phys 2006;16:5–17.
39. Chowdhury FU, Scarsbrook AF. The role of hybrid SPECT-CT in oncology: current and emerging clinical applications. Clin Radiol 2008;63:241–251.
40. Liang H, Yang Y, Yang K, et al. A microPET/CT system for in vivo small animal imaging. Phys Med Biol 2007;52:3881–3894.
41. Zaidi H, Montandon ML, Alavi A. The clinical role of fusion imaging using PET, CT, and MR imaging. Magn Reson Imaging Clin N Am 2010;18:133–149.
42. Kleiter MM, Yu D, Mohammadian LA, et al. A tracer dose of technetium-99m-labeled liposomes can estimate the effect of hyperthermia on intratumoral doxil extravasation. Clin Cancer Res 2006;12:6800–6807.
43. Head HW, Dodd GD 3rd, Bao A, et al. Combination radiofrequency ablation and intravenous radiolabeled liposomal Doxorubicin: imaging and quantification of increased drug delivery to tumors. Radiology 2010;255:405–414.
44. Ponce AM, Vujaskovic Z, Yuan F, et al. Hyperthermia mediated liposomal drug delivery. Int J Hyperthermia 2006;22:205–213.
45. Vujaskovic Z, Kim DW, Jones E, et al. A phase I/II study of neoadjuvant liposomal doxorubicin, paclitaxel, and hyperthermia in locally advanced breast cancer. Int J Hyperthermia 2010;26:514–521.
46. Ahmed M, Goldberg SN. Combination radiofrequency thermal ablation and adjuvant IV liposomal doxorubicin increases tissue coagulation and intratumoural drug accumulation. Int J Hyperthermia 2004;20:781–802.
47. Goldberg SN, Kamel IR, Kruskal JB, et al. Radiofrequency ablation of hepatic tumors: increased tumor destruction with adjuvant liposomal doxorubicin therapy. Am J Roentgenol 2002;179:93–101.
48. Yang W, Ahmed M, Elian M, et al. Do liposomal apoptotic enhancers increase tumor coagulation and end-point survival in percutaneous radiofrequency ablation of tumors in a rat tumor model? Radiology 2010;257:685–696.
49. Awasthi VD, Garcia D, Klipper R, et al. Neutral and anionic liposome-encapsulated hemoglobin: effect of postinserted poly(ethylene glycol)-distearoylphosphatidylethanolamine on distribution and circulation kinetics. J Pharmacol Exp Ther 2004;309:241–248.
50. Sou K, Goins B, Takeoka S, et al. Selective uptake of surface-modified phospholipid vesicles by bone marrow macrophages in vivo. Biomaterials 2007;28:2655–2666.

51. Sou K, Goins B, Oyajobi B, et al. Bone marrow-targeted liposomal carriers. Expert Opin Drug Deliv 2011;8:317–328.
52. Drouet M, Herodin F. Radiation victim management and the haematologist in the future: time to revisit therapeutic guidelines? Int J Radiat Biol 2010;86:636–648.
53. Ramakrishnan A, Deeg HJ. A novel role for the marrow microenvironment in initiating and sustaining hematopoietic disease. Expert Opin Biol Ther 2009;9:21–28.
54. Geddis AE. Megakaryopoiesis. Semin Hematol 2010;47:212–219.
55. Borregaard N. Neutrophils, from marrow to microbes. Immunity 2010;33:657–670.
56. Del Fattore A, Capannolo M, Rucci N. Bone and bone marrow: the same organ. Arch Biochem Biophys 2010;503:28–34.
57. Ahmed M, Lukyanov AN, Torchilin V, et al. Combined radiofrequency ablation and adjuvant liposomal chemotherapy: effect of chemotherapeutic agent, nanoparticle size, and circulation time. J Vasc Interv Radiol 2005;16:1365–1371.
58. Ahmed M, Liu Z, Lukyanov AN, et al. Combination radiofrequency ablation with intratumoral liposomal doxorubicin: effect on drug accumulation and coagulation in multiple tissues and tumor types in animals. Radiology 2005;235:469–477.
59. Ahmed M, Monsky WE, Girnun G, et al. Radiofrequency thermal ablation sharply increases intratumoral liposomal doxorubicin accumulation and tumor coagulation. Cancer Res 2003;63:6327–6333.
60. Goldberg SN, Girnan GD, Lukyanov AN, et al. Percutaneous tumor ablation: increased necrosis with combined radio-frequency ablation and intravenous liposomal doxorubicin in a rat breast tumor model. Radiology 2002;222:797–804.
61. Monsky WL, Kruskal JB, Lukyanov AN, et al. Radio-frequency ablation increases intratumoral liposomal doxorubicin accumulation in a rat breast tumor model. Radiology 2002;224:823–829.
62. Tashjian JA, Dewhirst MW, Needham D, et al. Rationale for and measurement of liposomal drug delivery with hyperthermia using non-invasive imaging techniques. Int J Hyperthermia 2008;24:79–90.
63. Toita T. Current status and perspectives of brachytherapy for cervical cancer. Int J Clin Oncol 2009;14:25–30.
64. Stock RG, Stone NN. Current topics in the treatment of prostate cancer with low-dose-rate brachytherapy. Urol Clin North Am 2010;37:83–96.
65. Niehoff P, Mundhenke C, Kimmig B, et al. Breast irradiation with brachytherapy: approved techniques and new concepts. Minerva Ginecol 2007;59:377–386.
66. Bao A, Zhao X, Phillips WT, et al. Theoretical study of the influence of a heterogeneous activity distribution on intratumoral absorbed dose distribution. Med Phys 2005;32:200–208.
67. Allard E, Hindre F, Passirani C, et al. 188Re-loaded lipid nanocapsules as a promising radiopharmaceutical carrier for internal radiotherapy of malignant gliomas. Eur J Nucl Med Mol Imaging 2008;35:1838–1846.
68. Khan MK, Minc LD, Nigavekar SS, et al. Fabrication of {198Au0} radioactive composite nanodevices and their use for nanobrachytherapy. Nanomedicine 2008;4:57–69.
69. Bao A, Phillips WT, Goins B, et al. Potential use of drug carried-liposomes for cancer therapy via direct intratumoral injection. Int J Pharm 2006;316:162–169.
70. Linninger AA, Somayaji MR, Mekarski M, et al. Prediction of convection-enhanced drug delivery to the human brain. J Theor Biol 2008;250:125–138.
71. Raghavan R, Brady ML, Rodriguez-Ponce MI, et al. Convection-enhanced delivery of therapeutics for brain disease, and its optimization. Neurosurg Focus 2006;20:E12.

72. Vogelbaum MA. Convection enhanced delivery for treating brain tumors and selected neurological disorders: symposium review. J Neurooncol 2007;83:97–109.
73. Wang SX, Bao A, Phillips WT, et al. Intraoperative therapy with liposomal drug delivery: retention and distribution in human head and neck squamous cell carcinoma xenograft model. Int J Pharm 2009;373:156–164.
74. Zavaleta CL, Goins BA, Bao A, et al. Imaging of 186Re-liposome therapy in ovarian cancer xenograft model of peritoneal carcinomatosis. J Drug Target 2008;16:626–637.
75. Gupta H, Weissleder R. Targeted contrast agents in MR imaging. Magn Reson Imaging Clin N Am 1996;4:171–184.
76. Mulder WJ, Strijkers GJ, van Tilborg GA, et al. Lipid-based nanoparticles for contrast-enhanced MRI and molecular imaging. NMR Biomed 2006;19:142–164.
77. Mulder WJ, van der Schaft DW, Hautvast PA, et al. Early in vivo assessment of angiostatic therapy efficacy by molecular MRI. FASEB J 2007;21:378–383.
78. Saito R, Bringas JR, McKnight TR, et al. Distribution of liposomes into brain and rat brain tumor models by convection-enhanced delivery monitored with magnetic resonance imaging. Cancer Res 2004;64:2572–2579.
79. Saito R, Krauze MT, Noble CO, et al. Tissue affinity of the infusate affects the distribution volume during convection-enhanced delivery into rodent brains: implications for local drug delivery. J Neurosci Methods 2006;154:225–232.
80. Chanda N, Kan P, Watkinson LD, et al. Radioactive gold nanoparticles in cancer therapy: therapeutic efficacy studies of GA-(198)AuNP nanoconstruct in prostate tumor-bearing mice. Nanomedicine 2010;6:201–209.
81. Early Breast Cancer Trialists' Collaborative Group. Effects of radiotherapy and surgery in early breast cancer. An overview of the randomized trials. N Engl J Med 1995;333: 1444–1455.
82. Van de Steene J, Vinh-Hung V, Cutuli B, et al. Adjuvant radiotherapy for breast cancer: effects of longer follow-up. Radiother Oncol 2004;72:35–43.
83. Arthur D. Accelerated partial breast irradiation: a change in treatment paradigm for early stage breast cancer. J Surg Oncol 2003;84:185–191.
84. Arthur DW, Vicini FA, Kuske RR, et al. Accelerated partial breast irradiation: an updated report from the American Brachytherapy Society. Brachytherapy 2003;2:124–130.
85. Hrycushko BA, Li S, Shi C, et al. Postlumpectomy focal brachytherapy for simultaneous treatment of surgical cavity and draining lymph nodes. Int J Radiat Oncol Biol Phys 2011;79:948–955.
86. Li S, Goins B, Phillips WT, et al. Post-lumpectomy intracavitary retention and lymph node targeting of (99m)Tc-encapsulated liposomes in nude rats with breast cancer xenograft. Breast Cancer Res Treat 2011;130:97–107.

CHAPTER 12

POLYMERIC MICELLES AS IMAGING AGENTS AND DRUG DELIVERY SYSTEMS

JUN ZHAO AND CHUN LI

12.1 INTRODUCTION

Polymeric micelles can be self-assembled from amphiphilic copolymers in aqueous solution and are usually characterized by a fairly narrow distribution of sizes and a unique core-shell structure [1]. Polymeric micelles have been widely used as carriers for drugs and diagnostic agents because of the similarities of micelles to natural transporters such as viruses and lipoproteins [2]. The uptake of polymeric micelles by the reticuloendothelial system (RES) can be minimized by optimizing their size and surface properties. Targeting moieties can be attached to micelles' surfaces to further increase the polymeric micelles' site specificity.

Among the numerous chemicals from which polymeric micelles can be built, amino acids exhibit some of the most attractive properties, including biocompatibility, nontoxicity, and versatile functionality [3]. In particular, poly(α,L-glutamic acid) (PG)-based polymeric micelles have been successfully used in many studies for the delivery of antitumor drugs and diagnostic imaging agents [4–6]. This chapter will discuss the synthesis and characterization of PG-based polymeric micelles and their use for delivering diagnostic imaging agents and drugs.

12.2 POLY(α,L-GLUTAMIC ACID)

Poly(α,L-glutamic acid) is prepared from poly(γ-benzyl-L-glutamate) (PBLG) by removing the protective benzyl group. PBLG, on the other hand, is generally synthesized via the amine-initiated ring-opening reaction of the carboxyanhydride (NCA) of γ-benzyl-L-glutamate (Fig. 12.1). Of note is that the amine initiator may deprotonate the nitrogen of the NCA ring. The formed NCA^- anion can also initiate NCA

Drug Delivery Applications of Noninvasive Imaging: Validation from Biodistribution to Sites of Action, First Edition. Edited by Chun Li and Mei Tian.

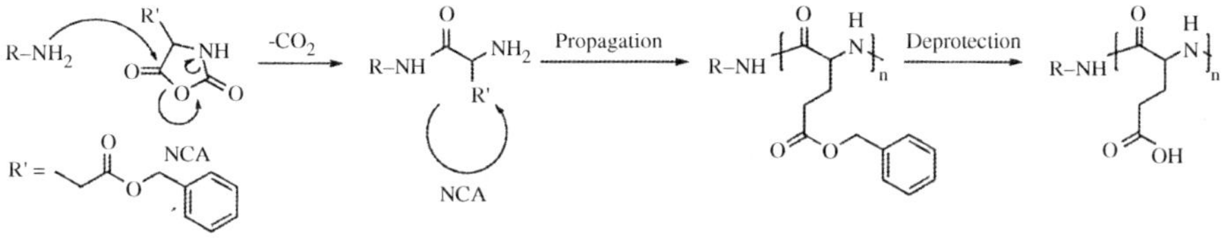

FIGURE 12.1 Amine-initiated polymerization of γ-benzyl-l-glutamate N-carboxyanhydride.

polymerization, and this side reaction broadens the distribution of molecular weight (MW) of the product polymer. One method that has been attempted to minimize such side reactions is using ammonium chloride as an alternative initiator. Because of the equilibrium between ammonium and amine, NCA polymerization can still be initiated by the amine molecules in the reaction mixture. However, if an NCA^- anion is formed, it is immediately quenched by the ammonium and thus cannot cause a side reaction. The use of ammonium chloride as an alternative initiator produces PG with a narrow distribution of MW. Unfortunately, the polymerization rate is also significantly reduced; only a moderate degree of polymerization was achieved even after several days of reaction in one study [7]. For this reason, amine initiators are still widely used for NCA polymerization.

Once the PBLG polymer is synthesized, PG is obtained by cleaving the benzyl groups from the side chains of PBLG via hydrolysis or hydrogenation. Some common combinations of the hydrolysis agents are HBr + $CHCl_2COOH$, CF_3COOH + CH_3SO_3H, and KOH + THF [7–10]. Hydrogenation is accomplished in an H_2 atmosphere using Pd/C as the catalyst [11].

Poly(α,L-glutamic acid) is nontoxic at the dose level required for drug delivery. It was well tolerated at a single dose of 800 mg/kg or a cumulative dose of 1.8 g/kg [12, 13]. PG itself is nonimmunogenic in humans, rabbits, and guinea pigs [14], and it does not tend to activate fibrinolysis [15]. A biodistribution study of PG (MW ~ 11,000) showed that most PG was recovered in the kidney and urine [16], probably because the hydrophilic PG had limited deposition in the RES and was primarily excreted via the renal route. PG is degradable by lysosomal enzymes [17]. The rate of degradation depends on pH and on PG's conformation. It has been suggested that the peptide bonds with adjacent uncharged side chains in a random-coil region of PG are particularly susceptible to enzyme attack [18]. On the other hand, the degradation may be altered by the drugs that are conjugated to the PG side chains. When a model drug, *p*-nitroanilide, was conjugated to PG via a Gly-Leu-Gly spacer, the overall degradation of the PG backbone was slower than unmodified PG. [8]. This reduced degradability was attributed to lower accessibility of the PG backbone to enzymes as well as alteration of the amino acid sequence adjacent to the degradation site.

Another form of poly(glutamic acid), poly(γ,L-glutamic acid), occurs naturally and can be manufactured by bacteria, most prominently Bacilli [19]. Although poly(γ,L-glutamic acid) is widely used for biomedical applications [20, 21], its MW and molecular architecture lack flexibility. In contrast, PG with well-controlled MWs and complicated architectures has been synthesized. For example, a star-shaped PG was prepared via NCA polymerization using a poly(amidoamine) dendrimer as the initiator. The MW of the product was accurately controlled by varying the monomer-to-initiator ratio. Narrow distribution of MW was demonstrated by gel permeation chromatography, and branched structure was confirmed by their Mark–Houwink α-values that are smaller than those associated with linear polymers [22].

12.3 PREPARATION AND CHARACTERIZATION OF POLYMERIC MICELLES FROM PG

12.3.1 Synthesis of Polymer Precursors Containing PG

Poly(α,L-glutamic acid)-containing amphiphilic polymer precursors can be categorized as block copolymers and graft copolymers. Block copolymers consist of sequences of PG blocks and non-PG block(s). PG blocks are introduced to the block copolymer by either NCA polymerization or direct conjugation with a preformed PG block. NCA polymerization can be initiated by the amine group at the terminal of a non-PG block [23, 24]. The length of PG blocks is determined by varying the NCA-to-initiator feeding ratio. However, NCA polymerization is less successful in the presence of tertiary amines, which also react with NCA. The positively charged polymer poly(2-(dimethylamino)ethyl methacrylate) is one example of a tertiary amine-containing polymer.

Direct conjugation with a preformed PG block requires a reaction with high yield and efficiency because reactions between two polymers are slower and more difficult to attain than those between small molecules. "Click" chemistry, for example, is well suited for such direct conjugation [25]. The click reaction is one type of Huisgen's 1,3-dipolar cycloadditions between azides and alkynes. Rao et al. used click chemistry to combine one block of poly(L-Z-lysine)-alkyne and two blocks of PBLG-azide into a Y-shaped miktoarm star polypeptide [26]. An excess of PBLG-azide was used to ensure complete conjugation, and the unreacted portion was removed by passing the reaction mixture through a resin that was premodified with the alkyne. Carbodiimide-assisted amide formation is another conjugation method. Yang et al. utilized the terminal NH_2 of PBLG to form the amide with the terminal carboxylate of the poly(lactic-co-glycolic acid) (PLGA) block [27]. To prevent interference of the PG carboxylates, the conjugation was performed before deprotection of the benzyl groups. After deprotection, the excess PG was removed by dialysis.

Graft copolymers, on the other hand, can be prepared from the pendent γ-carboxylates of the PG backbone. Simple alkylation of the γ-carboxylate increases the hydrophobicity of the polymer and enables polymer micellization in water [28]. Grafting with another amine-terminated polymer segment can be accomplished in the presence of carbodiimide. For example, He et al. grafted the amine-terminated poly(*N*-isopropylacrylamide) (PNIPAAm) to the PG backbone, thus incorporating thermoresponsiveness into the copolymer [29]. However, a long reaction time (72 h) was required. Unreacted PNIPAAm was removed by dialysis.

12.3.2 General Methods for Preparing Polymeric Micelles

Polymeric micelles (with or without drug loading) can be prepared by the self-assembly of amphiphilic polymers in aqueous solutions [30–33]. Four commonly used methods are direct dissolution, dialysis, nanoprecipitation, and film rehydration.

Direct dissolution is suitable for polymers that have relatively good water solubility. The drugs and polymers are codissolved in water or buffer solutions, in some cases with the help of stirring, heating, and/or sonicating. The polymers form micelles *in situ*, and the hydrophobic core of the micelles becomes the preferred location for the hydrophobic drugs.

Dialysis is commonly used for polymers that have limited water solubility. The drugs and polymers are first codissolved in a common solvent, e.g. dimethyl sulfoxide (DMSO). This mixture is then transferred to a preswollen dialysis bag and placed in a large volume of water to remove the nonwater solvent.

Nanoprecipitation starts with dissolving the polymer in a water-miscible solvent. This polymer solution is then slowly added into water under vigorous stirring to induce micelle formation. The precipitation can also be induced by slowly adding water into the polymer solution.

Film hydration starts with codissolving the polymer and drug in a volatile solvent, which is then dried into a polymer/drug film. The film is then immersed in water or buffer solutions and gradually dissolved with the help of heat or sonication.

12.3.3 Preparation of PG-Containing Polymeric Micelles

PG is responsive to pH because of its abundant γ-carboxylates. In basic solutions, γ-carboxylates are fully ionized, therefore PG has a random-coil conformation and is highly water soluble. In acidic solutions, γ-carboxylates are protonated, and PG takes a compact α-helix conformation and is insoluble in water. This pH-dependent shift of solubility in water makes PG suitable for both the micelle core and the micelle shell.

For micelles with a PG shell, hydrophobic polymers are needed to form the micelle core. PLGA is a widely used hydrophobic polymer because of its biodegradability and biocompatibility. Yang et al. prepared an AB-type copolymer with PLGA (MW ~ 8679) and PG (MW ~ 1976) blocks to formulate micelles [27]. The size and the shape of these micelles were pH dependent: 81.5-nm nanospheres were produced at pH 5 and 292.4-nm nanovesicles at pH 9. In another study, Kim et al. prepared micelles from PG-*b*-poly(L-phenylalanine). L-phenylalanine provided not only a hydrophobic interaction but also a hydrogen bond and π-π stacking to stabilize the micelle core [34]. The design was inspired by a previous study in which the π-π stacking stabilized the closed-caged nanostructures assembled from a simple dipeptide, diphenylglycine [35]. Above pH 6, the PG-*b*-poly(L-phenylalanine) self-assembled into 150-nm micelles with a critical micelle concentration of 0.11 mg/mL. The high critical micelle concentration was probably due to the short poly(L-phenylalanine) block. PG shells can be cross-linked to improve micelle stability. Rodríguez-Hernández et al. synthesized micelles from PG-*b*-poly(butadiene) and cross-linked the PG shell with 2,2′-(ethylenedioxy)diethylamine [36]. The cross-linked micelles were able to tolerate the addition of THF, which is a good solvent for the poly(butadiene) core. When the THF content was increased from 0 to 7.5% (v/v), the size of cross-linked micelles remained at 30 nm, while the size of micelles that were not cross-linked increased from 30 to 150 nm. However, it is noteworthy that most PG-shell micelles are stable only in media above a certain pH value because PG loses its water solubility in acidic solutions.

For micelles with a PG core, PG sometimes must be modified to increase its hydrophobicity. One method for doing this is to conjugate the γ-carboxylates with hydrophobic groups. The self-assembly of polyethylene glycol (PEG)-*b*-poly(stearyl α,L-glutamate) on highly oriented pyrolytic graphite has been studied in detail [37]. Kim et al. evaluated polymeric micelles from PEG-*b*-poly(alkyl α,L-glutamate) [38]. Hydrophobic antitumor drugs, such as cisplatin, have also been conjugated to the γ-carboxylates of PG. Kataoka et al. prepared micelles from PEG-*b*-(PG-*g*-cisplatin) and successfully use this formulation for drug delivery [39]. Another method for increasing PG's hydrophobicity is to form a polyelectrolyte complex between PG and a positively charged component. The concept of a polyelectrolyte complex was put forward by Kataoka and Kabanov in the 1990s [40, 41], and drug-loaded polyelectrolyte complex micelles have already been developed for cancer treatment [42, 43]. Luo et al. formulated a PG–chitosan polyelectrolyte complex micelle by slowly adding a PEG-*b*-PG solution (0.5 mg/mL) to a chitosan solution (0.5 mg/mL) [24]. Chitosan, a natural polymer that consists of poly[β-(1, 4)-2-amino-2-deoxy-D-glucopyranose], is an excellent candidate for biomedical applications because of its biodegradability, its nontoxicity, and its antibacterial properties [44, 45]. The amine functionality provides chitosan with a positive charge in acidic or neutral media. Dynamic light scattering and turbidity measurements found that stable micelles (200 nm in diameter) existed when the PEG-*b*-PG content in the mixture was between 20 and 55% and the solution pH between 2.5 and 4.5 [24]. No micelles were formed when PEG-*b*-PG content was lower than 20%, while a dramatic increase in micelle size was observed when PEG-*b*-PG content was higher than 55%. In a solution with pH below 2.5, PG was fully protonated and thus could not form complexes with chitosan; in a solution with pH above 4.5, loose large polymer aggregates persisted.

The pH responsiveness of PG led to the development of "smart" polymeric micelles that respond to environmental stimuli. Rodríguez-Hernández and Lecommandoux formulated a pH-sensitive micelle from PG_{15}-*b*-poly(L-lysine)$_{15}$ [46]. This polymer self-aggregated in both acidic and basic aqueous solutions. In acidic solution, PG was protonated and condensed into an α-helix conformation to form the hydrophobic core. In basic solution, the poly(L-lysine) block was insoluble in water and constituted the micelle core, while PG was flipped outside into the micelle shell. These two morphologies could reversibly transform into each other in response to pH changes. The hydrodynamic diameter of the aggregates was 110 nm at pH 3 and 150 nm at pH 15. This contrast between the low MW of the polymer and the large size of the aggregate suggested that the aggregates were hollow vesicles, which was confirmed by light-scattering and neutron-scattering experiments. This inside-out flipping of micelles might be useful for pH-triggered release of drugs.

Thermoresponsiveness has also been introduced in PG-containing polymeric micelles by incorporating PNIPAAm. PNIPAAm is a well-known polymer that undergoes a coil-to-globule transition at around 32°C in water [47]. The transition temperature is called the lower critical solution temperature (LCST). Below LCST, PNIPAAm has a random-coiled structure and thus is water soluble. Above LCST, PNIPAAm condenses into a globular structure and becomes water insoluble. Deng et al. studied micelles formulated from PG_{35}-*b*-$PNIPAAm_{55}$ [23]. pH-induced micellization occurred at 25°C, pH 5, and produced spherical micelles (40–60 nm in diameter) with a PG core and a PNIPAAm shell. Temperature-induced micellization occurred above 38°C in a solution with pH 10. The PNIPAAm block was water insoluble above 38°C, while the PG block was fully water soluble at pH 10. The micelles were rod shaped and measured 100 nm. A similar study was performed by Rao et al. using PG_{110}-*b*-$PNIPAAm_{65}$ [48]. In this study, transitions between the PG-core and PG-shell micelles upon changes in temperature and pH also were observed, but in contrast to the PG_{35}-*b*-$PNIPAAm_{55}$ micelles formed by Deng et al., both types of micelles formed from PG_{110}-*b*-$PNIPAAm_{65}$ were spherical with diameters below 50 nm. The well-controlled sizes may have been due to Rao et al.'s use of a much larger PG block, which could better stabilize the PG-shell micelles. Cai et al. studied the pH-responsive, thermoresponsive micelles formulated from the triblock copolymer PG_n-*b*-poly(propylene oxide)$_{69}$-*b*-PG_n (PG_n-*b*-PPO_{69}-*b*-PG_n) [49]. The length of the PPO block was 65 repeating units, and the length of the PG block (*n*) varied from 18 to 99 repeating units. The thermoresponsiveness arose from the dehydration of the PPO block at high temperatures, which resulted in the shrinkage of micelles. Transmission electron microscopy revealed self-assembled vesicles from polymers with 18-unit PG chains at high temperatures or low pH.

12.3.4 Characterization of PG-Containing Polymeric Micelles

Numerous publications have described methods for analyzing and characterizing polymeric micelles [50]. Only PG-related information is discussed in this section. Fourier transform infrared spectroscopy, which provides information about chemical bonds, shows that the characteristic N-H stretching peak of amide I is at ~3200–3400 cm^{-1} and the peaks for symmetric and asymmetric stretching of side chain $-CH_2-$ are at 2800–3000 cm^{-1}. The peak around 1740 cm^{-1} corresponds to the C=O stretching of the γ-carboxylic ester bond of the side chain. The amide I absorption peak is around 1650 cm^{-1}, while the amide II absorption peak is around 1550 cm^{-1} and corresponds to the α-helical structure of amide [51]. Nuclear magnetic resonance spectroscopy, routinely used to determine the chemical structures of monomers and polymers, can also elucidate the chemical composition of micelle shells when the micelles are dispersed in D_2O, since the shell components are soluble. Nuclear magnetic resonance spectroscopy has been used to monitor PG's transition from micelle core to micelle shell [10, 52]. It is well known that PG undergoes a coil-to-helix transition at pH 5.2 [53]. This transition can be monitored by circular dichroism [54]. PG's random-coil conformation at high pH exhibits a single maximum at 218 nm in circular dichroism spectra, while the α-helix conformation at lower pH exhibits two negative minima at 222 nm and 208 nm [10]. The ratio of random-coil conformation to α-helix conformation can be quantified from the absolute value of molar ellipticity [55].

12.4 PG-CONTAINING POLYMERIC MICELLES AS IMAGING AGENTS

Many imaging modalities are employed in modern medicine, including γ-scintigraphy, magnetic resonance imaging (MRI), computed tomography, and ultrasonography [56]. Imaging tracers are usually needed to differentiate the areas of interest from the background. Widely used low-MW tracers, however, have some drawbacks. One such drawback is that low-MW tracers (e.g., iodine, a contrast agent for computed tomography) are rapidly cleared from the circulation and therefore provide only a narrow time window for imaging. Repeated administration of such agents imposes heavy burdens on patients with comprised kidney function [57, 58]. Polymeric micelles are ideal carriers for imaging tracers for several reasons [59]. First, multiple imaging tracers can be incorporated into a single micelle to enable simultaneous visualization via different imaging techniques. Such multimodal imaging provides more accurate information since it can compensate for the deficiencies of individual imaging modalities. Micelle size and surface functionality can be finely tuned to optimize micelle performance for specific applications. For example, micelles with adequate size and surface protection can circulate for a prolonged time and passively accumulate in tumors via the enhanced permeation and retention effect [60]. The tumor topology can be mapped using polymeric micelles bearing imaging tracers. The prolonged circulation of polymeric micelles is still useful when they are conjugated with active targeting ligands because long-circulating micelles can increase the total quantity of targeting agents that passes through the tumor and becomes recognized by the targets [61].

The size of micelles plays an important role in their pharmacokinetics and biodistribution. Nanoparticles less than 100–400 nm in diameter are able to cross the wall of tumor vessels [62, 63]. Yang et al. studied the effects of the size of polymeric micelles on their pharmacokinetics and biodistribution [64]. The investigators used micelles of 20 nm, 30 nm, and 60 nm with similar chemical compositions and surface functionality: the micelle shell consisted of PEG, and the core was formed from poly(methoxyethylmethacrylate) using divinylbenzene as the core cross-linker. After conjugation with λ-imaging tracers, the micelles were intravenously injected into mice bearing xenografted MDA-MB-468 tumors. Pharmacokinetics data showed that both the systemic clearance and the volume of distribution significantly decreased as micelle size increased. This finding was consistent with the smaller percentage of larger micelles taken up by the liver: the values for percentage of injected dose per gram of tissue in liver (%ID/g) were 12% for the 30-nm and 60-nm micelles and 22% for the 20-nm micelles.

In addition to small micelle size, two other factors critical for achieving extended circulation are surface shielding of micelles from the RES and intrinsic micelle stability. Recently, our group developed 20-nm core-cross-linked polymeric micelles with a brushlike PEG shell and a silica-cross-linked core [65]. The brushlike PEG shell efficiently shielded the micelles from recognition and subsequent clearance by the RES, while the three-dimensional silica cross-linking network provided the micelles with excellent stability. The combined effects led to remarkably long-circulating micelles: the half-life of blood clearance was 46.18 ± 6.86 h.

12.4.1 PG-Polymer Conjugates for Imaging

PG is an ideal carrier for imaging tracers because of its biodegradability and abundant γ-carboxylic functionality. Our group has studied the chemistry of conjugating PG with imaging tracers such as those used for MRI and near-infrared fluorescence (NIRF). In one project, we studied the conjugation of PG to diethylene triamine pentaacetic acid (DTPA), an acyclic cation chelator for gadolinium (Gd) that can be used as an MRI contrast agent. PG-DTPA conjugate was synthesized via two routes (Fig. 12.2) [66].

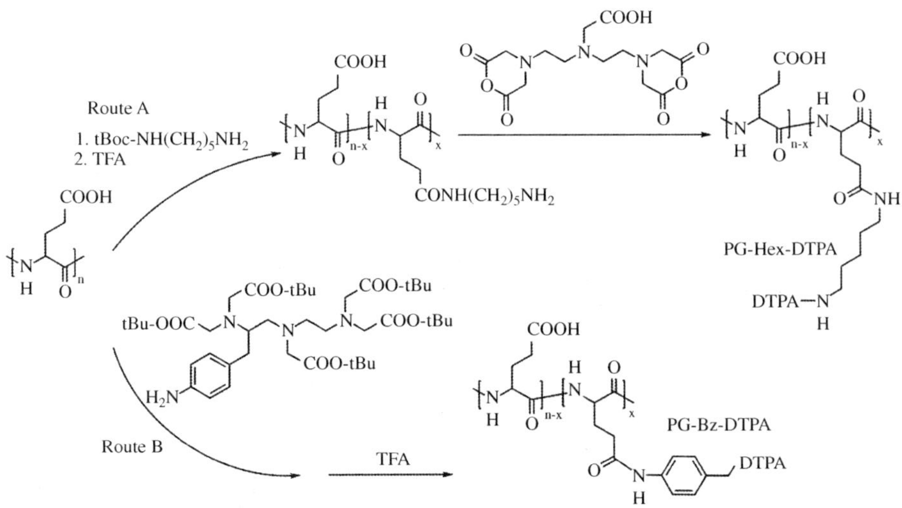

FIGURE 12.2 Synthesis of PG-DTPA conjugates.

Route A involved the use of DTPA anhydride and a 1,6-hexyldiamine spacer to form PG-Hex-DTPA. Route B involved the conjugation of *p*-aminobenzyl-DTPA t-butyl ester to PG and the subsequent removal of the t-butyl groups with TFA treatment to form PG-Bz-DTPA. Comparison between the two conjugates demonstrated that PG-Bz-DTPA was the better candidate for clinical applications for several reasons. First, route A produced PG-Hex-DTPA as a partially cross-linked polymer, which was revealed by gel permeation chromatography traces. Route B produced PG-Bz-DTPA as a linear polymer with well-controlled MW distribution. Second, the linear PG-Bz-DTPA-Gd complex exhibited significantly higher relaxivity than the cross-linked PG-Hex-DTPA-Gd complex (50% higher in R_1 and 75% in R_2) when the same amount of Gd was used. This difference was probably due to the better accessibility of Gd to water molecules in the complex with the linear polymer. Third, the degradation and clearance of PG-Bz-DTPA was faster than that of PG-Hex-DTPA. *In vivo* MRI scans (Fig. 12.3) showed that PG-Bz-DTPA-Gd provided remarkably enhanced vascular contrast but no significant loss of signal, probably because it had very limited diffusion into intravascular spaces. The clearance of PG-Bz-DTPA-Gd was confirmed as the signal intensity in blood gradually decreased over time. The signal intensities from the blood pool returned to the preinjection level by 24h after injection. In the meantime, signal from the tumor had actually increased, indicating a gradual retention of the PG-Bz-DTPA-Gd.

NIRF is another powerful tool for noninvasive imaging [67, 68]. One PG-conjugated NIRF dye, PG-NIR813, was prepared from a commercially available NIRF dye (Fig. 12.4) [69]. Because PG-NIR813 exhibited a degradation-induced increase of fluorescence, it was a useful tool for monitoring the *in vivo* PG degradation.

Prior to PG degradation, the dye was densely packed along the PG backbone and self-quenched. PG degradation caused the dye molecules to separate from each other. An increase in fluorescence was observed as a result. The *in vitro* results suggested that PG was selectively degraded by cysteine protease, particularly cathepsin B. Since cathepsin B correlates with tumor invasiveness

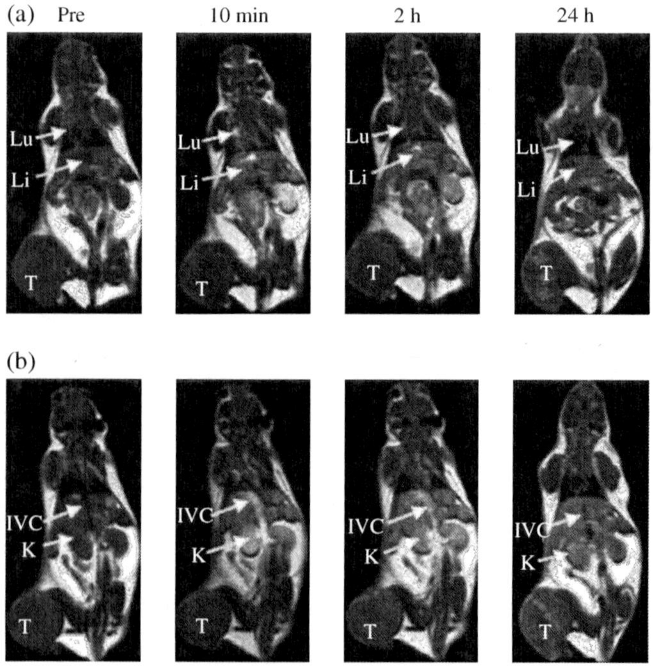

FIGURE 12.3 Whole-body MRI scans of mice before and at 10 min, 2 h, and 24 h after intravenous injection of PG-Bz-DTPA-Gd at a dose of 0.04 mmol Gd/kg. (a) Selected anterior coronal image after injection of contrast agent demonstrates pulmonary and hepatic perfusion. (b) Selected posterior coronal image after injection of contrast agent demonstrates renal perfusion as well as persistent enhancement of the vena cava. Note enhancement of tumor at 24 h post injection. Lu, lung; Li, liver; IVC, inferior vena cava; K, kidney; T, tumor. (Reprinted with permission from Bioconjugate Chemistry, 2004. 15[6]: p. 1408–1415 [66]. Copyright (2004) American Chemical Society.)

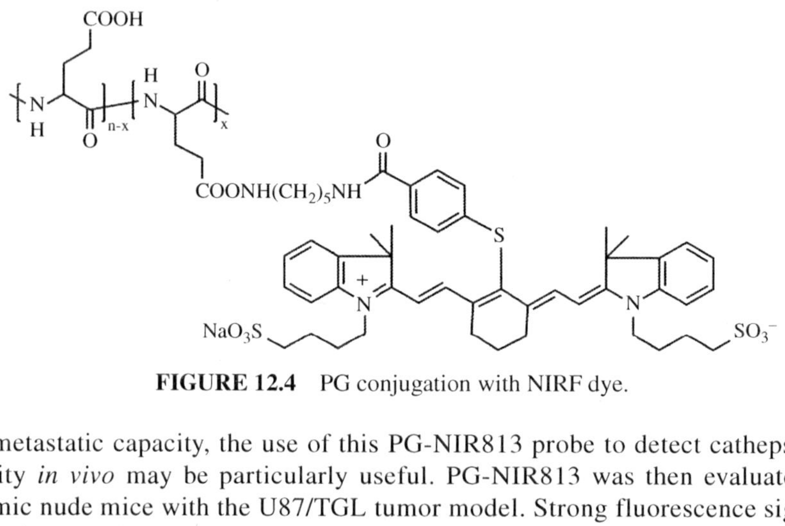

FIGURE 12.4 PG conjugation with NIRF dye.

and metastatic capacity, the use of this PG-NIR813 probe to detect cathepsin B activity *in vivo* may be particularly useful. PG-NIR813 was then evaluated in athymic nude mice with the U87/TGL tumor model. Strong fluorescence signals were observed in the blood pool 4 h after intravenous injection, indicating PG degradation (Fig. 12.5).

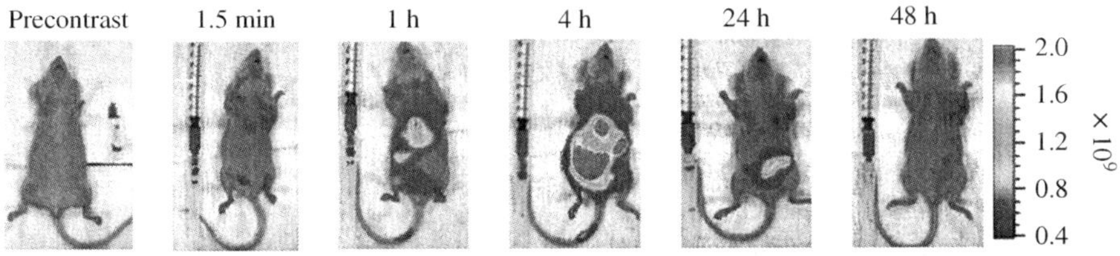

FIGURE 12.5 Whole-body NIFR images acquired after intravenous injection of PG-NIR813 at a dose of 10 nmol NIR813 per mouse. (Reprinted with permission from Pharmaceutical Research, 2007;24[6]:1217–1224 [69].)

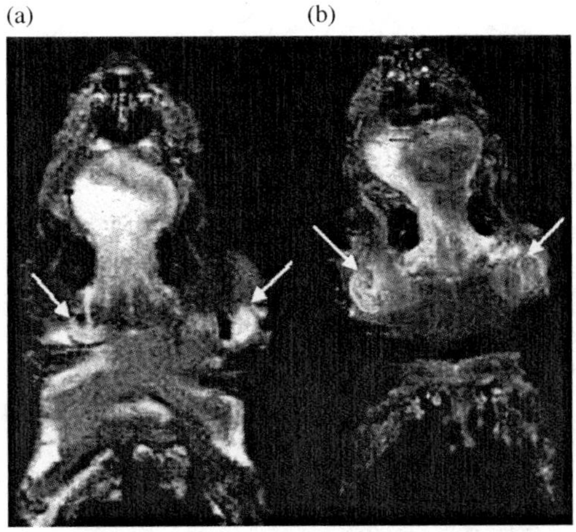

FIGURE 12.6 MRI of cervical lymph nodes after interstitial injection of PG-DTPA-Gd-NIR813 at a dose of 0.02 mmol Gd/kg into (a) the tongue of a normal mouse and (b) a mouse with a human DM14 squamous cell carcinoma in the tongue. (Modified with permission from Investigative Radiology, 2007;42[8]:569–578 [71].)

Recently, there has been increased interest in multimodal imaging [70]. PG has proved to be a convenient platform for simultaneously incorporating multiple imaging tracers. For example, PG-DTPA-Gd-NIR813, a PG derivative conjugated with Gd^{3+} and NIR831, was developed to map sentinel lymph nodes on MRI and NIRF [71]. Sentinel lymph node imaging has some challenging technical requirements; these include suitability for (i) high-resolution anatomic imaging prior to surgical dissection, (ii) real-time intraoperative visualization, and (iii) differentiation of normal sentinel lymph nodes from those with metastases. PG-DTPA-Gd-NIR813 proved to be a promising tool for meeting these requirements. With this imaging tool, the T_1-weighted MRI scans successfully differentiated normal lymph nodes from those with metastases. Compared with normal lymph nodes, lymph nodes with metastases showed fewer areas of enhancement; when enhancement occurred, it was located at the lymph node rims (Fig. 12.6). The differentiation was probably due to different macrophage uptake of PG-DTPA-Gd-NIR813 and impaired lymphatic drainage caused by micrometastases. PG-DTPA-Gd-NIR813 enabled NIRF imaging as well, which was more sensitive than MRI. Lymph nodes containing metastases were detected by NIRF as early as 10 min after the tracer was injected.

PG-DTPA-Gd-NIR813 has also been used to detect the density change of tumor-associated macrophages (TAM) [72], which frequently infiltrate solid tumors. Among such TAM, the classically activated (or M1) macrophages are commonly found during infection or inflammation. Another subset of TAM, M2 macrophages, promotes tumor growth and metastasis by secreting proangiogenic/growth factors. Therefore, selective depletion of M2 TAM is considered a novel approach to tumor treatment. However, a timely and noninvasive method for evaluating the effect of such treatment has been lacking. At present, imaging techniques either cannot show changes of TAM density early in therapy or cannot differentiate TAM from tumor cells. Recently, PG-DTPA-Gd-NIR813 was found to selectively accumulate in the necrotic/apoptotic regions of tumors and colocalize with CD68, an M2 TAM marker (Fig. 12.7). The colocalization was visualized *in vivo* by NIRF imaging and confirmed by immunohistochemical staining. Twenty-four hours after the tumor was treated with a TAM-depleting agent, PG-DTPA-Gd-NIR813 was injected intravenously and showed a decrease in TAM density. The simultaneous decrease of PG-DTPA-Gd-NIR813 and the CD68-positive cell population in the tumor following treatment with the depleting agent was confirmed by immunohistochemical staining.

12.4.2 PG-Based Polymeric Micelles for Imaging

Despite the extensive research on polymeric micelles as imaging agents, few of the polymeric micelles studied to date have included PG. In one study, a polymeric micelle system based on (PG-*g*-DTPA)-*b*-poly(L-lactic acid) was developed as the Gd^{3+} carrier for MRI. The DTPA chelator was grafted to the PG block that constituted the micelle shell (Fig. 12.8). Subsequently, the micelle-chelated Gd^{3+} was located in the hydrophilic environment.

Briefly, an amine-terminated poly(L-lactic acid) block was first synthesized via anionic polymerization of L-lactide. A PBLG block was built by NCA polymerization from the terminal amine of the poly(L-lactic acid) block. The benzyl groups were then removed under mild and anhydrous conditions (1 M trifluoromethanesulfonic acid and 1 M thioanisole in TFA) to yield PG-*b*-poly(L-lactic acid) [73]. The obtained PG-*b*-poly(L-lactic acid) polymers were grafted with p-amino-benzyl-DTPA t-butyl ester carbodiimide chemistry. The t-butyl groups were removed by TFA treatment to yield (PG-*g*-DTPA)-*b*-poly(L-lactic acid). Both deprotection steps were performed under anhydrous conditions so that the poly(L-lactic acid) block would not be degraded prematurely. Micelles were formed via the dialysis method. The hydrodynamic diameter of the micelles was pH dependent: 256 nm at pH 5.5, 230 nm at pH 7.4, and 185 nm at pH 8.5. The critical micelle concentration was 12 mg/L. Gd^{3+} was conjugated to the micelles in a pH 5.5 buffer solution, and the free Gd^{3+} ions were removed by dialysis against a dilute EDTA solution. The loading ratio of Gd^{3+} in the micelles was 5% by weight, which was twice as high as the loading ratio in a similar study in which the micelles were modified by a preformed DTPA-Gd complex [74]. After Gd^{3+} conjugation, the micelle diameter decreased to 210–230 nm at pH 7.4, probably because of the reduced anionic repulsion in the presence of Gd^{3+}.

The PG-micelle-Gd complex was used for MRI and provided T_1-weighted images of high quality because it exhibited high T_1 relaxivity due to its fast water-exchange rate and long rotational correlation lifetime. In the PG-micelle-Gd complex, the shell-located Gd exhibited a water-exchange rate as fast as that of its low-MW equivalents because the water's access to Gd was not impaired. In contrast, Gd-micelle

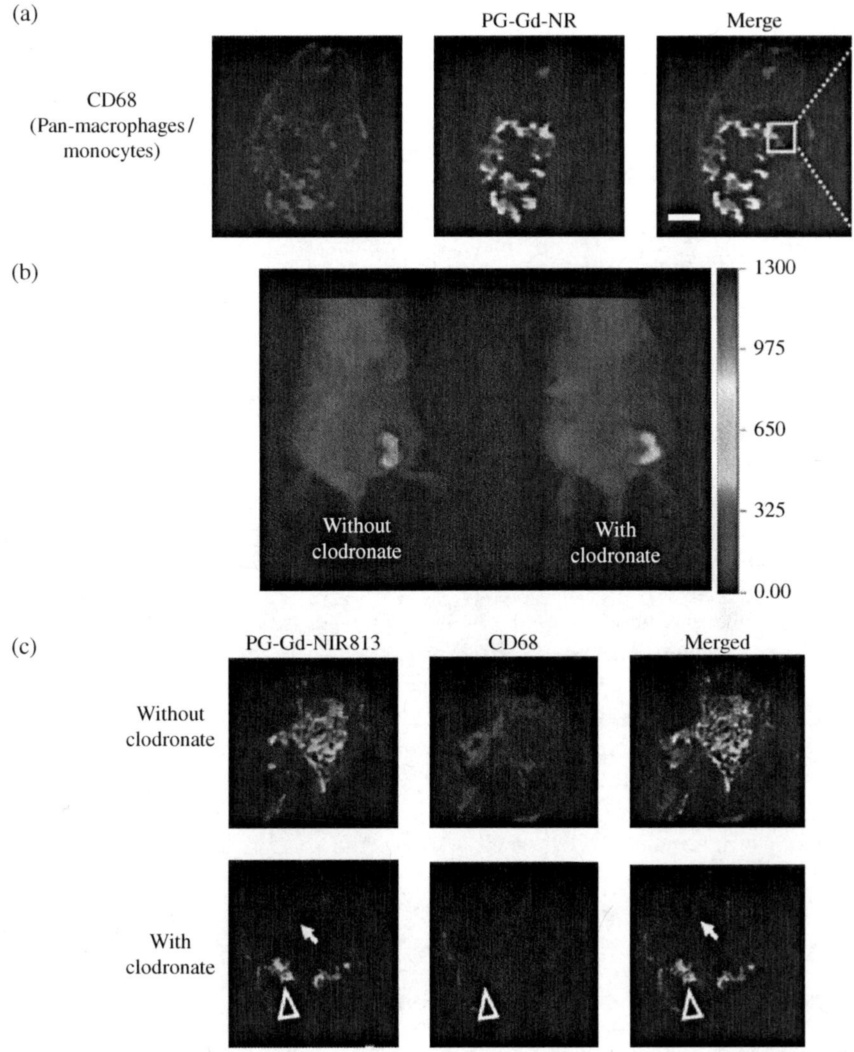

FIGURE 12.7 Noninvasive NIFR imaging of M2 macrophages using PG-DTPA-Gd-NIR813. (a) Co-localization of CD68 and PG-DTPA-Gd-NIR813 in C6 tumors 48 h after intravenous injection. (b) Fluorescence intensities of PG-Gd-NIR813 in A20 tumors in mice injected with clodronate, a macrophage-depleting agent, compared with saline control. (c) Fluorescence micrographs of C6 tumors in nude rats injected with PG-DTPA-Gd-NIR813 with or without macrophage depletion. (Reprinted with permission from Biomaterials, 2010;31[25]:6567–6573 [72].) For color details, please see color plate section.

complexes with Gd in the hydrophobic core exhibited low water-exchange rates [75]. Because the PG-micelle-Gd complex had a longer rotational correlation lifetime than its low-MW equivalents, the T_1 relaxivity of the PG-micelle-Gd complex (7.90/mM/s) was nearly twice as high that of the low-MW DTPA-Gd complex (4.26/mM/s). An *in vitro* cell study demonstrated that the PG-micelle-Gd complex had no cytotoxicity up to 100 μg/mL.

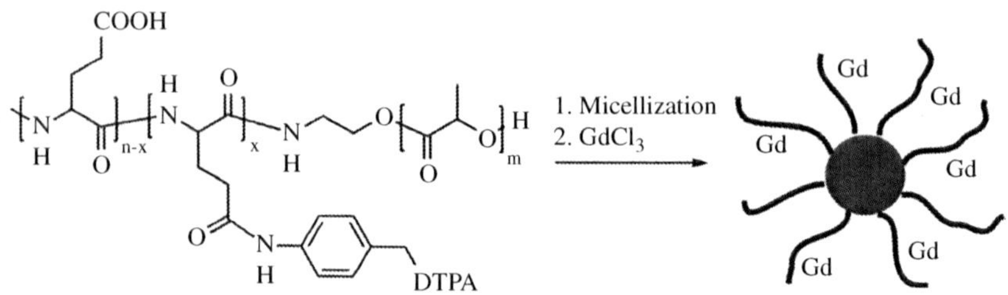

FIGURE 12.8 PG-based polymeric micelles with gadolinium labeling.

In general, the PG-micelle-Gd complex proved to be a promising contrast agent for MRI. Depending on the results of in-depth research on its applications *in vivo*, it may be an interesting candidate for multimodal tumor imaging with high quality, specificity, and sensitivity.

12.5 PG-BASED POLYMERIC MICELLES FOR ANTITUMOR DRUG DELIVERY

A common phenomenon observed with antitumor drugs is that a drug works well *in vitro* but not *in vivo*. The potential reasons for this phenomenon are multiple and include (i) poor bioavailability due to water insolubility, (ii) high systemic toxicity due to nonspecific action, and (iii) poor pharmacokinetics or biodistribution. Polymeric micelles are versatile drug carriers that can be engineered to overcome some of these issues; the interior of micelles provides an ideal shelter for the drugs against harsh *in vivo* microenvironments. Paclitaxel, doxorubicin, and platinum-based drugs are discussed as examples in this section to illustrate the use of PG-based polymeric micelles for drug delivery.

12.5.1 Paclitaxel

Paclitaxel is a potent inhibitor of cell replication that acts by promoting tubulin polymerization and microtubule assembly [12], thereby blocking the cell cycle at the late G2–mitosis phase [76]. Paclitaxel has been used to treat many malignancies, including primary epithelial ovarian carcinoma, nonsmall cell lung cancer, and cancers of the breast, colon, and head and neck [77]. However, the clinical use of paclitaxel has been hindered by the drug's poor solubility, as well as by the hypersensitivity, nephrotoxicity, and neurotoxicity caused by castor oil that is used as a delivery agent in some formulations [78]. In addition, resistance of tumor cells to paclitaxel, which is mediated by p-glycoprotein, has been documented [79, 80].

PG-conjugated paclitaxel (Fig. 12.9) was developed by Li et al. [12] with the goals of (i) improving the water solubility of paclitaxel, (ii) achieving a controlled paclitaxel release through the lysosomal degradation of PG, and (iii) using the PG degradation product sodium glutamate to alleviate paclitaxel-induced neuropathy [81]. PG-paclitaxel demonstrated significantly lower systemic toxicity than that of free paclitaxel and had remarkable antitumor efficacy, completely eliminating a well-established 13762F mammary adenocarcinoma *in vivo* [12]. Since then,

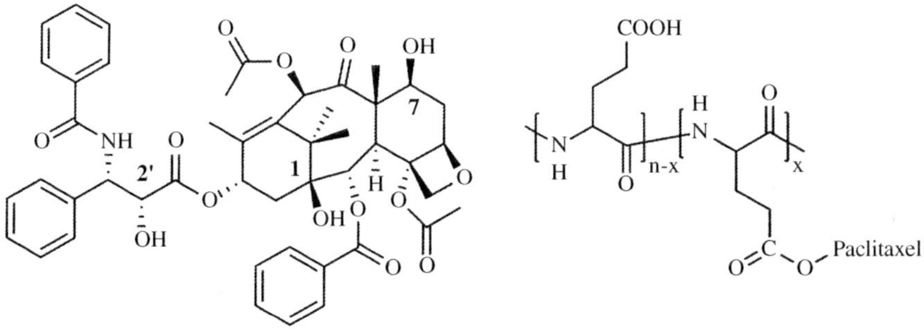

FIGURE 12.9 Paclitaxel and PG-conjugated paclitaxel.

PG-paclitaxel has been tested in a broad spectrum of tumor models and has exhibited antitumor activities superior to those of free paclitaxel in all these tests. PG-paclitaxel (paclitaxel poliglumex [PPX], also known as CT-2103, Opaxio™, and formerly as Xyotax™) has advanced to phase III clinical trials. The findings of previous clinical trials of this formulation were reviewed by Li and Wallace in 2008 [82]. PPX showed three safety-related advantages over conventional paclitaxel-based treatments: (i) alopecia was rare and there were no cases of complete hair loss, (ii) nausea and vomiting were uncommon, and (iii) hypersensitivity reactions were rare. PPX improved the overall survival rate of female patients in the combined log-rank analysis of the STELLAR 3 and 4 phase III studies. A confirmatory phase III trial was started in the United States in 2007 comparing PPX-carboplatin with paclitaxel–carboplatin for the treatment of nonsmall cell lung cancer in women ($n = 450$) [82]. It is noteworthy that the PG-paclitaxel conjugate might self-assemble into micelle structures at physiological pH (7.4) because of its amphiphilic nature.

Several new clinical studies have been conducted since 2008. In a phase I study of PPX and concurrent radiation for esophageal or gastric cancer, the maximum dose of PPX was 70 mg/m²/week [83]. In another phase I study, single-agent PPX at doses of 235 or 270 mg/m² every 21 days produced unexpectedly high rates of neutropenia and neuropathy in patients with advanced solid tumors [84]. In a phase II trial in patients with recurrent or persistent ovarian or primary peritoneal cancer, PPX at a dose of 175 mg/m² every 21 days showed modest activity of limited duration when it was given as second- or third-line therapy [85]. The median progression-free survival was 2.8 months, and the median overall survival was 15.4 months. A higher-than-predicted incidence of neuropathy, however, was observed in patients receiving this dose. Another phase II study tested the efficacy of PPX in patients with HER2-negative metastatic breast cancer [86]. Objective response was observed in 22% of the patients receiving a dose of 175 mg/m² over 10min every 21 days. Neurotoxicity and hypersensitivity reactions, however, were more frequent than expected. PPX was compared with docetaxel in a phase III trial as second-line treatment for non-small cell lung cancer [87]. PPX produced survival results similar to those of docetaxel: median overall survival was 6.9 months, the 1-year overall survival rate was 25%, and mean time to progression was 2 months. The toxicity profiles of the two drugs differed. Compared with docetaxel, PPX produced fewer incidences of alopecia and grade 3 or 4 neutropenia, had a shorter infusion time, and eliminated the routine use of medications to prevent hypersensitivity reactions. PPX at 210 mg/m^2 produced a higher rate of neurotoxicity than docetaxel. In a randomized phase III trial in patients with advanced nonsmall cell lung cancer and impaired performance status, PPX at 175 mg/m^2 was active and well tolerated [88]. The median survival was 7.3 months for patients receiving PPX and 6.6 months for the control group treated with vinorelbine or gemcitabine. Anemia and grade 3 or 4 neutropenia were less common in patients treated with PPX than in the control group. Although neuropathy was more common for PPX, grade 3 neuropathy occurred in only 3% of patients treated with PPX. In another study in patients with advanced nonsmall cell lung cancer, first-line treatment using PPX-carboplatin produced a comparable survival rate to paclitaxel–carboplatin [89]. Alopecia, arthralgias/myalgias, and cardiac events were significantly less frequent in the PPX-carboplatin group.

Research on paclitaxel-loaded polymeric micelles emerged in the late 1990s and has blossomed in recent years. The encapsulation of paclitaxel in polymeric micelles was mostly performed by physical entrapment using amphiphilic copolymers that form core-shell structures in aqueous solutions. The core is made of hydrophobic segments that serve as a reservoir for the water-insoluble paclitaxel, while the hydrophilic shell stabilizes the micelle–drug complex in water and shields the drug from the RES. The polymers used for the hydrophobic core have included biodegradable polyesters (poly(lactide), poly(ε-caprolactone), and poly(δ-valerolactone) [90–92]), the hydrophobic poly(amino acid)s (poly(phenylalanine) and poly(β-benzyl-L-aspartate) [93, 94]), polyether (poly(propylene oxide) [95]), poly(trimethylene carbonate) and its derivatives [96, 97], lipids [98], and poly(β-amino esters) [99]. Polymers used for the hydrophilic shell have included PEG (poly(N-vinylpyrrolidone) [100]) and the thermo- and pH-responsive polymers (poly(NIPAAm), poly(*N*-(2-hydroxypropyl) methacrylamide lactate), poly(acrylic acid), and poly(N,N-dimethylaminoethyl methacrylate) [101–106]). Some other amphiphilic polymers, such as modified chitosan, have also been studied for micellization and paclitaxel loading [107–109].

The chemical conjugation of paclitaxel to polymeric micelles was performed through an ester bond involving the –OH at C-2′ or C-7 of paclitaxel (Fig. 12.9) [110, 111]. For example, a polymeric micelle system containing L-glutamic acid was developed by Xie et al. for paclitaxel delivery [110]. The micelles were formulated from an ABA-type copolymer, poly((lactic acid)-*co*-[(glycolic acid)-*alt*-(L-glutamic acid)])-*b*-PEG-*b*-poly((lactic acid)-*co*-[(glycolic acid)-*alt*-(L-glutamic acid)]). Paclitaxel was conjugated to the glutamic acid units via an ester bond. The micelles were 119 nm in diameter with a paclitaxel loading of 27% by weight. The alternating sequence of lactic acid and L-glutamic acid rendered the micelles hydrolytically degradable in acidic solutions; therefore, the *in vitro* release followed a pH-dependent profile. The paclitaxel-micelles exhibited significant toxicity against RBG-6 tumor cells, although the toxicity was lower than that of free paclitaxel.

Recently, Alani et al. successfully conjugated paclitaxel to the micelles of PEG-*b*-poly(α, L-aspartic acid) via an acid-labile hydrazone linkage [111]. This approach could be extended to PG-based micelles without much difficulty. Briefly, PEG-*b*-poly(β-benzyl-L-aspartate) was first synthesized via NCA polymerization. The β-benzyl group was then substituted by hydrazine, which in turn reacted with ketone-containing levulinic acid or 4-acetylbenzoic acid to form an acid-labile hydrazone bond. Paclitaxel was subsequently conjugated to the polymer via an ester bond. Therefore, a hydrazone linkage was introduced between paclitaxel and the polymer backbone, enabling pH-sensitive drug release. The paclitaxel-loaded micelles exhibited hydrodynamic diameters between 41.9 and 137.0 nm. The *in vitro* release of paclitaxel followed a pH-sensitive profile, but no burst release was observed. In MCF-7 and SKOV-3 tumor cell lines, the concentration of paclitaxel causing 50% cell death (IC_{50}) was about three times that of free paclitaxel.

12.5.2 Doxorubicin

Doxorubicin is another widely used chemotherapy drug. Both physical entrapment and chemical conjugation have been used for loading doxorubicin into PG-based polymeric micelles. Physical encapsulation starts with deprotonating the commercially available doxorubicin hydrochloride salt using two to five equivalents of triethylamine. The resultant doxorubicin base is insoluble in water and thus can be preferentially encapsulated in the hydrophobic micelle core [112]. However, the chemical stability of doxorubicin remains a major concern during its micelle encapsulation. Acidic conditions (~pH 3–6) or low temperature (~4–5 °C) are usually required for the long-term storage of doxorubicin in aqueous solutions [113]. High-performance liquid chromatography analyses revealed a faster decomposition of doxorubicin in basic solutions or at high doxorubicin concentration [114], so special care is required.

Chemical conjugation of doxorubicin to polymeric micelles minimizes the potential for the burst release of doxorubicin. One conjugation method utilizes the amide bond between the doxorubicin amine functionality and carboxylate on the polymer backbone. This amide bond is less prone to degradation than the ester bond used for paclitaxel conjugation. A PG-doxorubicin conjugate in which doxorubicin was directly conjugated to PG via an amide bond exhibited practically no antitumor toxicity [115]. Therefore, additional degradable spacers (e.g., oligopeptides) were introduced between doxorubicin and the polymeric micelles [116]. Ulbrich et al. used a tetrapeptide, Gly-Phe-Leu-Gly, as the spacer between doxorubicin and poly(*N*-(2-hydroxypropyl)methacrylamide). This tetrapeptide was degradable by lysosomal enzymes [117]. It was later found that the antitumor activity was increased when the spacer was made of a longer oligopeptide or had a higher degradation rate [118]. Another conjugation method involved the use of an acid-labile hydrazone linkage. The γ-carboxylic acid of the glutamic unit was modified with a hydrazine, which then formed a hydrazone bond with the ketone of the 2-hydroxyacetyl group. The polymer–drug conjugation was carried out in absolute methanol in the presence of a catalytic amount of acetic acid [119, 120]. The hydrazone linkage was readily cleaved in acidic aqueous solutions, (pH 5 ~ 6) to release doxorubicin from the micelles (Fig. 12.10).

Lin et al. developed a doxorubicin-loaded hybrid micelle system in which the doxorubicin release rate could be tuned by changing temperature or pH [10]. The hybrid micelle was self-assembled from two copolymers: PG_{99}-*b*-PPO_{65}-*b*-PG_{99} and PEG_{31}-*b*-PPO_{10}. The micelle core consisted of a PPO block; the shell was composed of both PG and PEG chains. Doxorubicin was physically entrapped in the micelle core with loading ratios between 16.6 and 19.3% by weight. The doxorubicin-loaded micelles were around 90 nm in diameter. Accelerated doxorubicin release was

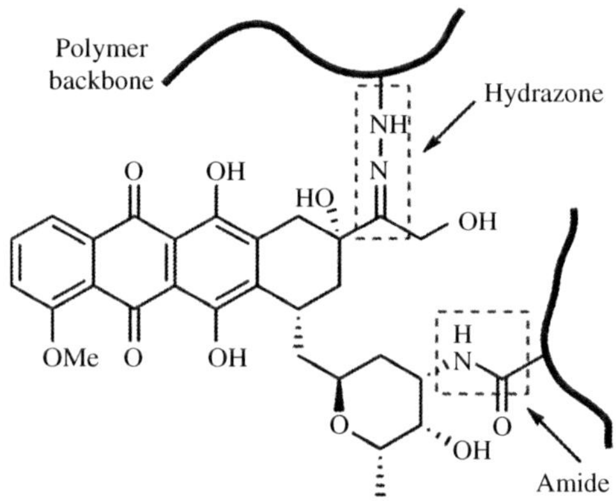

FIGURE 12.10 Chemical conjugation of doxorubicin to polymeric micelles.

triggered by either raising temperature or lowering pH. The influence of temperature was attributed to PPO dehydration and subsequent core shrinking at higher temperature, which resulted in doxorubicin's being squeezed out of the micelles. The influence of pH was attributed to a conformational change of PG from random coil to α-helix at lower pH. The collapse of the PG chain generated pressure on the doxorubicin-loaded core to accelerate the drug's release. The PEG chains were thought to form hydrophilic channels in the shell that may also have facilitated doxorubicin release [2]. Jeong et al. prepared micelles from a triblock copolymer, PLBG-*b*-PEG_{90}-*b*-PLBG [121]. PLBG, in this case, formed the micelle core because of the hydrophobicity of the benzyl functionality. The micelle diameter increased from 18 to 143 nm as the length of the PLBG block increased from 4 to 15 monomer units. Doxorubicin was loaded into micelles using the dialysis method, and the loading ratios ranged from 11.5 to 17.1% by weight. An *in vitro* release study in pH 7.4 phosphate-buffered saline found that 80% of doxorubicin was released by day 4. Na et al. prepared doxorubicin-loaded micelles from a triblock copolymer, PG-*b*-PEG-*b*-PG [122]. The drug loading was between 14.9 and 19.5%, and the micelles' diameter was around 80 nm. The micelles were injected intravenously at a dose of 5 mg/kg into mice bearing CT-26 tumors. The mean survival was 7 weeks. A synergistic effect was observed when doxorubicin-loaded micelles were coadministrated with etoposide- or paclitaxel-loaded micelles [122].

An exceptionally high loading of doxorubicin (47% w/w) was achieved by Sanson et al., who used the self-assembled polymersomes from PG_{16}-*b*-poly(trimethylene carbonate)$_{24}$ [123]. Poly(trimethylene carbonate) was used because this polycarbonate is biocompatible, enzymatically degradable, and nontoxic [124, 125]. Micelles were prepared by the nanoprecipitation method with the use of a solvent pair of DMSO and an aqueous buffer solution. DMSO was subsequently removed by dialysis. The optimal condition for drug loading was reached when carbonate buffer (pH 10.5) was used for nanoprecipitation. The diameters of doxorubicin-loaded polymersomes were below 64 nm at 5% doxorubicin loading and increased to 127–167 nm at 47% doxorubicin loading. An *in vitro* release study at 37°C in a pH 5.5 buffer solution showed a fast initial burst for the first 6 h and then slow release for up to 30 days, by which time 90% of doxorubicin had been released.

12.5.3 Platinum-Based Drugs

Cisplatin, or *cis*-dichlorodiammineplatinum (II), is used for the treatment of many malignant tumors, including head and neck, lung, breast, gastrointestinal, ovarian, and genitourinary cancers. However, the clinical use of cisplatin has been greatly hindered by intrinsic and acquired resistance to the drug and by adverse effects, including acute nephrotoxicity and chronic neurotoxicity [126–129]. A variety of cisplatin analogues have been developed to overcome these problems (Fig. 12.11). One analog, dichloro-(1,2-diamino-cyclohexane)platinum(II) (DACHPt), had lower toxicity than cisplatin. However, its water solubility was only one-fifth that of cisplatin [130]. Another analogue, oxaliplatin, was approved by the U.S. Food and Drug Administration in 2002. However, oxaliplatin exhibits dose-limiting cumulative peripheral distal neurotoxicity and acute dysesthesias [131].

Because of such problems with these low-MW platinum-based drugs, polymeric micelles have been developed as carriers for these drugs to improve their circulation in the blood, increase their accumulation in tumors, and minimize their adverse effects. A remarkable amount of work in this area has been accomplished by Kataoka et al. since the 1990s [5, 6, 39, 132]. This group successfully developed a polymeric micelle system with PEG-*b*-PG serving as the carrier for either cisplatin or DACHPt. The drugs were conjugated to the PG segment, rendering it hydrophobic. The PEG-*b*-(PG-*g*-drug) conjugates then gradually self-assembled into polymeric micelles with diameters less than 50 nm (Fig. 12.11). The polymeric micelles showed excellent stability in deionized water. The drug was released from the micelles in a chloride-rich medium, in which the chloride

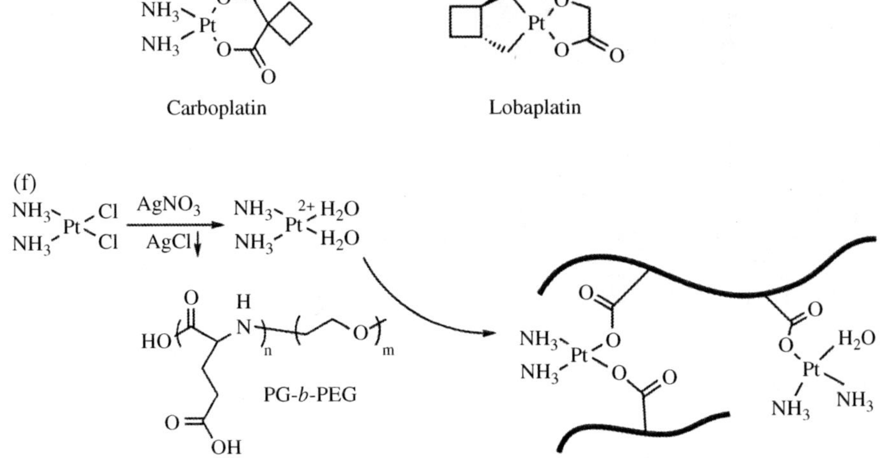

FIGURE 12.11 Structures of platinum-based antitumor drugs (a–e) and conjugation of cisplatin to PEG-*b*-PG block copolymer (f).

ions replaced the PG carboxylate and cleaved the drug from the micelles. When the micelles were placed in the releasing medium, a 15-h inducing phase was observed, probably due to the slow diffusion of chloride ions into the hydrophobic core. The absence of a burst release avoided premature drug leaking. Since polymeric micelles can passively accumulate in tumors due to the enhanced permeation and retention effect, this inducing phase helps these PEG-*b*-(PG-*g*-Pt) micelles achieve tumor-specific drug delivery.

Kataoka et al. performed both *in vitro* and in *vivo* studies using tumor cell lines LLC and C-26 [133, 134]. Because of the slow release of platinum-based drugs from the micelles, the IC_{50} values of drug-loaded micelles were about 10 times higher than those of free drugs. However, pharmacokinetics of the drug-loaded micelles were better than those of free drugs. Unlike free drugs, which are readily deactivated by opsonization, the drug-loaded micelles exhibited significantly longer circulation in blood. For example, 1 h after the injection of oxaliplatin, only 5–30% of the drug remained unbound by proteins. By 3 h after injection, there was no detectable oxaliplatin in plasma [133, 134]. Cisplatin was also rapidly cleared from plasma. In contrast, with the cisplatin-containing micelles, 60% of the injected dose remained in plasma at 8 h and 13% at 24 h after injection. The uptake of the drugs by tumors was improved by using the drug-loaded micelles. The uptake of cisplatin-containing micelles by tumors increased from 3% ID/g at 3 min after injection to 9% ID/g at 24 h after injection. It is noteworthy that free cisplatin showed high uptake (14% ID/g) in the kidneys shortly after injection, which may be responsible for its acute nephrotoxicity. In contrast, the highest kidney uptake of cisplatin-containing micelles was only 8% ID/g. Further investigation revealed that cisplatin-containing micelles can reduce both nephrotoxicity and neurotoxicity in rats compared with free cisplatin [132]. The reduced nephrotoxicity was attributed to the fact that the micelles are too large to penetrate the nephron as well as the lower cisplatin level in kidneys. Similarly, the lower level of neurotoxicity was attributed to the lower level of platinum accumulation in the neuro system when cisplatin-containing micelles were used.

The *in vivo* antitumor efficacy of cisplatin-containing micelles was evaluated in xenografted tumor models [132]. Significant tumor inhibition was observed in LLC-bearing mice that were treated with 4 mg/kg cisplatin-containing micelles five times at two-day intervals. Six of the 10 mice treated with micelles showed complete tumor regression. In contrast, no tumor cure was found in the mice treated with free cisplatin. Cisplatin-containing micelles were also less toxic than free cisplatin. No significant loss of body weight was found, and the blood urea nitrogen value remained within the normal range. Cabral et al. recently reported an interesting study of the tumor accumulation of polymeric micelles based on PEG-*b*-(PG-*g*-DACHPt) [135]. Four populations of micelles were prepared with average diameters of 30, 50, 70, and 110 nm. All the micelles penetrated highly permeable tumors (i.e., C26) in mice, but only the 30-nm micelles penetrated poorly permeable tumors (i.e., BxPC3).

12.6 TARGETED DELIVERY OF PG-BASED POLYMERIC MICELLES

Since most antitumor drugs are toxic not only to tumor cells but also to normal cells, tumor-specific drug delivery is highly preferred. Tumor-specific imaging agents are also favored since they allow for more accurate diagnosis [56]. Although polymeric micelles can passively accumulate in tumors via the enhanced permeation and retention effect, their tumor specificity can be further enhanced by tumor-targeting ligands. These ligands are molecules with high affinity for components that are highly expressed in tumors but not normal tissues, for example, small molecules such as folic acid, peptides such as RGD, and proteins [136–138].

Vega et al. used C225, an antibody against the epidermal growth factor receptor (EGFR), as the targeting moiety conjugated the PEG termini of PEG-*b*-(PG-*g*-doxorubicin) (PEG-*b*-(PG-*g*-Dox)) [4]. It was hypothesized, but not explicitly tested, that this block copolymer would self-assemble into micelles in aqueous solutions. C225-PEG-*b*-(PG-*g*-Dox) bound to EGFR-positive A431 cells within 5 min of co-incubation, but the control PEG-*b*-(PG-*g*-Dox) was slowly taken up via endocytosis after 24 h of coincubation. In addition, C225-PEG-*b*-(PG-*g*-Dox) co-localized with A431 nuclei, while the control did not. This contrast was not observed with the EGFR-negative cell line NIH3T3. The specific binding of C225-PEG-*b*-(PG-*g*-Dox) to A431 cells was completely blocked by free C225. In general, C225 appears to be a suitable targeting molecule to improve the tumor specificity of drug-loaded micelles.

Huang et al. prepared micelles from a block copolymer, PEG-*b*-PBLG, and tagged the PEG termini with glycyrrhetinic acid, a liver-targeting ligand [139]. Doxorubicin was loaded into micelles at loading ratios ranging from 7.1 to 7.5%. The micelle diameters were between 180 and 214 nm. Glycyrrhetinic acid-modified micelles were taken up by human hepatic carcinoma 7703 cells at 3.7 times the rate of nontargeted micelles. The targeted micelles killed 74% more 7703 cells than the nontargeted micelles did. An *in vivo* study in rats also found that the liver uptake of glycyrrhetinic acid-modified micelles was twice that of nontargeted micelles.

Recently, Yang et al. developed a PG-based polymeric vesicle capable of targeting the tumor, being visible on MRI, and delivering doxorubicin to the tumor [140]. This polymeric vesicle was loaded with superparamagnetic iron oxide (SPIO) nanoparticles and doxorubicin and was surface tagged with tumor-targeting folic acid. The polymeric vesicle was prepared from a triblock copolymer, PEG-b-PG-b-PEG; one terminal of PEG was tagged with folic acid and the other with acrylate to cross-link the micelles. The PG block was grafted with doxorubicin via an acid-sensitive hydrazone bond. Compared with nontargeted vesicles, the folic acid-conjugated vesicles had greater degrees of cellular uptake and cytotoxicity. The targeted vesicles also offered pH-dependent release of doxorubicin and greatly enhanced MRI sensitivity. However, no *in vivo* results have been reported so far for such polymeric micelles with multiple integrated functions.

12.7 CONCLUSION AND FUTURE PERSPECTIVES

In this chapter, we have systematically reviewed the synthesis, formulation, characterization, and applications for tumor imaging and drug delivery of polymeric micelles based on PG. While PG-based polymeric micelles have demonstrated strong potential utility in the fields of noninvasive imaging, tumor targeting, and drug delivery individually, a challenge for the future is the development of polymeric micelles that can achieve these multiple functions. In general, PG-based polymeric micelles with integrated targeting, diagnostic, and therapeutic functions would require only one dose of treatment to the target site. Such integration requires extensive effort and a comprehensive understanding of chemistry, medicine, biology, and imaging techniques. Meanwhile, the biodegradation and subsequent clearance of targeted micelles that have completed their missions would greatly improve the safety of polymeric micelles. It will be very exciting to see how such improvements affect future systems for imaging and drug delivery.

ACKNOWLEDGEMENT

This work was supported by the John S. Dunn Foundation. We thank Bryan Tutt for his editorial assistance.

REFERENCES

1. Kataoka K, Harada A, Nagasaki Y. Block copolymer micelles for drug delivery: design, characterization and biological significance. Adv Drug Deliv Rev 2001;47:113–131.
2. Lavasanifar A, Samuel J, Kwon GS. Poly(ethylene oxide)-block-poly(L-amino acid) micelles for drug delivery. Adv Drug Deliv Rev 2002;54:169–190.
3. Bae Y, Kataoka K. Intelligent polymeric micelles from functional poly(ethylene glycol)-poly(amino acid) block copolymers. Adv Drug Deliv Rev 2009;61:768–784.
4. Vega J, Ke S, Fan Z, et al. Targeting doxorubicin to epidermal growth factor receptors by site-specific conjugation of C225 to poly(L-glutamic acid) through a polyethylene glycol Spacer. Pharm Res 2003;20:826–832.
5. Nishiyama N, Okazaki S, Cabral H, et al. Novel cisplatin-incorporated polymeric micelles can eradicate solid tumors in mice. Cancer Res 2003;63:8977–8983.
6. Cabral H, Nishiyama N, Okazaki S, et al. Preparation and biological properties of dichloro(1,2-diaminocyclohexane)platinum(II) (DACHPt)-loaded polymeric micelles. J Control Release 2005;101:223–232.
7. Agut W, Brulet A, Taton D, et al. Thermoresponsive micelles from jeffamine-b-poly(L-glutamic acid) double hydrophilic block copolymers. Langmuir 2007;23:11526–11533.
8. Chiu HC, Kopeckova P, Deshmane SS, et al. Lysosomal degradability of poly(alpha-amino acids). J Biomed Mater Res 1997;34:381–392.
9. Zhang GD, Zhang R, Wen XX, et al. Micelles based on biodegradable poly(L-glutamic acid)-b-polylactide with paramagnetic gd ions chelated to the shell layer as a potential nanoscale IVIRI-visible delivery system. Biomacromolecules 2008;9:36–42.
10. Lin JP, Zhu JQ, Chen T, et al. Drug releasing behavior of hybrid micelles containing polypeptide triblock copolymer. Biomaterials 2009;30:108–117.
11. Deng C, Chen XS, Sun J, et al. RGD peptide grafted biodegradable amphiphilic triblock copolymer poly(glutamic acid)-b-poly(L-lactide)-b-poly(glutamic acid): synthesis and self-assembly. J Polym Sci A – Polym Chem 2007;45:3218–3230.
12. Li C, Yu DF, Newman RA, et al. Complete regression of well-established tumors using a novel water-soluble poly(L-glutamic acid) paclitaxel conjugate. Cancer Res 1998;58: 2404–2409.
13. Li C, Price JE, Milas L, et al. Antitumor activity of poly(L-glutamic acid)-paclitaxel on syngeneic and xenografted tumors. Clin Cancer Res 1999;5:891–897.

14. Kenny AD. Evaluation of sodium poly(alpha-L-glutamic acid) as a plasma expander. Exp Biol Med 1959;100:778–780.
15. Sumi H, Kawabe K, Nakajima N. Effect of various polyamino acids and D- and L-amino acids on the blood fibrinolytic system. Comp Biochem Physiol B: Biochem Mol Biol 1992;102:159–162.
16. Akamatsu K, Yamasaki Y, Nishikawa M, et al. Development of a hepatocyte-specific prostaglandin E(1) polymeric prodrug and its potential for preventing carbon tetrachloride-induced fulminant hepatitis in mice. J Pharmacol Exp Ther 1999;290:1242–1249.
17. Bayley D, Sancho MR, Brown J, et al. Soluble polymeric carriers for drug delivery - Part 6: preparation and biodistribution of N^5-hydroxyethyl-L-glutamine-co-L-glutamic acid copolymers in Rats. J Bioact Compat Polym 1993;8:51–68.
18. Miller WG. Degradation of synthetic polypeptides. II. Degradation of poly-a,L-glutamic acid by proteolytic enzymes in 0.20 M sodium chloride1. J Am Chem Soc 1964;86: 3913–3918.
19. Rodriguez-Carmona E, Villaverde A. Nanostructured bacterial materials for innovative medicines. Trends Microbiol 2010;18:423–430.
20. Buescher JM, Margaritis A. Microbial biosynthesis of polyglutamic acid biopolymer and applications in the biopharmaceutical, biomedical and food industries. Crit Rev Biotechnol 2007;27:1–19.
21. Feng ZL, Zhao G, Yu L, et al. Preclinical efficacy studies of a novel nanoparticle-based formulation of paclitaxel that out-performs Abraxane. Cancer Chemother Pharmacol 2010;65:923–930.
22. Tansey W, Ke S, Cao XY, et al. Synthesis and characterization of branched poly(L-glutamic acid) as a biodegradable drug carrier. J Control Release 2004;94:39–51.
23. Deng L, Shi K, Zhang YY, et al. Synthesis of well-defined poly(N-isopropylacrylamide)-b-poly(L-glutamic acid) by a versatile approach and micellization. J Colloid Interface Sci 2008;323:169–175.
24. Luo K, Yin JB, Song ZJ, et al. Biodegradable interpolyelectrolyte complexes based on methoxy poly(ethylene glycol)-b-poly(alpha,L-glutamic acid) and chitosan. Biomacromolecules 2008;9:2653–2661.
25. Kolb HC, Finn MG, Sharpless KB. Click chemistry: Diverse chemical function from a few good reactions. Angew Chem Int Ed Engl 2001;40:2004–2021.
26. Rao JY, Zhang YF, Zhang JY, et al. Facile preparation of well-defined AB(2) Y-shaped miktoarm star polypeptide copolymer via the combination of ring-opening polymerization and click chemistry. Biomacromolecules 2008;9:2586–2593.
27. Yang YN, Cai J, Zhuang XL, et al. pH-dependent self-assembly of amphiphilic poly(L-glutamic acid)-block-poly(lactic-co-glycolic acid) copolymers. Polymer 2010;51:2676–2682.
28. Portilla-Arias JA, Camargo B, Garcia-Alvarez M, et al. Nanoparticles made of microbial poly(gamma-glutamate)s for encapsulation and delivery of drugs and proteins. J Biomater Sci Polym Ed 2009;20:1065–1079.
29. He CL, Zhao CW, Guo XH, et al. Novel temperature- and pH-responsive graft copolymers composed of poly(L-glutamic acid) and poly(N-isopropylacrylamide). J Polym Sci A – Polym Chem 2008;46:4140–4150.
30. Riess G. Micellization of block copolymers. Prog Polym Sci 2003;28:1107–1170.
31. Motornov M, Roiter Y, Tokarev I, et al. Stimuli-responsive nanoparticles, nanogels and capsules for integrated multifunctional intelligent systems. Prog Polym Sci 2010;35:174–211.
32. Roy D, Cambre JN, Sumerlin BS. Future perspectives and recent advances in stimuli-responsive materials. Prog Polym Sci 2010;35:278–301.

33. Tyrrell ZL, Shen Y, Radosz M. Fabrication of micellar nanoparticles for drug delivery through the self-assembly of block copolymers. Prog Polym Sci 2010;35:1128–1143.
34. Kim MS, Dayananda K, Choi EY, et al. Synthesis and characterization of poly(L-glutamic acid)-block-poly(L-phenylalanine). Polymer 2009;50:2252–2257.
35. Reches M, Gazit E. Formation of closed-cage nanostructures by self-assembly of aromatic dipeptides. Nano Lett 2004;4:581–585.
36. Rodríguez-Hernández J, Babin J, Zappone B, et al. Preparation of shell cross-linked nano-objects from hybrid-peptide block copolymers. Biomacromolecules 2000;6: 2213–2220.
37. Gao G, Wang T, He J, et al. Hierarchical assembly of PEG-b-polypeptide hybrid block copolymers on graphite. Macromolecules 2007;40:2613–2619.
38. Kim G, Kim JY, Sohn D, et al. Synthesis and characterization of poly(alkyl alpha, L-glutamate-co-ethylene oxide). Macromole Res 2002;10:49–52.
39. Cabral H, Nishiyama N, Kataoka K. Optimization of (1,2-diamino-cyclohexane) platinum(II)-loaded polymeric micelles directed to improved tumor targeting and enhanced antitumor activity. J Control Release 2007;121:146–155.
40. Kabanov AV, Bronich TK, Kabanov VA, et al. Soluble stoichiometric complexes from poly(N-ethyl-4-vinylpyridinium) cations and poly(ethylene oxide)-block-polymethacry late anions. Macromolecules 1996;29:6797–6802.
41. Harada A, Kataoka K. Formation of polyion complex micelles in an aqueous milieu from a pair of oppositely-charged block copolymers with poly(ethylene glycol) segments. Macromolecules 1995;28:5294–5299.
42. Bontha S, Kabanov AV, Bronich TK. Polymer micelles with cross-linked ionic cores for delivery of anticancer drugs. J Control Release 2006;114:163–174.
43. Kim JO, Kabanov AV, Bronich TK. Polymer micelles with cross-linked polyanion core for delivery of a cationic drug doxorubicin. J Control Release 2009;138:197–204.
44. Rinaudo M. Chitin and chitosan: Properties and applications. Prog Polym Sci 2006;31: 603–632.
45. Sashiwa H, Aiba S-i. Chemically modified chitin and chitosan as biomaterials. Prog Polym Sci 2004;29:887–908.
46. Rodríguez-Hernández J, Lecommandoux S. Reversible inside-out micellization of pH-responsive and water-soluble vesicles based on polypeptide diblock copolymers. J Am Chem Soc 2005;127:2026–2027.
47. Zhang J, Pelton R, Deng Y. Temperature-dependent contact angles of water on poly(N-isopropylacrylamide) gels. Langmuir 1995;11:2301–2302.
48. Rao JY, Luo ZF, Ge ZS, et al. "Schizophrenic" micellization associated with coil-to-helix transitions based on polypeptide hybrid double hydrophilic rod-coil diblock copolymer. Biomacromolecules 2007;8:3871–3878.
49. Cai CH, Zhang LS, Lin JP, et al. Self-assembly behavior of pH- and thermosensitive amphiphilic triblock copolymers in solution: Experimental studies and self-consistent field theory simulations. J Phys Chem B 2008;112:12666–12673.
50. Rodríguez--Hernández J, Chécot F, Gnanou Y, et al. Toward "smart" nano-objects by self-assembly of block copolymers in solution. Prog Polym Sci 2005;30:691–724.
51. Balamurugan SS, Soto-Cantu E, Cueto R, et al. Preparation of organosoluble silica-polypeptide particles by "click" chemistry. Macromolecules 2010;43:62–70.
52. Sun J, Deng C, Chen XS, et al. Self-assembly of polypeptide-containing ABC-type triblock copolymers in aqueous solution and its pH dependence. Biomacromolecules 2007;8:1013–1017.

53. Chen Y, Dong CM. pH-sensitive supramolecular polypeptide-based micelles and reverse micelles mediated by hydrogen-bonding interactions or host-guest chemistry: characterization and in vitro controlled drug release. J Phys Chem B 2010;114:7461–7468.

54. Myer YP. pH-induced helix-coil transition of poly-L-lysine and poly-L-glutamic acid and 238-mμ dichroic band. Macromolecules 1969;2:624–628.

55. Kukula H, Schlaad H, Antonietti M, et al. The formation of polymer vesicles or "peptosomes" by polybutadiene-block-poly(L-glutamate)s in dilute aqueous solution. J Am Chem Soc 2002;124:1658–1663.

56. Torchilin VP. Multifunctional nanocarriers. Adv Drug Deliv Rev 2006;58:1532–1555.

57. Haller C, Hizoh I. The cytotoxicity of iodinated radiocontrast agents on renal cells in vitro. Invest Radiol 2004;39:149–154.

58. Hizoh I, Haller C. Radiocontrast-induced renal tubular cell apoptosis: hypertonic versus oxidative stress. Invest Radiol 2002;37:428–434.

59. Torchilin VP. PEG-based micelles as carriers of contrast agents for different imaging modalities. Adv Drug Deliv Rev 2002;54:235–252.

60. Maeda H, Wu J, Sawa T, et al. Tumor vascular permeability and the EPR effect in macromolecular therapeutics: a review. J Control Release 2000;65:271–284.

61. Li S-D, Huang L. Pharmacokinetics and biodistribution of nanoparticles. Mol Pharm 2008;5:496–504.

62. Kong G, Braun RD, Dewhirst MW. Hyperthermia enables tumor-specific nanoparticle delivery: effect of particle size. Cancer Res 2000;60:4440–4445.

63. Lu D, Wientjes MG, Lu Z, et al. Tumor priming enhances delivery and efficacy of nanomedicines. J Pharmacol Exp Ther 2007;322:80–88.

64. Yang Z, Leon J, Martin M, et al. Pharmacokinetics and biodistribution of near-infrared fluorescence polymeric nanoparticles. Nanotechnology 2009;20:165101 (11 pp).

65. Yang Z, Zheng SY, Harrison WJ, et al. Long-circulating near-infrared fluorescence core-cross-linked polymeric micelles: synthesis, characterization, and dual nuclear/optical imaging. Biomacromolecules 2007;8:3422–3428.

66. Wen XX, Jackson EF, Price RE, et al. Synthesis and characterization of poly(L-glutamic acid) gadolinium chelate: a new biodegradable MRI contrast agent. Bioconjug Chem 2004;15:1408–1415.

67. Gao J, Chen K, Xie R, et al. Ultrasmall near-infrared non-cadmium quantum dots for in vivo tumor imaging. Small 2010;6:256–261.

68. Guo R, Zhang L, Qian H, et al. Multifunctional nanocarriers for cell imaging, drug delivery, and near-IR photothermal therapy. Langmuir 2010;26:5428–5434.

69. Melancon MP, Wang W, Wang YT, et al. A novel method for imaging in vivo degradation of poly(L-glutamic acid), a biodegradable drug carrier. Pharm Res 2007;24:1217–1224.

70. Tan YF, Chandrasekharan P, Maity D, et al. Multimodal tumor imaging by iron oxides and quantum dots formulated in poly (lactic acid)-D-alpha-tocopheryl polyethylene glycol 1000 succinate nanoparticles. Biomaterials 2011;32:2969–2978.

71. Melancon MP, Wang YT, Wen XX, et al. Development of a macromolecular dual-modality MR-optical imaging for sentinel lymph node mapping. Invest Radiol 2007;42:569–578.

72. Melancon MP, Lu W, Huang QA, et al. Targeted imaging of tumor-associated M2 macrophages using a macromolecular contrast agent PG-Gd-NIR813. Biomaterials 2010;31: 6567–6573.

73. Arimura H, Ohya Y, Ouchi T. Formation of core-shell type biodegradable polymeric micelles from amphiphilic poly(aspartic acid)-block-polylactide diblock copolymer. Biomacromolecules 2005;6:720–725.

74. Turner JL, Pan D, Plummer R, et al. Synthesis of gadolinium-labeled shell-crosslinked nanoparticles for magnetic resonance imaging applications. Adv Funct Mater 2005;15: 1248–1254.
75. Nakamura E, Makino K, Okano T, et al. A polymeric micelle MRI contrast agent with changeable relaxivity. J Control Release 2006;114:325–333.
76. Foa R, Norton L, Seidman AD. Taxol (Paclitaxel) – a novel anti-microtubule agent with remarkable antineoplastic activity. Int J Clin Lab Res 1994;24:6–14.
77. Rowinsky EK. The development and clinical utility of the taxane class of antimicrotubule chemotherapy agents. Annu Rev Med 1997;48:353–374.
78. Rowinsky EK, Eisenhauer EA, Chaudhry V, et al. Clinical toxicities encountered with paclitaxel (Taxol(R)). Semin Oncol 1993;20:1–15.
79. Reinecke P, Schmitz M, Schneider EM, et al. Multidrug resistance phenotype and paclitaxel (Taxol (R)) sensitivity in human renal carcinoma cell lines of different histologic types. Cancer Invest 2000;18:614–625.
80. Sangrajrang S, Fellous A. Taxol resistance. Chemotherapy 2000;46:327–334.
81. Savarese DMF, Savy G, Vahdat L, et al. Prevention of chemotherapy and radiation toxicity with glutamine. Cancer Treat Rev 2003;29:501–513.
82. Li C, Wallace S. Polymer-drug conjugates: Recent development in clinical oncology. Adv Drug Deliv Rev 2008;60:886–898.
83. Dipetrillo T, Milas L, Evans D, et al. Paclitaxel poliglumex (PPX-Xyotax) and concurrent radiation for esophageal and gastric cancer: a phase I study. Am J Clin Oncol 2006;29: 376–379.
84. Veronese ML, Flaherty K, Kramer A, et al. Phase I study of the novel taxane CT-2103 in patients with advanced solid tumors. Cancer Chemother Pharmacol 2005;55:497–501.
85. Sabbatini P, Sill MW, O'Malley D, et al. A phase II trial of paclitaxel poliglumex in recurrent or persistent ovarian or primary peritoneal cancer (EOC): a gynecologic oncology group study. Gynecol Oncol 2008;111:455–460.
86. Lin NU, Parker LM, Come SE, et al. Phase II study of CT-2103 as first- or second-line chemotherapy in patients with metastatic breast cancer: unexpected incidence of hypersensitivity reactions. Invest New Drugs 2007;25:369–375.
87. Paz-Ares L, Ross H, O'Brien M, et al. Phase III trial comparing paclitaxel poliglumex vs docetaxel in the second-line treatment of non-small-cell lung cancer. Br J Cancer 2008;98:1608–1613.
88. O'Brien ME, Socinski MA, Popovich AY, et al. Randomized phase III trial comparing single-agent paclitaxel Poliglumex (CT-2103, PPX) with single-agent gemcitabine or vinorelbine for the treatment of PS 2 patients with chemotherapy-naive advanced non-small cell lung cancer. J Thorac Oncol 2008;3:728–734.
89. Langer CJ, O'Byrne KJ, Socinski MA, et al. Phase III trial comparing paclitaxel poliglumex (CT-2103, PPX) in combination with carboplatin versus standard paclitaxel and carboplatin in the treatment of PS 2 patients with chemotherapy-naive advanced non-small cell lung cancer. J Thorac Oncol 2008;3:623–630.
90. Kim SC, Kim DW, Shim YH, et al. In vivo evaluation of polymeric micellar paclitaxel formulation: toxicity and efficacy. J Control Release 2001;72:191–202.
91. Lin WJ, Juang LW, Lin CC. Stability and release performance of a series of pegylated copolymeric micelles. Pharma Res 2003;20:668–673.
92. Lee H, Zeng FQ, Dunne M, et al. Methoxy poly(ethylene glycol)-block-poly(delta-valerolactone) copolymer micelles for formulation of hydrophobic drugs. Biomacromolecules 2005;6:3119–3128.

93. Cho YW, Lee J, Lee SC, et al. Hydrotropic agents for study of in vitro paclitaxel release from polymeric micelles. J Control Release 2004;97:249–257.

94. Hamaguchi T, Kato K, Yasui H, et al. A phase I and pharmacokinetic study of NK105, a paclitaxel-incorporating micellar nanoparticle formulation. Br J Cancer 2007;97: 170–176.

95. Wang YZ, Yu L, Han LM, et al. Difunctional Pluronic copolymer micelles for paclitaxel delivery: synergistic effect of folate-mediated targeting and Pluronic-mediated overcoming multidrug resistance in tumor cell lines. Int J Pharm 2007;337:63–73.

96. Ould-Ouali L, Arien A, Rosenblatt J, et al. Biodegradable self-assembling PEG-copolymer as vehicle for poorly water-soluble drugs. Pharm Res 2004;21:1581–1590.

97. Liu JB, Zeng FQ, Allen C. Influence of serum protein on polycarbonate-based copolymer micelles as a delivery system for a hydrophobic anti-cancer agent. J Control Release 2005;103:481–497.

98. Sugin Z, Yuksel N, Baykara T. Preparation and characterization of polymeric micelles for solubilization of poorly soluble anticancer drugs. Eur J Pharm Biopharm 2006;64: 261–268.

99. Han JK, Kim MS, Lee DS, et al. Evaluation of the anti-tumor effects of paclitaxel-encapsulated pH-sensitive micelles. Macromol Res 2009;17:99–103.

100. Le Garrec D, Gori S, Karkan D, et al. Preparation, characterization, cytotoxicity and biodistribution of docetaxel-loaded polymeric micelle formulations. J Drug Deliv Sci Tech 2005;15:115–120.

101. Liu SQ, Tong YW, Yang YY. Thermally sensitive micelles self-assembled from poly(N-isopropylacrylamide-co-N,N-dimethylacrylamide)-b-poly(D,L-lactide-co-glycolide) for controlled delivers of paclitaxel. Mol Biosyst 2005;1:158–165.

102. Soga O, van Nostrum CF, Fens M, et al. Thermosensitive and biodegradable polymeric micelles for paclitaxel delivery. J Control Release 2005;103:341–353.

103. Seow WY, Xue JM, Yang YY. Targeted and intracellular delivery of paclitaxel using multi-functional polymeric micelles. Biomaterials 2007;28:1730–1740.

104. Liu BR, Yang M, Li XL, et al. Enhanced efficiency of thermally targeted taxanes delivery in a human xenograft model of gastric cancer. J Pharm Sci 2008;97:3170–3181.

105. You JO, Auguste DT. Feedback-regulated paclitaxel delivery based on poly(N,N-dimethylaminoethyl methacrylate-co-2-hydroxyethyl methacrylate) nanoparticles. Biomaterials 2008;29:1950–1957.

106. Zhang LY, Guo R, Yang M, et al. Thermo and pH dual-responsive nanoparticles for anti-cancer drug delivery. Adv Mater 2007;19:2988–2992.

107. Kim JH, Kim YS, Kim S, et al. Hydrophobically modified glycol chitosan nanoparticles as carriers for paclitaxel. J Control Release 2006;111:228–234. (Reprinted from Journal of Controlled Release, vol 109, pg 1, 2005).

108. Kim DG, Jeong YI, Nah JW. All-trans retinoic acid release from polyion-complex micelles of methoxy poly(ethylene glycol) grafted chitosan. J Appl Polym Sci 2007;105:3246–3254.

109. You J, Hu FQ, Du YZ, et al. Polymeric micelles with glycolipid-like structure and multiple hydrophobic domains for mediating molecular target delivery of paclitaxel. Biomacromolecules 2007;8:2450–2456.

110. Xie Z, Guan HL, Chen X, et al. A novel polymer-paclitaxel conjugate based on amphiphilic triblock copolymer. J Control Release 2007;117:210–216.

111. Alani AWG, Bae Y, Rao DA, et al. Polymeric micelles for the pH-dependent controlled, continuous low dose release of paclitaxel. Biomaterials 2010;31:1765–1772.

112. Kataoka K, Matsumoto T, Yokoyama M, et al. Doxorubicin-loaded poly(ethylene glycol)-poly(beta-benzyl-L-aspartate) copolymer micelles: their pharmaceutical characteristics and biological significance. J Control Release 2000;64:143–153.

113. Janssen MJH, Crommelin DJA, Storm G, et al. Doxorubicin decomposition on storage – effect of pH, type of buffer and liposome encapsulation. Int J Pharm 1985;23:1–11.

114. Beijnen JH, Vanderhouwen OAGJ, Underberg WJM. Aspects of the degradation kinetics of doxorubicin in aqueous-solution. Int J Pharm 1986;32:123–131.

115. Hoes CJT, Potman W, van Heeswijk WAR, et al. Optimization of macromolecular prodrugs of the antitumor antibiotic adriamycin. J Control Release 1985;2:205–213.

116. Krinick NL, Sun Y, Joyner D, et al. A polymeric drug-delivery system for the simultaneous delivery of drugs activatable by enzymes and or light. J Biomater Sci Polym Ed 1994;5:303–324.

117. Ulbrich K, Šubr V, Strohalm J, et al. Polymeric drugs based on conjugates of synthetic and natural macromolecules: I. Synthesis and physico-chemical characterisation. J Control Release 2000;64:63–79.

118. Hoes CJT, Grootoonk J, Duncan R, et al. Biological properties of adriamycin bound to biodegradable polymeric carriers. J Control Release 1993;23:37–53.

119. Guan H, McGuire MJ, Li S, et al. Peptide-targeted polyglutamic acid doxorubicin conjugates for the treatment of alpha(v)beta(6)-positive cancers. Bioconjug Chem 2008;19:1813–1821.

120. Sawant RM, Hurley JP, Salmaso S, et al. "SMART" drug delivery systems: double-targeted pH-responsive pharmaceutical nanocarriers. Bioconjug Chem 2006;17:943–949.

121. Jeong Y-I, Nah J-W, Lee H-C, et al. Adriamycin release from flower-type polymeric micelle based on star-block copolymer composed of poly(γ-benzyl L-glutamate) as the hydrophobic part and poly(ethylene oxide) as the hydrophilic part. Int J Pharm 1999;188:49–58.

122. Na HS, Lim YK, Jeong Y-I, et al. Combination antitumor effects of micelle-loaded anticancer drugs in a CT-26 murine colorectal carcinoma model. Int J Pharm 2010;383: 192–200.

123. Sanson C, Schatz C, Le Meins JF, et al. A simple method to achieve high doxorubicin loading in biodegradable polymersomes. J Control Release 2010;147:428–435.

124. Seow WY, Yang YY. Functional polycarbonates and their self-assemblies as promising non-viral vectors. J Control Release 2009;139:40–47.

125. Zhang Z, Kuijer R, Bulstra SK, et al. The in vivo and in vitro degradation behavior of poly(trimethylene carbonate). Biomaterials 2006;27:1741–1748.

126. Lebwohl D, Canetta R. Clinical development of platinum complexes in cancer therapy: an historical perspective and an update. Eur J Cancer 1998;34:1522–1534.

127. Timmerbosscha H, Mulder NH, Devries EGE. Modulation of cis-diamminedichloroplatinum(Ii) resistance – a review. Br J Cancer 1992;66:227–238.

128. Lieberthal W, Triaca V, Levine J. Mechanisms of death induced by cisplatin in proximal tubular epithelial cells: apoptosis vs. necrosis. Am J Physiol 1996;270:F700–F708.

129. Ries F, Klastersky J. Nephrotoxicity induced by cancer chemotherapy with special emphasis on cisplatin toxicity. Am J Kidney Dis 1986;8:368–379.

130. Kidani Y, Inagaki K, Iigo M, et al. Antitumor activity of 1,2-diaminocyclohexane--platinum complexes against sarcoma-180 ascites form. J Med Chem 1978;21:1315–1318.

131. Graham MA, Lockwood GF, Greenslade D, et al. Clinical pharmacokinetics of oxaliplatin: a critical review. Clin Cancer Res 2000;6:1205–1218.

132. Uchino H, Matsumura Y, Negishi T, et al. Cisplatin-incorporating polymeric micelles (NC-6004) can reduce nephrotoxicity and neurotoxicity of cisplatin in rats. Br J Cancer 2005;93:678–687.

133. Liu J, Kraut E, Bender J, et al. Pharmacokinetics of oxaliplatin (NSC 266046) alone and in combination with paclitaxel in cancer patients. Cancer Chemother Pharmacol 2002;49:367–374.

134. Massari C, Brienza S, Rotarski M, et al. Pharmacokinetics of oxaliplatin in patients with normal versus impaired renal function. Cancer Chemother Pharmacol 2000;45:157–164.

135. Cabral H, Matsumoto Y, Mizuno K, et al. Accumulation of sub-100 nm polymeric micelles in poorly permeable tumours depends on size. Nat Nanotechnol 2011;6:815–823.

136. Kim D, Lee ES, Park K, et al. Doxorubicin loaded pH-sensitive micelle: antitumoral efficacy against ovarian A2780/DOXR tumor. Pharma Res 2008;25:2074–2082.

137. Nasongkla N, Bey E, Ren JM, et al. Multifunctional polymeric micelles as cancer-targeted, MRI-ultrasensitive drug delivery systems. Nano Lett 2006;6:2427–2430.

138. Nasongkla N, Shuai X, Ai H, et al. cRGD-functionalized polymer micelles for targeted doxorubicin delivery. Angew Chem Int Ed Engl 2004;43:6323–6327.

139. Huang W, Wang W, Wang P, et al. Glycyrrhetinic acid-modified poly(ethylene glycol)-b-poly(γ-benzyl L-glutamate) micelles for liver targeting therapy. Acta Biomater 2010;6:3927–3935.

140. Yang X, Grailer JJ, Rowland IJ, et al. Multifunctional stable and pH-responsive polymer vesicles formed by heterofunctional triblock copolymer for targeted anticancer drug delivery and ultrasensitive MR imaging. ACS Nano 2010;4:6805–6817.

CHAPTER 13

PERFLUOROCARBON NANOPARTICLES: TRANSLATING BENCH OPPORTUNITIES TO THE CLINIC

G. M. LANZA, P. M. WINTER, S. D. CARUTHERS, A. H. SCHMIEDER, AND S. A. WICKLINE

13.1 RECOGNIZING AND ADDRESSING THE UNMET MEDICAL CHALLENGES

The first step in a translational research program is the identification of an intractable clinical problem worthy of the effort and resources that will be required. Understanding the practice of medicine helps to recognize these problems and aids in appreciating the clinical logistical boundaries that constrain the potential solutions. Beyond the critical issues of efficacy, safety (i.e., absorption, distribution, metabolism, and excretion (ADME)), and CMC (i.e., chemistry, manufacturing, and controls), infrastructural barriers such as cost, patient compliance, and the logistics of clinical use must be projected. Even the organizational structure of hospitals, which often dictates billing and institutional decisions, may become a barrier to new technologies if one department is poised to infringe on traditional "turf" of another. All questions cannot be solved at the outset, but a considered plan should recognize, prioritize, and resolve as many of the outstanding challenges as can be defined.

Although obvious challenges like "curing cancer" or "preventing heart attacks" can be the grand motivating themes, real progress always occurs incrementally through stepwise lesser achievements. Even small steps may constitute significant clinical advancement, as these accomplishments are always accompanied by new perspectives and deeper understandings about the project. Clinical translation of technology depends on successfully controlling the risk and setbacks. The old adage "crawl before walking and walk before running" is appropriately applied to biomedical research and reflects the need for a detailed logical development plan.

Perhaps the most important and expensive element in a clinical translation program is time. Timing relates to the limited longevity of intellectual property protection,

Drug Delivery Applications of Noninvasive Imaging: Validation from Biodistribution to Sites of Action, First Edition. Edited by Chun Li and Mei Tian.

the years for technology development and regulatory approval, and ultimately the expected market life as a proprietary product. Moreover, all innovations have discrete opportunity windows, which may be shortened by the emergence of alternative approaches or changing market needs.

Clinical translation of complicated new technologies is demanding, particularly for pharmaceutical development. Inefficient research and development due to inexperience or a lack of focus can lead to delays, unfruitful resource expenditures, and failure. Academic research is noted for the creative exploration of ideas often with complicated techniques, unusual materials, and high expense. For translational development, concepts must be simplified, the materials and processes must be pharmaceutically acceptable, and product candidates must have defined purity, potency, and proper dosages that achieve the desired benefit efficaciously and safely. Given the complexity and rigor required to generate new products, medical technology development requires an interdisciplinary team, preferably reinforced with an experienced private industry partner committed to completing the regulatory studies and marketing the product outcome. While academics often lead the early discovery and research with novel concepts and compelling preliminary data, the inclusion of experienced partners can focus select aspects of the research for commercial development and transition a concept into a bona fide product candidate. This approach essentially partners the inventiveness of academia with the creative development of industry, and this was the path pursued by Washington University, Kereos Inc., and Philips Healthcare for the development of targeted perfluorocarbon (PFC) nanoparticles, which are currently in phase I studies in the United States for magnetic resonance (MR) molecular imaging of angiogenesis in atherosclerosis with further plans to expand into other diseases. Upon this foundation of clinical safety and efficacy will be the rollout of various therapeutic products based on the same platform. The development of PFC nanoparticles has been an incremental but expanding plan of technology development intended to bring a family of product options to patients with intractable diseases.

13.2 PERFLUOROCARBON NANOPARTICLES: BACKGROUND

Perfluorocarbon nanoparticles are a multifunctional platform technology with versatile imaging and drug delivery potential demonstrated in a variety of preclinical animal models (Fig. 13.1). They are comprised of a phospholipid monolayer encapsulated around a PFC core (98% v/v). PFC nanoparticles differ distinctly from typical oil-in-water (O/W) emulsions by virtue of the physical–chemical properties of fluorine, the most electronegative of all elements. Fluorine has a high ionization potential and very low polarizability [1]. Larger than hydrogen, fluorine creates bulkier, stiffer compounds that typically adopt a helical conformation. The carbon–fluorine (C–F) bond is chemically and thermally stable and essentially biologically inert due to an encroachment barrier established by the dense electron cloud of fluorine in the chain [1]. Interestingly, PFCs are extremely hydrophobic and lipophobic simultaneously.

The biocompatibility of liquid fluorocarbons is well documented [1]. Even at large doses, most fluorocarbons are innocuous and physiologically inactive. No toxicity, carcinogenicity, mutagenicity, or teratogenic effects have been reported for pure fluorocarbons within the 460–520 MW range. PFCs have tissue half-life residencies ranging from

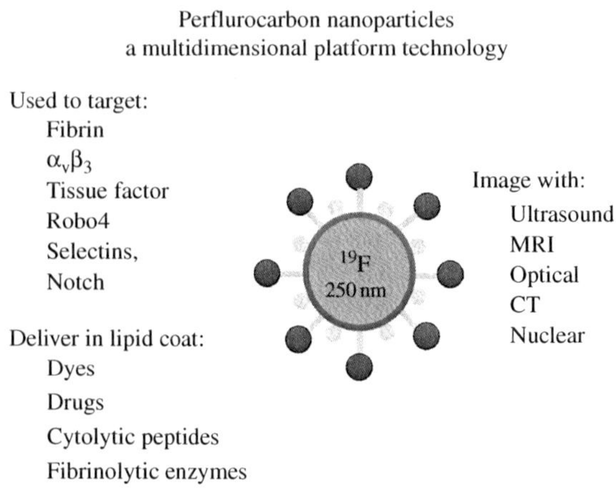

FIGURE 13.1 PFC nanoparticles are a multidimensional platform technology that has been functionalized for ultrasound, MRI, optical, and CT applications targeted against a variety of biosignatures incorporating a diverse group of payloads into the lipid coat.

4 days for perfluorooctylbromide up to 65 days for perfluorotripropylamine and are not metabolized but rather become slowly reintroduced from clearance tissues into the circulation in dissolved form by lipid carriers and ultimately expelled through the lungs [1].

Although the nominal size of PFC nanoparticles can be manipulated by chemical processing technique, the composition of the surfactant, and the PFC-to-surfactant ratio, the typical PFC nanoparticle size is ~200–300 nm, which confines their distribution to the vasculature. PFC nanoparticles have been considered for many biomedical applications [2–7], most notably as artificial blood substitutes. Unfortunately, the oxygen loading capacity of PFCs is linearly related to the partial pressure of oxygen in equilibrium with the emulsion. Thus, for almost any given partial pressure of oxygen, hemoglobin binds significantly more oxygen than can be dissolved in PFC. Moreover, PFCs demonstrate a nearly flat, linear oxygen dissociation curve as opposed to the sigmoidal dissociation curve of hemoglobin [8]. As a result, most of the oxygen bound to PFCs is released in the high-pressure atmosphere of the arteries, with little bound oxygen available for the capillary network where the partial pressure of oxygen is lower and the need is greatest.

The potential of PFCs for imaging was first studied by Mattrey and colleagues for blood pool diagnostic applications with ultrasound (US), MR, and computed tomography (CT) [9–12]. A distillation of this body of work reveals that the effectiveness of PFC emulsions for blood pool US and CT imaging was limited by a relatively weak contrast effect versus microbubble or iodinated contrasts, respectively. For MRI, ingestion of neat PFC provided excellent luminal negative contrast for gastrointestinal studies, but the added benefit to the patient was offset by higher material costs relative to the inexpensive barium-based agents.

13.3 IMAGE-GUIDED DRUG DELIVERY WITH PERFLUOROCARBON NANOPARTICLES

The concept of targeted drug delivery to improve the safety and efficacy of pharmaceuticals by optimizing therapeutic release at the site of disease is not a new concept. Implementation of this notion has been slowly evolving but is now a clinical reality.

Increasing dissatisfaction within the medical community with the inadequacy of our "one-size-fits-all" treatment paradigms has driven the discovery of new imaging-based technologies to biochemically characterize pathology noninvasively to enable a more personalized strategy for medical management. Today treatment protocols are based on the type and stage of disease, the presence of comorbidities, and likely side effects of the management approach. Unfortunately, few evidence-based options are available and the same blunt treatment approaches are employed without optimization for individual differences.

Research toward a newer approach to personalized targeted treatment of disease has been evolving, and the role of imaging for patient stratification of targeted therapy has become integral to this progress. Imaging offers the potential to noninvasively detect and biochemically characterize disease earlier, which is expected to lead to better risk stratification and more finely tuned patient management options. Molecular imaging agents combined with drugs can home to distinctive tumor biomarkers and deliver higher local levels of therapy with lower total body exposure. Moreover, the drug dose achieved at the target can be quantitatively confirmed and the adequacy of therapeutic response predicted. Imaging provides the essential tool that allows us to assess treatment response based on early biochemical changes, which enables timely re-intervention before clinical manifestations recrudesce, if needed. Although great technological progress has been made in recent years, the clinical implementation of image-guided drug delivery (IGDD) strategies has been rather slow.

The term theranostic is now commonly used to describe a nanotechnology platform designed for diagnostic molecular imaging in combination with specific delivery and release of drug into those targeted microanatomical sites. In most situations, the targeted imaging and therapeutic aspects of these platforms are being clinically developed independently. This incremental approach simplifies the clinical evaluation and approval process, by reducing the number of variables studied at one time.

Complementary to molecular imaging of angiogenesis with PFC nanoparticles in cancer and atherosclerosis [13–17], we have developed a novel approach to targeted drug delivery and termed it "contact facilitated drug delivery" [18]. Contact-facilitated drug delivery refers to the (ATP-independent) transfer of lipid surfactant components into the outer leaflet of the target cell membrane through a hemifusional complex [18–20]. It is an event that depends upon the persistent interaction of the lipid-encapsulated nanoparticle with the target cell membrane surface. In effect, theranostic PFC nanoparticles deliver a "kiss of death" without the typical requisite cellular internalization and subsequent drug payload escape from the endosomal compartment [19, 20]. Chemotherapeutics, such as doxorubicin [18], rapamycin [21], and fumagillin [22–25], as well as toxic peptides, such as melittin [26], and thrombolytic enzymes [27], have been effectively incorporated into the surfactant of PFC nanoparticles.

With respect to angiogenesis, fumagillin, a mycotoxin produced by *Aspergillus fumigatus*, suppresses endothelial proliferation by inhibition of methionine aminopeptidase 2 (MetAP2) [28, 29]. The antiangiogenic effectiveness of TNP-470, a water-soluble functional analogue of fumagillin, was widely demonstrated in rodents [30–33] and later studied in human clinical trials [34–38]. Anecdotal cases of complete remission of cervical cancer, regression of breast cancer, and stabilization of disease in patients with sarcoma, melanoma, and adenocarcinoma were reported. Unfortunately, TNP-470 has significant liabilities including a brief systemic half-life for itself (2 min) and its primary metabolite AGM-1883 (6 min) [34], and the dosage required to elicit therapeutic effects also induces sudden moderately severe symptoms of neurotoxicity, including weakness, nystagmus, diplopia, and ataxia [34, 35, 38]. As shown later in preclinical models of cancer and atherosclerosis, fumagillin incorporated into $\alpha_v\beta_3$-targeted PFC nanoparticle surfactant and delivered at less than a 10,000th of the dose applied clinically suppressed neovascular proliferation with no significant adverse effects in rabbits and rodents [22–25].

13.3.1 $\alpha_v\beta_3$-Targeted Fumagillin Nanoparticles in Early Atherosclerosis

$\alpha_v\beta_3$-targeted paramagnetic nanoparticles incorporating minute dosages of fumagillin were shown to diminish the development of neovasculature and to reduce V×2 tumor growth in rabbits [25]. Neither nontargeted fumagillin nanoparticles nor $\alpha_v\beta_3$-targeted nanoparticles without drug reduced angiogenesis or diminished tumor growth. Quantitative MR molecular imaging with $\alpha_v\beta_3$-targeted paramagnetic nanoparticles further revealed that the neovasculature was distributed predominately in the peripheral aspects of the tumor accounting for 7% of that volume. These measures were microscopically corroborated by the co-localization of $\alpha_v\beta_3$-targeted rhodamine nanoparticles with FITC-lectin exclusively in or adjacent to the tumor capsule (Fig. 13.2).

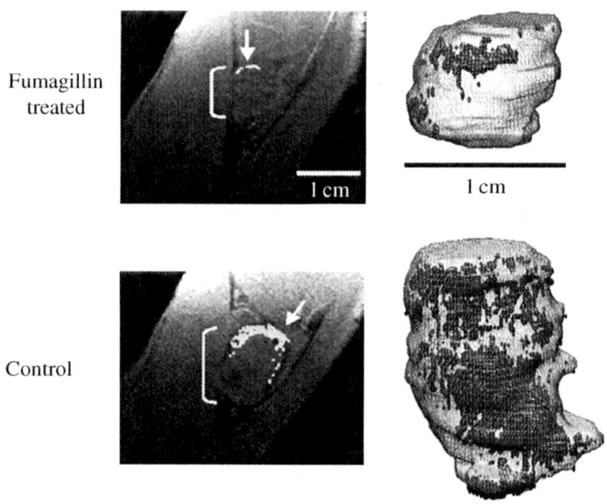

FIGURE 13.2 Diminished $\alpha_v\beta_3$-integrin contrast enhancement in T_1-weighted, fat-suppressed, 3D gradient-echo MR single slice images (250 × 250 μm, 500 μm slices, TR/TE = 40/5.6 ms, 65° flip angle, 1.5 T) in rabbits administered $\alpha_v\beta_3$-targeted fumagillin nanoparticles (top) versus those given $\alpha_v\beta_3$-targeted nanoparticles without drug (bottom). **Left:** Enhancing pixels, color coded in yellow (arrows), demonstrate sparse areas of angiogenesis in fumagillin-treated animal (top). **Right:** 3D neovascular maps of example V×2 tumors on day 16 following $\alpha_v\beta_3$-targeted fumagillin nanoparticles (top) versus $\alpha_v\beta_3$-nanoparticles without drug (bottom). Note the asymmetric distribution of angiogenic signal (color coded in blue) over the tumor surface in both the control and treated animals. Neovessel dense islands and the interspersed fine network of angiogenic proliferation over the tumor surface are diminished in rabbits receiving the targeted fumagillin treatment. Reproduced with permission from [25]. For color details, please see color plate section.

High-resolution three-dimensional (3D) neovascular maps illustrated the coherent asymmetric expression of angiogenesis as a few confluent regions of high-density neovascularity with an interspersed reticular network of enhancing voxels. Large regions of the tumor surface and the interior were devoid of $\alpha_v\beta_3$-targeted paramagnetic nanoparticle signal enhancement. Both the islands of dense neovasculature and the intervening network of neovessel formation were diminished by treatment with $\alpha_v\beta_3$-targeted fumagillin nanoparticles. Interestingly, non-necrotic regions of the tumor parenchyma were associated with increased accumulation of intratumoral cytotoxic lymphocytes following $\alpha_v\beta_3$-targeted fumagillin nanoparticle treatment versus controls.

In follow-up, the utility of $\alpha_5\beta_1$-integrin as a biomarker for angiogenesis in cancer was studied in the MDA-MB-435 xenograft mouse model [23]. $\alpha_5\beta_1$-integrin, like $\alpha_v\beta_3$-integrin, is an important adhesion molecule, which regulates endothelial cell migration and survival during neovascularization [39]. $\alpha_5\beta_1$-integrin is poorly expressed on normal quiescent blood vessels, but its expression is induced on tumor blood vessels and in response to angiogenic factors [40] including basic fibroblast growth factor, interleukin-8, tumor necrosis factor-α, and the angiomatrix protein Del-1. While the role of $\alpha_v\beta_3$-integrin is well documented in aggressive melanoma and breast cancer metastasis, $\alpha_5\beta_1$-integrin is frequently associated with tumor of lower malignant potential in addition to aggressive carcinomas.

The molecular imaging and antiangiogenic effects of fumagillin nanoparticles targeted by a dual $\alpha_5\beta_1(\alpha_v\beta_3)$-peptidomimetic versus the $\alpha_v\beta_3$-peptidomimetic alone were investigated using 3D MR neovascular mapping. Like the $\alpha_v\beta_3$-peptidomimetic described earlier, the $\alpha_5\beta_1\alpha_v\beta_3$-integrin antagonist was a quinolone nonpeptide (PCT/US1998/012010) with an IC_{50} of 390 pM for $\alpha_5\beta_1$and 87 nM for $\alpha_v\beta_3$ and no effective cross-reactivity with $\alpha_v\beta_5$ or GPIIb/IIIa. 3D reconstructions of the MR signal enhancement revealed the sparse spatial organization of neovessels, which predominantly were confined to the tumor periphery. $\alpha_5\beta_1(\alpha_v\beta_3)$-targeted nanoparticles containing fumagillin further decreased the already small amount of neovessel enhancement to an almost negligible level, but in contradistinction to the marked decrease in tumor volume in the V×2 model, fumagillin antiangiogenesis did not affect tumor volume in the MDA 435 animals. The effectiveness of targeted antiangiogenic therapy logically depends on the adequate expression of the neovascular biomarker and the dependence of tumor growth on the neovasculature. The MDA-MB-435 tumor model had a very sparse neovasculature, was relatively slow growing, and did not display a marked "angiogenic switch" phenotype as opposed to the V×2. From the clinical perspective, MR angiogenic characterization of small tumors, particularly breast, lung, or colon, may better stratify patients into medical strategies involving anti-VEGF treatment dependent on the prominence of neovasculature. Perhaps, such a patient enrichment approach would yield more cost-effective benefits with less unnecessary exposure to off-target risk.

13.3.2 $\alpha_v\beta_3$-Targeted Fumagillin Nanoparticles in Early Atherosclerosis

The effectiveness of $\alpha_v\beta_3$-targeted paramagnetic fumagillin nanoparticles in early atherosclerosis was demonstrated in the hyperlipidemic rabbit model (Fig. 13.3). In that experiment [22], three paramagnetic nanoparticle formulations (i.e., $\alpha_v\beta_3$-targeted

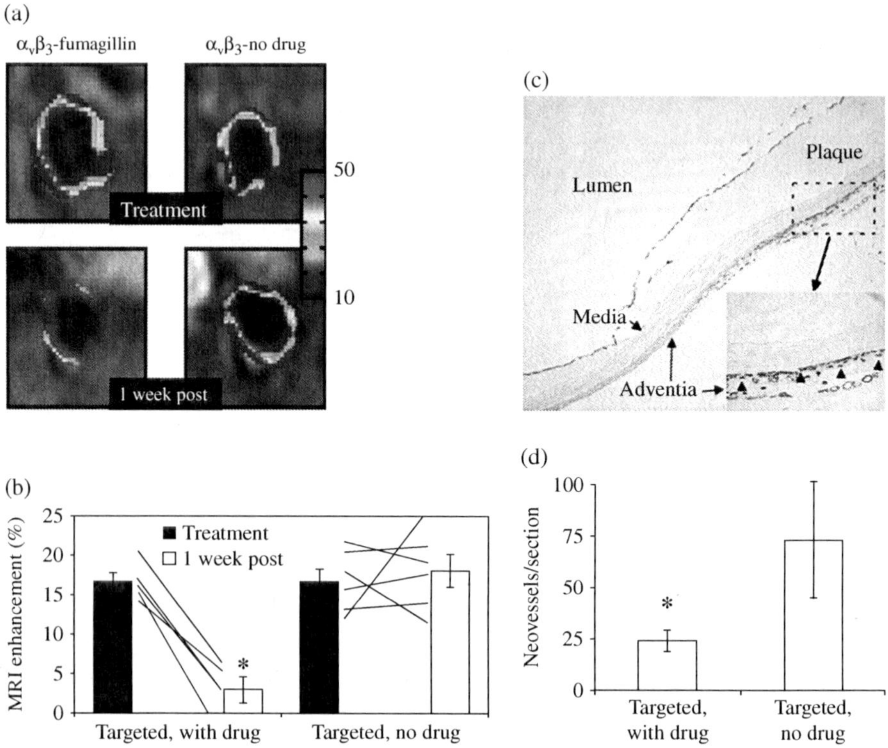

FIGURE 13.3 (a) MRI of abdominal aorta showing false-colored overlay of percent signal enhancement at the time of treatment (top) and one week post treatment (bottom). (b) Aortic MRI signal enhancement averaged over all imaged slices at the time of treatment (filled bars) and 1 week after treatment (open bars). Rabbits treated with $\alpha_v\beta_3$-targeted fumagillin nanoparticles had lower angiogenesis at 1 week compared with nanoparticles without fumagillin ($*p < 0.05$). The solid lines indicate individual animal response to treatment over the 7-day period. (c) PECAM-stained section (4×) of abdominal aorta from a hyperlipidemic rabbit displaying very early atherosclerotic changes in the adventitia, media, and plaque showing adventitial vasa vasorum expansion exclusively in region of plaque. Higher magnification inset (20×) shows that microvessels were predominantly located in the adventitia associated with thickening neointima. Arrowheads illustrate the type of PECAM microvessels counted within each section to assess fumagillin antiangiogenic effects. (d) The number of neovascular vessels within the adventitia was reduced ($*p < 0.05$) in fumagillin-treated rabbits over the proximal half of the aorta (i.e., renal artery to diaphragm), which correlated with the region of greatest MR signal and fumagillin response in the imaging studies. Reproduced with permission from [22]. For color details, please see color plate section.

fumagillin nanoparticles, $\alpha_v\beta_3$-targeted nanoparticles without fumagillin, and nontargeted fumagillin nanoparticles) were studied. During baseline scanning, T1-weighted black blood images showed no gross evidence of significant plaque development within the wall of the abdominal aorta in terms of luminal narrowing or wall thickening compared to previous experiments utilizing age-matched, nonatherosclerotic rabbits. As noted previously, MRI signal enhancement from the aortic walls of atherosclerotic rabbits injected with $\alpha_v\beta_3$-targeted nanoparticles, both with and without fumagillin, displayed a patchy distribution, with generally higher levels of angiogenesis near the diaphragm. MRI signal enhancement averaged over all imaged slices was identical for $\alpha_v\beta_3$-targeted nanoparticles with and without fumagillin.

One week after nanoparticle treatment, the residual angiogenic activity within the aortic wall was reassessed. The aortic wall signal intensity at baseline was identical for all groups at the time of treatment and at the 1-week follow-up, confirming that the paramagnetic nanoparticles administered 1 week prior were undetectable. MRI enhancement from angiogenic vasculature 1 week following $\alpha_v\beta_3$-targeted fumagillin nanoparticle treatment was markedly reduced in both spatial distribution and signal intensity. By comparison, the MR signal enhancement 1 week after treatment with $\alpha_v\beta_3$-targeted nanoparticles lacking fumagillin was undiminished. Treatment with nontargeted fumagillin nanoparticles did not significantly inhibit angiogenesis compared to $\alpha_v\beta_3$-targeted nanoparticles without fumagillin.

Abdominal aorta sections obtained 1 week after nanoparticle treatment revealed mild, heterogeneously distributed intimal thickening consistent with very early atherosclerotic neointimal development in all rabbits. The vast majority of neovessels within the aortic wall were located in the adventitia opposite regions of thickening intimal plaque. The total number of PECAM-positive microvessels per section when averaged across all aortic slices was greater ($p<0.05$) in control rabbits than in those given $\alpha_v\beta_3$-targeted fumagillin nanoparticles, particularly in the proximal half of the abdominal aorta, which displayed more prominent disease.

Next, the duration of a single dose of $\alpha_v\beta_3$-targeted fumagillin nanoparticles was examined as a function of fumagillin dose (30 μg/kg vs. 90 μg/kg) in animals continued and withdrawn from cholesterol diets following the start of treatment. At baseline, the average percent signal enhancement from $\alpha_v\beta_3$-targeted paramagnetic nanoparticles did not differ between treatment groups and was similar to earlier studies [24]. Again, $\alpha_v\beta_3$-targeted paramagnetic contrast was diffusely distributed across and within slices, and the enhancement calculated represented the average of all aortic voxels rather than a threshold subset. MR neovascular signal did not differ between fat-fed animals switched to standard chow and those continuing on the cholesterol-enriched diet over the 4-week study, which reflected the marked and persistent serum cholesterol elevation and fatty liver disease observed in all animals continuing throughout the study.

As noted previously, angiogenesis contrast among animals receiving $\alpha_v\beta_3$-targeted fumagillin nanoparticles was decreased by ~50–75% relative to the control animals ($p<0.05$). The decreased neovascular signal persisted through week 2. On week 3, aortic angiogenesis remained less than the control ($p<0.05$) in both dietary groups, but the difference was smaller, indicating that the effect of the single dose of fumagillin was waning. By week 4, there was no difference in the MR neovasculature signal between control and fumagillin-treated rabbits regardless of diet or fumagillin dose from 30 to 90 μg/ml.

HMG-CoA reductase inhibitors have been proven to reduce cardiovascular risk [41], but their benefits may exceed the primary lipid-lowering effects. The pleiotropic benefits of statins have been attributed to antioxidant effects [42] diminished leukocyte–endothelial cell adhesion [43], attenuated macrophage activation and cytokine release [44], increased endothelial nitric oxide activity [45], and atherosclerotic plaque stabilization.[46] Pathological data from excised carotid arteries of patients treated for 3 months with statins have revealed a reduction in microvascular density, which was proposed as an explanation for the additional benefit of statins [47].

The previous studies in hyperlipidemic rabbits demonstrated that $\alpha_v\beta_3$-targeted paramagnetic nanoparticles specifically home to angiogenic neovasculature in atherosclerotic aorta and deliver effective fumagillin therapy persisting for 3 weeks before recrudescing. The next but critical issue to be addressed was how to utilize and maintain the acute antiangiogenic benefits of fumagillin chronically in a clinically relevant care plan. The objective of this study was to determine whether the weak antiangiogenic response of statins would be adequate to sustain the acute antineovascular response of targeted fumagillin.

In a complicated 8-week experiment, the initial 4-week angiogenic response profile at baseline and following $\alpha_v\beta_3$-targeted fumagillin or control PFC nanoparticles did not differ from the previous 4-week experiment [24]. The neovascular integrated aortic signal of rabbits given atorvastatin alone did not differ from the nonstatin control counterparts. Animals receiving targeted fumagillin alone or in conjunction with oral statin had the same acute decrease in neovasculature at week 1, which returned to baseline levels at week 4.

However, following reinjection of targeted fumagillin in combination with continued dietary atorvastatin, the recrudescence of angiogenesis 4 weeks post treatment was not observed and a synergistic persistent antiangiogenesis effect, equivalent in magnitude to the 2-week response level, was observed. Dietary statin alone had no effect on angiogenesis in the second 4-week period (weeks 4–8 of study) nor did repeat treatment with fumagillin nanoparticles gain carry-over benefit from the prior treatment cycle. While atorvastatin alone exerted no significant effect on plaque angiogenesis in this short study, it is clear that dietary statin therapy synergistically extended the acute antineovascular benefits of $\alpha_v\beta_3^-$-targeted fumagillin PFC nanoparticles.

13.4 CONCLUSION: BENCH TO THE CLINIC

As described earlier, PFC nanoparticles are a versatile technology for multiple imaging modalities and for targeted drug delivery. In translating from these preclinical experiments to the clinic, a single indication must be presented to the FDA and tried in humans. Ultimately, molecular imaging of angiogenesis in carotids was selected as a means to identify patients with high stroke risk and only moderate disease (<70% stenosis). Preclinical data, both our own and in a growing literature, confirmed the critical need and relevance. Given the prevalence of disease, patient selection to validate the concept of homing and imaging during phase I in combination with safety was feasible. Further, carotid endarterectomy specimens could be available in some patients for pathological corroboration of targeted molecular imaging. We speculated that a phase II outcome trial in stroke prevention could be conducted with a reasonable number of patients over 3 years or less. Following a successful phase I study in atherosclerosis patients, molecular imaging of angiogenesis in cancer could be extended with a second IND. As the molecular imaging with MR rolled out in the clinic, providing an experience database in atherosclerosis and cancer, a second phase antiangiogenic agent development program using the same nanoplatform and targeting but with the new drug could be introduced perhaps first for atherosclerosis and then cancer. Moreover, we anticipated using the imaging agent, which would then be in phase II, to quantify the early antiangiogenic response, since morphological changes in plaques or tumors are to be expected to lag treatment consider ably.

As previously discussed, translation from the bench to the clinic requires a well-considered plan of incremental steps. Although this chapter provides only a flavor of the opportunities, issues, and decisions involved in PFC nanoparticle development, the considerations for almost any theranostic agent will follow a similar path.

REFERENCES

1. Krafft M. Fluorocarbons and fluorinated amphiphiles in drug delivery and biomedical research. Adv Drug Del Rev 2001;47:209–228.
2. Geyer RP. "Bloodless" rats through the use of artificial blood substitutes. Fed Proc 1975;34 (6):1499–1505.
3. Sloviter HA. Perfluoro compounds as artificial erythrocytes. Fed Proc 1975;34 (6): 1484–1487.
4. Mitsuno T, Ohyanagi H, Naito R. Clinical studies of a perfluorochemical whole blood substitute (Fluosol-DA) Summary of 186 cases. Ann Surg 1982;195 (1):60–69.
5. Kaufman R. Clinical development of perfluorocarbon-based emulsions as red cell substitutes. In: Sjöblom J, editor. *Emulsions and Emulsion Stability*. New York: Marcel Dekker; 1996. p 343–367.
6. Modell JH, Calderwood HW, Ruiz BC, et al. Liquid ventilation of primates. Chest 1976;69 (1):79–81.
7. Chang S, Zimmerman NJ, Iwamoto T, et al. Experimental vitreous replacement with perfluorotributylamine. Am J Ophthalmol 1987;103 (1):29–37.
8. Cohn CS, Cushing MM. Oxygen therapeutics: perfluorocarbons and blood substitute safety. Crit Care Clin 2009;25 (2):399–414.
9. Mattrey RF, Long DM, Multer F, et al. Perfluoroctylbromide: a reticuloendothelial-specific and tumor-imaging agent for computed tomography. Radiology 1982;145 (3):755–758.
10. Mattrey RF, Leopold GR, vanSonnenberg E, et al. Perfluorochemicals as liver- and spleen-seeking ultrasound contrast agents. J Ultrasound Med 1983;2 (4):173–176.
11. Mattrey RF, Hajek PC, Gylys-Morin VM, et al. Perfluorochemicals as gastrointestinal contrast agents for MR imaging: preliminary studies in rats and humans. Am J Roentgenol 1987;148 (6):1259–1263.
12. Mattrey RF, Long DC. Potential role of PFOB in diagnostic imaging. Invest Radiol 1988;23 Suppl 1:S298–S301.
13. Schmieder A, Winter P, Caruthers S, et al. MR molecular imaging of melanoma angiogenesis with avb3-targeted paramagnetic nanoparticles. Magn Reson Med 2005;53:621–627.
14. Winter PM, Caruthers SD, Kassner A, et al. Molecular imaging of angiogenesis in nascent Vx-2 rabbit tumors using a novel alpha(nu)beta3-targeted nanoparticle and 1.5 tesla magnetic resonance imaging. Cancer Res 2003;63:5838–5843.
15. Winter PM, Morawski AM, Caruthers SD, et al. Molecular imaging of angiogenesis in early-stage atherosclerosis with alpha(v)beta3-integrin-targeted nanoparticles. Circulation 2003;108:2270–2274.
16. Cyrus T, Abendschein DR, Caruthers SD, et al. MR three-dimensional molecular imaging of intramural biomarkers with targeted nanoparticles. J Cardiovasc Magn Reson 2006;8 (3):535–541.

17. Cyrus T, Caruthers S, Allen J, et al. Intramural delivery of rapamycin with avb3-integrin-targeted paramagnetic nanoparticles inhibits stenosis following angioplasty. Circulation 2005;112 Suppl 17U224–U225.
18. Lanza GM, Yu X, Winter PM, et al. Targeted antiproliferative drug delivery to vascular smooth muscle cells with a magnetic resonance imaging nanoparticle contrast agent: implications for rational therapy of restenosis. Circulation 2002;106 (22): 2842–2847.
19. Soman N, Lanza G, Heuser J, et al. Synthesis and characterization of stable fluorocarbon nanostructures as drug delivery vehicles for cytolytic peptides. Nano Lett 2008;8: 1131–1136.
20. Crowder KC, Hughes MS, Marsh JN, et al. Sonic activation of molecularly-targeted nanoparticles accelerates transmembrane lipid delivery to cancer cells through contact-mediated mechanisms: implications for enhanced local drug delivery. Ultrasound Med Biol 2005;31 (12):1693–1700.
21. Cyrus T, Zhang H, Allen JS, et al. Intramural delivery of rapamycin with alphavbeta3-targeted paramagnetic nanoparticles inhibits stenosis after balloon injury. Arterioscler Thromb Vasc Biol 2008;28 (5):820–826.
22. Winter P, Neubauer A, Caruthers S, et al. Endothelial alpha(nu)beta(3)-Integrin targeted fumagillin nanoparticles inhibit angiogenesis in atherosclerosis. Arterioscler Thromb Vasc Biol 2006;26:2103–2109.
23. Schmieder AH, Caruthers SD, Zhang H, et al. Three-dimensional MR mapping of angiogenesis with {alpha}5{beta}1({alpha}{nu}{beta}3)-targeted theranostic nanoparticles in the MDA-MB-435 xenograft mouse model. FASEB J 2008;22:4179–4189.
24. Winter P, Caruthers S, Zhang H, et al. Antiangiogenic synergism of integrin-targeted fumagillin nanoparticles and atorvastatin in atherosclerosis. J Am Coll Cardiol Img 2008;1:624–634.
25. Winter PM, Schmieder AH, Caruthers SD, et al. Minute dosages of alpha(nu)beta3-targeted fumagillin nanoparticles impair Vx-2 tumor angiogenesis and development in rabbits. FASEB J 2008;22:2758–2767.
26. Soman N, Baldwin S, Hu G, et al. A platform of molecularly targeted nanostructures for anticancer therapy with cytolytic peptides. J Clin Invest 2009;119:2830–2842.
27. Marsh J, Senpan A, Hu G, et al. Fibrin-targeted perfluorocarbon nanoparticles for targeted thrombolysis. Nanomedicine 2007;2:533–543.
28. Liu S, Widom J, Kemp CW, et al. Structure of human methionine aminopeptidase-2 complexed with fumagillin. Science 1998;282 (5392):1324–1327.
29. Sin N, Meng L, Wang MQ, et al. The anti-angiogenic agent fumagillin covalently binds and inhibits the methionine aminopeptidase, MetAP-2. Proc Natl Acad Sci USA 1997;94 (12):6099–6103.
30. Bergers G, Javaherian K, Lo KM, et al. Effects of angiogenesis inhibitors on multistage carcinogenesis in mice. Science 1999;284 (5415):808–812.
31. Castronovo V, Belotti D. TNP-470 (AGM-1470): mechanisms of action and early clinical development. Eur J Cancer 1996;32A (14):2520–2527.
32. Konno H, Tanaka T, Kanai T, et al. Efficacy of an angiogenesis inhibitor, TNP-470, in xenotransplanted human colorectal cancer with high metastatic potential. Cancer 1996;77 (8 Suppl):1736–1740.
33. Shusterman S, Grupp SA, Barr R, et al. The angiogenesis inhibitor tnp-470 effectively inhibits human neuroblastoma xenograft growth, especially in the setting of subclinical disease. Clin Cancer Res 2001;7 (4):977–984.

34. Bhargava P, Marshall JL, Rizvi N, et al. A Phase I and pharmacokinetic study of TNP-470 administered weekly to patients with advanced cancer. Clin Cancer Res 1999;5 (8):1989–1995.
35. Kudelka AP, Verschraegen CF, Loyer E. Complete remission of metastatic cervical cancer with the angiogenesis inhibitor TNP-470. N Engl J Med 1998;338 (14):991–992.
36. Kudelka AP, Levy T, Verschraegen CF, et al. A phase I study of TNP-470 administered to patients with advanced squamous cell cancer of the cervix. Clin Cancer Res 1997;3 (9):1501–1505.
37. Logothetis CJ, Wu KK, Finn LD, et al. Phase I trial of the angiogenesis inhibitor **TNP-470** for progressive androgen-independent prostate cancer. Clin Cancer Res 2001;7 (5): 1198–1203.
38. Offodile R, Walton T, Lee M, et al. Regression of metastatic breast cancer in a patient treated with the anti-angiogenic drug TNP-470. Tumori 1999;85 (1):51–53.
39. Kim S, Harris M, Varner J. Regulation of integrin alpha vbeta 3-mediated endothelial cell migration and angiogenesis by integrin alpha5beta1 and protein kinase A. J Biol Chem 2000;275 (43):33920–33928.
40. Boudreau N, Varner J. The homeobox transcription factor Hox D3 promotes integrin alpha5beta1 expression and function during angiogenesis. J Biol Chem 2004;279 (6):4862–4868.
41. Pasternak RC, Smith SC Jr, Bairey-Merz CN, et al. ACC/AHA/NHLBI clinical advisory on the use and safety of statins. Circulation 2002;106 (8):1024–1028.
42. Girona J, La Ville AE, Sola R, et al. Masana, Simvastatin decreases aldehyde production derived from lipoprotein oxidation. Am J Cardiol 1999;83 (6):846–851.
43. Kimura M, Kurose I, Russell J, et al. Effects of fluvastatin on leukocyte-endothelial cell adhesion in hypercholesterolemic rats. Arterioscler Thromb Vasc Biol 1997;17 (8):1521–1526.
44. Verhoeven BA, Moll FL, Koekkoek JA, et al. Statin treatment is not associated with consistent alterations in inflammatory status of carotid atherosclerotic plaques: a retrospective study in 378 patients undergoing carotid endarterectomy. Stroke 2006;37 (8):2054–2060.
45. Laufs U, La Fata V, Plutzky J, et al. Upregulation of endothelial nitric oxide synthase by HMG CoA reductase inhibitors. Circulation 1998;97 (12):1129–1135.
46. Sukhova GK, Williams JK, Libby P. Statins reduce inflammation in atheroma of nonhuman primates independent of effects on serum cholesterol. Arterioscler Thromb Vasc Biol 2002;22 (9):1452–1458.
47. Koutouzis M, Nomikos A, Nikolidakis S, et al. Statin treated patients have reduced intraplaque angiogenesis in carotid endarterectomy specimens. Atherosclerosis 2007;192 (2):457–463.

CHAPTER 14

CANCER THERANOSTICS WITH HOLLOW GOLD NANOSPHERES

WEI LU AND CHUN LI

14.1 INTRODUCTION

Combining diagnosis and therapy in one process is an emerging biomedical paradigm referred to as theranostics. This new approach tailors treatment to the individual patient and has the potential to offer improved therapeutic efficacy, reduced side effects, and improved cost-effectiveness. Recent advances in nanotechnology have provided unprecedented opportunities for development of theranostic technologies. In particular, multifunctional nanomaterials incorporating both diagnostic and therapeutic functions have been created and evaluated for cancer theranostic applications [1].

Among those nanomaterials, gold nanostructures exhibit a unique and tunable optical property, termed surface plasmon resonance (SPR) [2, 3]. SPR is a collective oscillation of conduction-band electrons within the structures induced by the oscillating dipole of a resonant wavelength of light. For solid gold particles, the resonance peak appears at ~520 nm, and the peak varies slightly depending on the size of the particle and the medium in which the particle is embedded. With core-shell gold nanostructures—such as silica-cored gold nanoshells, which are regarded as the first generation of gold nanoshells—the SPR peaks can be tuned from the visible to the near-infrared (NIR) region by formulating nanostructures with different particle size and composition. These unique optical properties combined with excellent biocompatibility of gold make gold nanostructures well suited for various biomedical applications [4–9].

Hollow gold nanospheres (HAuNS) are considered the second generation of gold nanoshells and have the unique combination of small size (~40 nm in diameter), spherical shape, and hollow interior with tunable SPR. These nanoparticles have physicochemical, optical, and pharmacokinetic characteristics that make them extremely well suited for molecular imaging and drug delivery. HAuNS have been shown to be potentially useful for a variety of potential applications in biomedicine, including photoacoustic tomography (PAT), photothermal ablation (PTA), and light-controlled

Drug Delivery Applications of Noninvasive Imaging: Validation from Biodistribution to Sites of Action, First Edition. Edited by Chun Li and Mei Tian.

release of chemotherapeutic agents and small interfering RNA (siRNA). In this chapter, we will provide an up-to-date review of the synthesis, characterization, and preclinical evaluation of HAuNS, with emphasis on the development of multifunctional HAuNS for cancer theranostic applications.

14.2 HOLLOW GOLD NANOSPHERES

14.2.1 Synthesis and Characterization

Gold core-shell nanoparticles are generally synthesized by a template method, that is, the core nanoparticle is coated with a gold shell by reduction of chloroauric acid ($HAuCl_4$) onto the surface of the core, as is the case for silica-cored gold nanoshells [10]. HAuNS, the second generation of gold nanoshells, consist of only a thin gold wall with a hollow interior. HAuNS are synthesized using a reductive metal as a sacrificed template to form the core. When the $HAuCl_4$ is reduced, the template core is consumed, and the resulting metal ions diffuse into bulk solution, leaving a hollow interior.

Liang et al. first reported the formation of HAuNS with tunable interior-cavity sizes using cobalt (Co) nanoparticles as a reductive template [11]. The Co nanoparticles were prepared according to the method reported by Kobayashi et al. [12]. In brief, $CoCl_2$ solution was added into deaerated aqueous solution and reduced by sodium borohydride ($NaBH_4$) in the presence of citric acid as a stabilizer. The pink color of Co^{2+} solution changed to dark brown as Co colloids were formed. Hydrogen was evolved, while $NaBH_4$ was oxidized, a process that continued for several minutes. When the gas evolution ceased, the Co nanoparticle colloidal solution was transferred to stirred $HAuCl_4$ aqueous solution. Since the reduction potential of the $AuCl_4^-/Au$ redox couple (0.935 V vs. standard hydrogen electrode (SHE)) was much higher than that of the Co^{2+}/Co redox couple (−0.377 V vs. SHE), $AuCl_4^-$ was reduced to Au atoms as soon as Co nanoparticles were added into the $HAuCl_4$ solution:

$$3Co + 2AuCl_4^- = 2Au + 3Co^{2+} + 8Cl^- \qquad (14.1)$$

The reduced Au atoms nucleated, grew into very small particles, and eventually evolved into a thin shell around the Co nanoparticles. The shells presumably had an incomplete porous structure at an early stage when Co^{2+} and $AuCl_4^-$ were able to continuously diffuse across the shell in reverse directions [11, 13]. When $AuCl_4^-$ was completely consumed, the remaining Co "core" was continually oxidized by H^+ in aqueous solution (the reduction potential of the H^+/H_2 redox couple is 0.0 vs. SHE). Because the reduced gold atoms were largely confined to the vicinity of the sacrificial Co template's outer surface, the size of the resulting HAuNS was governed by the initial size of the Co nanoparticles (Fig. 14.1).

Prevo et al. [14] reported another approach to the synthesis of HAuNS. These authors modified the method of gold nanocage preparation [15–17] and described a galvanic replacement reaction between silver nanoparticles and an aqueous $HAuCl_4$ solution as a simple and convenient way to synthesize HAuNS [14]. Silver nanoparticles can be synthesized by reducing silver nitrate with $NaBH_4$, resulting in nanoparticles 15–25 nm in diameter. These silver nanoparticles acted as nuclei for further growth

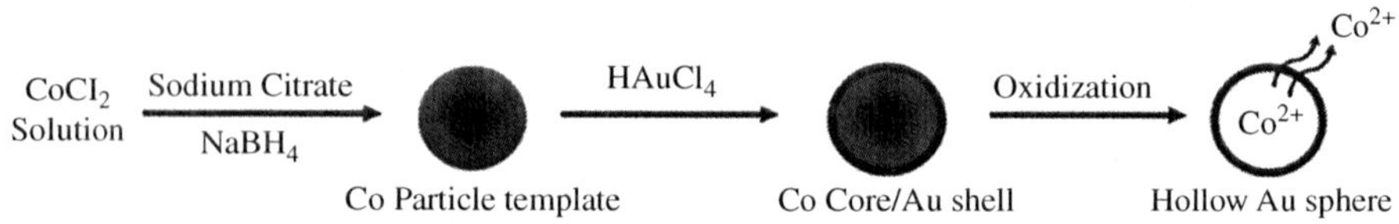

FIGURE 14.1 Schematic drawing of synthesis of HAuNS. From reference [83]. Copyright American Association for Cancer Research. Reprinted with permission.

with addition of hydroxylamine hydrochloride and additional silver nitrate. The silver nanoparticles were converted into gold nanoshells by the addition of $HAuCl_4$ directly into the as-grown nanoparticle solution. Since the reduction potential of the $AuCl_4^-/Au$ redox couple was higher than that of the Ag^+/Ag redox couple (0.8 V vs. SHE), the replacement reaction was [15]:

$$3Ag + AuCl_4^- = Au + 3Ag^+ + 4Cl^- \tag{14.2}$$

Upon the addition of $HAuCl_4$, the solution turned from yellow/orange to gray/yellow to blue/gray to blue/turquoise within seconds as the silver and gold were oxidized and reduced, respectively [14]. The gold deposited on the sliver template surface, nucleated, grew into very small particles, and eventually evolved into a thin shell around the template. Similar to what happens during synthesis of cobalt-cored gold nanoshells, the shell on the silver template had an incomplete structure in the initial stage because both $HAuCl_4$ and AgCl continuously diffused across this layer until the silver nanotemplate was completely consumed. However, as the reaction proceeded with continuous refluxing at a temperature of ~100 °C, the gold shells reconstructed their walls into highly crystalline structures via processes such as Ostwald ripening [18]. Therefore, annealing of the resultant gold shells generated seamless hollow structures [17]. This new synthesis approach of Prevo et al. [14] is rapid, stable, highly scalable, and in many cases a true "one-pot" synthesis.

14.2.2 Tuning SPR

SPR is an optical phenomenon arising from the interaction between an electromagnetic wave and the conduction-band electrons in a metal [19]. When exposed to light at a resonance frequency, the conduction-band electrons in a metal nanostructure are driven by the electric field to collectively oscillate at a resonance frequency relative to the lattice of positive ions. At this resonance frequency, the incident light is absorbed by the nanostructure. Some of these absorbed photons are released with the same frequency in all directions, a process known as scattering. At the same time, a fraction of these absorbed photons are converted into phonons (vibrations of the lattice), a process known as absorption. In general, the SPR peak of a metal nanostructure includes both scattering and absorption components [20].

The optical properties of gold nanospheres are quantified in terms of their calculated absorption and scattering efficiencies (Q_{abs} and Q_{sca}) and their optical resonance wavelength (λ_{max}). The efficiencies Q_i for the interaction of radiation with a scattering sphere of radius r are cross sections σ_i normalized to the particle cross section, πr^2, where i stands for extinction (i = ext), absorption (i = abs), or scattering (i = sca); thus [21]

$$Q_i = \frac{\sigma_i}{\pi r^2} \tag{14.3}$$

Energy conservation requires that

$$Q_{\text{ext}} = Q_{\text{sca}} + Q_{\text{abs}}, \tag{14.4}$$

or

$$\sigma_{\text{ext}} = \sigma_{\text{sca}} + \sigma_{\text{abs}}. \tag{14.5}$$

In the case of metallic nanoshells, when the real part of a dielectric permittivity of a nanoshell has a negative value, the spectra of optical absorption and scattering have specific resonance peaks, which correspond to the excitation of surface plasmons in nanoshells [22]. To calculate these spectra of absorption and scattering of light in HAuNS with a water core (HAuNS in water), the incident light was assumed as a plane wave propagating along the z axis of a Cartesian system of coordinates. This assumption allowed application of standard Mie theory [23] to calculations, as well as use of the following expressions for scattering and absorption cross sections [24]:

$$\sigma_{\text{sca}} = \frac{2\pi\sqrt{\varepsilon_m}}{k_m^2}\sum_{n=1}^{\infty}(2n+1)\cdot\left(|a_n|^2 + |b_n|^2\right) \tag{14.6}$$

$$\sigma_{\text{abs}} = \frac{2\pi\sqrt{\varepsilon_m}}{k_m^2}\sum_{n=1}^{\infty}(2n+1)\cdot\text{Re}(a_n + b_n) - \sigma_{\text{sca}} \tag{14.7}$$

where $\sqrt{\varepsilon_m}$ is a dielectric constant of a host medium, k_m is a wave number of an incident plane wave (in a host medium), and a_n, b_n are Mie coefficients of scattered field. It should be noted here that in the cross sections of Equations 14.6 and 14.7, the physical meaning of scattered and absorbed power is divided by the energy density of the incident laser radiation in a vacuum. To calculate Mie coefficients, boundary conditions for the Maxwell equations are used for the considered geometry. In the case of a metal nanoshell with a dielectric core, Mie coefficients are calculated by using the method described by Kerker [25].

Jain et al. used Mie theory and the discrete dipole approximation method to calculate the absorption and scattering efficiencies and optical resonance wavelengths for solid gold nanoparticles [21]. The calculated spectra clearly reflected the well-known dependence of nanoparticles' optical properties (i.e., the resonance wavelength), extinction cross section, and ratio of scattering to absorption on nanoparticle dimension. When the diameter of the gold nanoparticles was increased from 20 to 80 nm, the magnitude of extinction and the relative contribution of scattering to the extinction rapidly increased, and the maximum plasmon wavelength of the nanospheres increased slightly from ~520 to 550 nm. Gold nanoshells were found to have extinction cross sections similar to or even higher than those of solid gold nanoparticles [20]. However, the resonance wavelength of gold nanoshells can be increased rapidly by increasing the particle size or by increasing the ratio of the core thickness to the shell radius. The relative contribution of scattering to the extinction can be increased rapidly by increasing the particle size or decreasing the ratio of the core thickness to the shell radius [21].

Prodan et al. [26] and Halas [19] used "plasmon hybridization" theory to explain the remarkable tunability of the optical resonance of nanoshells. The highly geometry-dependent plasmon response can be seen as an interaction or "hybridization" between the essentially fixed frequency plasmon response of a solid nanosphere and the plasmon response of a nanocavity. The sphere and cavity plasmons are electromagnetic excitations that induce surface charges at the inner and outer interfaces of the metal shell. Because of the finite thickness of the shell layer, the sphere and cavity plasmons interact with each other. The strength of the interaction between the sphere and cavity plasmons is controlled by the thickness of the metal shell layer. This interaction results in the splitting of the plasmon resonances into two new resonances: the lower-energy symmetric, or "bonding," plasmon and the higher-energy antisymmetric, or "antibonding," plasmon [26]. Only the lower-energy plasmon interacts strongly with an incident optical field. That is the reason why the nanoshell plasmon shifts to the lower energies (longer wavelengths) as the shell thickness is reduced. The theory of plasmon hybridization provides a general principle to guide the design of metallic nanostructures and prediction of their plasmon response qualitatively and quantitatively [19].

Liang et al. [11] and Schwartzberg et al. [13] showed that the SPR peak of HAuNS can be controlled by adjusting the interior-cavity sizes using Co as the displacement template. With this approach, Co colloidal particles reduced $AuCl_4^-$ to form the gold shell outside. The shell formation was an inward growth process, in which the thickness of the gold shell increased inward as the replacement between Co nanoparticles and $HAuCl_4$ continued. Therefore, the outer diameters were controlled by the diameter of Co nanoparticles, whereas interior-cavity sizes were controlled by changing the stoichiometric ratio of $HAuCl_4$ over the reducing agents [11]. For example, the SPR peak of HAuNS increased from 526 to 628 nm as the interior-cavity size was increased from 0 to 40 nm while the exterior size was fixed at ~60 nm [11]. Schwartzberg et al. modified the synthetic method and expanded the tunable peak of the surface plasmon band absorption to 550–820 nm at a smaller exterior size of around 40 nm [13].

The SPR of HAuNS can also be tuned by using silver as a core template. When different amounts of $HAuCl_4$ were added to silver nanoparticles grown from an original 0.2-mmol $AgNO_3$ stock solution, the gold peak gradually red shifted during the reaction from 400 to about 620 nm [14]. This suggested that the silver template dissolved with decreasing SPR peak (~400 nm) of silver nanoparticle and finally disappeared. On the other hand, gold shells formed as a result of the increase of independent SPR peak at longer wavelength. However, no peak around 530 nm (the position of the SPR peak for gold solid colloids) was found in the reaction system, suggesting that no gold solid colloids formed in the solution [17]. In addition, the resulting final absorbance spectra for HAuNS could be controlled by adjusting the size of the silver template. When the stoichiometric ratio of gold to silver was kept constant at 0.37–0.38:1 to ensure complete reaction, the SPR of the final HAuNS could be precisely manipulated within the wavelength range of 600–800 nm by controlling the initial silver concentration (template size). The experimental result from this approach correlated well with Mie theory calculations [14].

Not only can monoparticular SPR be tuned but also the internanoparticular SPR of dimeric HAuNS can be tuned, through adjustment of surface medication. Chandra et al. [27] showed that altering particle aspect ratio, interparticle gap, and nanosphere inner surface spatial separation in aggregated HAuNS led to control over the interparticle plasmon resonance. Ethanedithiol-mediated aggregation resulted in dimeric structures in which monomer subunits were spatially separated by less than 3 Å. In this case, the dimeric aggregates exhibited an obvious blueshift of internanoparticle SPR derived from a charge-transfer plasmon resonance at the dimer interface. The extent of the blueshift was dependent on shell thickness. Dimers comprised of HAuNS with thin shells exhibited the largest shift; aggregates containing HAuNS with thick shells (≥7 nm) did not display a significant SPR shift when the individual particles were in contact. In contrast, all cysteine-induced aggregates examined in the study displayed large interparticle gaps (>1 nm) and a redshifted SPR, regardless of particle dimension [27].

14.2.3 Photothermal Effect

When the incident light frequency matches the intrinsic electron oscillation frequency of a gold core-shell nanostructure, light is absorbed, resulting in SPR absorption. SPR absorption in gold nanoparticles is followed by energy relaxation through nonradioactive decay channels. This results in an increase in kinetic energy, leading to overheating of the local environment around the light-absorbing species. This phenomenon of rapid conversion of absorbed light into heat is known as the photothermal effect [3, 22, 28].

Using femtosecond transient absorption spectroscopy, Link and El-Sayed [29] showed that the photoexcitation of metal nanostructures results in the formation of a heated electron gas that subsequently cools rapidly—within ~1 ps—as a result of exchange of energy with the nanoparticle lattice. This process is followed by phonon–phonon interactions in which the nanoparticle lattice cools rapidly (within ~100 ps) as a result of exchange of energy with the surrounding medium [29]. This mechanism has also been confirmed in gold nanocages using ultrafast laser spectroscopy or time-resolved spectroscopy as well as vibrational spectroscopy [30, 31].

The photothermal effect in gold nanostructures is highly efficient. Electron temperatures of several thousand Kelvin are easily reached in gold nanoparticles, even with laser excitation powers as low as 100 nJ, and lattice temperatures on the order of a few tens of degrees can be achieved [29]. The absorption cross section of gold nanoparticles is several orders of magnitude stronger than that of the strongest-absorbing rhodamine 6G dye molecules [21, 32]. Conventional NIR dyes like indocyanine green possess an absorption cross section of $\sim 1.66 \times 10^{-20}$ m^2 at wavelengths of ~800 nm, while silica-cored gold nanoshells possess absorption cross sections on the order of 3.8×10^{-14} m^2. Thus, such nanoshells are over one million times more likely than the comparable dye to encounter an absorbing event and convert that light into thermal energy [7].

Besides occurring with silica-cored gold nanoshells and gold nanocages, the photothermal effect occurs with HAuNS, as reported by Lu et al., who observed photothermal-related conversion of HAuNS morphology [33]. These authors found that the SPR absorbance and peak wavelength of HAuNS were significantly reduced upon pulsed NIR laser irradiation. The reduced SPR absorbance as well as considerable blueshift was reversely proportional to the laser light in a dose- and time-dependent manner. At higher laser dose, the SPR band in the NIR region of HAuNS completely disappeared, while an absorption peak at ~520 nm resembling an SPR band of spherical solid gold nanoparticles appeared. Transmission electron microscopy confirmed the spectral change of HAuNS related to the loss of their structural integrity upon NIR irradiation. The laser-induced heat caused increasing numbers of HAuNS to melt and collapse into solid droplets and even transformed HAuNS into solid spherical particles in a dose-dependent manner [33]. The phenomenon of shape change as a result of NIR light irradiation was in line with observations with other gold-based nanostructures [34–38].

Near-infrared laser-induced photothermal conversion has also been termed hole-burning effect. Wheeler et al. investigated the effect of pulsed femtosecond laser irradiation on the persistent spectral hole-burning characteristics of HAuNS and the mechanism underlying this effect [39]. On the basis of the correlation between UV/Vis spectroscopy and transmission electron microscopy data, the mechanism was explained as two processes: (i) conversion of NIR-absorbing HAuNS to bluer-absorbing HAuNS upon NIR irradiation and (ii) breakdown of NIR-absorbing HAuNS into solid gold nanoparticles directly. The branching ratio between these two processes was dependent on the peak power of the femtosecond laser pulses; lower peak power favored process and higher peak power favored process. This finding had important implications for various applications of HAuNS, including PTA of cancer and drug delivery [39].

Prevo et al. also observed that pulsed-laser light could heat HAuNS to sufficiently high temperatures so that they were quickly melted and annealed into more stable solid spherical nanoparticles [14]. These authors further determined that not only optimization of HAuNS characteristics for maximum absorption of femtosecond NIR pulses but also laser type played an important role in heating individual nanoshells to the highest temperatures during irradiation. For continuous-wave laser irradiation, the nanoshell temperature did not increase significantly over the temperature of the solution as the energy input to the particle was dissipated quickly into the surrounding medium [40, 41]. The instantaneous continuous-wave power input was of the order of 10^{10} lower than for femtosecond pulses, even though the average power input to the system was the same [14].

14.3 PHOTOACOUSTIC TOMOGRAPHY

Photoacoustic tomography, also referred to as optoacoustic tomography, is a hybrid technology that visualizes the internal distribution of optical energy deposition in biological tissues on the basis of the detection of laser-induced ultrasonic waves (photoacoustic or optoacoustic waves) [42, 43]. PAT provides higher spatial resolution than traditional optical imaging in deep biological tissues because ultrasonic scattering in such tissues is two orders of magnitude less than optical scattering

[44, 45]. This lower scattering significantly reduces the signal intensity of background scattering and improves image quality. PAT has been successfully applied to the visualization of different structures in biological tissues and has been especially useful in imaging the cerebral cortex of small animals [46, 47]. In addition, PAT has high sensitivity and specificity in monitoring of exogenous optical contrast agents [48, 49]. Optical contrast allows spectroscopic separation of signal contributions from multiple optical absorbers, such as oxyhemoglobin, deoxyhemoglobin, and molecular contrast agents, thus enabling simultaneous molecular and functional imaging [49].

Optical contrast agents with strong NIR light absorption are preferable for PAT. Hemoglobin and water, the major absorbers in humans and animals of visible and infrared light, respectively, have their lowest absorption coefficient in the NIR region around 650–900 nm [50]. Therefore, NIR light permits deep optical penetration into biological samples [51]. Gold nanoparticles such as gold nanoshells [51], gold nanocages [52, 53], gold nanorods [54–60], and HAuNS [61] have been employed as molecular contrast agents in the NIR range. These gold nanoparticles can increase photoacoustic signals because of the light excitation in the NIR spectral region, where the signal ratio of gold nanoparticles to hemoglobin is higher and contrast with endogenous chromophores is great [62, 63]. Moreover, gold nanoparticles are not susceptible to photobleaching, a problem commonly associated with the use of organic dyes. Gold nanoparticles have been successfully used to enhance photoacoustic mapping of cerebral vasculature [51, 52], sentinel lymph nodes [53, 55, 64], and tumors *in vivo* [62, 65]; joint tissue *ex vivo* [66]; and macrophages in atherosclerotic plaques [67]. In addition, the progressive extravasation of gold nanoshells through solid tumor vasculature *in vivo* has been noninvasively imaged with PAT [68].

Lu et al. investigated HAuNS-enhanced PAT of the living mouse brain vasculature [61]. Methoxy-polyethylene glycol-SH (MW 5000) was conjugated with HAuNS through an Au-S bond to form pegylated HAuNS (PEG-HAuNS). The resulting nanoparticles had average outer diameter of 43–46 nm and a mean shell thickness of 2–4 nm (Fig. 14.2a). The resonance absorbance of these nanoparticles peaked at 800 nm (Fig. 14.2b). According to Equations 14.6 and 14.7, the theoretically calculated inner core radius and gold shell thickness of HAuNS with an extinction spectrum that peaked at 800 nm matched well with the measured dimensions of HAuNS on transmission electron microscopy (Fig. 14.2c). The molar extinction coefficient associated with the optical absorption cross section of HAuNS was calculated to exceed 1.4×10^{11}/M/cm at 800 nm based on Figure 14.2c. In comparison, the molar extinction coefficient of hemoglobin is $\sim 1 \times 10^{3}$/M/cm [69]. As a result, PEG-HAuNS concentrations of only ~20 pM yielded optical absorption approximately as strong as that of hemoglobin at the typical concentration of 2.3 mM in blood. According to the Mie theory and the plasmon resonance theory, the optical absorption coefficient of gold nanoshells increases, and the optical scattering coefficient decreases with decreasing diameter of the core. Compared with silica-cored gold nanoshells (145-nm outer diameter, 10-nm shell thickness) with larger particle size [52], HAuNS had an absorption coefficient 32 times higher and a similar scattering coefficient [61]. Since PAT is an absorption-based technique, this high absorption coefficient makes HAuNS an excellent contrast agent for photoacoustic imaging.

Photoacoustic tomography with PEG-HAuNS allowed imaging of the mouse brain vasculature with high spatial resolution and satisfactory sensitivity [61]. On live

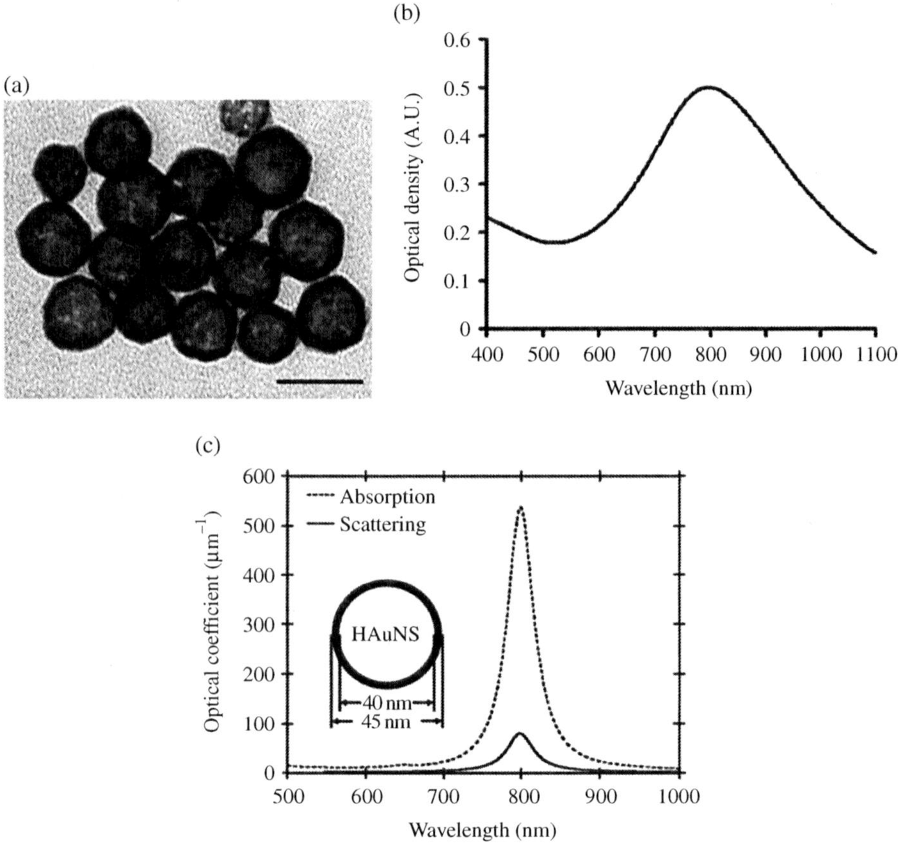

FIGURE 14.2 (a) Transmission electron micrograph of PEG-HAuNS (bar = 50 nm). (b) Experimental absorbance spectrum of PEG-HAuNS in water. The peak was at 800 nm. A.U. indicates arbitrary unit. (c) Theoretically calculated absorption and scattering spectra of PEG-HAuNS with water core having dielectric permittivity of 1.77. The inner core radius was 20 nm, and the gold shell thickness was 2.5 nm, both of which values were consistent with the measured dimensions of the HAuNS on transmission electron microscopy. From reference [61]. Copyright Elsevier. Reprinted with permission.

imaging of nude mice, the photoacoustic signal of the brain vasculature was strong, and the vasculature was imaged with great clarity, especially small blood vessels at 5 min and 2 h after PEG-HAuNS injection. The depiction of the brain vessels on PEG-HAuNS-enhanced PAT matched well with anatomical photographs. However, without PEG-HAuNS, the photoacoustic signal of the brain vasculature was weak because of the low intrinsic absorption of oxyhemoglobin and deoxyhemoglobin at 800 nm. Only large vessels distributed along the mouse brain cortical surface, such as the middle cerebral artery, were visualized. The contrast enhancement deriving from the use of PEG-HAuNS was well explained through differential images obtained by subtracting the preinjection image from the postinjection image pixel by pixel (Fig. 14.3) [61].

Polyethylene glycol-HAuNS proved to be ideal optical contrast agents because of their excellent pharmacokinetics and lack of acute toxicity [61]. After intravenous

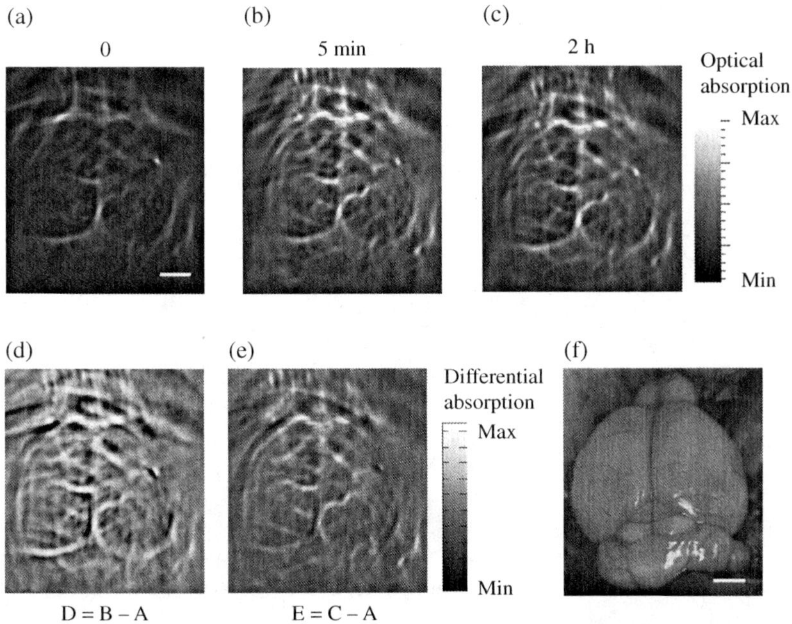

FIGURE 14.3 Noninvasive PAT of a mouse brain *in vivo* with PEG-HAuNS and NIR light at a wavelength of 800 nm. (a–b) Photoacoustic images acquired (a) before, (b) 5 min after, and (c) 2 h after IV injection of PEG-HAuNS. (d and e) Differential images obtained by subtracting the preinjection image from the postinjection images (image d = image b – image a; image e = image c – image a). Bar = 2 mm. (f) Open-skull photograph of the mouse brain cortex obtained after data acquisition for PAT. Bar = 2 mm. From reference [61]. Copyright Elsevier. Reprinted with permission.

injection, PEG-HAuNS displayed long blood circulation profile with a distribution half-life of 1.38 ± 0.38 h and an elimination half-life of 71.82 ± 30.46 h. The long circulation time of the nanoparticles meant that the photoacoustic signal of the brain vessels remained strong for a long period. Quantitative analysis showed that the photoacoustic absorption signal in the brain vessels at 5 min after contrast agent injection was on average 2.3 times the preinjection value and at 2 h after contrast agent injection was on average 1.9 times the preinjection value [61]. The fact that the signal intensity at 2 h was not much less than the signal intensity at 5 min indicated that PEG-HAuNS are well suited for sequential collection of three-dimensional images in clinical applications.

The kidneys, liver, and spleen were the major organs taking up a significant amount of PEG-HAuNS. Uptake of PEG-HAuNS in all major organs except the kidneys and bone had decreased by day 7 after injection, suggesting metabolism or disposition of HAuNS. Since gold is an inert element, it is supposed to be nonbiodegradable. After intravenous injection, the only metabolic pathway is kidney excretion. This fact could partially explain why accumulation of gold nanoparticles in the kidney did not decrease by day 7 after injection. The analysis of mouse serum biochemical parameters proved that a single dose of PEG-HAuNS did not impair liver or kidney function. Histological examination of the liver, spleen, and kidneys revealed no apparent change in cellular structures seven days after PEG-HAuNS injection [61].

14.4 PTA OF CANCER

PTA is a therapeutic modality in which photothermal agents are employed to achieve selective heating of the local environment [70–73]. The heat produced can lead to localized temperatures far above the threshold temperature (~330 K) that causes irreversible cell death [20]. This method is very appealing because it is a physical treatment and therefore associated with fewer side effects than conventional cancer treatment strategies such as radiotherapy and chemotherapy. PTA can also be performed repeatedly without the accumulation of toxic side effects. Mild increases in temperature can also boost the efficacy of radiotherapy and chemotherapy [74].

Gold nanostructures are ideally suited for use as photothermal coupling agents. The enhanced absorption cross sections of gold nanostructures ensure effective laser therapy at relatively low laser output energies; in addition, the SPR absorption band of core-shell-structured nanoparticles is readily tunable in the NIR region, which makes it possible to deliver thermal dose to deeper tissues. Metal nanostructures also display much higher photostability than organic dyes, and the surface of metal nanostructures can be readily modified with receptor ligands to direct nanoparticles to the target sites, allowing selective thermal ablation of targeted tumor cells.

The efficacy of PTA for cancer treatment was initially studied in human prostate cancer cells *in vitro* with laser-activated silica-cored gold nanoshells that were taken up by cells via nonspecific phagocytosis or pinocytosis. It was shown that a ratio of at least 5000 nanoshells per human prostate cancer cell was critical for achieving efficient cell killing effect [75–77].

Later studies showed that HAuNS were more efficient than silica-cored gold nanoshells for PTA of cancer. HAuNS are less than one-third the size of silica-cored gold nanoshells (40 nm in outer diameter vs. 145 nm in outer diameter). The smaller size of HAuNS is of utmost importance for efficient cellular uptake [78]. Particles larger than 100 nm may not readily enter cells and may interrupt some cellular functions. Similarly, particles smaller than 20 nm may not be retained by cells. Thus, nanostructures with sizes between 20 and 100 nm may be favorable [13]. In passive tumor targeting through the enhanced permeability and retention effect, particle size is also a critical characteristic. It was shown that smaller gold nanoparticles (20–40 nm) coated with PEG had longer blood circulation time than larger gold nanoparticles (80 nm). The 20-nm PEG-coated gold nanoparticles exhibited the lowest uptake by reticuloendothelial cells and the slowest clearance from the body [79]. Moreover, smaller particles have a better chance of extravasating from blood vessels into extravascular fluid space. In certain tumors, such as glioma and ovarian cancer, pore cutoff size was 7–100 nm [80, 81]. Therefore, to improve targeted delivery of gold nanostructures, it is desirable that the size of these nanoparticles be less than 80 nm. In light of this guideline, HAuNS with a diameter of ~40 nm have been identified as a promising photothermal mediator [82, 83].

Cheng et al. compared three gold-based nanomaterials (silica–Au nanoshells [composed of a dielectric core (silica) coated with gold], hollow gold or silver nanospheres, and gold nanorods) for their photothermal efficiencies in killing three types of malignant cells (A549 lung cancer cells, HeLa cervix cancer cells, and transitional cell carcinoma bladder cancer cells) *in vitro* using an NIR laser. Photodestructive efficiency was evaluated as a function of the number of nanoparticles required to destroy the cancer cells under 808-nm laser wavelength at fixed laser power. Of the three nanomaterials, silica–Au nanoshells needed the smallest number of particles to produce effective photodestruction, whereas gold nanorods needed the largest number of particles. The photothermal efficiency rankings were silica–Au nanoshells > hollow gold or silver nanospheres > gold nanorods [84].

Melancon et al. investigated HAuNS-mediated PTA with C225, a monoclonal antibody directed at epidermal growth factor receptor (EGFR), covalently attached to HAuNS (C225-HAuNS) [82]. When an aqueous solution of C225-HAuNS (7.3×10^{10} nanoshells/ml) was exposed to an NIR laser at an output power of $8\,W/cm^2$, the temperature of the solution increased 16.5 °C after ~4 min of light exposure. Without HAuNS, little temperature change was observed at the same laser power and duration of exposure [82]. The *in vitro* results showed EGFR-mediated selective uptake of C225-HAuNS in EGFR-positive A431 tumor cells but no uptake of IgG-HAuNS control. Treatment of A431 cells with C225-HAuNS plus NIR laser resulted in selective destruction of these cells. In contrast, treatment with C225-HAuNS alone, laser alone, or IgG-HAuNS plus laser did not result in any observable effect on cell viability [82].

Attachment of targeting molecules that possess high affinity for unique molecular signatures found in malignant cells has been found to improve the residence time of HAuNS in tumors. Melancon et al. used C225-HAuNS to target EGFR-overexpressing A431 human squamous carcinoma in mice. Radiotracer counting for tumor biodistribution showed that C225-HAuNS had a 48% higher uptake value in the A431 tumors than did IgG-conjugated HAuNS (control) at 24 h after injection. The number of nanoshells and/or nanoshell aggregates per field observed under dark-field microscope in the perivascular area of the tumor was more than three times as high in mice injected with C225-HAuNS as it was in mice injected with IgG-HAuNS, suggesting that the interaction between C225-HAuNS and EGFR may have facilitated extravasation of the nanoshells into the interstitial space [82].

In addition to being conjugated with an antibody, HAuNS have been conjugated with a small molecular weight peptide, [Nle^4,D-Phe^7]α-MSH (NDP-MSH), as a targeting moiety through a PEG linker [83]. NDP-MSH is a potent agonist of melanocortin type 1 receptor, which is overexpressed in many melanoma cells [85], and binds to the receptor with high affinity ($IC_{50} = 0.21$ nM) [86]. The tumor uptake of NDP-MSH-conjugated PEG-HAuNS was significantly higher than that of control PEG-HAuNS without peptide conjugation (12.6 ± 3.1 percentage of injected dose per gram of tissue (%ID/g) vs. 4.3 ± 1.2%ID/g) at 4 h after intravenous injection in B16/F10 melanoma-bearing nude mice [83]. This suggested that the targeted HAuNS exhibited enhanced extravasation from tumor blood vessels and dispersion into the tumor matrix. Furthermore, the PTA effect of the nanoparticles was confirmed both histologically, by examination of excised tissue, and functionally, by [^{18}F]fluorodeoxyglucose positron emission tomography ([^{18}F]FDG-PET) (Fig. 14.4). The successful active targeting of NDP-MSH-conjugated HAuNS to melanoma suggested their potential application in targeted PTA for melanoma [83].

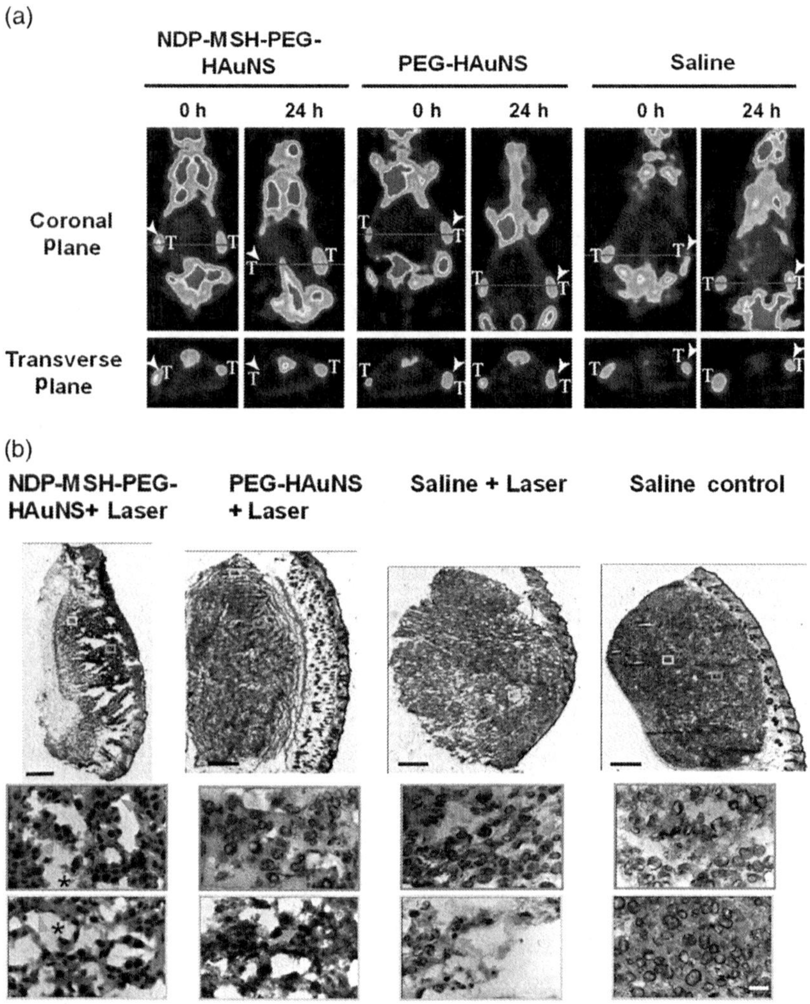

FIGURE 14.4 *In vivo* PTA-induced selective destruction of B16/F10 melanoma in nude mice. (a) [^{18}F]FDG-PET shows significantly reduced tumor metabolic activity after PTA only in mice pretreated with NDP-MSH-PEG-HAuNS. [^{18}F]FDG-PET was conducted before (0 h) and 24 h after NIR laser irradiation, which was commenced 4 h after intravenous injection of NDP-MSH-PEG-HAuNS, PEG-HAuNS, or saline control. For each mouse, tumors on one flank were randomly selected for laser treatment (0.5 W/cm^2 at 808 nm for 1 min). T: tumor; Arrowheads: tumors irradiated with NIR light. (b) Histological assessment of tumor necrosis. The whole tumor from each mouse 24 h after NIR irradiation was removed, fixed, and stained with hematoxylin–eosin (bar = 500 μm). Representative images show that in the group treated with NDP-MSH-PEG-HAuNS plus laser, most of tumor cells were necrotic as indicated by pyknosis (arrows), karyolysis (arrowheads), cytoplasmic acidophilia, and degradation and corruption of the extracellular matrix of the tumor (asterisks). Only a small fraction of necrotic zone was found in the tumors treated with PEG-HAuNS plus laser, laser alone, and tumor without any treatment. Bar = 50 μm. From reference [83]. Copyright American Association for Cancer Research. Reprinted with permission. For color details, please see color plate section.

14.5 NIR LIGHT-TRIGGERED DRUG DELIVERY

HAuNS with high photothermal coupling efficiency can be incorporated into drug delivery vehicles such as liposomes and microspheres to produce NIR light-controlled drug-release systems. Wu et al. synthesized small HAuNS that were either encapsulated within liposomes (by an interdigitation–fusion method) or tethered to the liposome membrane with a Au-S-PEG-lipid linker [87]. HAuNS with a maximum absorption at 820 nm were coated with 750-Da PEG-thiol to enhance particle stability and were concentrated by ultracentrifugation. The diameter of the HAuNS was 33 ± 13 nm, and shell thickness was 3.4 ± 0.9 nm. The HAuNS were encapsulated within dipalmitoylphosphatidylcholine liposomes together with a fluorescent dye, 6-carboxyfluorescein, used as a soluble model drug. The pulsed NIR light absorbed by HAuNS triggered near-instantaneous release of liposome contents, providing precise spatial and temporal control. The laser-heated HAuNS acted as optically triggered nano-"sonicators" to temporally disrupt the lipid membrane. HAuNS tethered to the liposome membrane, HAuNS encapsulated within liposomes, and HAuNS suspended freely outside liposomes all induced liposome disruption; however, tethered HAuNS achieved the highest release efficacy owing to their proximity to the lipid membrane. With this new NIR-activated release, disease cells can be synergistically targeted by combining drug-carrying particles (liposomes) and energy-absorbing particles (HAuNS). Continued irradiation of the HAuNS can induce localized hyperthermia or permeabilize cell membranes, both of which can facilitate the cellular uptake of large macromolecules, including proteins and DNA [87].

Near-infrared laser-modulated release of anticancer agents was also reported with biodegradable and biocompatible microspheres containing HAuNS [88]. In this system, poly(lactide-co-glycolide) (PLGA) copolymers were used to formulate 1–15-μm-diameter microsphere using a water-in-oil-in-water double-emulsion solvent evaporation method, incorporating HAuNS and water-insoluble paclitaxel. It was anticipated that paclitaxel dissolved in the organic phase together with the PLGA would be uniformly distributed within the polymer, whereas HAuNS dissolved in the original aqueous phase would be dispersed in the inner water phase inside the microspheres. Upon NIR laser irradiation, rapid and repetitive paclitaxel release from the paclitaxel/HAuNS-containing microspheres was observed, whereas when the NIR light was switched off, paclitaxel release was insignificant (Fig. 14.5). The extent of paclitaxel release from the microspheres was dependent on the output power of the NIR laser, duration of irradiation, treatment frequency, and concentration of HAuNS embedded inside the microspheres. HAuNS encapsulated in microspheres had the same photothermal coupling efficiency as plain HAuNS. *In vitro* results showed that cancer cells incubated with paclitaxel/HAuNS-loaded microspheres and subjected to NIR light irradiation displayed significantly greater cytotoxic effects than cells incubated with the microspheres alone or cells irradiated with NIR light alone, owing to NIR light-triggered drug release. In nude mice bearing human U87 gliomas and MDA-MB-231 mammary tumor xenografts, intratumoral injection of paclitaxel/HAuNS-loaded microspheres followed by NIR irradiation caused significantly greater tumor growth delay than treatment with HAuNS-loaded microspheres (no paclitaxel) and NIR irradiation or paclitaxel/HAuNS-loaded microspheres alone. The superimposed antitumor efficacy of paclitaxel/HAuNS-loaded microspheres was attributed to the simultaneous modulation of drug release and induction of photothermal cell killing triggered by NIR laser [88].

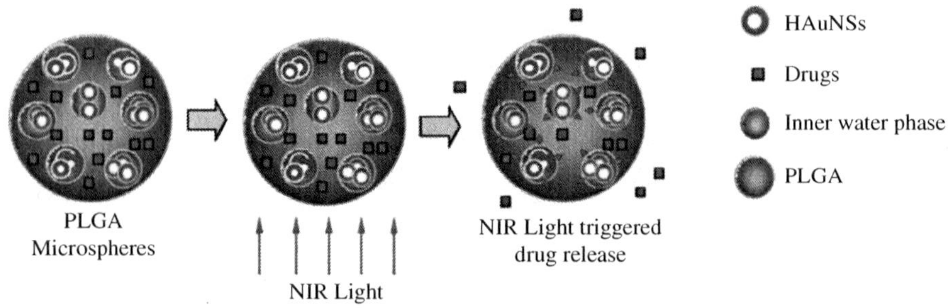

FIGURE 14.5 Proposed mechanism of NIR light-triggered drug release from PLGA microspheres encapsulating both HAuNS and paclitaxel. Paclitaxel is dispersed uniformly in the PLGA matrix, whereas HAuNS are primarily dispersed in the inner water phase of the microspheres. From reference [88]. Copyright Wiley-VCH. Reprinted with permission.

For laser-controllable drug delivery systems, HAuNS not only can serve as photothermal coupling elements within the system but also can act as delivery vehicles carrying therapeutic agents. Because of their hollow interior, HAuNS can offer super drug-payload capacity, which can allow excellent NIR laser-controlled release of therapeutic agents. The anticancer drug doxorubicin can be absorbed onto HAuNS through the electrostatic interaction between the positively charged drug molecules and the negatively charged citrate-stabilized HAuNS [89]. HAuNS have been shown to have extremely high payload capacity, resulting in as high as 63% loading of doxorubicin by weight, which is at least 3.5 times the doxorubicin payload of solid gold nanoparticles of the same size, surface charge, and weight. HAuNS with high photothermal conversion efficiency can be exploited for controlled release of doxorubicin from the particles using NIR light as the external stimulus to trigger drug release (Fig. 14.6). *In vitro* results showed that doxorubicin-loaded HAuNS permitted two modes of cell killing with a single nanodevice, that is, PTA mediated by HAuNS and cytotoxicity mediated by doxorubicin released from HAuNS upon NIR laser irradiation [89].

14.6 PHOTOTHERMAL TRANSFECTION

Traditional nanoparticulate polyelectrolyte-mediated delivery of siRNA molecules does not allow temporal and spatial control of RNA interference (RNAi) [90–92]. Once siRNA is taken up by cells, gene silencing becomes an irreversible process. Unlike this traditional approach, light-induced siRNA delivery techniques have enormous advantages in terms of limiting siRNA delivery to the illuminated area, thereby reducing nonspecific effects. However, a major limitation of these techniques is that UV/Vis light (commonly 350 ~ 450 nm) cannot penetrate tissues [93, 94].

Recently, Lu et al. reported NIR light-inducible silencing of NF-κB p65 through the use of photothermal coupling agents, that is, HAuNS, as siRNA carriers [33]. The authors achieved highly efficient siRNA conjugation through the functionalized sulfhydryl group at the 5′ end of the sense chain in siRNA. HAuNS' combination of optimal light absorption capacity in the NIR region and small size enabled targeted delivery of siRNA to cancer cells. Micro-positron emission tomography/computed

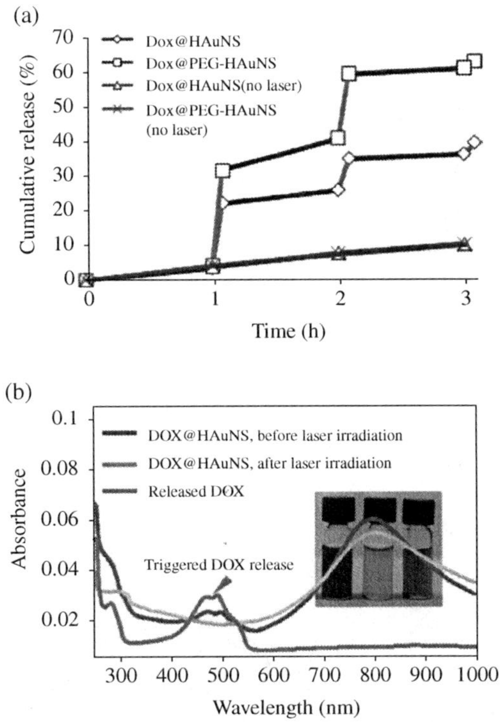

FIGURE 14.6 Mechanism NIR light-triggered release of doxorubicin (DOX) from DOX-loaded HAuNS (DOX–HAuNS) and DOX-loaded PEG-HAuNS (DOX–PEG-HAuNS). (a) DOX release profiles in the presence and absence of NIR laser irradiation. Irradiation with NIR laser caused rapid DOX release during NIR exposure (5 min, mid gray), and the release was turned off when laser was switched off. (b) Absorption spectra of DOX–HAuNS before and after NIR laser irradiation. Inset: Photograph of aqueous solutions of (a) DOX–HAuNS before laser irradiation, (b) released DOX collected in the supernatant, and (c) DOX–HAuNS after complete release of DOX. From reference [89]. Copyright American Chemical Society. Reprinted with permission.

tomography (micro-PET/CT) proved that folate-targeted nanoconstructs exhibited significantly higher tumor uptake than their nontargeted counterparts after intravenous injection. Efficient downregulation of NF-κB p65 and enhanced apoptotic response to irinotecan were demonstrated both *in vitro* and *in vivo* after systemic administration of the HAuNS nanoconstructs and NIR irradiation.

This form of RNAi technique was termed "photothermal transfection" [33]. Unlike traditional nanoparticulate polyelectrolyte-mediated RNAi, photothermal transfection can be confined to the illuminated area, which reduces nonspecific side effects. Furthermore, because photothermal transfection is induced by NIR light and because tissue absorption of NIR light is minimal, photothermal transfection can cause less skin or normal tissue damage than UV/Vis light-induced siRNA delivery techniques.

Photothermal transfection mediated by HAuNS and NIR light involves three steps: First, siRNA-coated HAuNS are internalized by the target cells through receptor-mediated endocytosis. Second, irradiation of the target cells by pulsed NIR laser induces selective thermal disruption of the endolysosomal membrane and escape of

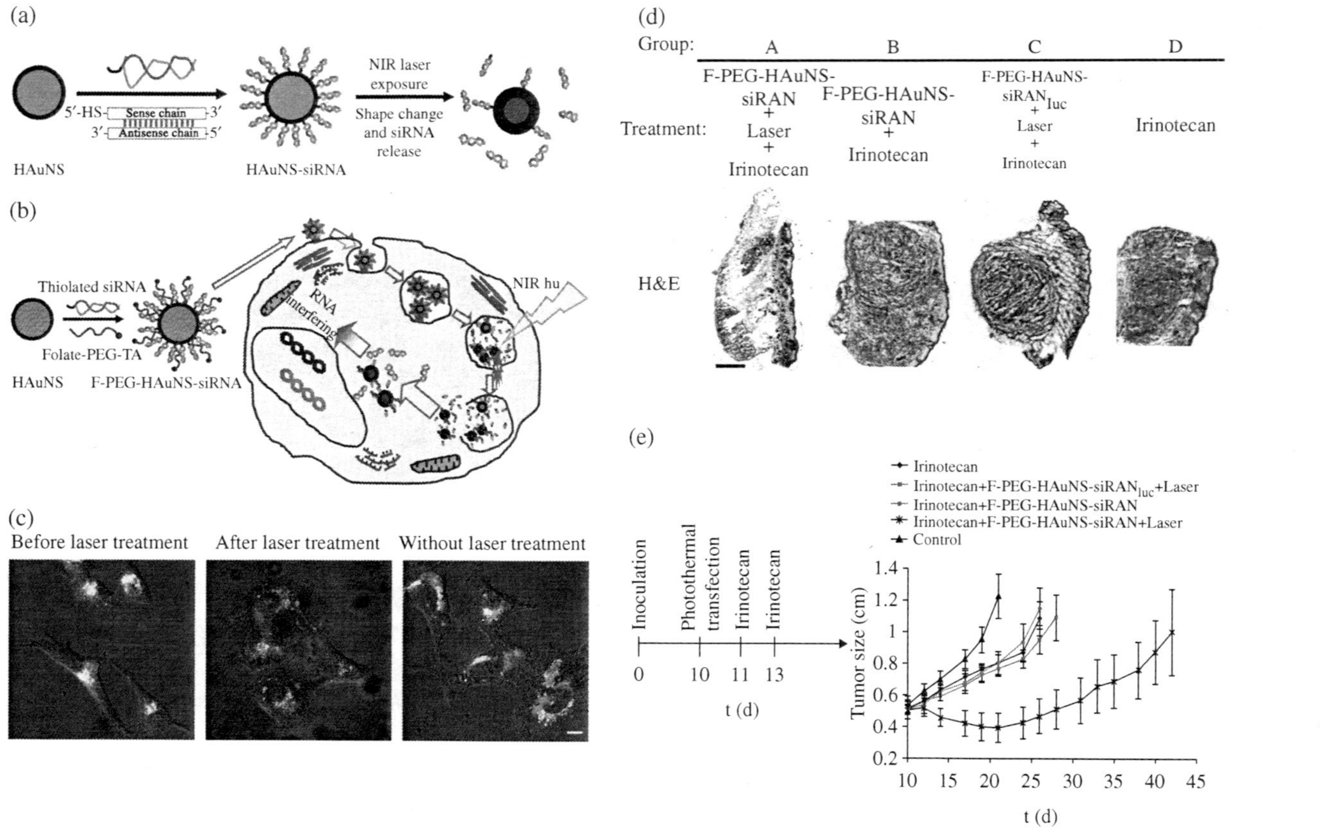

FIGURE 14.7 Photothermal transfection. (a) Scheme for the conjugation of HAuNS-siRNA and photothermal-induced release of siRNA. (b) Scheme for the synthesis of folate-modified HAuNS-siRNA conjugates (F-PEG-HAuNS-siRNA) and their proposed intracellular itinerary following NIR laser irradiation. TA, thioctic acid. (c) Photothermal-induced endolysosomal escape of Dy547-labeled siRNA. Green represents LysoTracker green-labeled endolysosomes; red represents Dy547-labeled siRNA. Bar, 10 μm. (d and e) Effect of p65 siRNA photothermal transfection combined with irinotecan on nude mice bearing HeLa cancer xenografts. (d) Representative micrographs of tumors stained with hematoxylin–eosin. Bar, 1 mm. (e) Tumor size versus time curve. Control, tumor-bearing mice that did not receive any treatment. From reference [33]. Copyright American Association for Cancer Research. Reprinted with permission. For color details, please see color plate section.

nanoparticles from endolysosomes. Third, NIR light simultaneously induces decoupling of siRNA from the nanoparticle carrier and release of free siRNA into the cytosol (Fig. 14.7) [33]. The compelling *in vivo* results provided evidence that photothermal transfection with siRNA targeting NF-κB p65 could potentially sensitize tumors to chemotherapeutic agents in the clinic. Photothermal transfection may be used in rational design of nanoscale delivery system for improving success rates of cancer chemotherapies. Because NIR light can penetrate deep into tissues and can be delivered at predetermined sites and because many target siRNAs have overlapping/redundant features in biological systems, RNAi regulated by NIR irradiation should offer a significant advantage in avoiding unwanted side effects.

Another research group showed evidence of pulsed NIR laser-dependent release of siRNA from siRNA-coated 40-nm HAuNS *in vitro* [95]. In that system, HAuNS were conjugated with thiolated siRNA to form a HAuNS-siRNA assembly. Tat-lipid conjugate was then coated on this assembly, creating a lipid bilayer. The resulting complex was prepared through electrostatic assembly of highly cationic Tat-lipid around the anionic HAuNS-siRNA assembly, with membrane sealing to stabilize the complex forming the inner layer, which enabled the outer Tat layer to stretch outward for cellular binding. This design was also expected to protect RNA from rapid degradation by enzymes. Given endocytosis of the earlier complex, a direct endosome release mechanism was supposed to be activated by pulsed NIR laser treatment. The escape of the siRNA from endosomes was attributed to local heating and cavitation. *In vitro* results showed that the NIR laser-induced release of siRNA from the HAuNS was power and time dependent, through surface-linker bond cleavage. This HAuNS-activated release of siRNA using pulsed NIR laser achieved gene silencing in a temporally and spatially controlled manner, which may facilitate more sophisticated developmental biology and therapeutic studies [95].

14.7 PERSPECTIVES ON USE OF MULTIFUNCTIONAL HAuNS AS CANCER THERANOSTIC AGENTS

Theranostic nanomedicines useful for diagnosis, drug delivery, and monitoring of therapeutic response are expected to play a significant role in the dawning era of personalized medicine [1]. The main challenges in developing theranostic agents are achieving both multifunctionality and selectivity. The criteria for successful nanoparticle-based theranostics include (i) intrinsic imaging contrast properties, (ii) high payload capacity for both drugs and contrast agents, and (iii) feasibility of surface modification for selective and specific delivery at the cell and molecular levels.

As outlined earlier in this chapter, HAuNS have been established to be ideal candidate nanocarriers. By tuning the particle size and the dimensions of the core and shell of the nanostructure, one can fabricate HAuNS with light absorption and scattering in the NIR regions of the electromagnetic spectrum. As NIR light is harmless and has maximum tissue penetration, HAuNS with absorption and scattering in the NIR region have been successfully tailored for PAT, PTA, and controlled release of chemotherapeutic agents and siRNA.

Hollow gold nanospheres that integrate diagnostic and therapeutic capabilities in a single entity can readily be explored for a myriad of potential applications in cancer

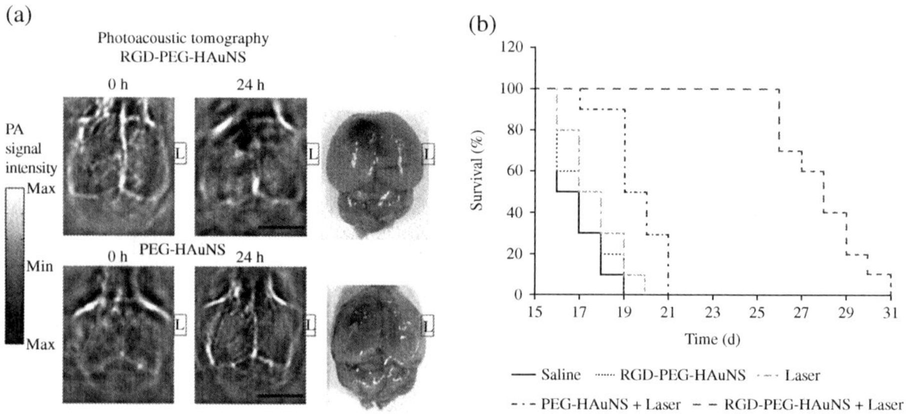

FIGURE 14.8 (a) PAT images of U87 human glioma in mouse brains before (0 h) and 24 h after intravenous injection of nanoparticles (bar, 5 mm). Photographs of corresponding mouse brains were used to confirm tumor location. Arrows, locations of tumors; L, left. (b) PAT of U87 human glioma in mouse brains following different treatments. The life spans of the tumor-bearing mice are presented by Kaplan–Meier survival curves. From reference [96]. Copyright American Association for Cancer Research. Reprinted with permission.

theranostics. HAuNS will enable acquisition of *in vivo* imaging data with PAT in the preoperative, intraoperative, and postoperative settings, as well as image-guided surgery or treatment, such as PTA and NIR-triggered drug treatment. In the current study, Lu et al. [96] developed cyclic(KRGDf) peptide-conjugated HAuNS targeting integrins that are overexpressed in both glioma and angiogenic blood vessels. The targeted HAuNS permitted accurate photoacoustic imaging of orthotopically inoculated U87 glioma in nude mice and mediated selective antitumor effect when mice were irradiated with NIR laser (Fig. 14.8). The findings suggest potential applications of HAuNS as a novel theranostic platform. HAuNS can serve as (i) efficient optical contrast agents for photoacoustic imaging for pretreatment diagnosis, real-time monitoring of treatment, as well as assessment of treatment outcome and (ii) efficient photothermal coupling agents for PTA of tumor cells under the guidance of photoacoustic imaging. In clinic, human malignant gliomas are characterized by their ability to infiltrate and invade surrounding normal brain tissues. Complete tumor resection is often limited by the surgeon's ability to distinguish residual tumor tissue from surrounding brain tissue. Tumors often recur because it is extremely difficult to remove infiltrating tumor cells without affecting vital brain functions. We envision with this technique NIR laser treatment of the surgical bed through optical fiber for the eradication of residual brain tumor cells.

Furthermore, HAuNS labeled with different types of radioisotopes, such as ^{64}Cu and ^{111}In, could be sensitively detected with positron emission tomography (PET) or single photon emission computed tomography, respectively. Such radioisotope-labeled HAuNS could be used for a "see-and-treat" strategy—for example, imaging after injection of HAuNS would be used to test individual patients for drug delivery efficiency and to tailor an innovative treatment for each patient based on the imaging results. This strategy is expected to have significant impact on the management of cancer. For instance, given that micro-PET/CT using ^{64}Cu-labeled c(KRGDf)-PEG-HAuNS confirmed selective uptake of the nanoparticles in U87 tumors (Fig. 14.9)

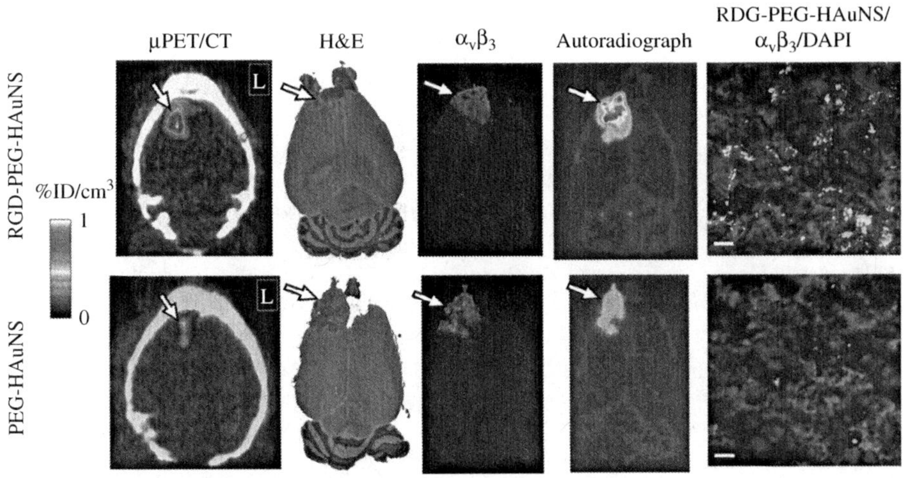

FIGURE 14.9 Targeted delivery of ^{64}Cu-labeled c(KRGDf)-PEG-HAuNS to U87 human glioma in mouse brains. Shown are micro-PET/CT images (obtained 24 h after intravenous injection of nanoparticles), photographs of mouse brains stained with hematoxylin–eosin and $\alpha_v\beta_3$ and autoradiographs of mouse brains, and immunofluorescence micrographs of tumor tissue at high magnification. Red, $\alpha_v\beta_3$; green, scattering signal of HAuNS under dark field; blue, DAPI-stained cell nuclei; arrows, tumors; bar, 10 μm. From reference [96]. Copyright American Association for Cancer Research. Reprinted with permission. For color details, please see color plate section.

[96], PET/CT with ^{64}Cu-labeled targeted HAuNS may provide effective imaging method for HAuNS-mediated PTA therapy.

We expect that in the future, development of HAuNS with multimodal cell killing techniques will provide the synergy needed for a breakthrough in the treatment of accessible advanced cancers. For example, integration of PTA with laser-controlled release of chemotherapeutic agents into a single nanodevice would allow synergistic anticancer treatment. Integration of photothermal transfection into HAuNS mediating laser-controlled release of siRNA or PTA would sensitize cancer cells to the effects of chemotherapy or heating. Such "two-punch" approaches to concurrent cancer therapy are expected to increase the likelihood of cell killing with precise temporal and spatial control and are thus promising approaches to improving therapeutic efficacy.

To date, most research on nanoparticle-based theranostics has reached only the proof-of-concept stage. Detailed preclinical studies should be conducted, especially in the areas of pharmacokinetics and *in vivo* tumor targeting and therapeutic effect. Furthermore, the metabolism of the gold nanoparticles and acute and long-term toxicity need to be examined thoroughly, although preliminary studies suggest that gold nanoparticles are nontoxic. HAuNS should be required to meet the safety, effectiveness, and quality control standards required for all new drugs.

ACKNOWLEDGEMENTS

We would like to thank Stephanie Deming for editing the manuscript. This work was supported in part by the National Institutes of Health through MD Anderson Cancer

Center support grant CA016672 and by the National Institutes of Health grant R01 CA119387 and the John S. Dunn Foundation.

REFERENCES

1. Xie J, Lee S, Chen X. Nanoparticle-based theranostic agents. Adv Drug Deliv Rev 2010;62 (11):1064–1079.
2. Jain PK, Huang X, El-Sayed IH, et al. Review of some interesting surface plasmon resonance-enhanced properties of noble metal nanoparticles and their applications to biosystems. Plasmonics 2007;2 (3):107–118.
3. Huang XH, Jain PK, El-Sayed IH, et al. Plasmonic photothermal therapy (PPTT) using gold nanoparticles. Laser Med Sci 2008;23 (3):217–228.
4. Hirsch LR, Jackson JB, Lee A, et al. A whole blood immunoassay using gold nanoshells. Anal Chem 2003;75 (10):2377–2381.
5. West JL, Halas N, Sershen SR. Optically-responsive nanoshell composites. Abstr Papers Am Chem Soc 2003;225:U522.
6. Gobin AM, Lee MH, Drezek RA, et al. Nanoshells for combined cancer therapy and imaging in vivo. Clin Cancer Res 2005;11 (24):9095s.
7. Hirsch LR, Stafford RJ, Bankson JA, et al. Nanoshell-mediated near-infrared thermal therapy of tumors under magnetic resonance guidance. Proc Natl Acad Sci USA 2003;100 (23):13549–13554.
8. West JL, Halas NJ. Engineered nanomaterials for biophotonics applications: improving sensing, imaging, and therapeutics. Annu Rev Biomed Eng 2003;5:285–292.
9. Lal S, Clare SE, Halas NJ. Nanoshell-enabled photothermal cancer therapy: impending clinical impact. Acc Chem Res 2008;41 (12):1842–1851.
10. Oldenburg SJ, Averitt RD, Westcott SL, et al. Nanoengineering of optical resonances. Chem Phys Lett 1998;288 (2–4):243–247.
11. Liang HP, Wan LJ, Bai CL, et al. Gold hollow nanospheres: tunable surface plasmon resonance controlled by interior-cavity sizes. J Phys Chem B 2005;109 (16):7795–7800.
12. Kobayashi Y, Horie M, Konno M, et al. Preparation and properties of silica-coated cobalt nanoparticles. J Phys Chem B 2003;107 (30):7420–7425.
13. Schwartzberg AM, Olson TY, Talley CE, et al. Synthesis, characterization, and tunable optical properties of hollow gold nanospheres. J Phys Chem B 2006;110 (40):19935–19944.
14. Prevo BG, Esakoff SA, Mikhailovsky A, et al. Scalable routes to gold nanoshells with tunable sizes and response to near-infrared pulsed-laser irradiation. Small 2008;4 (8):1183–1195.
15. Sun YG, Mayers BT, Xia YN. Template-engaged replacement reaction: a one-step approach to the large-scale synthesis of metal nanostructures with hollow interiors. Nano Lett 2002;2 (5):481–485.
16. Sun YG, Xia YN. Shape-controlled synthesis of gold and silver nanoparticles. Science 2002;298 (5601):2176–2179.
17. Sun Y, Xia Y. Increased sensitivity of surface plasmon resonance of gold nanoshells compared to that of gold solid colloids in response to environmental changes. Anal Chem 2002;74 (20):5297–5305.
18. Roosen AR, Carter WC. Simulations of microstructural evolution: anisotropic growth and coarsening. Phys Stat Mech Appl 1998;261 (1–2):232–247.

19. Halas N. Playing with plasmons. Tuning the optical resonant properties of metallic nanoshells. MRS Bulletin 2005;30 (5):362–367.

20. Hu M, Chen JY, Li ZY, et al. Gold nanostructures: engineering their plasmonic properties for biomedical applications. Chem Soc Rev 2006;35 (11):1084–1094.

21. Jain PK, Lee KS, El-Sayed IH, et al. Calculated absorption and scattering properties of gold nanoparticles of different size, shape, and composition: applications in biological imaging and biomedicine. J Phys Chem B 2006;110 (14):7238–7248.

22. El-Sayed MA. Some interesting properties of metals confined in time and nanometer space of different shapes. Acc Chem Res 2001;34 (4):257–264.

23. Mie G. Beiträge zur optik trüber medien, speziell kolloidaler metallösungen. Ann Phys 1908;25 (3):377–445.

24. Stratton JA. *Electromagnetic Theory*. New York: McGraw-Hill; 1941. pp 569.

25. Kerker M. *Scattering of Light and Other Electromagnetic Radiation*. Physical Chemistry. New York: Academic Press; 1997. 666 pp.

26. Prodan E, Radloff C, Halas NJ, et al. A hybridization model for the plasmon response of complex nanostructures. Science 2003;302 (5644):419–422.

27. Chandra M, Dowgiallo AM, Knappenberger KL, Jr. Controlled plasmon resonance properties of hollow gold nanosphere aggregates. J Am Chem Soc 2010;132 (44): 15782–15789.

28. Jain PK, El-Sayed IH, El-Sayed MA. Au nanoparticles target cancer. Nano Today 2007; 2 (1):18–29.

29. Link S, El-Sayed MA. Shape and size dependence of radiative, non-radiative and photothermal properties of gold nanocrystals. Int Rev Phys Chem 2000;19 (3):409–453.

30. Hu M, Petrova H, Chen JY, et al. Ultrafast laser studies of the photothermal properties of gold nanocages. J Phys Chem B 2006;110 (4):1520–1524.

31. Hartland GV. Measurements of the material properties of metal nanoparticles by time-resolved spectroscopy. Phys Chem Chem Phys 2004;6 (23):5263–5274.

32. Du H, Fuh RCA, Li JZ, et al. PhotochemCAD: a computer-aided design and research tool in photochemistry. Photochem Photobiol 1998;68 (2):141–142.

33. Lu W, Zhang G, Zhang R, et al. Tumor site-specific silencing of NF-kappaB p65 by targeted hollow gold nanosphere-mediated photothermal transfection. Cancer Res 2010;70 (8):3177–3188.

34. Link S, Burda C, Nikoobakht B, et al. Laser-induced shape changes of colloidal gold nanorods using femtosecond and nanosecond laser pulses. J Phys Chem B 2000;104 (26):6152–6163.

35. Chen JY, Wiley B, Li ZY, et al. Gold nanocages: engineering their structure for biomedical applications. Adv Mater 2005;17 (18):2255–2261.

36. Chen CC, Lin YP, Wang CW, et al. DNA-gold nanorod conjugates for remote control of localized gene expression by near infrared irradiation. J Am Chem Soc 2006;128 (11):3709–3715.

37. Takahashi H, Niidome Y, Yamada S. Controlled release of plasmid DNA from gold nanorods induced by pulsed near-infrared light. Chem Commun 2005;(17):2247–2249.

38. Akchurin G, Khlebtsov B, Akchurin G, et al. Gold nanoshell photomodification under a single-nanosecond laser pulse accompanied by color-shifting and bubble formation phenomena. Nanotechnol 2008;19 (1):1–8.

39. Wheeler DA, Newhouse RJ, Wang HN, et al. Optical properties and persistent spectral hole burning of near infrared-absorbing hollow gold nanospheres. J Phys Chem C 2010;114 (42):18126–18133.

40. Richardson HH, Hickman ZN, Govorov AO, et al. Thermooptical properties of gold nanoparticles embedded in ice: characterization of heat generation and melting. Nano Lett 2006;6 (4):783–788.

41. Roper DK, Ahn W, Hoepfner M. Microscale heat transfer transduced by surface plasmon resonant gold nanoparticles. J Phys Chem C Nanomater Interfaces 2007;111 (9):3636–3641.

42. Zhang HF, Maslov K, Wang LHV. In vivo imaging of subcutaneous structures using functional photoacoustic microscopy. Nat Protoc 2007;2 (4):797–804.

43. Zhang HF, Maslov K, Stoica G, et al. Functional photoacoustic microscopy for high-resolution and noninvasive in vivo imaging. Nat Biotechnol 2006;24 (7):848–851.

44. Oleary MA, Boas DA, Chance B, et al. Experimental images of heterogeneous turbid media by frequency-domain diffusing-photon tomography. Opt Lett 1995;20 (5):426–428.

45. Boas DA, Brooks DH, Miller EL, et al. Imaging the body with diffuse optical tomography. IEEE Signal Proc Mag 2001;18 (6):57–75.

46. Wang XD, Pang YJ, Ku G, et al. Noninvasive laser-induced photoacoustic tomography for structural and functional in vivo imaging of the brain. Nat Biotechnol 2003;21 (7):803–806.

47. Hoelen CGA, de Mul FFM, Pongers R, et al. Three-dimensional photoacoustic imaging of blood vessels in tissue. Opt Lett 1998;23 (8):648–650.

48. Wang XD, Ku G, Wegiel MA, et al. Noninvasive photoacoustic angiography of animal brains in vivo with near-infrared light and an optical contrast agent. Opt Lett 2004;29 (7):730–732.

49. Li ML, Oh JT, Xie XY, et al. Simultaneous molecular and hypoxia imaging of brain tumors in vivo using spectroscopic photoacoustic tomography. Proc IEEE 2008;96 (3):481–489.

50. Weissleder R. A clearer vision for in vivo imaging. Nat Biotechnol 2001;19 (4):316–317.

51. Wang YW, Xie XY, Wang XD, et al. Photoacoustic tomography of a nanoshell contrast agent in the in vivo rat brain. Nano Lett 2004;4 (9):1689–1692.

52. Yang XM, Skrabalak SE, Li ZY, et al. Photoacoustic tomography of a rat cerebral cortex in vivo with au nanocages as an optical contrast agent. Nano Lett 2007;7 (12):3798–3802.

53. Song KH, Kim CH, Cobley CM, et al. Near-infrared gold nanocages as a new class of tracers for photoacoustic sentinel lymph node mapping on a rat model. Nano Lett 2009; 9 (1):183–188.

54. Agarwal A, Huang SW, O'Donnell M, et al. Targeted gold nanorod contrast agent for prostate cancer detection by photoacoustic imaging. J Appl Phys 2007;102 (6):064701.

55. Song KH, Kim C, Maslov K, et al. Noninvasive in vivo spectroscopic nanorod-contrast photoacoustic mapping of sentinel lymph nodes. Eur J Radiol 2009;70 (2):227–231.

56. Li PC, Wang CRC, Shieh DB, et al. In vivo photoacoustic molecular imaging with simultaneous multiple selective targeting using antibody-conjugated gold nanorods. Opt Express 2008;16 (23):18605–18615.

57. Li PC, Wei CW, Liao CK, et al. Photoacoustic imaging of multiple targets using gold nanorods. IEEE Trans Ultrason Ferroelectr Freq Control 2007;54 (8):1642–1647.

58. Kim K, Huang SW, Ashkenazi S, et al. Photoacoustic imaging of early inflammatory response using gold nanorods. Appl Phys Lett 2007;90 (22):223901.

59. Eghtedari M, Oraevsky A, Copland JA, et al. High sensitivity of in vivo detection of gold nanorods using a laser optoacoustic imaging system. Nano Lett 2007;7 (7):1914–1918.

60. Eghtedari M, Liopo AV, Copland JA, et al. Engineering of hetero-functional gold nanorods for the in vivo molecular targeting of breast cancer cells. Nano Lett 2009;9 (1):287–291.

61. Lu W, Huang Q, Ku G, et al. Photoacoustic imaging of living mouse brain vasculature using hollow gold nanospheres. Biomaterials 2010;31 (9):2617–2626.

62. De La Zerda A, Zavaleta C, Keren S, et al. Carbon nanotubes as photoacoustic molecular imaging agents in living mice. Nat Nanotechnol 2008;3 (9):557–562.

63. Copland JA, Eghtedari M, Popov VL, et al. Bioconjugated gold nanoparticles as a molecular based contrast agent: Implications for imaging of deep tumors using optoacoustic tomography. Mol Imaging Biol 2004;6 (5):341–349.

64. Pramanik M, Song KH, Swierczewska M, et al. In vivo carbon nanotube-enhanced non-invasive photoacoustic mapping of the sentinel lymph node. Phys Med Biol 2009;54 (11):3291–3301.

65. Xiang LZ, Yuan Y, Xing D, et al. Photoacoustic molecular imaging with antibody-functionalized single-walled carbon nanotubes for early diagnosis of tumor. J Biomed Opt 2009;14 (2):021008.

66. Chamberland DL, Agarwal A, Kotov N, et al. Photoacoustic tomography of joints aided by an Etanercept-conjugated gold nanoparticle contrast agent—an ex vivo preliminary rat study. Nanotechnology 2008;19 (9):095101.

67. Wang B, Yantsen E, Larson T, et al. Plasmonic intravascular photoacoustic imaging for detection of macrophages in atherosclerotic plaques. Nano Lett 2009;9 (6):2212–2217.

68. Slomiany MG, Dai L, Tolliver LB, et al. Inhibition of functional hyaluronan-CD44 interactions in CD133-positive primary human ovarian carcinoma cells by small hyaluronan oligosaccharides. Clin Cancer Res 2009;15 (24):7593–7601.

69. Prahl S. 1998. Optical absorption of hemoglobin. Available at http://omlc.ogi.edu/spectra/hemoglobin/. Accessed 2013 Apr 23.

70. Anderson RR, Parrish JA. Selective photothermolysis—precise microsurgery by selective absorption of pulsed radiation. Science 1983;220 (4596):524–527.

71. Jori G, Spikes JD. Photothermal sensitizers—possible use in tumor-therapy. J Photochem Photobiol B 1990;6 (1–2):93–101.

72. Camerin M, Rello S, Villanueva A, et al. Photothermal sensitisation as a novel therapeutic approach for tumours: studies at the cellular and animal level. Eur J Cancer 2005;41 (8):1203–1212.

73. Soncin M, Busetti A, Fusi F, et al. Irradiation of amelanotic melanoma cells with 532 nm high peak power pulsed laser radiation in the presence of the photothermal sensitizer Cu(II)-hematoporphyrin: a new approach to cell photoinactivation. Photochem Photobiol 1999;69 (6):708–712.

74. Moyer HR, Delman KA. The role of hyperthermia in optimizing tumor response to regional therapy. Int J Hyperthermia 2008;24 (3):251–261.

75. Stern JM, Hsieh JT, Park S, et al. Successful gold nanoshell assisted laser ablation of prostate carcinoma in vitro. J Urol 2007;177 (4):52.

76. Stern JM, Hsieh JT, Park S, et al. Gold nanoshell assisted laser ablation of a prostate cancer cell line. J Endourol 2006;20:A193.

77. Stern JM, Stanfield J, Lotan Y, et al. Efficacy of laser-activated gold nanoshells in ablating prostate cancer cells in vitro. J Endourol 2007;21 (8):939–943.

78. Chithrani BD, Ghazani AA, Chan WCW. Determining the size and shape dependence of gold nanoparticle uptake into mammalian cells. Nano Lett 2006;6 (4):662–668.

79. Zhang G, Yang Z, Lu W, et al. Influence of anchoring ligands and particle size on the colloidal stability and in vivo biodistribution of polyethylene glycol-coated gold nanoparticles in tumor-xenografted mice. Biomaterials 2009;30 (10):1928–1936.

80. Yuan F, Dellian M, Fukumura D, et al. Vascular permeability in a human tumor xenograft: molecular size dependence and cutoff size. Cancer Res 1995;55 (17):3752–3756.
81. Hobbs SK, Monsky WL, Yuan F, et al. Regulation of transport pathways in tumor vessels: role of tumor type and microenvironment. Proc Natl Acad Sci USA 1998;95 (8):4607–4612.
82. Melancon MP, Lu W, Yang Z, et al. In vitro and in vivo targeting of hollow gold nanoshells directed at epidermal growth factor receptor for photothermal ablation therapy. Mol Cancer Ther 2008;7 (6):1730–1739.
83. Lu W, Xiong C, Zhang G, et al. Targeted photothermal ablation of murine melanomas with melanocyte-stimulating hormone analog-conjugated hollow gold nanospheres. Clin Cancer Res 2009;15 (3):876–886.
84. Cheng FY, Chen CT, Yeh CS. Comparative efficiencies of photothermal destruction of malignant cells using antibody-coated silica–Au nanoshells, hollow Au/Ag nanospheres and Au nanorods. Nanotechnology 2009;20 (42):425104.
85. Siegrist W, Solca F, Stutz S, et al. Characterization of receptors for alpha-melanocyte-stimulating hormone on human melanoma cells. Cancer Res 1989;49 (22):6352–6358.
86. Chen J, Cheng Z, Hoffman TJ, et al. Melanoma-targeting properties of (99 m)technetium-labeled cyclic alpha-melanocyte-stimulating hormone peptide analogues. Cancer Res 2000;60 (20):5649–5658.
87. Wu G, Mikhailovsky A, Khant HA, et al. Remotely triggered liposome release by near-infrared light absorption via hollow gold nanoshells. J Am Chem Soc 2008;130 (26):8175–8177.
88. You J, Shao RP, Wei X, et al. Near-infrared light triggers release of paclitaxel from biodegradable microspheres: photothermal effect and enhanced antitumor activity. Small 2010;6 (9):1022–1031.
89. You J, Zhang G, Li C. Exceptionally high payload of doxorubicin in hollow gold nanospheres for near-infrared light-triggered drug release. ACS Nano 2010;4 (2):1033–1041.
90. Medina-Kauwe LK, Xie J, Hamm-Alvarez S. Intracellular trafficking of nonviral vectors. Gene Ther 2005;12 (24):1734–1751.
91. Bartlett DW, Su H, Hildebrandt IJ, et al. Impact of tumor-specific targeting on the biodistribution and efficacy of siRNA nanoparticles measured by multimodality in vivo imaging. Proc Natl Acad Sci USA 2007;104 (39):15549–15554.
92. Thomas M, Klibanov AM. Conjugation to gold nanoparticles enhances polyethylenimine's transfer of plasmid DNA into mammalian cells. Proc Natl Acad Sci USA 2003;100 (16):9138–9143.
93. Boe S, Longva AS, Hovig E. Evaluation of various polyethylenimine formulations for light-controlled gene silencing using small interfering RNA molecules. Oligonucleotides 2008;18 (2):123–132.
94. Oliveira S, Fretz MM, Hogset A, et al. Photochemical internalization enhances silencing of epidermal growth factor receptor through improved endosomal escape of siRNA. Biochim Biophys Acta 2007;1768 (5):1211–1217.
95. Braun GB, Pallaoro A, Wu GH, et al. Laser-activated gene silencing via gold nanoshell-siRNA conjugates. Acs Nano 2009;3 (7):2007–2015.
96. Lu W, Melancon MP, Xiong C, et al. Effects of photoacoustic imaging and photothermal ablation therapy mediated by targeted hollow gold nanospheres in an orthotopic mouse xenograft model of glioma. Cancer Res 2011;71 (19):6116–6121.

CHAPTER 15

IMAGING PULMONARY DRUG DELIVERY

STEPHEN P. NEWMAN

15.1 PULMONARY DRUG DELIVERY

15.1.1 The Development of Pulmonary Drugs

The history of pulmonary drug delivery dates back thousands of years, with records being found in the writings of many ancient cultures [1]. Its application in traditional Ayurvedic medicine in India around 2000 B.C. may mark the origin of using inhalation to treat respiratory conditions. Most ancient remedies involved inhaling the smoke or vapors from burning plants, and this practice continued into modern times, with asthma cigarettes containing *Datura stramonium* and other substances being popular in the early twentieth century [2]. Vapor inhalers were available in Europe from around 1780, the first of these being a pewter tankard used to deliver opium vapor for treatment of cough [3]. The first atomizers did not appear until the mid-nineteenth century, but this development marked an important transition to the use of aerosol particles and droplets (rather than vapor) as a delivery medium for drug [4]. The twentieth century saw the development of modern drugs and of modern inhaler devices to deliver them.

Pulmonary delivery mainly involves topical therapy of some pulmonary conditions but also offers potential for systemic therapy using the lungs as a portal of entry to the body (Fig. 15.1). The most commonly used topical therapies are inhaled bronchodilators and corticosteroids, given to patients with asthma and chronic obstructive pulmonary disease (COPD). These two distinct, although related, conditions are very common in many modern societies. Asthma has a country-dependent prevalence of 1–18% of the population [6]. COPD is a major cause of morbidity and mortality worldwide, and its prevalence is linked to cigarette smoking [7]. Inhaled bronchodilators are either beta-2 receptor agonists (e.g., salbutamol, terbutaline sulfate, or salmeterol) or anticholinergic compounds intended to block muscarinic receptors (e.g., ipratropium bromide or tiotropium bromide). Bronchodilators may be short

Drug Delivery Applications of Noninvasive Imaging: Validation from Biodistribution to Sites of Action, First Edition. Edited by Chun Li and Mei Tian.

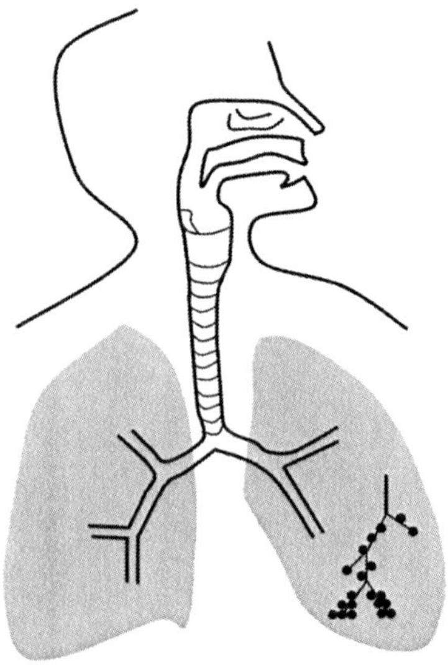

FIGURE 15.1 Drugs are given by the pulmonary route to treat certain lung diseases (notably asthma, COPD, and CF). The pulmonary route also has potential for systemic therapy using the lungs as a portal of entry to the body. However, of the systemic therapies listed, only one insulin product and an inhaler containing ergotamine tartrate for treatment of migraine have so far been marketed (Adapted from Anderson and Newman [5]).

acting (up to 6 h) or long acting (up to 12 h), and drugs with 24-h duration of action have recently been developed [8]. Inhaled corticosteroids (e.g., beclomethasone dipropionate, budesonide, or fluticasone propionate) are used widely as first-line therapy in asthma, because of their anti-inflammatory action [9], but their use in COPD is more controversial [10]. Formulations for inhalation containing both a bronchodilator and a corticosteroid are becoming increasingly common. At least in asthma patients, there seem to be advantages in delivering both "reliever" and "controller" drugs in a single inhaler [11].

Other drugs given by the pulmonary route to treat lung diseases include antibiotics, antiviral agents, and mucoactive agents, for instance, in the treatment of patients with cystic fibrosis (CF) who often have infected pulmonary mucus [12]. Delivery of drug directly to its required site by the pulmonary route offers the potential advantages of minimizing systemic side effects while using a relatively small dose, and in the case of some drugs attaining a rapid onset of action [13].

The potential of the lungs as a portal of entry to the body has long been recognized, and current developments involve three types of drug: proteins and peptides, rapidly acting small molecules, and vaccines. Delivery of peptides and proteins by the pulmonary route can avoid the need for injections of drugs that may be poorly absorbed orally [14, 15]. Remarkably, the first record of delivering insulin by the pulmonary route dates from 1925 [16], although the first pulmonary insulin product (Exubera®) did not reach the market until the twenty-first century [17]. This product was not a commercial success [18], but other inhaled insulin products may follow, and possible roles are seen for other inhaled peptides and proteins, including interferons, calcitonin, and parathyroid hormone [19]. Small molecules with high pulmonary bioavailability and rapid onset of action include nicotine, ergotamine, apomorphine, and fentanyl [20]. The use of the pulmonary route to deliver measles vaccine to children in the developing world is a goal that has attracted much interest, because this would avoid some major problems associated with the use of needles [21].

15.1.2 Modern Inhaler Devices

Today's inhaler devices may be divided into three main categories (i) pressurized metered dose inhalers (pMDIs), (ii) dry powder inhalers (DPIs), and (iii) nebulizers [22].

In the Western world, most pMDIs are now powered by hydrofluoroalkane (HFA) propellants [23], although ozone-depleting chlorofluorocarbons (CFCs) are still used widely in some developing countries [24]. pMDIs contain between 100 and 400 doses and may be used alone or in conjunction with a spacer or "add-on" device placed between the nozzle of the pMDI and the patient's mouth. The drug in a pMDI may be present either as a micronized suspension or as a solution dissolved in a formulation comprising propellants and sometimes other excipients [25]. Patients may not find pMDIs easy to use correctly, particularly because they need to "press and breathe" simultaneously [26].

DPIs contain micronized drug particles, which may be mixed with carrier particles generally composed of lactose [27]. Many different brands of DPI are available, and these may contain single doses in gelatin capsules, a series of individually dispensed doses in capsules or blisters, or multiple doses stored in a powder reservoir [28]. Most DPIs are breath actuated, because the patient's inhalation through the device disperses the drug powder and delivers some of the dose into the lungs [29]. The strength of the patient's inhalation may determine how effectively the powder is dispersed and hence how much drug is deposited in the lungs [30]. However, this effect is highly dependent on the characteristics of both device and formulation. Some novel DPIs are "active" devices in which the powder is dispersed by a mechanism other than the patient's inhalation [31, 32]. Active DPIs may prove particularly suitable for achieving efficient and reproducible pulmonary delivery of novel inhaled therapies, especially those with narrow therapeutic windows.

Nebulizers contain the drug in solution, or occasionally as a suspension of particles. The aerosol may be generated by the passage of compressed air through a venturi (jet nebulizer), by the vibrations in a piezoelectric crystal underneath the nebulizer cup (ultrasonic nebulizer), or by the piezoelectric oscillations in a vibrating mesh or membrane (vibrating membrane nebulizer) [33, 34]. Nebulizers are generally the least convenient inhalers, because they contain single doses, they may have long relatively treatment times, and they may be relatively bulky. However, they are probably the most reliable inhalers, because patients can simply adopt normal relaxed breathing when using them [35]. They can be used to deliver virtually any drug, including off-label therapies [36]. The liquid formulations used in nebulizers may also be used in an additional category of device, metered dose liquid inhalers or "soft mist" inhalers, some of which could become competitors for pMDIs and DPIs in the multidose inhaler market [37]. Currently, only one such device (Respimat® Soft Mist™ Inhaler) has been marketed, but the introduction of other devices in this category is anticipated [38].

15.1.3 Basic Pulmonary Anatomy and Physiology

A basic understanding of pulmonary anatomy and physiology is useful if the challenges of pulmonary drug delivery are to be addressed and if imaging studies of pulmonary drug delivery are to be correctly interpreted. The respiratory tract as a whole consists of three main regions: (i) the upper (extrathoracic) airways, (ii) the conducting (tracheobronchial) airways of the lungs, and (iii) the alveolated (respiratory) airways of the lungs where gas exchange takes place. The upper airways comprise the nasal cavity, the mouth, and the pharynx. The upper airways are potentially a major deposition site of inhaled particles and droplets [39, 40], especially for the rapidly moving "ballistic" sprays produced by many pMDIs [41, 42]. Drugs intended for pulmonary applications are usually inhaled via the mouth rather than the nose in order to avoid aerosol losses in the nasal cavity.

The airways of the lungs may be described to a first approximation by the model of Weibel [43], who considered the lungs as branching system of airway generations in which each airway of a given generation divides into two airways in the next generation. Weibel considered the lungs to consist of 24 branching generations, extending from the trachea (generation 0) to the alveolar sacs (generation 23). The first 17 generations, extending down to the terminal bronchioles (generation 16), are the conducting airways. The larger conducting airways are known as bronchi and smaller airways as bronchioles. The walls of the conducting airways are covered with cilia which move a layer of mucus and deposited materials to the pharynx, from where they may be swallowed or expectorated.

The alveolated airways begin at generation 17 (respiratory bronchioles). There are 300–600 million alveoli in the healthy human lungs [44], and estimates of the average alveolar surface area range from 100 to 190 m^2 [45]. Clearance of deposited insoluble materials from the alveolated region of the lungs takes place much more slowly than from the conducting airways and mainly involves particle engulfment by alveolar macrophages [46]. The human alveoli are lined with a thin layer of epithelial cells (thickness $< 1\,\mu m$). The close proximity of the alveolar air spaces to the pulmonary capillaries facilitates gas exchange and may also allow absorption of deposited drugs into the bloodstream. Many small molecules pass unchanged through the air–blood barrier, whereas some large peptides can undergo marked enzymatic degradation [19].

The conducting airways are thought to be completely cleared of deposited insoluble material within 24 h, at least in healthy subjects. For this reason, the 24-h retention of a deposited insoluble radiolabel (e.g., ^{99m}Tc-labeled colloid particles) may be regarded as a measure of the fraction of the lung dose delivered to the alveolated airways [47, 48].

15.1.4 Aerosol Deposition Processes

Imaging studies are used mainly to quantify pulmonary deposition of drug aerosols in the lungs and in the upper airways (Fig. 15.2). Deposition of pharmaceutical aerosols takes place mainly by inertial impaction (upper airways and larger conducting airways) and gravitational sedimentation (smaller conducting airways and alveoli), supplemented by Brownian diffusion (for submicronic particles) and in some situations by the effects of electrostatic forces [50, 51].

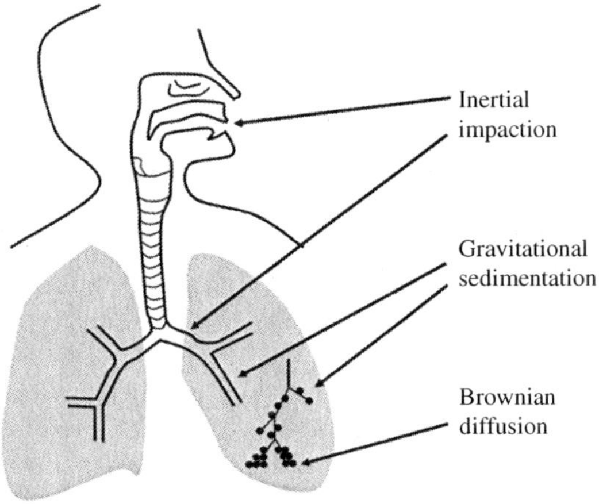

FIGURE 15.2 The deposition of pharmaceutical aerosols in the respiratory tract occurs mainly by inertial impaction (in the upper airways and at bifurcations between major airways) and by gravitational sedimentation (in smaller airways and alveoli). Submicronic particles may deposit by Brownian diffusion in the alveoli (Adapted from Newman [49]).

The probability of inertial impaction is proportional to an "impaction parameter" ($D_a^2 \times Q$), where D_a is the particle aerodynamic diameter and Q is the inhaled flow rate. For aerosols released by many pMDIs, deposition is dominated by inertial impaction in the upper airways [52]. Avoidance of deposition in the upper airways, allowing penetration into the lungs, favors relatively small particles and relatively low inhaled flow rates. Most pharmaceutical aerosols are heterodisperse in character and contain particles with a spectrum of sizes. The aerodynamic particle size distributions (APSDs) of such aerosols are often bimodal in nature. The APSD may be described by the mass median aerodynamic diameter (MMAD) and by the geometric standard deviation (GSD), which are useful measures of the average aerosol size and the spread of sizes, respectively [53]. The "fine particle mass" from an inhaler is defined as the mass of drug contained within the smallest particle size fractions, e.g., particles with $D_a < 5\,\mu m$ [53].

For pMDIs [54] and nebulizers [55], lung deposition is usually enhanced by slow deep inhalation. For breath-actuated DPIs, the situation is more complex, because the aerodynamic diameter of inhaled particles may be determined by the inhaled flow rate; fast inhalation with maximal inspiratory effort is often needed through DPIs in order to de-aggregate the formulation and hence to provide a sufficiently high yield of fine particles [28]. Breath holding at the end of inhalation may augment lung deposition, by allowing time for particles to sediment onto the airway surfaces.

In addition to APSD and inhalation mode, deposition also depends on the airway dimensions of individual subjects. There is high intersubject variability in aerosol deposition resulting partly from random variations in airway dimensions between individuals [56]. In addition, there is a systematic effect caused by the presence of airway narrowing in asthma, COPD, and other diseases, which may cause aerosol particles to deposit in more central airways of the lungs, so that a reduced amount of drug is available to the lung periphery. If an airway is totally blocked with mucus in severe disease, then the area of lung downstream will not be ventilated and no drug aerosol can penetrate [13].

15.2 OVERVIEW OF IMAGING METHODS USED TO QUANTIFY PULMONARY DRUG DELIVERY

Three radionuclide imaging methods may be used to quantify pulmonary drug delivery (i) gamma scintigraphy, (ii) single photon emission computed tomography (SPECT), and (iii) positron emission tomography (PET) [57, 58]. Gamma scintigraphy is a two-dimensional (2D) or planar imaging method using a gamma camera with one or two detector heads and collects data in a 2D array of picture elements (pixels). SPECT and PET are three-dimensional (3D) imaging methods, using a gamma camera with two or three rotating heads and a fixed detection system, respectively. These two imaging methods collect data in a 3D array of volumetric picture elements (voxels).

The anatomy of the human head, thorax, and abdomen lends itself quite well to planar imaging of a radioactive marker deposited in the upper airways and in lungs, as well as any swallowed radioactive marker present in the stomach. However, the sheer complexity of the lung anatomy is a barrier to the interpretation of imaging studies. If the lungs are viewed in 2D, it is inevitable that almost any region of interest will contain an overlapping mixture of conducting airways and alveoli, and relating deposition patterns to anatomical structures will be challenging [59]. The situation is better if the lungs are viewed in 3D, but the inherent resolution of the imaging method used will restrict the airways that can be visualized and the level of detail that can be distinguished. These issues will be discussed further in later sections of this chapter.

15.3 GAMMA SCINTIGRAPHIC IMAGING

15.3.1 Origins

Following the invention of the gamma camera in the late 1950s by Hal Anger [60], 2D radionuclide imaging of various gamma-labeled radiotracers became the cornerstone of nuclear medicine. The ability of 2D imaging to visualize the structure and function of various organs was a significant step forward in medical diagnosis, often providing information that was not previously available. In due course, it was recognized that 2D imaging could also be used to visualize and to quantify drug delivery by a variety of routes [61, 62], and the first studies to quantify pulmonary deposition from a nebulizer were undertaken in the 1970s [63, 64]. Studies to assess deposition from pMDIs [41, 65] and DPIs [66] followed a few years later, once appropriate radiolabeling methods had been developed. Most studies in man have involved the subject sitting or standing in front of a single-headed gamma camera, or between the detectors of a twin-headed gamma camera.

15.3.2 Radiolabeling for Gamma Scintigraphy

Radiolabeling for gamma scintigraphic studies is a major technical issue. The first nebulizer studies used ^{99m}Tc-labeled placebo solutions, and it was recognized that solution formulations in nebulizers are relatively straightforward to radiolabel, because the radioactive tracer can simply be mixed with drug solution [67]. Early studies involving pMDIs used either ^{99m}Tc-labeled solutions [41] or ^{99m}Tc-labeled Teflon particles [65] which were added to placebo pressurized canisters. In the 1980s, it was shown that it is possible to add a ^{99m}Tc radiolabel to a pMDI canister containing drug in such a way that the radiolabel is distributed throughout different particle size bands in an identical or very similar fashion to the drug [68]. There is a reasonable theoretical basis for believing that ^{99m}Tc can associate mainly with the drug particles and not with any other constituent of the formulation, although the quality of the drug–radiolabel association in pMDIs formulated as suspensions is to some extent dependent on the type of surfactant used [69]. Formulation-specific radiolabeling methods have been developed for DPIs [70–72], often involving the addition of ^{99m}Tc to drug particles and then blending these radiolabeled particles with lactose carrier.

A program of radiolabeling validation is an important part of any scintigraphic study, and it must be demonstrated both that the characteristics of the drug product have not been changed significantly by the radiolabeling procedure and that the radiolabel distribution matches the drug distribution as closely as possible across all particle size bands. Comparison of drug and radiolabel size distributions in a cascade impactor is the accepted way of validating a radiolabeling method that is to be used for assessing drug deposition [73, 74]. It is important to recognize that a ^{99m}Tc radiolabel is a marker for drug particles and droplets and not for the drug molecule. Since the radiolabel is not bonded to the drug molecule, it is anticipated that drug and radiolabel will almost certainly be absorbed at different rates through the lung epithelium. Hence ^{99m}Tc labeling is suitable for monitoring drug deposition, but not its subsequent fate.

When used in deposition studies involving pMDIs and DPIs, ^{99m}Tc is usually in the form of sodium pertechnetate, which is absorbed via the epithelium with a half-time of only 10–15 min [75]. This short half-time necessitates rapid administration and imaging in order to avoid errors associated with loss of aerosol from the lungs [76]. A wider range of ^{99m}Tc-labeled radiopharmaceuticals can be used in nebulizers, including ^{99m}Tc-diethylene triamine pentaacetic acid (DTPA) or ^{99m}Tc-labeled colloid particles. These radiopharmaceuticals are cleared more slowly from the lungs [77]; colloid particles are cleared by the mucociliary mechanism. Since administration times using nebulizers are usually much longer than those required for pMDI and DPI, the use of these more slowly cleared radiopharmaceuticals is more appropriate than pertechnetate.

The lack of suitable alternative radiolabels has resulted in virtually all studies of pulmonary drug deposition using ^{99m}Tc. This radionuclide has some ideal characteristics for gamma camera imaging, including a gamma ray energy of 140 keV, which can be detected efficiently by gamma camera, and yet which needs only a few millimeters of lead shielding. The physical half-life of ^{99m}Tc is 6 h, which is long enough for most practical requirements, but short enough that long-term exposure is not an issue. Apart from gamma rays, ^{99m}Tc emits few other particles, and it is among the safest radionuclides administered to man for medical or pharmaceutical purposes. Surprisingly little use has been made of ^{111}In (gamma rays energies of 173 and 247 keV, half-life of 2.8 days) in studies of pulmonary drug delivery, but this radionuclide could have a role in assessing delivery from combination inhalers containing a bronchodilator and a corticosteroid, where one drug might be traced by ^{99m}Tc and the other by ^{111}In. Only occasional studies have attempted to radiolabel a drug molecule for a gamma scintigraphic study. One study used ^{77}Br to radiolabel ipratropium bromide in a pMDI [78], while another study used ^{123}I to radiolabel pentamidine in a nebulizer [79].

The major constituents of organic molecules are hydrogen, carbon, nitrogen, and oxygen. As will be discussed later, ^{11}C, ^{13}N, and ^{15}O are positron emitters, and their high photon energy (511 keV) renders them unsuitable for gamma camera imaging, while their short half-lives (≤20 min) create a number of practical or logistical problems. The radionuclides ^{3}H and ^{14}C are commonly used in medical research but are beta-emitters that cannot be detected by gamma camera.

15.3.3 Quantification of Lung Dose

Gamma scintigraphy studies provide striking visual images of pulmonary deposition patterns (Fig. 15.3), which have often been used as a means of promoting one inhaler product over another. However, as described some years ago, "it has been a challenge to investigators to develop methods to quantitate what is observed visually" [81]. The two major issues in quantification of whole lung deposition or "lung dose" are (i) converting gamma camera counts into deposition data, while taking into account the attenuation losses that occur due to interaction of gamma rays with body tissues [82, 83], and (ii) mass balance, in other words ensuring that the amount of radioactivity present in a dose can be fully accounted for once it has been delivered [84, 85]. These two issues are related, since achievement of an accurate mass balance provides evidence that appropriate correction has been made for tissue attenuation losses.

Correction for tissue attenuation of gamma rays is virtually essential in any gamma scintigraphy study if lung deposition is to be quantified accurately and if reliable comparisons are to be made between different regimens. The image of the lungs is restricted to one plane in gamma scintigraphy, and this makes it impossible to know the thickness of body tissues through which gamma rays have passed and hence impossible to make precise corrections for gamma ray attenuation [86, 87]. However, several tissue attenuation correction methods making a variety of assumptions and

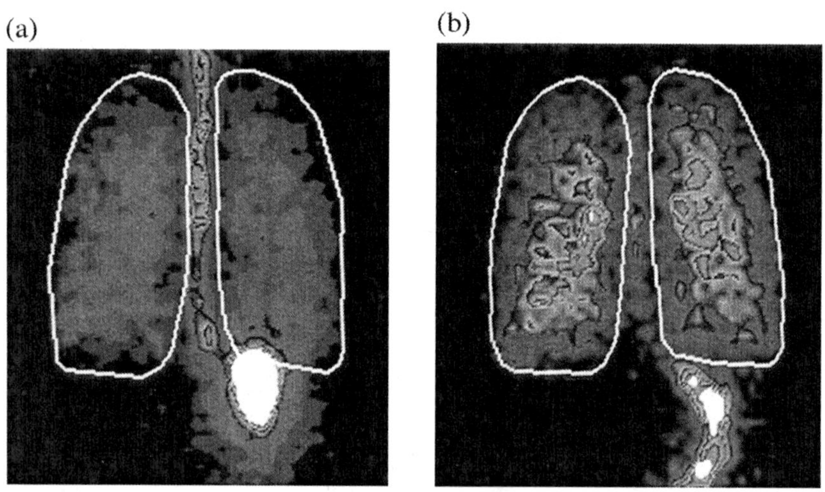

FIGURE 15.3 Examples of gamma scintigraphic images of lung deposition. (a) pMDI and (b) pMDI plus spacer device. The spacer device has increased lung deposition and has reduced the amount of swallowed radioactivity in the stomach. The lung outlines were defined from the contours of a ventilation scan using a radioactive inert gas (Adapted from Davis et al. [80]).

TABLE 15.1 Methods of Correction for Tissue Attenuation and Scatter in Gamma Scintigraphic Studies

Method	Reference
Comparison of lung count versus known amount of radioactivity in lung phantom	89
Comparison of lung count versus count from injected ^{99m}Tc-macroaggregated albumin (assumed to be trapped in lung capillaries)	90
Calculation of attenuation correction factors, from assumed values of narrow beam or broad beam linear attenuation coefficients	88, 91
Determination of attenuation correction factors based on previously determined correlations with body mass index	92
Determination of separate attenuation and scatter corrections	86, 93
Transmission scanning using large flood field source	94, 95
Transmission scanning coupled with use of phantoms	96

approximations have been used, and several of these seem to be satisfactory, enabling whole lung deposition to be quantified to within ±10% of its true value [88]. Some of these methods are listed on Table 15.1. Perhaps the two most widely used methods involve either the calculation of attenuation factors from the known linear attenuation coefficients of ^{99m}Tc gamma rays in various tissues or transmission scanning using a large flood field source containing ^{99m}Tc or ^{57}Co [95]. The latter radionuclide has similar gamma ray energy to ^{99m}Tc, but a half-life of 270 days, so that it can be incorporated into a commercial flood field source. To a close approximation, an attenuation correction factor for the lungs can be defined by the term $(C_0/C_1)^{1/2}$, where C_0 and C_1 are counts from a flood field source without and with a subject present [94]. However, this simple correction factor works less well for other organs, where the tissue is denser than that in the lungs.

In addition to gamma ray attenuation, gamma ray scatter is also a factor potentially influencing quantification [86, 93]. When gamma rays undergo Compton scatter by body tissues, scattered photons with energy less than that of the original gamma ray are produced, and some of these will be detected by the gamma camera, even if it is set to record the main gamma ray peak for the radionuclide being used. While methods are available for making a dedicated scatter correction [86, 93, 97], both the importance of scatter correction and the way in which scatter should be handled are more controversial than attenuation correction. It has been reported that an effective way of dealing with the problem of scatter is to make a correction to the observed data using a broad beam attenuation coefficient, which allows for both attenuation and scatter in a single operation [88].

For some drugs, a measure of absolute pulmonary bioavailability, or of relative pulmonary bioavailability, can be obtained from pharmacokinetic studies. The charcoal block pharmacokinetic method has been used to quantify absolute pulmonary bioavailability for several drugs including terbutaline and budesonide [98]. Pulmonary bioavailability is not identical to whole lung deposition, although the numerical values of whole lung deposition from gamma scintigraphy and pulmonary bioavailability from the charcoal block method are similar [99, 100]. This similarity has provided a form of validation for scintigraphic data.

15.3.4 Regional Lung Deposition

Generally, quantification of the whole lung deposition by gamma scintigraphy to a sufficient degree of accuracy is not a major problem [83]. However, meaningful quantification of regional lung deposition is more difficult. The lungs may be divided into a variety of regions of interest (ROIs), as shown in Figure 15.4, and it is usual to define a peripheral/central zone (P/C) or central/peripheral zone (C/P) deposition ratio [101, 103, 104]. Sometimes these ratios are normalized by dividing by the P/C or C/P ratio of a radioactive inert gas, e.g., ^{81m}Kr. The terms "outer" and "inner" are sometimes preferred to "peripheral" and "central" when describing lung regions and may be recommended in future guidelines. Since the lungs are not homogeneous structures, it may be desirable to apply different tissue attenuation correction factors to central and peripheral regions [105]. Either a ventilation scan with a radioactive inert gas or a transmission scan using a flood field source may be used to define the lung borders. As described above, it is difficult to relate these indices to anatomical structures because of the 2D nature of the images. A further problem is that different groups or centers undertaking gamma scintigraphy have chosen to define lung regions in different ways. Hence it is difficult, and often impossible, to compare the regional lung deposition data for the same inhaler or formulation between study centers.

Despite its limitations, a simple P/C ratio has been shown to respond in the expected way to changes in aerodynamic particle size, inhaled flow rate, and airway patency. P/C ratio was shown to be higher for 2-μm particles compared to 6-μm particles, for slow inhalation compared to fast inhalation, and in healthy subjects compared to patients with obstructive lung disease [106]. P/C or C/P ratios were

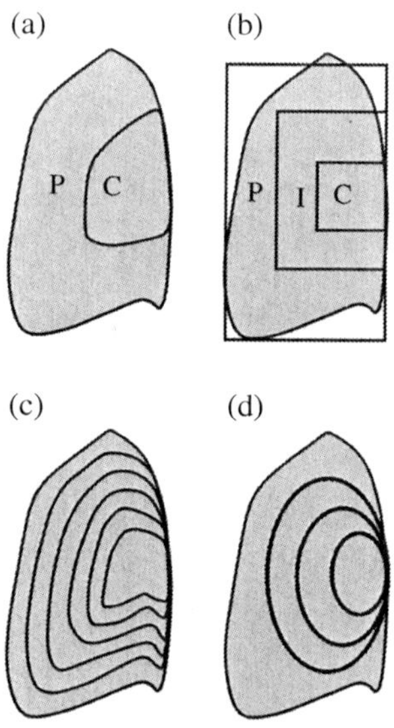

FIGURE 15.4 Examples of the division of the lungs into regions in gamma scintigraphy. (a) Two zones, from which a peripheral/central (P/C) or central/peripheral (C/P) ratio may be defined, (b) three zones, which also includes an intermediate (I) zone, (c) six concentric lung-shaped zones, and (d) several oval-shaped zones (Adapted from Agnew [101] and Newman [102]).

shown to correlate significantly with the 24-h whole lung retention of an insoluble radiolabel [106, 107], which may be considered as a functional measure of alveolar deposition [47]. These correlations suggest that simple indices of regional lung deposition contain at least some information about the anatomical site of deposition in the lungs. However, as will be described, both SPECT and PET overcome the limitation of viewing the complex structure of the lungs in 2D and can be expected to provide markedly superior regional lung deposition data. A method has been proposed for recovering the missing third dimension in gamma scintigraphy studies of pulmonary drug delivery [108].

15.3.5 Uses of Gamma Scintigraphy

A major use of gamma scintigraphy applied to pulmonary drug delivery has been to improve our fundamental understanding of the drug delivery process and of the manner in which this is influenced by lung physiology and pathology [109–111]. Scintigraphic data have been used to provide *in vivo* proof of concept data for new inhaler devices or formulations, often in comparison with existing products [112–114]. An example of gamma scintigraphic data comparing two inhalers is given in Figure 15.5, showing the fractionation of the dose between lungs and upper airways, the retention on the device and recovery from an exhaled air filter, as well as the distribution between three lung regions. When applied to DPIs, gamma scintigraphy has demonstrated the wide variation in lung deposition between different brands and the extent to which lung deposition from different inhalers varies with the patient's inhaled flow rate [116–118]. Gamma scintigraphy has also demonstrated that only a small fraction of the dose reaches the lungs from many pMDIs, and that this fraction can be zero, or virtually zero, in patients with poor inhaler technique [119].

Lung deposition data have been used in order to set the drug doses for phase II clinical comparisons between inhalers more realistically than might be possible using *in vitro* data alone [120]. Lung deposition is a useful study end point for a wide range of drugs delivered by the pulmonary route whose clinical efficacy may not be straightforward to measure. An example is the quantification of whole lung and peripheral lung deposition of alpha-1 antitrypsin to patients with a deficiency of this enzyme, enabling a rational choice of nebulizer system to be made for replacement therapy by the pulmonary route [121].

Although the use of a ^{99m}Tc radiolabel does not permit drug absorption to be monitored, it is possible to combine the measurement of lung deposition by gamma scintigraphy with the assessment of systemic drug levels using conventional pharmacokinetic methods. Examples of this "pharmacoscintigraphic" approach have included the measurement of lung deposition from a novel inhaler device, coupled with assessment of plasma concentrations of morphine sulfate versus subcutaneous injection [122], and of ergotamine tartrate versus another inhaler [123].

15.3.6 Data from Meta-Analyses

Over 100 gamma scintigraphy studies investigating aspects of pulmonary drug delivery have now been published in peer-reviewed journals [83]. This has allowed the pooling of data in such a way as to enhance our understanding of the drug delivery process.

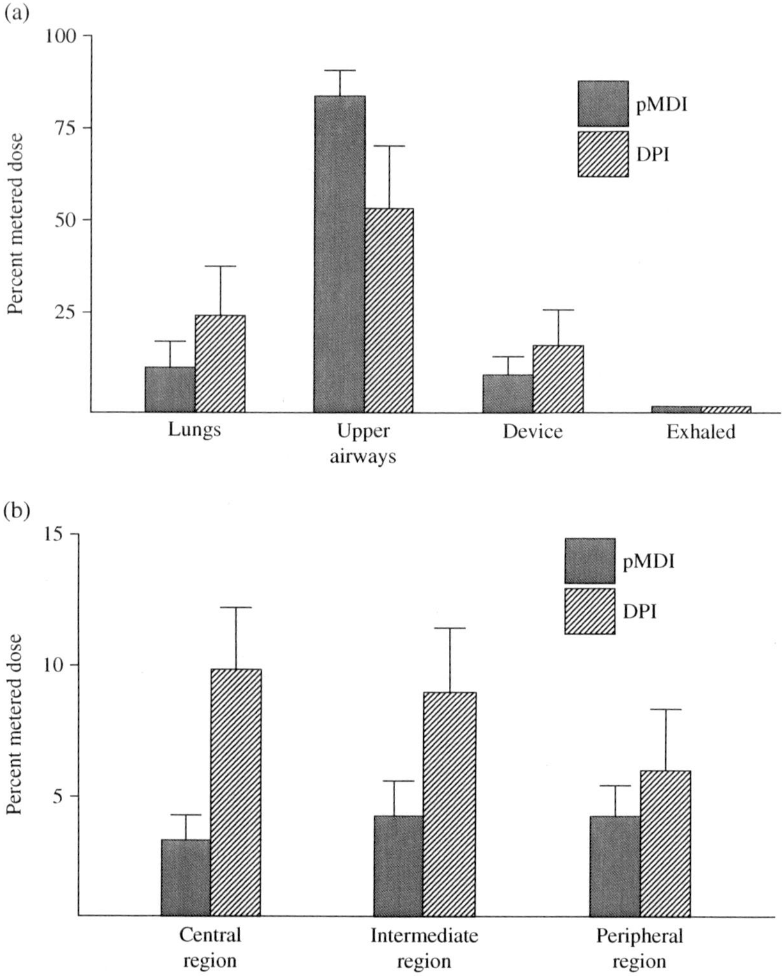

FIGURE 15.5 Data provided by a typical gamma scintigraphy study comparing deposition from a pMDI and a DPI. (a) Percentages of the metered (ex-valve) dose deposited in the lungs and upper airways, retained on the device, and recovered from an exhaled air filter. (b) Percentages of the metered (ex-valve) dose deposited in central, intermediate, and peripheral lung regions, as defined in Figure 15.4(b). Despite more than doubling the percentage of the dose deposited in the lungs, the deposition pattern for the DPI was markedly more central (mean P/C ratio of 0.64) compared to that for the pMDI (mean P/C ratio of 1.24). Mean and SD data adapted from Thorsson et al. [115].

Borgström et al. [124] analyzed the data from 187 study legs and showed that there is a broad inverse correlation between the mean lung deposition and its variability, expressed as a coefficient of variation. This phenomenon was ascribed to the way in which inhaled particles interact with the human upper airways, with high upper airway deposition being associated with low (and highly variable) lung deposition. This analysis predicted that if an inhaler can achieve lung deposition $>30\%$ of the ex-valve dose, then a coefficient of variation of lung deposition $<30\%$ is virtually guaranteed, assuming the inhaler is used correctly.

In an analysis of 33 study legs, Newman and Chan [125] showed that fine particle fraction measured by cascade impactor was correlated significantly with whole lung deposition measured by gamma scintigraphy, although whole lung deposition was overestimated by the *in vitro* measurement. A correlation closely matching the line of identity was found between fraction of the drug dose contained in particles $<3\,\mu m$ diameter and whole lung deposition.

Perhaps predictably, the conclusions reached from meta-analyses involving regional lung deposition are less clear. Clark [107] assessed the results of more than 100 study legs from studies undertaken at a single center and found that whole lung deposition and a P/C ratio showed little or no correlation. The reasons for this lack of correlation are not yet fully understood but may relate in part to the ballistic component of most pMDI sprays and the very broad or bimodal size distributions seen for DPI aerosols. Large or rapidly moving particles impact in the upper airways, and this process will determine whole lung deposition. However, since few of these large or rapidly moving particles penetrate into the lungs, they have little influence on regional lung deposition.

15.4 SPECT IMAGING

15.4.1 Principles of Data Acquisition and Processing

The major limitation of gamma scintigraphy is that it compresses a 3D distribution of radiolabel in the lungs into a 2D image. SPECT overcomes this limitation by collecting multiple images from different angles around a supine subject and then using a reconstruction algorithm to determine the original 3D distribution [59, 62, 126]. Images are acquired either with the detector heads stepped between multiple fixed imaging positions (step-and-shoot acquisition) or with continuous acquisition during which the detector heads move at a constant rotational speed [127]. Detector systems for SPECT are gamma cameras with two or three heads, the former being more common. Since the detectors are essentially the same as those used for gamma scintigraphy, the image resolutions of SPECT and gamma scintigraphy are similar (~10–14 mm full width at half maximum (FWHM)). 3D imaging using SPECT is an important tool in nuclear medicine, and its use was extended to study pulmonary drug delivery, first from a nebulizer [128], and subsequently from pMDIs [129–131].

There are two major approaches to data collection in SPECT studies used to assess pulmonary drug delivery. Some centers have undertaken concurrent planar imaging and have used the planar images in order to fractionate the dose into amounts deposited in the upper airways, lungs, and stomach. In these studies, SPECT has been used only to assess regional lung deposition [59, 132]. Repeat planar imaging of the lungs at the end of the SPECT sequence allows correction to be made for radiotracer losses from the lungs during SPECT data collection [133]. Other centers have used SPECT imaging sequences of the upper airways, lungs, and stomach in order to determine the fractions of the dose located at each of these sites [126, 134]. The axial field of view is typically 35–40 cm for twin-headed cameras and 20–25 cm for triple-headed cameras, so that it may be necessary to use multiple bed positions in order to record the radiotracer present at all three sites. The theoretical basis for quantification in SPECT is stronger than in gamma scintigraphy, and whole lung deposition can be estimated with greater accuracy [135]. Hence whole lung deposition data assessed by SPECT has been used as a "gold standard," against which to validate methods for assessing whole lung deposition by gamma scintigraphy [88].

The technique of filtered back projection has been used widely for data reconstruction in nuclear medicine and has also been used in SPECT lung studies. However, an iterative method such as ordered subset expectation maximization (OSEM) has advantages over filtered back projection, including the minimization of streak artifacts on the reconstructed images [126]. In addition to recording gamma rays emitted from the body, it is also necessary to use either a transmission source integral to the camera or a second imaging method, in order to define the lung outline, and to provide an attenuation map which enables attenuation correction to be performed [136]. Twin-headed cameras may be equipped with an integral transmission source of ^{153}Gd (gamma ray energy of 103 keV, half-life of 236 days). With triple-headed cameras, it may be possible to record transmission of gamma rays from an integral ^{99m}Tc source with one of the three detectors while recording emission of gamma rays with the other two detectors. In addition to correcting for tissue attenuation, corrections can also be made to the data for gamma ray scatter and the partial volume effect [137]. The latter may result in a gamma ray being recorded in a region of interest adjacent to the one from which it has originated. Not all groups working in the field consider scatter correction to be of major importance [126].

Co-registration with images acquired by X-ray computed tomography (CT) or magnetic resonance imaging (MRI) allows SPECT data to be related to anatomical structures and provides a basis for tissue attenuation correction [59, 126, 138]. Acquisition of these images on a separate scanner may lead to problems with accurate image co-registration. Imaging systems are now available that allow combined SPECT and CT scanning using a single instrument, which should allow more accurate co-registration [138, 139]. If CT is used as an adjunct to SPECT, then the added radiation dose may be significant, and this may restrict the number of study legs in which any subject can participate.

15.4.2 Regional Lung Deposition Analysis in SPECT

Regional lung deposition in SPECT may be determined from the analysis of transverse, coronal, or sagittal sections. Another approach is to treat the lungs as a series of concentric shells, radiating from the hilum (Fig. 15.6); the major bronchi will be located in the more central shells, while small airways and alveoli will be located in the more peripheral shells [59]. A further step in regional deposition analysis is to estimate the deposition per airway generation [136, 137], although this requires assumption of a specific morphometric model in terms of both the number of airways and their location in space. The distribution of deposition by airway generation is potentially useful to clinicians seeking to optimize inhalation therapy and to scientists seeking to validate their lung deposition models in human subjects [140]. An example of deposition expressed per airway generation for two aerosols of different sizes is given in Figure 15.7. Validation of this approach has been demonstrated using simulated lung images generated from a known distribution per airway generation, with reasonable agreement being found between actual and estimated depositions in various airways [140]. In another study, the deposition per generation of a ^{99m}Tc-labeled human serum albumin (HSA) aerosol was determined, and the conducting airway deposition fraction

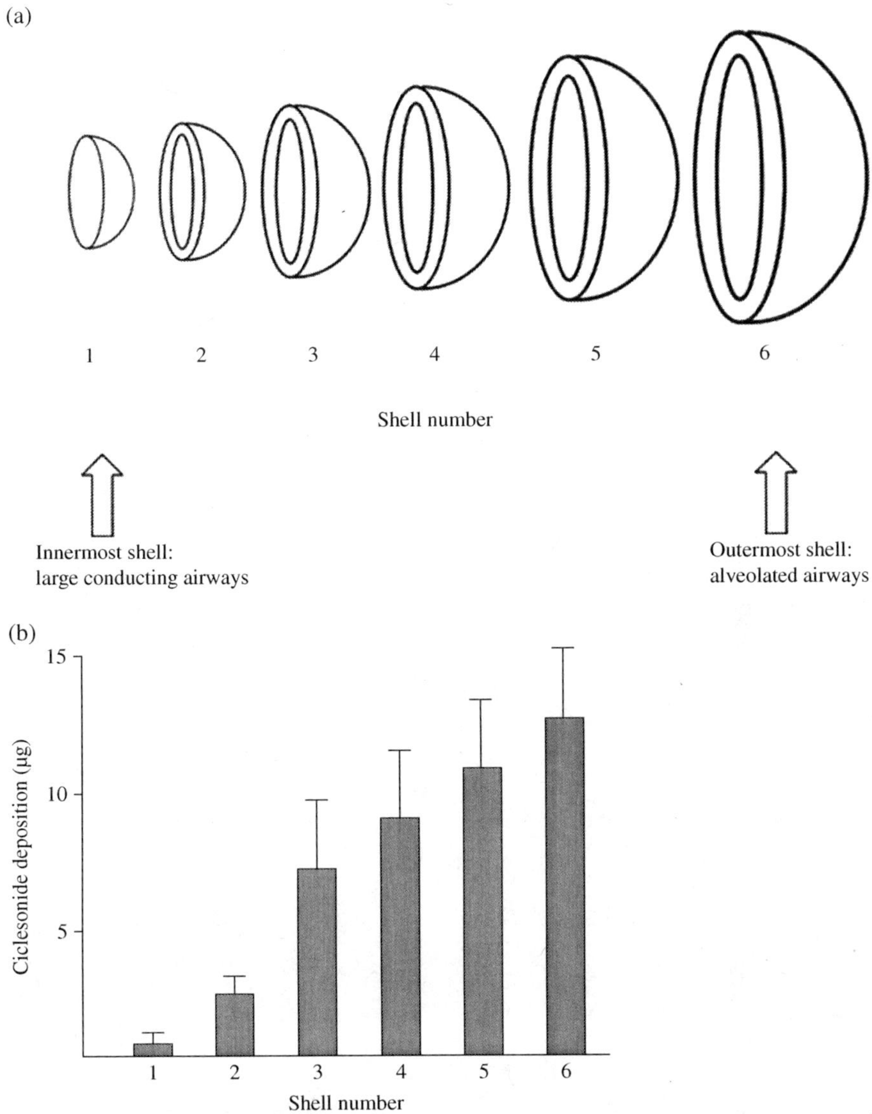

FIGURE 15.6 SPECT data analysis. (a) Schematic showing division of lungs into six concentric or nested shells (adapted from Newman and Pitcairn [62] and from Newman et al. [132]). (b) Mass of ciclesonide deposited in six concentric shells when delivered by pMDI. Mean and SD data adapted from Newman et al. [132].

was calculated by summing the depositions of appropriate generations. The mean value of conducting airway deposition agreed with the mean value given by the 24-h whole lung clearance of ^{99m}Tc-HSA, but the correlation between these two methods for individual subjects was not statistically significant [141].

SPECT data have been shown in several studies to detect differences between treatment regimens that either were missed or were demonstrated only weakly, by

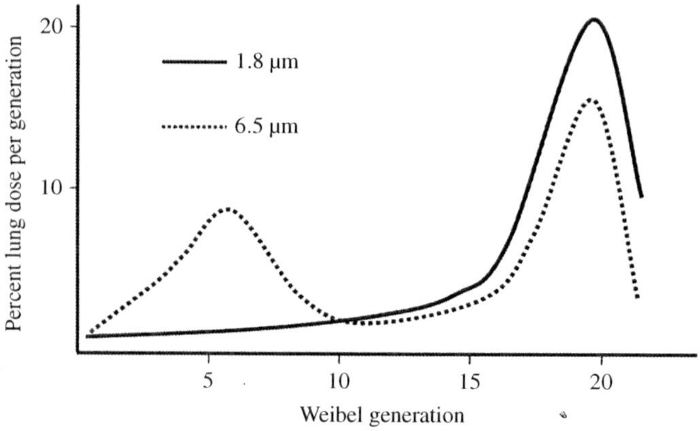

FIGURE 15.7 Transform of SPECT data to estimate deposition in 24 airway generations for aerosols of diameters 1.8 and 6.5 μm. The data show peaks around generation 20 for both aerosols and a peak around generation 6 for the larger aerosol. Mean data adapted from Fleming [137].

TABLE 15.2 Comparison of Regional Lung Deposition Indices Using 2D Gamma Scintigraphy and 3D SPECT

	2D C/P ratio	SPECT C/P ratio	Conducting airway/alveolated airway deposition ratio
Fine aerosol	2.03	0.93	2.43
Coarse aerosol	2.61	2.19	5.27

The SPECT C/P ratio was determined by approximating the lung as 10 concentric shells and then calculating the ratio of deposition in the innermost three shells to that in the outermost three shells. The conducting airway-to-alveolated airway ratio was determined from calculations of deposition per airway generation. Fine and coarse aerosols had MMAD of 1.5 and 6.5 μm, respectively.

Note: Mean data adapted from Fleming et al. [143].

gamma scintigraphy. For instance, the use of C/P ratios from mid-lung transverse or coronal slices obtained by SPECT was better able to detect a difference in deposition between two differently sized aerosols (MMADs 2.6 vs. 5.5 μm) than C/P ratios obtained by gamma scintigraphy [142]. In another study [143], superior discrimination between fine and coarse aerosols compared to 2D imaging was shown by a 3D C/P ratio based on concentric shells and by the ratio of conducting airway to alveolar deposition from assessment of deposition per airway generation (Table 15.2).

15.4.3 Radiolabeling for SPECT

Radiolabeling issues in SPECT studies are mostly similar to those in gamma scintigraphy, with the majority of studies using ^{99m}Tc as a marker for the drug. As in gamma scintigraphy, the ^{99m}Tc radiolabel is not chemically bonded to a drug molecule. Therefore, the radiolabeling validation testing procedures appropriate to SPECT are the same as those in gamma scintigraphy. However, data acquisition times in SPECT are usually longer than those in gamma scintigraphy, because of the time required for multiple images taken from different angles [144]. Therefore, the tracers used in SPECT studies should ideally be cleared relatively slowly from the lungs. Where possible, radiotracers such as ^{99m}Tc-DTPA or ^{99m}Tc-labeled colloid particles are preferable, because their clearance half-times (~1 h for DTPA) are markedly longer than those of pertechnetate [75, 77]. It may not be possible to use slowly cleared radiolabels with all formulations; for instance, ^{99m}Tc-pertechnetate is the radiolabel most commonly used for radiolabeling the contents of pMDIs and DPIs [73, 113]. A novel radiotracer method for conducting SPECT studies on DPIs (Technecoat) was developed [145], in which drug particles were radiolabeled by depositing carbon nanoparticles containing ^{99m}Tc (Technegas) on their surface. Technegas particles used to label drug particles were shown to have a clearance half-time of ~3 h [146]. It has been suggested that more new radiolabeling methods for SPECT lung studies are required to provide radiolabels that remain in the lungs longer than pertechnetate [83].

15.4.4 Fast SPECT

Long image acquisition times (≥15 min) are a common in SPECT studies. For instance, the acquisition of 32 pairs of images, each of 30 s duration, would result in a total acquisition time of 16 min. Depending on the radiopharmaceutical used, losses of radiotracer from the lungs owing to absorption and clearance could be significant during this imaging period. Recognition of this limitation has led to development of "fast SPECT" or "rapid SPECT" imaging methods [147, 148]. Using modern triple-headed cameras, it may be possible to complete a rotation of the thorax in as little as 1 min, enabling a clockwise and counterclockwise rotation to be combined into a frame with total duration of 2 min.

Fast SPECT overcomes the problem of radiotracer clearance during long imaging times, but as with any procedure involving radioactive materials, sufficient radioactivity must be present for good counting statistics. The selection of an appropriate amount of radiotracer for SPECT lung studies involves making a compromise between the conflicting requirements of low radiation dose, short imaging time, and good counting statistics. Ideally a lung activity > 10 MBq is required with an imaging time of 2 min [126]. Since many inhaler devices only deposit about 10% of the drug dose in the lungs, it may be necessary to deliver 100 MBq to the body. This is several times higher than the lung activity and total body activity required for planar imaging and hence is associated with a higher radiation dose. However, the effective (whole body) dose from SPECT lung studies has been estimated as 0.3–0.6 mSv for ^{99m}Tc-DTPA and 0.6–1.3 mSv for ^{99m}Tc-pertechnetate [126]. These doses are relatively low compared to those from many X-ray or nuclear medicine diagnostic procedures. An increase in radiation dose compared with planar imaging is generally acceptable, assuming that the SPECT data are superior. For inhalers or formulations which are known or suspected to deposit markedly more than 10% of the drug dose in lungs, it may be possible to reduce the administered radioactivity and hence the radiation dose.

15.5 PET IMAGING

15.5.1 Principles of Data Acquisition and Processing

PET is the third imaging method used to assess pulmonary drug delivery and has potential advantages over both gamma scintigraphy and SPECT. As a 3D imaging method, it overcomes the limitations of gamma scintigraphy that are associated with reducing a 3D distribution in the body to a single plane [58]. SPECT provides 3D deposition data but generally uses ^{99m}Tc, which is not chemically bound to a drug molecule. Since several positron-emitting radionuclides (including ^{11}C, ^{13}N, and ^{15}O) are isotopes of the constituent elements of drug molecules, it becomes theoretically possible to include the radiolabel within the structure of the drug molecule itself to produce a molecule that is chemically and pharmacologically identical to the molecule before labeling [149]. This provides assurance that the images produced are genuinely images of drug deposition, rather than deposition a radiolabel that has only a loose physical association with drug particles. PET imaging permits not only the initial deposition of drug to be assessed but also its *in vivo* kinetics and clearance [150, 151]. However, the short half-life of these radionuclides (20 min for ^{11}C and shorter for ^{13}N and ^{15}O) limits the period over which drug kinetics can be followed.

PET imaging is based on the detection, by a ring-shaped detector array, of pairs of 511 keV photons traveling in opposite directions from the original decay event. These photons are created after emission of a positron during radioactive decay, by the annihilation of that positron when it encounters an electron. Positrons typically have ranges of a few millimeters in water or soft tissues before they are annihilated [149]. The resolution of PET is superior to that of SPECT (4–6 mm FWHM vs. 10–14 mm), although the resolution of PET is slightly degraded in lung tissue where the density of matter is lower and the range of positrons is consequently higher [152]. Small-animal PET scanners, with reduced detector element size, provide sufficient resolution for high quality images to be obtained in rodents and hence make PET feasible for preclinical investigations [153]. Radiolabeling validation testing for PET studies similar to that in gamma scintigraphy and SPECT has been suggested [154], but this may need to be supplemented by other tests to demonstrate that the radiolabeled drug has the same physicochemical properties, and is likely to have the same pharmacokinetics, as the unlabeled drug.

Corrections for tissue attenuation in PET studies may be made using either an integral transmission source (e.g., ^{68}Ge/^{68}Ga, where the long-lived radionuclide ^{68}Ge decays into the short-lived positron-emitting radionuclide ^{68}Ga) or a second imaging method, such as high resolution CT [58]. Co-registration with CT not only allows precise delineation of the lung borders but also provides an attenuation map of individual voxels, allowing tissue attenuation corrections to be made. Data presentation is similar to SPECT, for instance, as transverse, sagittal, and coronal sections through the lungs, or by treating the lungs as a series of concentric shells. It is also possible to estimate deposition per airway generation [155], assuming a lung model such as that of Weibel. PET is considered to be the most accurate of the three radionuclide imaging methods for quantifying lung deposition [156] and was even claimed as "the only generally reliable quantification in nuclear imaging" [157].

In nuclear medicine, a possible simple imaging option for positron-emitting radionuclides is to use a twin-headed gamma camera to detect coincidences between pairs of 511 keV photons [158, 159], but this approach does not seem to have been applied to the assessment of pulmonary drug delivery.

15.5.2 Functional Imaging

The scanning agent most widely used in PET studies is ^{18}F-fluorodeoxyglucose (^{18}F-FDG) [160]. ^{18}F has a half-life of 109 min, which allows it to be traced in the body for many hours. This agent was first introduced in the late 1970s and has had a major influence in medical diagnosis and monitoring of treatment, because it concentrates in areas that are metabolically active, including cancers, regions of infection, and regions of inflammation. PET scanning with ^{18}F-FDG is thus considered as a functional imaging method, unlike gamma scintigraphy and SPECT [58, 154, 161]. It has been suggested that PET imaging following inhalation of ^{18}F-FDG could be used as a biomarker to monitor the treatment of inflammation in some lung diseases [162], especially if combined with CT scanning using the same imaging device [163]. Metabolic processes in the body can be followed using positron-emitting radionuclides other than ^{18}F. For instance, the metal ion ^{64}Cu (half-life of 12.7 h) has been radiolabeled to a peptide molecule in order to monitor inflammation in experimentally induced pulmonary injury [164].

15.5.3 Examples of PET Applied to Pulmonary Drug Delivery

PET is well established in pulmonary medicine, in order to assess a range of parameters including blood flow and ventilation [165, 166] and the manner in which these parameters change with disease or treatment. Its use has been extended by several groups to the assessment of pulmonary drug delivery, both in preclinical studies in animals and in man, using a range of positron-emitting radionuclides (Table 15.3). To date, there have been relatively few pulmonary drug delivery studies in man, compared with the wide range of studies involving gamma scintigraphy, and the increasing number of studies using SPECT. Deposition and *in vivo* kinetics of a variety of drugs have been assessed using PET, including ^{11}C-labeled triamcinolone acetonide [169], ^{11}C-labeled nicotine [171, 172], and ^{18}F-labeled fluticasone propionate [170]. Berridge et al. [169] prepared a formulation of ^{11}C-labeled triamcinolone acetonide (TAA) in a pMDI and administered it to five healthy human subjects. Repeat imaging sequences of the head, lungs, and stomach were performed over

TABLE 15.3 Selected Preclinical and Clinical Pulmonary Studies Using PET Imaging

Species	Radiopharmaceutical	Subject	Reference
Rat	^{18}F-FDG	Monitoring progression of experimentally induced idiopathic pulmonary fibrosis	161
Mouse	^{64}Cu peptide	Neutrophil-specific imaging agent shown to detect inflammation in lung injury	164
Man	^{18}F-FDG	Effect of inhaled corticosteroid on ventilation, as assessed by distribution of ^{18}F-FDG aerosol	154
Man	^{18}F-FDG	Deposition of particles delivered by DPI	167
Man	^{11}C-adrenoceptor antagonist	Drug deposition and clearance, showing relationship to adrenoceptor blockade	168
Man	^{11}C-TAA	Deposition and regional clearance of inhaled corticosteroid	169
Man	^{18}F-fluticasone propionate	Deposition and regional clearance of inhaled corticosteroid	170
Man	^{11}C-nicotine	Clearance kinetics of inhaled nicotine from cigarette and novel inhaler	171

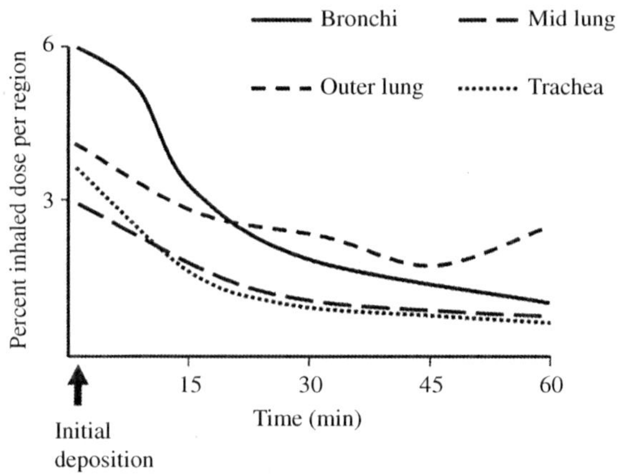

FIGURE 15.8 Example of PET data analysis. Initial deposition and subsequent clearance of triamcinolone acetonide for selected regions following inhalation by pMDI coupled to a spacer device. Mean data adapted from Berridge et al. [169].

2.5 h, enabling the initial deposition pattern of ^{11}C-TAA and its clearance from several pulmonary regions to be quantified. The PET lung images were co-registered with lung images obtained by fast spiral CT. The study confirmed the results of gamma scintigraphy studies, in that a mean of 10–15% of the dose reached the lungs, with the majority detected in the upper airways and stomach, but the data provided relating to ^{11}C-TAA clearance from the lungs could not have been obtained by either gamma scintigraphy or SPECT (Fig. 15.8). In addition to studies where a positron-emitting radionuclide has been incorporated into the structure of a drug molecule, ^{18}F-FDG has been used in solution to assess delivery from a nebulizer [154, 156] and tagged to powder particles to assess deposition from a DPI [167]. However, the use of ^{18}F-FDG in studies of pulmonary drug delivery negates one of the major advantages of PET, namely, the opportunity to study the deposition of actual radiolabeled drug molecules.

Using appropriate inhaled radioligands, the interaction of drugs with airway beta-adrenoceptors may be visualized and quantified by PET. van Waarde et al. [168] used a nebulizer to administer an ^{11}C-radiolabeled antagonist (CGP12388) to eight healthy volunteers, at baseline, and then after pretreatment with a beta-adrenergic bronchodilator. CGP12388 had been shown previously to bind to beta-adrenoceptors *in vivo*. Drug treatment accelerated pulmonary clearance of CGP12388, an effect which was ascribed to the blockade of beta-adrenoceptors in the lungs.

15.5.4 Practical Challenges

The advantages of PET as a means of assessing pulmonary drug delivery are offset by its practical challenges. Radiolabeled drug analogs for use in pulmonary studies are a significant advance, but radiolabeling is unlikely to be straightforward. According to one estimate, it may take more than a year and cost more than one million U.S. dollars to develop a radiolabeled drug analog suitable for a PET study in man [154]. Specialized radiochemical expertise is virtually essential [173]. The logistics of performing the radiolabeling on study days and then administering the radiolabeled drug may also be challenging, because of the short half-lives of the most suitable radionuclides (e.g., 20 min for ^{11}C). Following radiolabeling, the formulation may need to be incorporated into an inhaler device, and quality control checks should be carried out before administration, while the radionuclide is rapidly decaying. The longer half-life of ^{18}F renders it more convenient, but few drugs contain a fluorine atom, fluticasone propionate being a notable exception [170].

There are also issues with hardware associated with PET studies. PET scanners are expensive [156], and access to scanners in busy clinical departments may be difficult. The imaging center must be located close to a cyclotron which is the source of positron-emitting radionuclides, and for the most short-lived radionuclides used in PET studies, the cyclotron needs to be on-site. Most PET cameras have axial fields of view that are too short (≤20 cm) to include the entire lungs, so that lung scans taken at different levels may have to be performed and joined together to make a composite image [126]. More modern PET scanners have longer axial fields of view that allow the lungs to be covered in a single image.

15.6 CHOICE OF IMAGING METHODS

The three radionuclide imaging methods used to assess pulmonary drug delivery (gamma scintigraphy, SPECT, and PET) all have advantages and disadvantages [58, 59, 102]. In the case of systemically acting drugs, it could be argued that PET is the only method that actually quantifies pulmonary drug delivery, as opposed to pulmonary drug deposition, because it alone is able to monitor the fate of inhaled drugs once they have been deposited.

Several factors need to be considered when choosing a radionuclide imaging method to assess pulmonary drug delivery. Gamma scintigraphy is a relatively unsophisticated method, and compared to SPECT or PET, the fundamental basis for quantification is less solid. The quality of regional lung deposition data obtained by gamma scintigraphy is relatively poor. It has been suggested that the only real advantage of gamma scintigraphy is that it is simple in terms of both data acquisition and data analysis [157], but in practice, this advantage will often make it the method of choice, rather than a more sophisticated alternative. Perhaps for this reason, together with the fact that it has been so extensively used, gamma scintigraphy was described some years ago as "the industry standard" [144]. Arguably, gamma scintigraphy still holds this position as an imaging method for assessing drug delivery for new inhaler devices or formulations.

It is generally considered inappropriate to use SPECT or PET merely in order to quantify deposition from a novel inhaler device or formulation [83], although it has been suggested that SPECT could be used to obtain a bank of regional lung deposition data from different inhalers, analogous to the bank of whole lung deposition data that has been obtained using gamma scintigraphy. Instead, SPECT or PET is better used in order to provide information that gamma scintigraphy cannot provide. In the case of PET, this includes the intrapulmonary distribution of actual drug, changes in biodistribution with time, responses to treatment, or interaction of radiolabeled ligands with receptors in the lungs [174]. Either SPECT or PET should be considered when assessment of regional lung deposition is a major study objective. SPECT

TABLE 15.4 Advantages and Disadvantages/Challenges of PET Imaging for Assessing Pulmonary Drug Delivery

Advantages	Disadvantages/challenges
• 3D imaging method, allowing better assessment of regional lung deposition than gamma scintigraphy • Basis for quantification better than gamma scintigraphy • Most accurate deposition data • Better image resolution than gamma scintigraphy or SPECT • Radiolabeled drug molecules can be used • Drug clearance and *in vivo* kinetics can be monitored, as well as drug deposition • Can be used to assess interactions between radiolabeled drug ligands and receptors	• Imaging hardware is expensive, and access to it may be an issue • Cyclotron needed either nearby or on-site • Developing a radiolabeled drug analog may be expensive and time consuming • The most appropriate radionuclides for radiolabeling drug molecules have half-lives $\leq$20 min • Timelines of studies likely to be longer, and costs higher, than for gamma scintigraphy or SPECT

offers the possibility to assess regional lung deposition in a way that can be related to lung anatomy, but using a conventional gamma camera and a ^{99m}Tc radiolabel.

The challenges of PET when applied to radiolabeled drugs delivered from inhaler devices should not be underestimated. Although PET is the most sophisticated of the methods, with the best resolution and probably the greatest accuracy, it can only be undertaken at centers possessing the necessary hardware and radiochemical expertise (Table 15.4). Not only are the costs of PET likely to be high, but perhaps even more important in the context of a drug development program, the timelines of completing a study involving a labeled drug analog in an inhaler device may be inconveniently long.

15.7 REGULATORY ISSUES

Measurement of drug deposition or drug delivery, using radionuclide imaging methods, holds an intermediate position between *in vitro* tests (emitted dose, APSD, and spray pattern) and measures of drug safety and efficacy from pharmacodynamic studies or controlled clinical trials. *In vitro* tests provide key quality control data and may be undertaken with great precision [175]. However, they do not reflect the way in which patients use inhalers or in which the spray from an inhaler device interacts with the patient's upper airways.

Measurements of lung deposition or delivery from imaging studies are not accepted by the U.S. Food and Drug Administration (FDA) as the source of pivotal data in New Drug Applications (NDAs), because the critical requirement is the demonstration of safety and efficacy [83]. Nevertheless, imaging studies may drive development programs in individual companies and may be seen by product developers as a useful way of obtaining supporting data in regulatory applications about the drug delivery characteristics of their products. There is anecdotal evidence that regulators find lung images compelling and that they recognize imaging studies to provide evidence of the delivery of radiolabeled particles into the lungs.

In the case of Abbreviated New Drug Applications (ANDAs), there has been hope that the FDA might accept radionuclide imaging studies as a means of showing bioequivalence between two or more products. However, there is a problem which could well prove insurmountable. In order to undertake a radionuclide imaging study, it is necessary to add a radiolabel to the product. Even if this is done in such a way that the product after radiolabeling is indistinguishable from the product before labeling, the act of radiolabeling could be construed in a legal sense as having changed or "adulterated" the product. This is not permitted under the U.S. Code of Federal Regulations [176]. In a comparison of generic and innovator inhalers, it could be argued that the radiolabeled innovator reference product is not the same as true innovator reference product and that the innovator–generic comparison is not valid. Another potential legal issue is the "Least Burdensome Provision" of the Code of Federal Regulations [177] which could be interpreted to mean that sophisticated radionuclide imaging methods are unacceptable to assess bioequivalence because they are either not widely available or too expensive [83]. However, these concerns are not apparently shared by all regulatory authorities. For instance, recent European guidelines mentioned gamma scintigraphic studies as a possible method for showing equivalence between two products [178] and suggested statistical criteria that could be applied to whole lung and regional lung deposition data.

In addition to legal issues, regulators also have expressed several scientific concerns for radionuclide imaging data. Radiolabeling may sometimes cause small changes in the APSD, or the drug and radiolabel size distributions may not match perfectly. The possible effects of these small changes in APSD on drug safety and efficacy are poorly understood, partly because there is a lack of data clearly relating aerodynamic particle size of inhaled drugs to their clinical effects [179]. Another regulatory concern involves the link between lung deposition of inhaled drugs and their clinical effects. It is obvious that respiratory drugs must be deposited in the lungs if a therapeutic effect is to be achieved, but the manner in which therapeutic response changes with variations in lung deposition is poorly understood. This situation may not have been helped by several high profile cases where lung deposition and clinical response appeared to correlate poorly. A further regulatory concern is the presence of marked interlaboratory variations in methodology for conducting radionuclide imaging studies [180].

These scientific issues can be addressed. The need to develop new and better radiolabeling methods and to undertake more studies to show the relationship between lung deposition (especially regional lung deposition) and clinical effects has been recognized [83]. Strenuous efforts have been made via the auspices of the International Society for Aerosols in Medicine (ISAM) to standardize radionuclide imaging methodology. New guidelines were recently published relating to radiolabeling and its validation [181] and to gamma scintigraphy [182], as well as SPECT and PET. The use of such methodology would increase the likelihood that future radionuclide imaging studies are conducted according to an agreed quality standard and that the results of studies undertaken at different centers can be compared more readily than is presently the case.

15.8 LOOKING TO THE FUTURE

There is likely to be a continued requirement for radionuclide imaging studies to assess pulmonary drug delivery. During the 1990s, there was a surge in development of novel inhaler devices, resulting in part from the desire to eliminate the use of CFC propellants in pMDIs. It was also recognized that drug delivery needed to be become both more efficient and more reproducible if inhalers were to prove suitable for delivering many of the new molecules being proposed for pulmonary delivery [183]. In the 2000s, a number of novel sophisticated respiratory formulations were developed, and these too were assessed by radionuclide imaging [184, 185]. Gamma scintigraphy proved to be an ideal imaging method for assessing the plethora of new devices and formulations [124, 186].

The rate at which novel inhalers and formulations were developed has subsequently decreased, as companies have sought to develop appropriate drug molecules to be used with the new inhaler technologies. The future could see an increase in SPECT and PET studies of pulmonary drug delivery, as a means of investigating regional lung deposition, and its relationship to pharmacokinetic and pharmacodynamic end points, in greater detail. The increased use of PET to monitor both deposition and clearance of drugs from inhalers is possible, but it is perhaps surprising how few studies have been published concerning the use of PET to assess the delivery of radiolabeled drugs from pMDIs and DPIs. Some of the practical challenges of PET described in this chapter may account for the lack of published data.

Multimodality imaging, combining a radionuclide imaging method with CT or MRI, as a means of providing anatomical information in drug delivery studies is likely to increase. Scanners combining either SPECT and CT or PET and CT in the same hardware are now available [152, 187] and are playing an important role in medical diagnosis and staging of disease. The accuracy of co-registration between radionuclide image and anatomical image is likely to be significantly enhanced if subjects do not have to be moved between scanners. Of course, it is essential to ensure that the radiation dose from such procedures is carefully monitored.

MRI itself may offer future possibilities in the assessment of pulmonary drug delivery. Ventilation scanning using hyperpolarized ^{3}He or ^{129}Xe provides striking high resolution images, which can demonstrate the presence of ventilation defects [188]. It may also be possible to define indices from these MRI scans that can distinguish between different patient groups or different types of abnormality better than established pulmonary function tests [189]. Studies in animal models have shown that it is possible to quantify regional deposition in the lungs after inhalation of iron oxide nanoparticles [190, 191], but whether such methods could be safely and reliably used as a surrogate for deposited drugs in man is as yet unknown.

Finally, a long-term goal must be to avoid undertaking radionuclide imaging studies at all and to rely on deposition calculations from mathematical modeling studies. Deposition modeling is based on assumptions about lung morphology, aerosol transport in the airways, and deposition processes and generally uses computational fluid dynamic calculations [192]. Modeling techniques can be used to just estimate whole lung and upper airway depositions [193], but have become increasingly sophisticated, allowing calculation of deposition in specific airway generations and/or at individual airway bifurcations [194]. Pharmaceutical aerosols often have complex

APSDs, which may change with time owing to evaporation and other processes, making the calculation of deposition more challenging than for stable monodisperse particles of known size. For the time being at least, studies using radionuclide imaging methods are likely provide more reliable data concerning pulmonary drug delivery. SPECT and PET studies may be able to assist in the validation of lung deposition computer models [83].

REFERENCES

1. Anderson PJ. History of aerosol therapy; liquid nebulization to MDIs and DPIs. Respir Care 2005;50:1139–1150.
2. Grossman J. The evolution of inhaler technology. J Asthma 1994;31:55–64.
3. Mudge J. *A Radical and Expeditious Cure for a Recent Catarrhous Cough*. London: Allen; 1778.
4. Sanders M. Inhalation therapy: an historical review. Primary Care Respir J 2007;16:71–81.
5. Anderson P, Newman S. Drugs for topical and systemic delivery by the pulmonary route. In: Newman S, Anderson P, Byron P, et al. *Respiratory Drug Delivery: Essential Theory and Practice*. Richmond: RDD Online/Virginia Commonwealth University; 2009. p 337–382.
6. Masoli M, Fabian D, Holt S, et al. The global burden of asthma: executive summary of the GINA dissemination committee report. Allergy 2004;59:469–478.
7. Rabe KF, Hurd S, Anzueto A, et al. Global strategy for the diagnosis, management, and prevention of chronic obstructive pulmonary disease: GOLD executive summary. Am J Respir Crit Care Med 2007;176:532–555.
8. Tashkin DP. Indacaterol maleate for the treatment of chronic obstructive pulmonary disease. Expert Opin Pharmacother 2010;11:2077–2085.
9. Barnes PJ, Pedersen S, Busse WW. Efficacy and safety of inhaled corticosteroids: new developments. Am J Respir Crit Care Med 1998;157:S1–S53.
10. Barnes PJ. Inhaled corticosteroids in COPD: a controversy. Respiration 2010;80:89–95.
11. Barnes PJ. Scientific rationale for using a single inhaler for asthma control. Eur Respir J 2007;29:587–595.
12. Garcia-Contrares L, Hickey AJ. Aerosol treatment of cystic fibrosis. Crit Rev Ther Drug Carrier Syst 2003;20:317–356.
13. Labiris NR, Dolovich MB. Pulmonary drug delivery. Part I: physiological factors affecting therapeutic effectiveness of aerosolized medications. Br J Clin Pharmacol 2003; 56:588–599.
14. Patton JS, Bukar J, Nagarajan S. Inhaled insulin. Adv Drug Deliv Revs 1999;35:235–247.
15. Kohlhäufl M, Scheuch G. Systemic aerosol therapy—the clinician's perspective. In: Bechtold-Peters K, Luessen H, editors. *Pulmonary Drug Delivery: Basics, Applications and Opportunities for Small Molecules and Biopharmaceutics*. Berlin: Cantor Verlag; 2007. p102–120.
16. Gannslein M. Uber inhalation von insulin. Klin Wochenschr 1925;4:71–72.
17. Harper NJ, Gray S, De Groot J, et al. The design and performance of the Exubera pulmonary insulin delivery system. Diabetes Technol Ther 2007;9 Suppl 1:S16–S27.
18. Siekmeier R, Scheuch G. Inhaled insulin–does it become reality? J Physiol Pharmacol 2008;59 Suppl 6:81–113.

19. Adjei AL, Gupta PK. *Inhalation Delivery of Therapeutic Peptides and Proteins*. New York: Marcel Dekker; 1997.
20. Patton JS. Market trends in pulmonary therapies. In: Bechtold-Peters K, Luessen H, editors. *Pulmonary Drug Delivery: Basics, Applications and Opportunities for Small Molecules and Biopharmaceutics*. Berlin: Cantor Verlag; 2007. p 372–391.
21. Laube BL. The expanding role of aerosols in systemic drug delivery, gene therapy and vaccination. Respir Care 2005;50:1161–1176.
22. Dolovich MB, Dhand R. Aerosol drug delivery: developments in device design and clinical use. Lancet 2010;377:1032–1045.
23. Atkins PJ. Transition to HFA MDIs: how did we get here and where are we going?. In: Dalby RN, Byron PR, Peart J, et al., editors. *Respiratory Drug Delivery 2006*. River Grove: Davis Healthcare International; 2006. p 83–90.
24. Woodcock A. Responsible inhaler development in Article 5 countries: navigating the scientific, regulatory, political and commercial challenges. In: Dalby RN, Byron PR, Peart J, et al., editors. *Respiratory Drug Delivery 2010*. River Grove: Davis Healthcare International; 2010. p 121–129.
25. Bell J, Newman SP. The rejuvenated pressurised metered dose inhaler. Expert Opin Drug Deliv 2007;4:215–234.
26. Laube BL, Janssens HM, de Jongh FH, et al. ERS/ISAM task force consensus statement. Eur Respir J 2011;37:1308–1417.
27. Smyth HDC, Hickey AJ. Carriers in drug powder delivery: implications for inhalation system delivery. Am J Drug Deliv 2005;3:117–132.
28. Smith IJ, Parry-Billings M. The inhalers of the future? A review of dry powder devices on the market today. Pulm Pharmacol Ther 2003;16:79–95.
29. Telko MJ, Hickey AJ. Dry powder inhaler formulation. Respir Care 2005;50:1209–1227.
30. Newman SP. Dry powder inhalers for optimal drug delivery. Expert Opin Biol Ther 2004;4:23–33.
31. Tobyn M, Staniforth JN, Morton D, et al. Active and intelligent inhaler device development. Int J Pharm 2004;277:31–37.
32. Schumacher JD, Harper NJ. Exubera: planning successful product use—educating the patient and healthcare professional. In: Dalby RN, Byron PR, Peart J, et al., editors. *Respiratory Drug Delivery Europe 2007*. River Grove: Davis Healthcare International; 2007. p 11–20.
33. Rau JL. Design principles of nebulization devices currently in use. Respir Care 2002;47: 1257–1278.
34. Dhand R. Nebulizers that use a vibrating mesh or plate with multiple apertures to generate aerosol. Respir Care 2002;47:1406–1418.
35. Geller DE. Comparing clinical features of the nebulizer, metered dose inhaler and dry powder inhaler. Respir Care 2005;50:1313–1322.
36. Shirk M, Donahue KR, Shirvani J. Unlabeled uses of nebulized medications. Am J Health Syst Pharm 2006;63:1704–1716.
37. Geller DE. New liquid aerosol generation devices: systems that force pressurized liquids through nozzles. Respir Care 2002;47:1392–1405.
38. Dhand R. Aerosol plumes: slow and steady wins the race. J Aerosol Med 2005;18: 261–263.
39. Swift DL. The oral airway—a conduit or collector for pharmaceutical aerosols? In: Byron PR, Dalby RN, Farr SJ, editors. *Respiratory Drug Delivery IV*. Buffalo Grove: Interpharm Press; 1994. p 187–195.

40. Finlay WH, Golshahi L, Noga M, et al. Choosing 3D mouth-throat dimensions: a rational merging of medical imaging and aerodynamics. In: Dalby RN, Byron PR, Peart J, et al., editors. *Respiratory Drug Delivery 2010*. River Grove: Davis Healthcare International; 2010. p 185–193.
41. Dolovich MB, Ruffin R, Corr D, et al. Clinical evaluation of a simple demand-inhalation MDI aerosol delivery device. Chest 1983;84:36–41.
42. Clark AR. MDIs: physics of aerosol formation. J Aerosol Med 1996;9 (Suppl 1):S19–S26.
43. Weibel ER. *Morphometry of the Human Lung*. New York: Academic Press; 1963.
44. Stocks J, Hislop AA. Structure and function of the respiratory system: development aspects and their relevance to aerosol therapy. In: Bisgaard H, O'Callaghan C, Smaldone GC, editors. *Drug Delivery to the Lung*. New York: Marcel Dekker; 2002. p 47–104.
45. Brain JD. Inhalation, deposition and fate of insulin and other therapeutic proteins. Diabetes Technol Ther 2007;9 Suppl 1:S4–S15.
46. Bur M, Bock U, Haltner-Ukomadu E, et al. In vitro models for pulmonary drug absorption. In: Bechtold-Peters K, Luessen H, editors. *Pulmonary Drug Delivery: Basics, Applications and Opportunities for Small Molecules and Biopharmaceutics*. Berlin: Cantor Verlag; 2007. p 58–81.
47. Camner P, Philipson K. Human alveolar deposition of 4 μm Teflon particles. Arch Environ Health 1978;36:171–175.
48. Bailey MR, Fry FA, James AC. The long-term clearance kinetics of insoluble particles from the human lung. Ann Intern Med 1982;26:273–290.
49. Newman S. Aerosol properties and deposition principles. In: Newman S, Anderson P, Byron P, et al. *Respiratory Drug Delivery: Essential Theory and Practice*. Richmond: RDD Online/Virginia Commonwealth University; 2009. p 29–58.
50. Brain JD, Blanchard JD. Mechanisms of particle deposition and clearance. In: Morén F, Dolovich MB, Newhouse MT, et al., editors. *Aerosols in Medicine: Principles, Diagnosis and Therapy*. Amsterdam: Elsevier; 1993. p 117–156.
51. Scheuch G, Zimlich WC, Siekmeier R. Biophysical parameters determining pulmonary drug delivery. In: Bechtold-Peters K, Luessen H, editors. *Pulmonary Drug Delivery: Basics, Applications and Opportunities for Small Molecules and Biopharmaceutics*. Berlin: Cantor Verlag; 2007. p 46–54.
52. Newman S, Peart J. Pressurized metered dose inhalers. In: Newman S, Anderson P, Byron P, et al. *Respiratory Drug Delivery: Essential Theory and Practice*. Richmond: RDD Online/Virginia Commonwealth University; 2009. p 177–215.
53. Mitchell JP, Nagel MW. Cascade impactors for size characterization of aerosols from medical inhalers: their uses and limitations. J Aerosol Med 2003;16:341–377.
54. Dolovich M, Ruffin RE, Roberts R, et al. Optimal delivery of aerosols from metered dose inhalers. Chest 1981;80 (6 Suppl):911–915.
55. Bennett WD. Controlled inhalation of aerosolised therapeutics. Expert Opin Drug Deliv 2005;2:763–767.
56. Yu CP, Nicolaides P, Soong TT. Effect of random airway sizes on aerosol deposition. Am Ind Hyg Assoc J 1979;40:999–1005.
57. Dolovich MB. Measuring total and regional lung deposition using inhaled radiotracers. J Aerosol Med 2001;14 Suppl 1:S35–S44.
58. Berridge MS, Lee Z, Heald DL. Regional distribution and kinetics of inhaled pharmaceuticals. Curr Pharm Des 2000;6:1631–1651.
59. Newman SP, Fleming JS. Challenges in assessing regional distribution of inhaled drug in the human lungs. Expert Opin Drug Deliv 2011;8:841–855.

60. Wagner HJ. Hal Anger: nuclear medicine's quiet genius. J Nucl Med 2003;44:26–34.

61. Meseguer G, Gurny R, Buri P. In vivo evaluation of dosage forms: application of gamma scintigraphy to non-enteral routes of administration. J Drug Targeting 1994;2:269–288.

62. Newman SP, Pitcairn GR. Gamma scintigraphy and SPECT. In: Lappin G, Temple S, editors. *Radiotracers in Drug Development*. Boca Raton: Taylor and Francis; 2006. p 269–288.

63. Asmundsson T, Johnson RF, Kilburn KH, et al. Efficiency of nebulizers for depositing saline in human lung. Am Rev Respir Dis 1973;108:506–512.

64. Dolovich MB, Killian D, Wolff RK, et al. Pulmonary aerosol deposition in chronic bronchitis: intermittent positive pressure breathing versus quiet breathing. Am Rev Respir Dis 1977;115:397–402.

65. Newman SP, Pavia D, Morén F, et al. Deposition of pressurised aerosols in the human respiratory tract. Thorax 1981;36:52–55.

66. Newman SP, Morén F, Trofast E, et al. Deposition and clinical efficacy of terbutaline sulphate from Turbuhaler, a new multi-dose powder inhaler. Eur Respir J 1989;2:247–252.

67. Ruffin RE, Kenworthy MC, Newhouse MT. Response of asthmatic patients to fenoterol inhalation: a method of quantifying the airway bronchodilator dose. Clin Pharmacol Ther 1978;23:338–345.

68. Köhler D, Fleischer W, Matthys H. A new method for easy labelling of beta-2 agonists in the metered dose inhaler with technetium 99m. Respiration 1988;53:65–73.

69. Clarke JG, Farr SJ, Wicks SR. Technetium-99m labelling of suspension type pressurised metered dose inhalers comprising various drug/surfactant combinations. Int J Pharm 1992;80:R1–R5.

70. Arppe J, Vidgren M. Practical gamma labelling method for metered-dose inhalers and inhalation powders. STP Pharma Sciences 1994;4:19–22.

71. Pitcairn GR, Lunghetti G, Ventura P, et al. A comparison of the lung deposition of salbutamol inhaled from a new dry powder inhaler, at two inhaled flow rates. Int J Pharm 1994;102:11–18.

72. Pitcairn GR, Lim J, Hollingworth A, et al. Scintigraphic assessment of drug delivery from the Ultrahaler dry powder inhaler. J Aerosol Med 1997;10:295–306.

73. Walker PS, Petterson GL, Bondesson E, et al. An advanced and detailed in vitro validation procedure for the radiolabelling of carrier-free terbutaline sulphate dry powder. J Aerosol Med 2001;14:227–236.

74. Roller CM, Schaefer NC, Zhang G, et al. In vitro validation of ^{99m}Tc-HFA-FP delivered via pMDI-spacer. J Aerosol Med 2006;19:254–260.

75. Walker PS, Conway JH, Fleming JS, et al. Pulmonary clearance rate of two chemically different forms of inhaled pertechnetate. J Aerosol Med 2001;14:209–215.

76. Hasani A, Agnew JE, Toms N, et al. Scintigraphic assessment of lung drug deposition distribution using a radiolabelled drug surrogate: significance of radiolabel transport post deposition as a source of error. In: *Drug Delivery to the Lungs X*. Portishead: The Aerosol Society; 1999. p 96–99.

77. Brådvik I, Wollmer P, Evander E, et al. Kinetics of lung clearance of 99mTc-DTPA in smoking patients with sarcoidosis compared to healthy smokers. Respir Med 2002;96: 317–321.

78. Short MD, Singh CA, Few JD, et al. The labelling and monitoring of lung deposition of an inhaled synthetic anticholinergic bronchodilating agent. Chest 1981;80 (6 Suppl):918–921.

79. Sinclair AJ, Signore A, Bomanji J, et al. In-vivo kinetics of I123-labelled insulin: studies in normal subjects and patients with diabetes mellitus. Nucl Med Comm 1987;8:779–786.

80. Davis SS, Hardy JG, Newman SP, et al. Gamma scintigraphy in the evaluation of pharmaceutical dosage forms. Eur J Nucl Med 1992;19:971–986.
81. Gerrity TR. Aerosol deposition using new radioaerosol imaging techniques. In: Byron PR, Dalby RN, Farr SJ, editors. *Respiratory Drug Delivery IV*. Buffalo Grove: Interpharm Press; 1994. p 179–186.
82. Gonda I. Absolute and relative quantitation of deposition and clearance by scintigraphy. J Biopharm Sci 1992;3:191–197.
83. Scheuch G, Bennett W, Borgström L, et al. Deposition, imaging and clearance: what remains to be done? J Aerosol Med Pulm Drug Deliv 2010;23 Suppl 2:S39–S57.
84. Coates AL, Dinh L, MacNeish CF, et al. Accounting for radioactivity before and after nebulization of tobramycin to ensure accuracy of quantification of lung deposition. J Aerosol Med 2000;13:169–178.
85. Everard ML, Dolovich MB. In vivo measurement of lung dose. In: Bisgaard H, O'Callaghan C, Smaldone GC, editors. *Drug Delivery to the Lung*. New York: Marcel Dekker; 2002. p 173–209.
86. Lee Z, Berridge MS, Nelson AD, et al. The effect of scatter and attenuation on aerosol deposition as determined by gamma scintigraphy. J Aerosol Med 2001;14:167–183.
87. Lee Z, Ljungberg M, Muzic RF, et al. Usefulness and pitfalls of planar gamma scintigraphy for measuring aerosol deposition in the lungs: a Monte Carlo investigation. J Nucl Med 2001;42:1077–1083.
88. Fleming JS, Conway JH, Bolt L, et al. A comparison of planar scintigraphy and SPECT measurement of total lung deposition of inhaled aerosol. J Aerosol Med 2003; 16:9–19.
89. Messina MS, Smaldone GC. Evaluation of quantitative aerosol techniques for use in bronchoprovocation studies. J Allergy Clin Immunol 1985;75:252–257.
90. Cross CE, Hornhof WJ, Koblik PD, et al. Aerosol deposition: practical considerations of methodology for the direct measurement of aerosol delivery to the lung bronchiolar-alveolar surfaces. J Aerosol Med 1992;5:39–45.
91. De Backer W, Devolder A, Poli G, et al. Lung deposition of BDP - formoterol HFA pMDI in healthy volunteers, asthmatic, and COPD patients. J Aerosol Med Pulm Drug Deliv 2010;23:137–148.
92. Fleming JS, Conway JH, Holgate ST, et al. Evaluation of the accuracy and precision of lung aerosol deposition measurements from planar radionuclide imaging using simulation. Phys Med Biol 1998;43:2423–2429.
93. Forge NI, Mountford PJ, O'Doherty MJ. Quantification of technetium-99m lung radioactivity from planar images. Eur J Nucl Med 1993;20:10–15.
94. Macey DJ, Marshall R. Absolute quantitation of radiotracer uptake in the lungs using a gamma camera. J Nucl Med 1982;23:731–735.
95. Pitcairn GR, Newman SP. Tissue attenuation corrections in gamma scintigraphy. J Aerosol Med 1997;10:187–198.
96. Taylor, G, Warren SJ. Methods for tissue attenuation correction in inhalation pharmaceutical scintigraphy studies. In: Dalby RN, Byron PR, Peart J, et al., editors. *Respiratory drug delivery 2010*. River Grove: Davis Healthcare International; 2010. p 929–933.
97. Buvat I, Benali H, Todd-Pokropek A, et al. Scatter correction in scintigraphy: the state of the art. Eur J Nucl Med 1994;21:675–694.
98. Borgström L, Nilsson M. A method for determination of the absolute pulmonary bioavailability of inhaled drugs: terbutaline. Pharm Res 1990;7:1068–1070.

99. Borgström L, Newman SP, Weisz A, et al. Pulmonary deposition of inhaled terbutaline: comparison of scanning gamma camera and urinary excretion methods. J Pharm Sci 1992;81:753–755.

100. Newman SP, Steed K, Hooper G, et al. Comparison of gamma scintigraphy and a pharmacokinetic technique for assessing pulmonary deposition of terbutaline sulphate delivered by pressurized metered dose inhaler. Pharm Res 1995;12:231–236.

101. Agnew JE. Characterizing lung aerosol penetration. J Aerosol Med 1991;4:237–249.

102. Newman S. In vivo assessment of pulmonary drug delivery. In: Newman S, Anderson P, Byron P, et al. *Respiratory Drug Delivery: Essential Theory and Practice*. Richmond: RDD Online/Virginia Commonwealth University; 2009. p 97–133.

103. Smaldone GC. Deposition patterns of nebulized drugs: is the pattern important? J Aerosol Med 1994;7 Suppl 1:S25–S31.

104. Biddiscombe MF, Meah SN, Underwood SR, et al. Comparing lung regions of interest in gamma scintigraphy for assessing inhaled therapeutic aerosol deposition. J Aerosol Med Pulm Drug Deliv 2011;24:165–173.

105. Möller W, Felten K, Meyer G, et al. Errors in dose determination of radiolabelled particles deposited in central airways due to inhomogeneities in gamma ray absorption. J Aerosol Med 2007;20:193 [abstract].

106. Newman SP, Hirst PH, Pitcairn GR, et al. Understanding regional lung deposition in gamma scintigraphy. In: Dalby RN, Byron PR, Farr SJ, editors. *Respiratory Drug Delivery VI*. Buffalo Grove: Interpharm Press; 1998. p 9–15.

107. Clark AR. Beyond the throat: preferred methods of assessing regional distribution in the lung. In: Dalby RN, Byron PR, Peart J, et al., editors. *Respiratory Drug Delivery 2010*. River Grove: Davis Healthcare International; 2010. p 235–244.

108. Tossici-Bolt L, Fleming JS, Conway JH, et al. Analytical technique to recover the third dimension in planar imaging of inhaled aerosols: (1) impact on spatial quantification. J Aerosol Med 2006;19:565–579.

109. Ruffin RE, Dolovich MB, Wolff RK, et al. The effects of preferential deposition of histamine in the human airway. Am Rev Respir Dis 1978;117:485–492.

110. Newman SP, Pitcairn GR, Hirst PH. A brief history of gamma scintigraphy. J Aerosol Med 2001;14:139–145.

111. Smaldone GC. Respiratory physiology and disease effects on aerosol drug deposition. In: Dalby RN, Byron PR, Peart J, et al., editors. *Respiratory Drug Delivery IX*. River Grove: Davis Healthcare International; 2004. p 179–186.

112. Newman SP, Wilding IR, Hirst PH. Human lung deposition data: the bridge between in vitro and clinical evaluations for inhaled drug products? Int J Pharm 2000;208:49–60.

113. Newman SP, Pitcairn G, Hirst PH, et al. Radionuclide imaging technologies and their use in evaluating asthma drug deposition in the lungs. Adv Drug Deliv Revs 2003;55:851–867.

114. Warren S, Taylor G. Scintigraphic analysis in the clinical research of pharmaceuticals. EPC 2001;2001:54–61.

115. Thorsson L, Kenyon CJ, Newman SP, et al. Lung deposition of budesonide in asthmatics: a comparison of different formulations. Int J Pharm 1998;168:119–127.

116. Borgström L, Bondesson E, Morén F, et al. Lung deposition of budesonide inhaled via Turbuhaler: a comparison with terbutaline sulphate in normal subjects. Eur Respir J 1994;7:69–73.

117. Pitcairn GR, Lankinen T, Seppälä O-P, et al. Pulmonary drug delivery from the Taifun dry powder inhaler is relatively independent of the patient's inspiratory effort. J Aerosol Med 2000;13:97–104.

118. Hirst PH, Newman SP, Clark DA, et al. Lung deposition of budesonide from the novel dry powder inhaler Airmax. Respir Med 2002;96:389–396.

119. Newman SP, Weisz AWB, Talaee N, et al. Improvement of drug delivery with a breath actuated pressurised aerosol for patients with poor inhaler technique. Thorax 1991;46:712–716.

120. Newman SP. Use of gamma scintigraphy to evaluate the performance of new inhalers. J Aerosol Med 1999;12 Suppl 1:S25–S31.

121. Brand P, Schulte M, Wencker M, et al. Lung deposition of inhaled alpha1-proteinase inhibitor in cystic fibrosis and alpha1-antitrypsin deficiency. Eur Respir J 2009;34:354–360.

122. Hirst PH, Bacon RE, Newman SP, et al. Deposition, absorption and bioavailability of aerosolized morphine sulfate delivered by a novel hand-held device, the metered solution inhaler (MSI). In: Dalby RN, Byron PR, Farr SJ, et al., editors. *Respiratory Drug Delivery VII*. Raleigh: Serentec Press; 2000. p 467–469.

123. Newman SP, Hirst PH, Bacon RE, et al. Pulmonary delivery of ergotamine tartrate using a new breath-actuated pressurized aerosol device, the aerosol drug delivery system (ADDS). In: Dalby RN, Byron PR, Farr SJ, et al., editors. *Respiratory Drug Delivery VII*. Raleigh: Serentec Press; 2000. p 471–473.

124. Borgström L, Olsson B, Thorsson L. Degree of throat deposition can explain the variability in lung deposition of inhaled drugs. J Aerosol Med 2006;19:473–483.

125. Newman SP, Chan H-K. In vitro/in vivo comparisons in pulmonary drug delivery. J Aerosol Med Pulm Drug Deliv 2008;21:77–84.

126. Eberl S, Chan H-K, Daviskas E. SPECT imaging for radioaerosol deposition and clearance studies. J Aerosol Med 2006;19:8–20.

127. Cao Z-J, Maunoury C, Chen CC, et al. Comparison of continuous step-and-shoot versus step-and-shoot acquisition SPECT. J Nucl Med 1996;37:2037–2040.

128. Logus JW, Trajan M, Hooper HR, et al. Single photon emission tomography of the lung imaged with 99mTc-labeled aerosols. J Can Assoc Radiol 1984;35:133–138.

129. Perring S, Summers Q, Fleming JS, et al. A new method of quantification of the pulmonary regional distribution of aerosols using combined CT and SPECT and its application to nedocromil sodium administered by metered dose inhaler. Br J Radiol 1994;67:46–53.

130. Summers QA, Fleming JS, Dai Y, et al. The pulmonary deposition of two aerosol preparations of nedocromil sodium delivered by MDI assessed by single photon emission computed tomography. J Aerosol Med 1996;9 Suppl 1:S93–S109.

131. Conway JH, Walker P, Fleming JS, et al. Three-dimensional description of the deposition of inhaled terbutaline sulphate administered via Turbuhaler. In: Dalby RN, Byron PR, Farr SJ, et al, editors. *Respiratory Drug Delivery VII*. Raleigh: Serentec Press; 2000. p 607–609.

132. Newman SP, Salmon A, Nave R, et al. High lung deposition of ^{99m}Tc-labeled ciclesonide administered via HFA-MDI to patients with asthma. Respir Med 2006;100:375–384.

133. Newman SP, Pitcairn GR, Kalirai CS, et al. Using gamma scintigraphy and single photon emission computed tomography to assess the pulmonary deposition of a new inhaled formulation of flunisolide. In: *Drug Delivery to the Lungs XIII*. Portishead: The Aerosol Society; 2002. p 87–90.

134. Glover W, Chan H-K, Eber S, et al. Lung deposition of mannitol powder aerosol in healthy subjects. J Aerosol Med 2006;19:522–533.

135. Fleming JS, Sauret V, Conway JH, et al. Evaluation of the accuracy and precision of lung aerosol deposition measurements from single-photon emission computed tomography using simulation. J Aerosol Med 2000;13:187–198.

136. Fleming JS, Hashish AH, Conway JH, et al. Assessment of deposition of inhaled aerosol in the respiratory tract of man using three-dimensional multimodality imaging and mathematical modeling. J Aerosol Med 1996;9:317–327.
137. Fleming JS. Measuring drug deposition by airway generation in humans. In: Dalby RN, Byron PR, Peart J, et al., editors. *Respiratory Drug Delivery IX*. River Grove: Davis Healthcare International; 2004. p 187–193.
138. Fleming JS, Conway J, Majoral C, et al. The use of combined single photon emission computed tomography and X-ray computed tomography to assess the fate of inhaled aerosol. J Aerosol Med Pulm Drug Deliv 2011;24:49–60.
139. Bocher M, Balan A, Krausz Y, et al. Gamma camera-mounted anatomical X-ray tomography: technology, system characteristics and first images. Eur J Nucl Med 2000;27:619–627.
140. Fleming JS, Conway JH, Holgate ST, et al. Comparison of methods for deriving aerosol deposition by aerosol generation from three-dimensional radionuclide imaging. J Aerosol Sci 2000;31:1251–1259.
141. Fleming JS, Quint M, Bolt L, et al. Comparison of SPECT aerosol deposition data with twenty-four-hour clearance measurements. J Aerosol Med 2006;19:261–267.
142. Phipps PR, Gonda I, Bailey DL, et al. Comparisons of planar and tomographic gamma scintigraphy to measure the penetration index of inhaled aerosols. Am Rev Respir Dis 1989;139:1516–1523.
143. Fleming JS, Halson P, Conway J, et al. Three-dimensional description of pulmonary deposition of inhaled aerosol using data from multimodality imaging. J Nucl Med 1996;37:873–877.
144. Smaldone GC. Radionuclide scanning, respiratory physiology and pharmacokinetics. J Aerosol Med 2001;14:135–137.
145. Pitcairn G, James J, Joyson A, et al. TechneCoat: a novel method for radiolabelling dry powder formulations in radionuclide imaging studies. In: Dalby RN, Byron PR, Peart J, et al, editors. *Respiratory Drug Delivery VIII*. Raleigh: Davis Horwood; 2002. p 553–556.
146. Ito H, Pitcairn G, Hirst P, et al. An in vivo study to assess the suitability of Technegas as a SPECT imaging agent for pulmonary drug formulations. In: Dalby RN, Byron PR, Peart J, et al., editors. *Respiratory Drug Delivery VIII*. Raleigh: Davis Horwood; 2002. p 439–442.
147. Eberl S, Chan H-K, Daviskas E, et al. Aerosol deposition and clearance measurement: a novel technique using dynamic SPET. Eur J Nucl Med 2001;28:1365–1372.
148. Chan H-K, Eberl S, Daviskas E, et al. Changes in lung deposition of aerosols due to hygroscopic growth. A fast SPECT study. J Aerosol Med 2002;15:307–311.
149. Koetz BS, Price PM. Positron emission tomography. In: Lappin G, Temple S, editors. *Radiotracers in Drug Development*. Boca Raton: Taylor and Francis; 2006. p 251–267.
150. Jones T. New opportunities in molecular imaging using PET. Drug Information J 1997;31:991–995.
151. Aboagye EO, Price PM, Jones T. In vivo pharmacokinetics and pharmacodynamics in drug development using positron-emitting tomography. Drug Discov Today 2001;6:293–302.
152. Bailey DL. Imaging the airways in 2006. J Aerosol Med 2006;19:1–7.
153. Sanchez-Crespo A, Andreo P, Larsson SA. Positron flight in human tissues and its influence on PET image spatial resolution. Eur J Nucl Med Mol Imaging 2004;31:44–51.
154. Dolovich MB. ^{18}F-fluorodeoxyglucose positron emission tomographic imaging of pulmonary functions, pathology, and drug delivery. Proc Am Thorac Soc 2009;6:477–485.
155. Lee Z, Berridge MS, Finlay WH, et al. Mapping PET-measured triamcinolone acetonide aerosol distribution into deposition by airway generation. Int J Pharm 2000;199:7–16.

156. Dolovich M, Nahmias C, Coates G. Unleashing the PET: 3D imaging of the lung. In: Dalby RN, Byron PR, Farr SJ, et al., editors. *Respiratory Drug Delivery VII*. Raleigh: Serentec Press; 2000. p 215–230.

157. Berridge MS, Heal, DL, Lee Z. Imaging studies of biodistribution and kinetics in drug development. Drug Dev Res 2003;59:208–226.

158. Van Lingen A, Huijgens PC, Visser FC, et al. Performance characteristics of a 511 KeV collimator for imaging positron emitters with a standard gamma camera. Eur J Nucl Med 1992;19:315–321.

159. Lin P, Delaney G, Chu J, et al. Fluorine-18 FDG dual-headed gamma camera coincidence imaging of radiation pneumonitis. Clin Nucl Med 2000;25:866–869.

160. Basu S, Alavi A. Unparalleled contribution of 18F-FDG PET to medicine over 3 decades. J Nucl Med 2008;49:17N–21N, 37N.

161. Labiris R, Ask K, Kol M, et al. Pulmonary disease models with lung imaging outcome measures: making the black box more transparent. In: Dalby RN, Byron PR, Peart J, et al, editors. *Respiratory Drug Delivery 2006*. River Grove: Davis Healthcare International; 2006. p.191–204.

162. Chen DL, Schuster DP. Imaging pulmonary inflammation with positron emission tomography: a biomarker for drug development. Mol Pharm 2006;3:488–495.

163. Dolovich MB, Schuster DP. Positron emission tomography and computed tomography versus positron emission tomography-computed tomography: tools for imaging the lung. Proc Am Thorac Soc 2007;4:328–333.

164. Locke LW, Chordia MD, Zhang Y, et al. A novel neutrophil specific PET imaging agent: CFLFLFK-PEG-64Cu. J Nucl Med 2009;50:790–797.

165. Rhodes CG, Valind SO, Brudin LH, et al. Quantification of regional V/Q ratios in humans by use of PET. II. Procedure and normal values. J Appl Physiol 1989;66:1905–1913.

166. Rhodes CG, Hughes JMB. Pulmonary studies using positron emission tomography. Eur Respir J 1995;8:1001–1017.

167. Yanai M, Hatazawa J, Ojima F, et al. Deposition and clearance of inhaled ^{18}FDG powder in patients with chronic obstructive pulmonary disease. Eur Respir J 1998;11:1342–1348.

168. van Waarde A, Maas B, Doze P, et al. Positron emission tomography studies of human airways using an inhaled β-adrenoceptor antagonist, S-11C-CGP 12388. Chest 2005;128:3020–3027.

169. Berridge MS, Lee Z, Heald DL. Pulmonary distribution and kinetics of inhaled 11C triamcinolone acetonide. J Nucl Med 2000;41:1603–1611.

170. Constantinou M, Waters SL, Steel CJ, et al. Flixotide pressurized metered dose inhalers loaded with ^{18}F-fluticasone dipropionate particles for drug deposition studies in humans with PET—formulation and analysis. J Labelled Compd Radiopharm 2004;47:55–70.

171. Bergström M, Nordberg A, Lunnell E, et al. Regional deposition of inhaled ^{11}C-nicotine vapor in the human airway as visualized by positron emission tomography. Clin Pharmacol Ther 1995;57:309–317.

172. Lunnell E, Bergström M, Antoni G, et al. Nicotine deposition and body distribution from a nicotine inhaler and a cigarette studied with positron emission tomography. Clin Pharmacol Ther 1996;58:593–594 [letter].

173. Pike VW. The status of PET radiochemistry for drug development and evaluation. Drug Information J 1997;31:997–1013.

174. Dolovich M, Labiris R. Imaging drug delivery and drug responses in the lung. Proc Am Thorac Soc 2004;1:329–337.

175. Dalby R, Byron P. In vitro assessment of inhaled products. In: Newman S, Anderson P, Byron P, et al. *Respiratory Drug Delivery: Essential Theory and Practice*. Richmond: RDD Online/Virginia Commonwealth University; 2009. p 59–96.

176. Code of Federal Regulations. Adulterated Drugs and Devices. Title 21, Chapter 9, Section 351; 2009.

177. Code of Federal Regulations. Classification of Devices Intended for Human Use. Title 21, Chapter 9, Section 360; 2009.

178. European Medicines Agency. Guideline on the requirements for clinical documentation for orally inhaled products (OIP) including the requirements for demonstration of therapeutic equivalence between two inhaled products for use in the treatment of asthma and chronic obstructive pulmonary disease (COPD) in adults and for use in the treatment of asthma in children and adolescents. Doc Ref CPMP/EWP/4151/00; 2009.

179. Mitchell JN, Newman SP, Chan H-K. In vitro and in vivo aspects of cascade impactor testing and inhaler performance: a review. AAPS PharmSciTech 2007;8 (4):237–248.

180. Adams W., Ahrens RC, Chen M-L, et al. Demonstrating bioequivalence of locally acting orally inhaled drug products (OIPs): workshop summary report, cosponsored by Product Quality Research Institute (PQRI) and Inhalation and Nasal Technology Focus Group (INTFG). J Aerosol Med Pulm Drug Deliv 2010;23:1–29.

181. Devadason SG, Chan H-K, Haeussermann S, et al. Validation of radiolabelling of drug formulations for aerosol deposition assessment of orally inhaled products. J Aerosol Med Pulm Drug Deliv 2012;25 Suppl 1: S6–S9.

182. Newman SP, Bennett WD, Biddiscombe M, et al. Standardization of techniques for using planar (2D) imaging for aerosol deposition assessment of orally inhaled products. J Aerosol Med Pulm Drug Deliv 2012;25 Suppl 1: S10–S28.

183. Newman SP. Improving drug metering to the lung: making the patient the most important factor. In: Dalby RN, Byron PR, Peart J, et al., editors. *Respiratory Drug Delivery 2006*. River Grove: Davis Healthcare International; 2006. p 1–12.

184. Dunbar C, Scheuch G, Sommerer K, et al. In vitro and in vivo dose delivery characteristics of large porous particles for inhalation. Int J Pharm 2002;245:179–189.

185. Newhouse MT, Hirst PH, Duddu SP, et al. Inhalation of a dry powder tobramycin PulmoSphere formulation in healthy volunteers. Chest 2003;124:360–366.

186. Pauwels R, Newman SP, Borgström L. Airway deposition and airway effects of anti-asthma drugs delivered from metered dose inhalers. Eur Respir J 1997;10:2127–2138.

187. Townsend DW. Multimodality imaging of structure and function. Phys Med Biol 2008;53:R1–R39.

188. Altes TA, Powers PL, Knight-Scott J, et al. Hyperpolarized 3He lung ventilation imaging in asthmatics: preliminary findings. J Magn Reson Imaging 2001;13:378–384.

189. Casselbrant I, Diaz S, Petterson G, et al. Apparent diffusion coefficient (ADC) of hyperpolarized 3-He magnetic resonance imaging (MRI) in patients with emphysema and healthy volunteers. Eur Respir J 2004;24 Suppl 48:235s [abstract].

190. Martin AR, Thompson RB, Finlay W. MRI measurement of regional lung deposition in mice exposed nose-only to nebulized superparamagnetic iron oxide particles. J Aerosol Med Pulm Drug Deliv 2008;21:335–342.

191. Thompson RB, Finlay WH. Using MRI to measure aerosol deposition. J Aerosol Med Pulm Drug Deliv 2012;25:55–62.

192. Martonen TB, Smyth HD, Isaacs KK, et al. Issues in drug delivery: concepts and practice. Respir Care 2005;50:1228–1252.

193. Finlay WH, Martin AR. Recent advances in predictive understanding of respiratory tract deposition. J Aerosol Med Pulm Drug Del 2008;21:189–206.

194. Martonen TB, Musante CJ, Segal RA, et al. Lung models: strengths and limitations. Respir Care 2000;45:712–736.

CHAPTER 16

IN VIVO IMAGING FOR VALIDATION OF ORAL DOSE PERFORMANCE

FIONA J. MCINNES AND HOWARD N. E. STEVENS

16.1 INTRODUCTION

The oral route remains the most widely used and accepted method of administering therapeutic compounds. As the challenges of modern drug delivery drive the progression toward increasingly complex dosage forms, the importance of confirming successful performance *in vivo* cannot be underestimated. Realization of the importance of biorelevant testing has, in recent years, resulted in a significant expansion in the literature surrounding *in vitro* assessment of formulations, with a view to predicting *in vivo* behavior. While *in vitro* testing can provide valuable predictive information, it is only the process of evaluating modern formulations in the clinic that will ensure confidence in performance. Examples can be found in the literature of formulations that did not perform as predicted by *in vitro* tests [1–3] emphasizing the importance of progressing to *in vivo* evaluation as soon as it is safe and practical to do so.

While pharmacokinetic and pharmacodynamic data are invaluable in determining the therapeutic efficacy of a compound, they do not necessarily confirm that a formulation has behaved in accordance with its design or provide information on the location in the gastrointestinal (GI) tract at any particular time. It may, for example, be desirable to achieve maximum spread of the formulation in the gut in order to increase absorption of poorly soluble compounds, or to prolong gastric residence in order to achieve increased bioavailability of a compound with a narrow absorption window in the upper small intestine. Alternatively, the aim may be to target a formulation to release the drug only in the colon, for local effect or improved bioavailability, while avoiding exposure of the patient to the drug in regions of the GI tract where it is not required. In many cases such formulations will be designed using multiple components or complex physical control of drug release, which must be robust and reproducible enough to perform in the varying physiological and physicochemical conditions that will be experienced in different regions of the gut.

Drug Delivery Applications of Noninvasive Imaging: Validation from Biodistribution to Sites of Action, First Edition. Edited by Chun Li and Mei Tian.

In vivo imaging therefore remains crucial in determining desirable formulation performance. In this chapter, we will outline the commonly used imaging techniques that have been reported and give examples of their use for validation of oral dose performance in the different sections of the GI tract.

16.2 IMAGING TECHNIQUES

16.2.1 Gamma Scintigraphy

Gamma scintigraphy is perhaps the most widely used *in vivo* imaging technique for validation of oral dose performance. In this technique, a key component of the formulation is labeled with a gamma-emitting radioisotope, in order to track the progress and behavior of that aspect of the dose throughout the GI tract. The dose is administered to a volunteer, who is then placed in front of the gamma camera at various intervals in order for images of the formulation to be acquired. Images can be acquired as static images taken over a period of 30–60 s or as a series of dynamic images if the process to be investigated is expected to be rapid. The camera used to detect the gamma emissions (Fig. 16.1) consists of a detector head, over which a lead plate collimator is placed in order to screen out unnecessary interference.

A parallel hole collimator is typically used in oral studies, a lead plate with a series of parallel holes that filter out any gamma rays not traveling directly toward the detector, creating a clear image of the radioactive source. When gamma rays pass through the collimator, they interact with the sodium iodide crystal that lies behind it, producing light, which is transformed by a series of photomultiplier tubes into an electrical signal that can be interpreted by the system software to form an image (Fig. 16.2).

As it is gamma emissions that are detected, the data that are obtained are quantitative, with each pixel representing the number of gamma rays detected in that area [4], which for many is a key benefit of this technique over other *in vivo* imaging options. The analytical software then allows further analysis of the images, with the user drawing "regions of interest" (ROIs) around areas such as the stomach or the colon, allowing determination of the proportion of the radiolabeled dose that was in that particular anatomical region at the time the image was acquired.

Commonly used isotopes are of low energy, safely allowing crossover studies in volunteers, and include technetium-99m (^{99m}Tc), indium-111 (^{111}In), and samarium-153 (^{153}Sm), with others such as erbium-171 (^{171}Er) used less frequently. When studying the performance of an oral formulation, a key aspect of the molecule used for labeling is that it should not be in a form that can be absorbed to the systemic circulation. The objective is to visualize the distribution of the formulation in the GI tract, and so the isotope is commonly complexed with a molecule such as diethylene triamine pentaacetic acid (DTPA) in order to ensure that it cannot cross the epithelial barrier. When it has been decided which aspect of the formulation will be labeled (e.g., should it be homogeneously distributed throughout, or is it a "layer" or "shell" that is of interest?), it is also crucial to undertake *in vitro* validation of the labeling process, ensuring that the radiolabel does indeed remain with that component of the

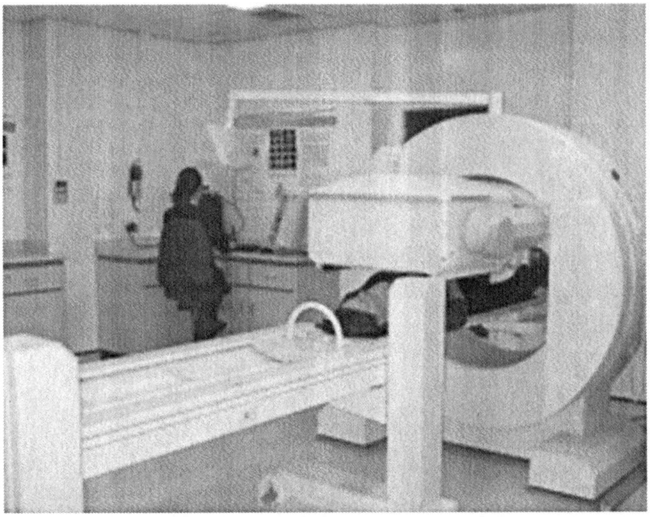

FIGURE 16.1 A typical gamma camera.

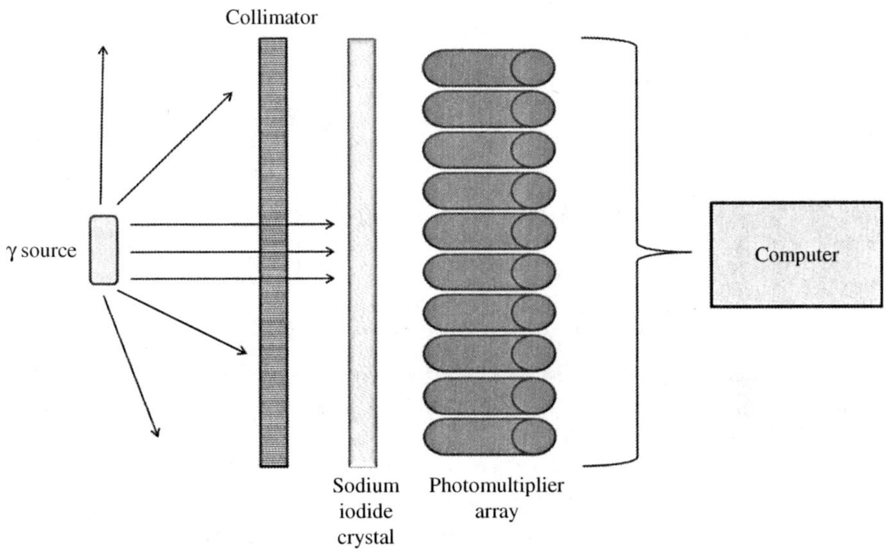

FIGURE 16.2 Schematic diagram of a gamma camera.

formulation for the duration of the study and that the radiolabel does not influence the properties of the formulation. In a multicomponent formulation, it is possible to visualize the behavior of each component independently; a bilayer tablet or multicomponent capsule, for example, could be manufactured with ^{99m}Tc in one layer and ^{111}In in the other (Fig. 16.3).

Pharmacokinetic blood sampling can be carried out in volunteers simultaneously to scintigraphic imaging in order to correlate key pharmacokinetic events with location in the GI tract, a technique that is referred to as pharmacoscintigraphy. The

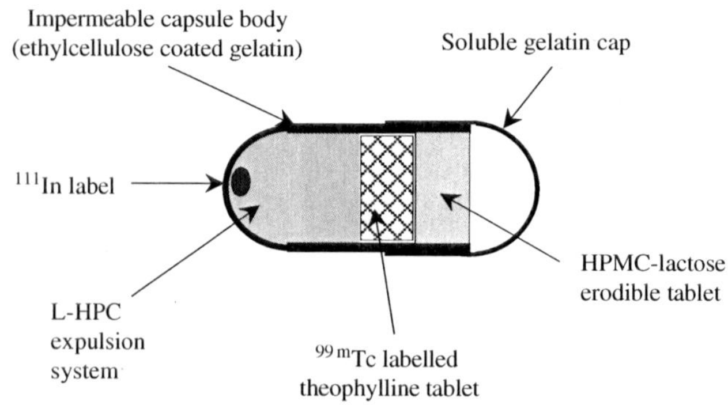

FIGURE 16.3 Use of different isotopes to radiolabel a multicomponent delivery system [89]. (With permission from John Wiley and Sons.)

great majority of oral formulation studies reported in the literature utilize the technique of gamma scintigraphy, and therefore, the major focus of this chapter will be the use of this technique in assessing oral dose performance; however, there are a variety of other *in vivo* imaging techniques that are also used, and the relative merits of these should also be discussed.

16.2.2 Magnetic Moment Imaging

Magnetic moment imaging (MMI) is an imaging method that circumvents the use of radioactive isotopes for noninvasive *in vivo* imaging of dosages. The food colorant, ferrimagnetic iron oxide magnetite (Fe_3O_4), is commonly used as an insoluble marker, meaning that it will not be absorbed from the GI tract [5]. A small quantity of this marker (0.1–0.2 mg) is incorporated into the dosage form to be administered to the subject, which is subsequently magnetized with a strong bar magnet or electromagnet with a field strength of at least 0.6 T, in order to align the magnetic orientations of the particles [5]. The dipole field is measured using instrumentation incorporating sensitive magnetic flux sensors, superconductive quantum interference devices (SQUIDS). Bergstrand et al. demonstrated the potential of this technique in dosage form development, in a study that modeled food interactions with felodipine tablets [6]. It was found that both felodipine release and absorption were affected by the location of the tablet in the GI tract. A benefit of this technique is the high special resolution and sensitivity provided, along with data on the three-dimensional movements of the dose (x, y, and z planes, Fig. 16.4); however, the sensors are large and must be housed in magnetically shielded rooms in order to eliminate interference from the magnetic field of the earth [7].

Alternative instrumentation has been designed that does not require such housing, and is more portable, although this requires the use of larger quantities of more strongly magnetic materials in order to compensate for the lack of shielding from the environment and reduced sensitivity of the instrumentation [5, 7]. In both cases, the magnetic moment is lost as the tablet disintegrates, allowing the instrumentation to provide a certain level of detail and quantification in this respect. Confirmation of the performance of this portable system was obtained by Goodman et al., who studied

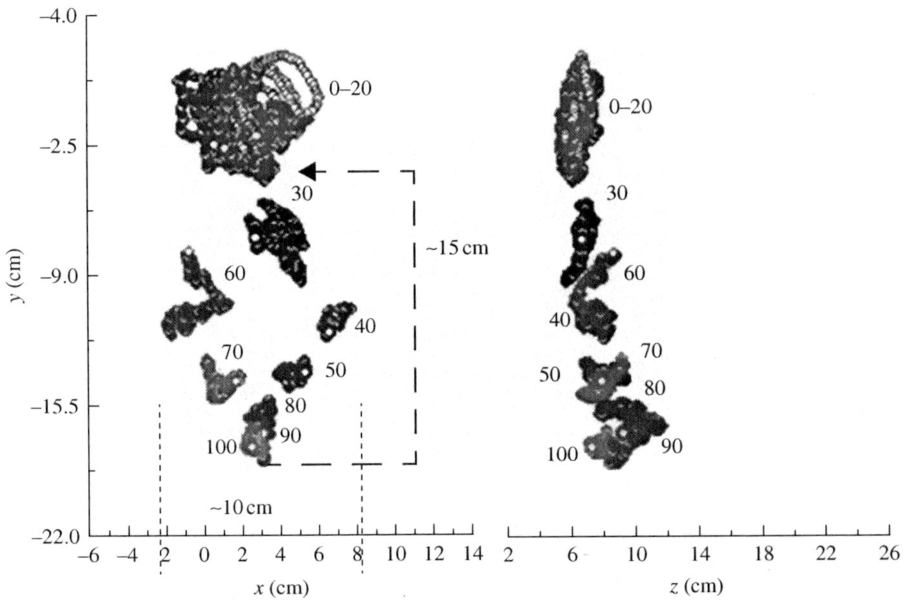

FIGURE 16.4 The localization of an orally administered tablet in the frontal view (*x*,*y*) and lateral view (*z*,*y*), determined by MMI (Reprinted from reference [7], with permission from Elsevier.)

the transit of immediate release, enteric coated, and non-disintegrating tablets using both MMI and scintigraphy and found good correlation between the two techniques [7].

16.2.3 AC Biosusceptometry

AC biosusceptometry (ACB) is another technique using magnetic sensors to track dosage forms; however, in this case the magnetic particles in the dose are not previously magnetized, therefore allowing both marker movement and dispersion to be monitored [8]. A magnetically susceptible material, ferrite, is used as a tracer, and induction coils are used to measure flux variations in response to the application of an alternating magnetic field. A more detailed description of the instrumentation of this technique can be found elsewhere in the literature [8]. Similarly to the alternative MMI instrumentation described in Section 16.2.2, ACB does not require a magnetically shielded room and is portable, unlike the biomagnetic measurement system reported by Hu et al. [9].

16.2.4 Magnetic Resonance Imaging

Magnetic resonance imaging (MRI) is a medical imaging technique that has also been applied to the evaluation of oral pharmaceuticals. A strong magnetic field and radio wave irradiation are applied, resulting in a nuclear magnetic resonance signal that can be detected from nuclei such as ^{1}H, ^{19}F, ^{31}P, and ^{13}C [10, 11]. MRI is predominantly used to detect the presence of water, as the high ^{1}H concentration results in a stronger signal. One benefit of the use of MRI for *in vivo* investigation of pharmaceuticals is that the technique provides anatomical detail and high-resolution information on the dosage. Recently, for example, it has been shown that this technique can allow detection of individual 1-mm pellets in the human stomach (labeled with magnetite), along with differentiation from food particles present in the same region [12]. Given the near continuous contractions of the stomach in the fed state and the propensity for patient movement to interfere with image quality, this level of detail represents a remarkable achievement. This qualitative advantage must, however, be balanced against the less quantitative nature of the data obtained, which can confirm the ability of formulations such as floating rafts to perform appropriately *in vivo* [13], but not necessarily provide detailed information on the kinetics of emptying. In addition, availability of this type of instrumentation is often limited to medical premises, with clinical investigations taking priority over research.

16.3 IMAGING OF ORAL DOSES IN THE GI TRACT

16.3.1 The Buccal Cavity

In formulations administered to the buccal cavity, the primary interest generally lies in the rate of clearance of the dose from the mouth. Gamma scintigraphy was used to validate the *in vivo* performance of the Expidet formulation, which was designed to dissolve rapidly in the mouth in order to achieve rapid therapeutic onset via sublingual absorption or to circumvent the problem of patients swallowing tablets without water [14]. Micronized Amberlite resin was labeled with ^{99m}Tc and incorporated into the Expidet dose, which was placed on the tongue of volunteers. Scintigraphic images demonstrated that within 15 s the formulation had totally dispersed and that by 30 s activity had appeared in the esophagus (Fig. 16.5), and the technique was used to demonstrate that the inclusion of salivary stimulants in the formulation did not have any effect on the rate of clearance from the buccal cavity. It was noted by the authors that while the formulation dispersed rapidly, there was no lateral spread across the floor of the buccal cavity.

In many instances drug delivery to the buccal cavity involves preparations such as lozenges for topical delivery or chewing gum for systemic absorption. In a study

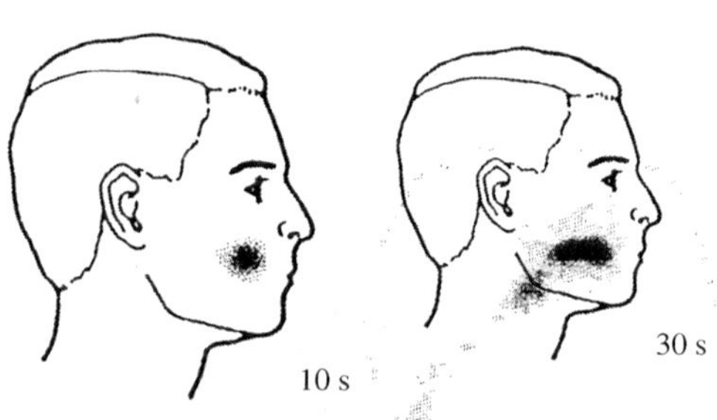

FIGURE 16.5 Scintigraphic images of the distribution of Expidet formulations containing 10 mg micronized ion exchange resin labeled with technetium-99 m. (Reprinted from reference [14], with permission from Elsevier.)

using gamma scintigraphy, the efficacy of lozenge, chewing gum, and tablet formulations was compared to investigate the resultant spread of the drug within the buccal cavity and the time at which the drug was swallowed [15]. While it was shown that the release of a model radiolabeled compound was similar for each formulation, by evaluating the clearance kinetics of the radioactivity from the buccal cavity and esophagus, it was concluded that the longer clearance half-life and higher area under the curve of chewing gum should make it more effective for local targeting, while sublingual tablets should perform better than chewing gum when the target is systemic absorption through the oral mucosa. In addition, the authors here suggested that the difference in clearance rates observed may be a function of the differing formulation effects on salivary flow, indicating a behavior that should be taken into account when designing buccal doses.

The concept of analyzing residence time of buccal formulations is of critical importance for those that are designed to be bioadhesive and deliver a drug over a prolonged period of time, and *in vivo* imaging can provide crucial information on this without the need to disturb the dosage form *in situ*. Indeed, this technique has also been investigated to confirm residence time of bioadhesive formulations in the periodontal pocket for delivery of antimicrobials [16]. In another study, it was shown that gel formulations composed of hydrophilic excipients exhibited a biphasic clearance from the buccal cavity with a rapid rate of clearance in the first 5 min (Fig. 16.6), while formulations composed of hydrophobic excipients exhibited a much slower, zero-order clearance [17].

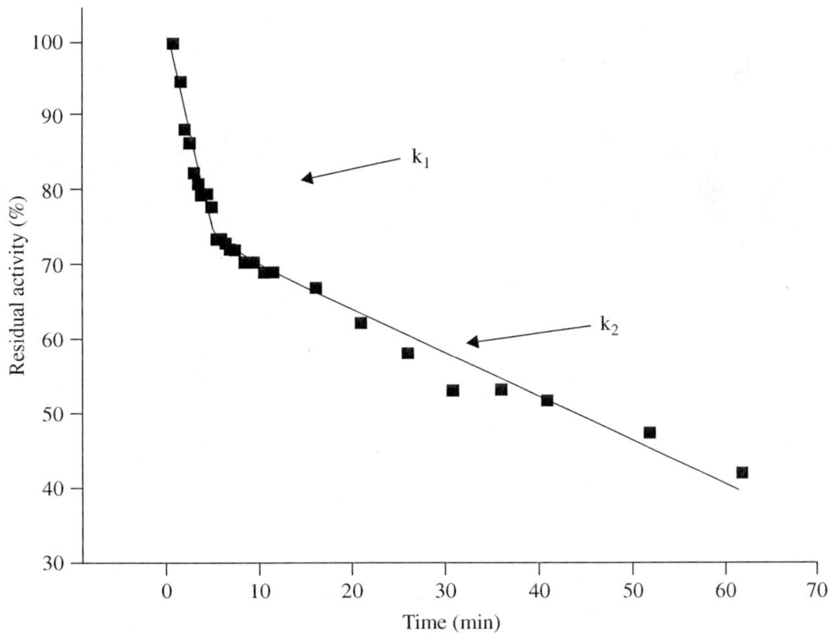

FIGURE 16.6 Elimination kinetics of a hydrophilic bioadhesive gel from the buccal cavity (Reprinted from reference [17], with permission from Elsevier.).

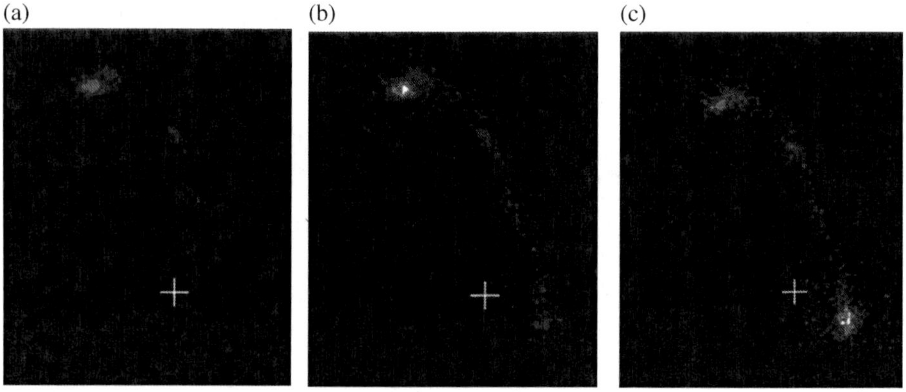

FIGURE 16.7 Gamma scintigraphic images showing the evaluation of esophageal transit of film-coated tablet formulations, with (a) the tablet in the mouth, (b) transit through the esophagus, and (c) in the stomach.

16.3.2 Esophageal Transit

The esophagus is not generally used as a site for drug delivery; transit through the esophagus is rapid (around 10–14 s [18]), and the striated squamous mucosal surface is not ideal for systemic drug absorption. A major issue concerning the esophagus is the tendency of solid dose oral formulations to adhere to the epithelial surface and cause damage to the mucosa, particularly in the case of elderly patients where the impaired esophageal clearance of this demographic increases the risk significantly [19]. The extent of damage caused will be largely dependent on the physicochemical nature of the drug contained within the formulation; thus in the case of compounds that are known to cause significant mucosal damage, the esophageal transit of the formulation should be evaluated *in vivo*. Gamma scintigraphy can allow visualization of transit rate (Fig. 16.7) and has been used to demonstrate the propensity of gelatin capsules to reside for longer periods of time in the esophagus of healthy volunteers aged 50–79 years, in comparison with enteric coated tablets [20].

The authors suggested that this emphasized the importance of considering this factor when designing formulations for compounds that have the potential to cause mucosal injury. This problem is not restricted to the above age group; studies using barium-labeled tablets with radiological imaging have also demonstrated the occurrence of esophageal adhesion in the general population [21]. Other important factors in esophageal clearance determined using gamma scintigraphy include the position of the patient at the time of dosing, with supine patients shown to have a significantly higher risk of formulation adherence [22]; the imbibition of water at the time of dosing, with tablets showing a greater tendency to adhere when taken with only a small volume of liquid [21]; and the size and shape of the tablet [21, 23].

The effect of modern film coatings on esophageal transit time of oval- and caplet-shaped tablets was investigated using gamma scintigraphy, and in both cases it was demonstrated that Opaglos® and Opadry® coatings reduced the tendency of tablets to adhere [24]. In fact, it has also been suggested that it is not only the outer shell formulation characteristics that should be assessed for mucoadhesive propensity in the esophagus but that the excipients contained within should also be considered. A case study was presented by Säkkinen et al., where the authors proposed that adherence of a capsule formulation to the esophagus of a healthy male volunteer for a period of 1.75 h was a result of not only the tendency of gelatin capsules to adhere but also the microcrystalline chitosan granules included as an excipient within the capsule [25]. In this case, the inclusion of chitosan in the formulation was designed to achieve gastric retention for therapeutic benefit but clearly demonstrated the need to incorporate *in vivo* imaging in the development process to identify unanticipated problems.

Despite the undesired propensity for solid dose formulations to become stuck in the esophagus, in situations where this behavior is desirable for local treatment, it can be understandably difficult to achieve long-term retention. Esophageal adhesion may, however, be advantageous in the treatment of localized diseases such as Barrett's esophagus. In this case it has been proposed that polymer gels could be used to achieve adhesion for prolonged localized drug delivery, and MRI has been used demonstrate the feasibility of this concept [26].

16.3.3 Delivery to the Stomach

16.3.3.1 Gastric Emptying Rate Quantification of gastric emptying rates can be important for formulation development, both in terms of the behavior of the dosage form itself and also the contents of the stomach at the time of dosing. Gamma scintigraphy has been used to demonstrate the effect of different types of food on the gastric emptying of oral formulations [27, 28], with the quantity of fat consumed known to exert a significant delay in gastric emptying. This will impact on the performance of disintegrating and particulate systems, slowing down the rate of gastric emptying as it exits the stomach along with food, and for monolithic systems designed to remain intact (e.g., enteric coated or matrix systems), gastric emptying of the dosage will not occur until all of the food has left the stomach and the migrating myoelectric complex (MMC) cycle has reached phase III [29, 30]. Any change in gastric emptying rate may impact on onset of drug action and/or bioavailability where absorption primarily occurs in the small intestine. Gamma scintigraphy has also been used to demonstrate that gastric emptying rates can vary in accordance with circadian rhythm, with the rate being higher in the morning than in the afternoon [18]. Conversely, the formulation or the drug contained within may have an influence on normal gastric emptying of food. One study suggested that administration of lipids in quantities that might be contained in pharmaceutical formulations resulted in a slight delay in gastric emptying relative to a placebo (10 ml water) [31].

In a study where GI transit was increased or delayed in healthy volunteers by the use of metoclopramide or propantheline, it was found that the bioavailability of metformin was enhanced in the case of delayed gastric emptying of the radiolabeled solution [32]. This finding is important in terms of both therapy and formulation. It is known that diabetic patients often suffer from gastroparesis [33], meaning that the performance of an oral metformin formulation may be different in these patients in comparison to healthy volunteers. Secondly, the authors concluded that the findings had significant implications for the design of a modified release oral dose (immediate

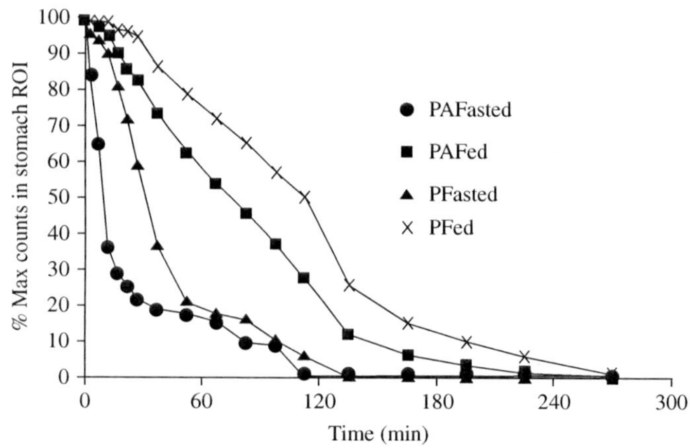

FIGURE 16.8 Mean gastric emptying of Panadol Actifast (PA) and Panadol (P) in the fed and fasted state ($n=11$) [35]. (With kind permission from Springer Science+Business Media B.V.)

release metformin is often administered three times a day), providing key evidence for the strategy required to maintain adequate absorption. Pharmacoscintigraphic techniques also confirmed an effect of gastric emptying rate on the bioavailability of an investigational hypolipidemic agent, despite the apparent observation that absorption was higher in the ileal region [34].

Radiolabeling has been used to demonstrate that rapid T_{max} in a commercial paracetamol (acetaminophen) formulation containing bicarbonate, Panadol Actifast®, in comparison with conventional paracetamol tablets resulted from a combination of faster disintegration and gastric emptying of the formulation (Fig. 16.8) [35].

This study demonstrated a direct influence of formulation strategy on the therapeutic effect of an active compound. Gamma scintigraphy has also demonstrated unexpected formulation effects on gastric residence. The Expidet rapid disintegrating buccal formulation with micronized resin embedded in the matrix was found to coat the stomach and achieve a longer residence time, when compared to liquid and tablet formulations [36]. Many imaging studies have reported the influence of formulation size, shape, and density on their time of gastric emptying [37–39].

16.3.3.2 *In Vivo Erosion and Formulation Robustness* Robustness in the presence of the varying hydrodynamic and physical forces in the different regions of the GI tract is of paramount importance in the development of oral controlled release matrix formulations. If the matrix fails *in vivo*, then the active drug that was incorporated at a dose designed to be slowly released over a prolonged period of time can be "dumped" in a single place, resulting in a potentially dangerous overdose. The forces in the stomach can be much higher in the fed state [40], and there are reports in the literature of formulation failure in this region [1, 41]. Recently, efforts to understand dose dumping have increased dramatically, particularly in light of the suggestion that not only food components but also alcohol may have an impact on dosage form failure [42–44], as highlighted by the FDA alert on an alcohol interaction with the Palladone (hydromorphone) extended release formulation.

TABLE 16.1 Complete *In vivo* Disintegration Data for an MR Tablet Following Fasted and Fed Dosing[a]

	Fasted		Fed	
Subject number	Time	Anatomical site	Time	Anatomical site
1	9.64	Colon	6.98	Small intestine
2	7.88	Colon	2.73	Stomach
3	12.67	Colon	6.70	Stomach
4	6.48	Colon	4.04	Stomach
5	9.13	Colon	4.88	Stomach
6	8.78	Colon	6.21	Stomach
7	10.11	Colon	5.68	Stomach
8	11.79	Colon	5.36	Small intestine
9	7.98	Colon	10.49	Stomach
Mean	9.38		5.90	
SD	1.95		2.18	
Median	9.13		5.68	
n	9		9	

[a]Ref. 1, with permission from John Wiley and Sons.

An extended release hydroxypropyl methylcellulose (HPMC) matrix formulation of an experimental compound for treatment of prostatic hyperplasia was found in pharmacokinetic studies to result in variable absorption in volunteers following consumption of a high-fat breakfast, while performance in the fasted state reflected the expected 18-h release profile predicted by *in vitro* experimentation. The investigators suspected that this may have been a result of dose dumping from the matrix in the fed state and, by labeling the matrix with ^{153}Sm, were able to confirm this theory, showing significantly faster erosion in the fed state and complete disintegration predominantly occurring in the stomach (Table 16.1) [1].

This study demonstrated the usefulness of gamma scintigraphy in the identification of formulation failure and the need to further understand the erosion and performance of oral matrix systems. In contrast, MMI has been used in the case of an extended release felodipine HPMC matrix tablet to demonstrate that the increased C_{max} observed was due to poor mixing in the proximal stomach, rather than failure of the formulation and dose dumping [45].

Even without the occurrence of dose dumping, matrix erosion rates *in vivo* could be markedly different from the simulated *in vitro* environment. Ghimire et al. have reported the use of gamma scintigraphy for quantitative assessment of erosion *in vivo* (Fig. 16.9) [46, 47].

An HPMC hydrophilic matrix tablet and a glyceryl behenate placebo wax matrix tablet were radiolabeled by incorporating a small amount of ^{99m}Tc-radiolabeled charcoal into the matrix at the time of manufacture. As charcoal is an insoluble material, the homogeneously distributed particles would only be able to separate from the matrix following erosion of the polymer layer around it (rather than by diffusion through the matrix, which would occur for a water soluble particle). This method had previously been reported as a method of labeling a swelling gastroretentive formulation

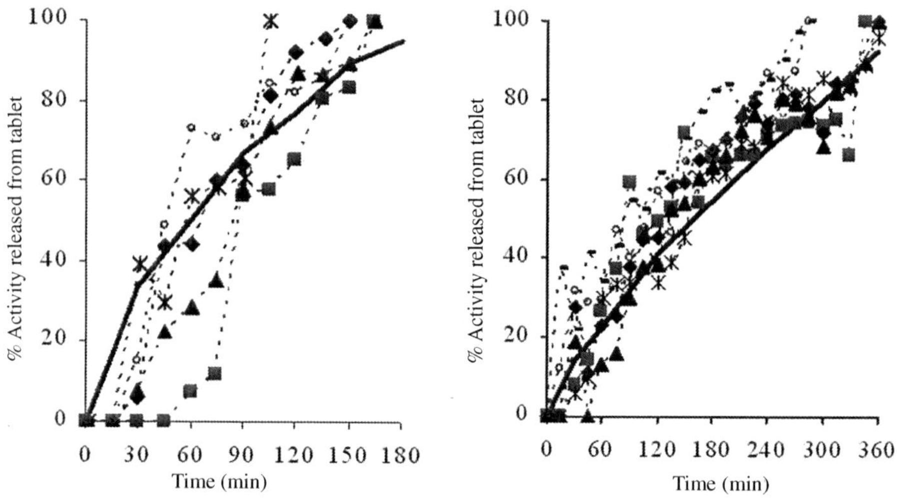

FIGURE 16.9 Individual *in vivo* erosion profiles of two different HPMC matrix tablets in six subjects (S1,■; S3,♦; S4,▲; S5,○; and S6,*) and comparison to mean *in vitro* gravimetric erosion profile (solid line). (Reprinted from reference [46], with permission from Elsevier.)

that had proven difficult to label with "conventional" methods due to leaching of the isotope [48]. MMI has also been reported as a suitable *in vivo* method for visualization and evaluation of oral dose disintegration [5, 7], as has the magnetic technique ACB [49]. The study of disintegration is not, however, limited to matrix tablet formulations; scintigraphic imaging has also aided in confirming performance of hard-shell capsules for ubiquitous formulation use [50]. HPMC capsules (QUALI-V®) prepared from plant sources rather than gelatin are available as an alternative to conventional gelatin capsules, for cases where gelatin is not suitable, either for ethical/dietary reasons or perhaps because of stability issues related to their higher moisture content. By labeling the contents of an HPMC and a gelatin capsule with different isotopes (^{99m}Tc and ^{111}In), the authors were able to demonstrate that the disintegration times of both capsules were "rapid and comparable."

16.3.3.3 Floating Raft Formulations Raft-forming formulations are one approach to the prevention of acid reflux and operate by the formation of a physical barrier over the contents of the stomach in order to prevent gastric juices from refluxing into and irritating/damaging the esophagus. The performance of one such system has been monitored using gamma scintigraphy and demonstrated the correlation between physical presence of the formulation and suppression of acid reflux (Fig. 16.10) [51].

The key to reliable therapeutic performance of such systems is therefore the ability to maintain this barrier for an appreciable length of time, in order to prevent this occurrence and allow healing of any damaged tissue. Probably the most well-known example of such formulation is the alginate raft-forming formulation, Gaviscon®. The ability of such alginate formulations to form a suitable raft has been demonstrated using gamma scintigraphy, where the raft was labeled with ^{113m}In, and

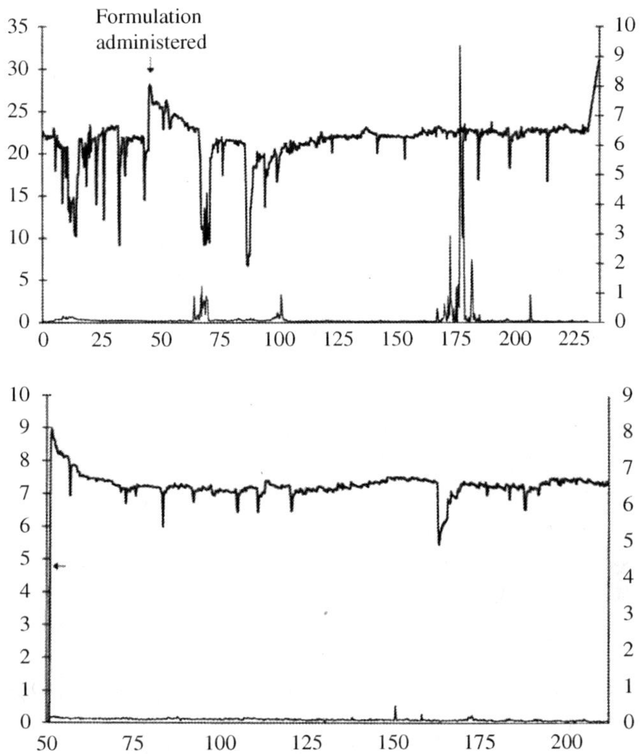

FIGURE 16.10 Correlation of floating raft performance with acid reflux, showing a typical trace after radiolabeled food with placebo (top) and Aflurax (bottom). pH trace, showing acid reflux (upper line); counts trace, showing food reflux (lower line). (Reprinted from reference [51], with permission from Elsevier.)

administered to volunteers [52–54]. The alginate formulations were found to float on the contents of the stomach and empty slowly, in comparison with antacid formulations that emptied with other stomach contents. More detailed information on the spatial distribution of the rafts was obtained using an MRI variant, echo planar imaging, which could detect when the raft was present in larger fragments in individuals, rather than as a continuous mass [55]. Furthermore, subsequent gamma scintigraphic studies demonstrated that despite the reliance of the Gaviscon® formulations on interaction with acid in the stomach to form the alginate raft, pretreatment with the acid-suppressing proton pump inhibitor omeprazole did not have a detrimental effect on performance [56]. A similar conclusion was also reached on the administration of the H_2 receptor antagonist cimetidine [57]. This important finding using an imaging technique provides data that can directly inform best practice in clinical prescribing.

16.3.3.4 Gastroretentive Formulations The aim of gastroretentive formulations is to achieve prolonged gastric residence, in order to attain more efficacious treatment of a local condition such as *Helicobacter pylori*, to achieve prolonged absorption to the systemic circulation by slowly releasing the active compound to the small intestine, or to increase bioavailability of drugs with a narrow absorption window in the proximal intestine by a similar mechanism. There are a variety of strategies that have been reported in order to achieve this, and imaging techniques have been intricately involved in the evaluation of their performance. The concept of a formulation floating on the gastric contents has been described in the context of alginate rafts in Section 16.3.3.3; however, other types of floating formulations have also been evaluated using gamma scintigraphy. These include confirmation of gastric retention of multiple unit alginate systems designed to circumvent the "all-or-nothing" gastric emptying of unit dose formulations [58] and their further formulation development to include coadministration of citric acid solution as a means of further delaying gastric emptying [59].

Floating tablets have also been evaluated using techniques such as gamma scintigraphy and MRI, with particular attention paid to the effect of fed status and meal type [60, 61], as a concern with floating formulations is the need for continued presence of fluid in the stomach for them to float on. Various studies have been performed on resins such as the anionic exchange resin cholestyramine and appear to show mucoadhesive properties with distribution and spread over the different sections of the stomach [62, 63] and retention of at least 20% of the dose for up to 6 h [64, 65]. Spread in the stomach was also demonstrated to be formulation specific in a scintigraphic study involving sucralfate, where a gel was found to achieve better spread and retention (for up to 5 h) in comparison with a suspension formulation [66].

The size and shape of oral formulations has also been investigated as a means of achieving gastro-retention, generally by one of two different mechanisms: by swelling or expansion in some way in order to render the formulation too large to pass through the pylorus or by utilizing unusual shapes or dimensions, which again prevent exit from the stomach. A study was performed using gamma scintigraphy to visualize the *in vivo* performance of different shaped silicone formulations, and gastric residence times of around 3–5 h were achieved [67].

16.3.4 The Small Intestine

As many compounds are absorbed predominantly in the small intestine, it is important to establish parameters such as transit time through this region and correct formulation actuation. Recently, for example, radiolabeling has been used to confirm performance of a novel double enteric coating technique, which was designed to accelerate release in the proximal small intestine [68]. The transit of the contents of the gut is generally regarded as being constant at around 3–5 h and unaffected by the intake of food. Furthermore, it has also been reported that the transit of dosage forms through this region is consistent, with transit of solutions, pellets, and tablets being independent of the fed state [69]. A pharmacoscintigraphic study of a radiolabeled xanthan gum matrix controlled release tablet containing diclofenac sodium in healthy volunteers reported that while food intake had the effect of delaying gastric emptying time, the transit of the tablets through the small intestine was just over 3 h and unaffected by food [70]. Despite this, the use of gamma scintigraphy has produced recent evidence that small intestinal transit of non-disintegrating dosage forms may not be as constant as once thought. While administration of tablets either simultaneously with food or in the fasted state produced similar transit times of around 204–210 min, administration of the same quantity of food 45 min after the tablet resulted in a significantly shorter small intestinal mean transit time of 141 min [71]. Furthermore, the authors reported that in 6 of the 10 volunteers, the tablet had emptied from the stomach by the time food was administered, and in these six cases, the small intestinal transit time was only 100 min. These results may have significant implications for the design of dosage forms and differences in bioavailability of the active drug administered.

It is not only the effect of food on the small intestinal transit of dosage forms that has been evaluated using *in vivo* imaging; gamma scintigraphy has also been used to identify an effect of formulation excipients on transit rate. Generally, it is expected that the excipients contained within an oral formulation are present to manipulate drug release in some manner, but should be "neutral" in terms of any pharmacological effect. Polyethylene glycol (PEG) is often incorporated into oral formulations in order to improve solubility of the drug; however, it has been determined that the presence of 10 g of PEG 400 reduced the small intestinal transit of liquid by 35% [72] and as a result significantly reduced the absolute bioavailability of ranitidine by 31% [73]. These findings demonstrate that careful consideration must be given to excipient choice in oral formulations, in particular where the drug is known to be absorbed primarily from the small intestine. This effect was also shown to be concentration dependent [74] and interestingly is also gender specific, with bioavailability of ranitidine enhanced in male, but not female subjects [75].

Of course, when absorption of a compound occurs predominantly in the small intestine, it may be of interest to prolong intestinal residence in order to enhance bioavailability. Mucoadhesive formulations have been developed for this purpose, although results to date have been variable. ^{153}Sm-labeled chitosan granules were studied using gamma scintigraphy, and it was concluded that despite promising *in vitro* data, they could not reliably achieve mucoadhesion in the upper GI tract *in vivo* [76].

16.3.5 Colon Targeting

The colon is a region of interest for the delivery of compounds to treat local inflammatory conditions and also for oral delivery of peptides, which may benefit from the reduced peptidase activity in this area [77]. In such instances, formulations are designed to delay drug release until a specific trigger occurs in the region of the colon, in order to avoid unnecessary exposure to active compound where it is not required. One method of triggering release in the colon is to coat the formulation in a substance such as pectin, which is only digested in the gut by the bacteria that reside predominantly in the colon [77–79]. One such formulation consisted of a core tablet (where the active drug would be contained) compression coated with a pectin–HPMC mix, in order to delay drug release until the time of arrival at the colon [78]. By distributing ^{99m}Tc-DTPA-labeled lactose in the inner core tablet, the transit and site of release could be followed using gamma scintigraphy, validating that in each volunteer the formulation coat did not break down in the stomach or small intestine and reached the colon with the inner core intact. Despite the small number of subjects included in this study, the authors were able to further analyze the scintigraphic data obtained and demonstrate the critical influence of position in the gut and water availability on the performance of the system, allowing feedback into the development process.

Another method for triggering drug release in the colon is to utilize the pH changes that occur along the different sections of the GI tract [80–82]. Eudragit® S100 is an anionic copolymer based on methacrylic acid and methyl methacrylate, which is designed to dissolve only at a pH of > 7. By incorporating a radiolabel in beads coated in this polymer, the authors of one study were able to demonstrate site-specific release in the colon for delivery of a bee venom peptide [83].

A biomagnetic measurement system was also used to evaluate a pressure-controlled colon delivery capsule and demonstrated that the capsule could be visualized in the ascending colon at the same time as caffeine was detected in the saliva, thus suggesting that the capsules did indeed "burst" on entry into the colon and exposure to a high-pressure zone [9]. In addition, scintigraphic studies have also provided evidence of a high-pressure zone around the ileocaecal area resulting in the disruption of a pectin–HPMC hydrogel layer [78].

The relatively consistent transit time of dosage forms through the small intestine has also been exploited as a means of targeting drug release in the colon *in vivo* [84]. In such systems, the mechanism of action is generally dependent on the gradual ingress of water into the system, which ultimately triggers release at a predetermined time after dosing. The hydrophilic-sandwich (HS) capsule system consists of a "capsule within a capsule" concept, where a "sandwich" layer of HPMC polymer is placed between an inner drug-containing capsule and the outer enteric coated capsule [85]. Following emptying from the stomach and dissolution of the enteric coating, the composition and density of the HPMC layer control hydration and erosion rates, generating a lag time before drug release. The performance of this system has been monitored and validated using gamma scintigraphy, with ^{111}In representing drug release from the inner capsule [86].

In each of these methods, one of the main hindrances to drug release and absorption in the colonic region is the relatively low availability of water in this region. Using MMI, it was shown that the water in the colon tends to be available only in pockets, with the total free available water being only 13 ml [87]. This will not only have a significant effect on the extent to which the drug can be solubilized in order to be absorbed but will also impact on the behavior of modified release formulations that predominantly rely on water ingress to elicit their desired behavior. This has been demonstrated in imaging studies, where spread of the formulation was observed to be limited in the colonic region [88, 89].

16.4 EXPLORATORY DEVICES

A further use of *in vivo* imaging techniques in the development of oral doses is to verify the performance of devices that are used to deliver a drug to a specific region in the GI tract [90]. Such devices are generally used in regional absorption studies, delivering the active compound to different target areas such as the proximal small intestine or colon [91]. In such studies, it is essential to verify the location of the device at the time of drug release in order to ensure that the absorption data for that region are correct. Equally, reliable actuation and release of drug should also be confirmed to avoid false negatives.

16.5 CONCLUSIONS

In vivo imaging techniques have proven invaluable in oral dosage form development, with gamma scintigraphy the most widely reported method across the literature. Imaging studies have been used not only to confirm behavior that was designed and predicted during *in vitro* development but also to identify cases where *in vivo* performance was not as expected. Early-stage evaluation of performance can help inform key decisions in the development of oral products and help avoid costly problems due to formulation failure. Equally, the imaging studies described here demonstrate how the use of imaging techniques *in vivo* can be utilized to understand the mechanisms of behavior and explain pharmacokinetic profiles.

While the focus in this section has been maintained on oral dose performance in clinical studies, it should be remembered that early stages of formulation development often involve large animal models, and with the use of carefully considered experimental setup, gamma scintigraphy and MRI have also been used to understand performance of oral systems and the relevance of the model to man [92–96].

REFERENCES

1. Davis J, Burton J, Connor AL, et al. Scintigraphic study to investigate the effect of food on a HPMC modified release formulation of UK-294,315. J Pharm Sci 2009;98:1568–1576.
2. McInnes FJ, O'Mahony B, Lindsay B, et al. Nasal residence of insulin containing lyophilised nasal insert formulations, using gamma scintigraphy. Eur J Pharm Sci 2007;31:25–31.
3. Hendeles L, Weinberger M, Milavetz G, et al. Food-induced dose-dumping from a once-a-day theophylline product as a cause of theophylline toxicity. Chest 1985;87:758–765.
4. Perkins AC. Instrumentation, imaging, data analysis and display. In: Perkins AC, Frier M, editors. *Nuclear Medicine in Pharmaceutical Research*. London: Taylor and Francis; 1999. p 15–30.
5. Weitschies W, Blume H, Monnikes H. Magnetic marker monitoring: high resolution real-time tracking of oral solid dosage forms in the gastrointestinal tract. Eur J Pharm Biopharm 2010;74:93–101.
6. Bergstrand M, Soderlind E, Weitschies W, et al. Mechanistic modeling of a magnetic marker monitoring study linking gastrointestinal tablet transit, in vivo drug release, and pharmacokinetics. Clin Pharmacol Ther 2009;86:77–83.
7. Goodman K, Hodges LA, Band J, et al. Assessing gastrointestinal motility and disintegration profiles of magnetic tablets by a novel magnetic imaging device and gamma scintigraphy. Eur J Pharm Biopharm 2010;74:84–92.
8. Cora LA, Americo MF, Romeiro FG, et al. Pharmaceutical applications of AC biosusceptometry. Eur J Pharm Biopharm 2010;74:67–77.
9. Hu ZP, Mawatari S, Shibata N, et al. Application of a biomagnetic measurement system (BMS) to the evaluation of gastrointestinal transit of intestinal pressure-controlled colon delivery capsules (PCDCs) in human subjects. Pharm Res 2000;17:160–167.

10. Richardson JC, Bowtell RW, Mader K, et al. Pharmaceutical applications of magnetic resonance imaging (MRI). Adv Drug Del Rev 2005;57:1191–1209.
11. Melia CD, Rajabi-Siahboomi AR, Bowtell RW. Magnetic resonance imaging of controlled release pharmaceutical dosage forms. Pharm Sci Technol Today 1998;1:32–39.
12. Knorgen M, Spielmann RP, Abdalla A, et al. Non-invasive MRI detection of individual pellets in the human stomach. Eur J Pharm Biopharm 2010;74:120–125.
13. Paterson RS, Eccleston GM, O'Mahoney B, et al. *Magnetic resonance imaging (MRI) permits visualisation of floating raft formation in man.* International Symposium on Controlled Release of Bioactive Materials; 2001 23–28 June; San Diego: Controlled Release Society; 2001. p 794–795.
14. Wilson CG, Washington N, Peach J, et al. The behavior of a fast-dissolving dosage from (Expidet) followed by gamma-scintigraphy. Int J Pharm 1987;40:119–123.
15. Christrup LL, Davis SS, Frier M, et al. Deposition of a model substance, 99m Tc E-HIDA, in the oral cavity after administration of lozenges, chewing gum and sublingual tablets. Int J Pharm 1990;66:169–174.
16. Needleman I, Mather SJ, Martin GP, et al. Periodontal pocket clearance by gamma scintigraphy in human volunteers. J Clin Periodontol 2000;27:904–909.
17. Jacques Y, NguyenXuan T, Ionescu E, et al. In vivo evaluation of hydrophilic and hydrophobic mucoadhesive semi-solid formulations containing sucralfate and lidocaine for intraoral use. Eur J Pharm Biopharm 1997;43:59–63.
18. Wilson C, Washington N. *Physiological Pharmaceutics*. 2nd ed. London: Taylor and Francis; 2003.
19. Ekberg O, Feinberg MJ. Altered swallowing function in elderly patients without dysphagia: radiologic findings in 56 cases. Am J Roentgenol 1991;156:1181–1184.
20. Perkins AC, Wilson CG, Blackshaw PE, et al. Impaired esophageal transit of capsule versus tablet formulations in the elderly. Gut 1994;35:1363–1367.
21. Hey H, Jorgensen F, Sorensen K, et al. Esophageal transit of six commonly used tablets and capsules. Br Med J 1982;285:1717–1719.
22. Wilson CG, Washington N, Norman S, et al. A gamma scintigraphic study to compare esophageal clearance of expidet formulations, tablets and capsules in supine volunteers. Int J Pharm 1988;46:241–246.
23. Channer KS, Virjee JP. The effect of size and shape of tablets on their esophageal transit. J Clin Pharmacol 1986;26:141–146.
24. Wilson CG, O'Mahoney B, Farrell T, et al. Modern tablet film coatings and influence on ease of swallowing. AAPS J 2006;8:T2221.
25. Sakkinen M, Marvola J, Kanerva H, et al. Scintigraphic verification of adherence of a chitosan formulation to the human oesophagus. Eur J Pharm Biopharm 2004;57:145–147.
26. Potts AM, O'Mahony B, Foster J, et al. The use of magnetic resonance imaging (MRI) to visualise the oesophageal transit of liquid and gel formulations. J Pharm Pharmacol 2000;52:18.
27. Holt S, Reid J, Taylor TV, et al. Gastric emptying of solids in man. Gut 1982; 23:292–296.
28. Meyer JH, Ohashi H, Jehn D, et al. Size of liver particles emptied from the human stomach. Gastroenterology 1981;80:1489–1496.
29. Coupe AJ, Davis SS, Evans DF, et al. Correlation of the gastric-emptying of nondisintegrating tablets with gastrointestinal motility. Pharm Res 1991;8:1281–1285.

30. Mojaverian P, Reynolds JC, Ouyang A, et al. Mechanism of gastric-emptying of a non-disintegrating radiotelemetry capsule in man. Pharm Res 1991;8:97–100.

31. Kossena GA, Charman WN, Wilson CG, et al. Low dose lipid formulations: effects on gastric emptying and biliary secretion. Pharm Res 2007;24:2084–2096.

32. Marathe PH, Wen YD, Norton J, et al. Effect of altered gastric emptying and gastrointestinal motility on metformin absorption. Brit J Clin Pharmacol 2000;50:325–332.

33. Kong MF, Horowitz M. Gastric emptying in diabetes mellitus: relationship to blood-glucose control. Clin Geriatr Med 1999;15:321–338.

34. Sun JX, Walter B, Sandefer EP, et al. Explaining variable absorption of a hypolipidemic agent (CGP 43371) in healthy subjects by gamma scintigraphy and pharmacokinetics. J Clin Pharmacol 1996;36:230–237.

35. Kelly K, O'Mahony B, Lindsay B, et al. Comparison of the rates of disintegration, gastric emptying, and drug absorption following administration of a new and a conventional paracetamol formulation, using gamma scintigraphy. Pharm Res 2003;20:1668–1673.

36. Washington N, Wilson CG, Greaves JL, et al. A gamma scintigraphic study of gastric coating by Expidet, tablet and liquid formulations. Int J Pharm 1989;57:17–22.

37. Davis SS, Stockwell AF, Taylor MJ, et al. The effect of density on the gastric-emptying of single-unit and multiple-unit dosage forms. Pharm Res 1986;3:208–213.

38. Davis SS, Norringchristensen F, Khosla R, et al. Gastric-emptying of large single unit dosage forms. J Pharm Pharmacol 1988;40:205–207.

39. Clarke GM, Newton JM, Short MD. Gastrointestinal transit of pellets of differing size and density. Int J Pharm 1993;100:81–92.

40. Kamba M, Seta Y, Kusai A, et al. Comparison of the mechanical destructive force in the small intestine of dog and human. Int J Pharm 2002;237:139–149.

41. Schug BS, Brendel E, Wolf D, et al. Formulation-dependent food effects demonstrated for nifedipine modified-release preparations marketed in the european union. Eur J Pharm Sci 2002;15:279–285.

42. Fadda HM, Mohamed MAM, Basit AW. Impairment of the in vitro drug release behaviour of oral modified release preparations in the presence of alcohol. Int J Pharm 2008;360:171–176.

43. Walden M, Nicholls FA, Smith KJ, et al. The effect of ethanol on the release of opioids from oral prolonged-release preparations. Drug Dev Ind Pharm 2007;33:1101–1111.

44. Levina M, Vuong H, Rajabi-Siahboomi AR. The influence of hydro-alcoholic media on hypromellose matrix systems. Drug Dev Ind Pharm 2007;33:1125–1134.

45. Weitschies W, Wedemeyer RS, Kosch O, et al. Impact of the intragastric location of extended release tablets on food interactions. J Control Release 2005;108:375–385.

46. Ghimire M, Hodges LA, Band J, et al. In-vitro and in-vivo erosion profiles of hydroxypropylmethylcellulose (HPMC) matrix tablets. J Control Release 2010;147:70–75.

47. Ghimire M, Hodges LA, Band J, et al. Correlation between in vitro and in vivo erosion behaviour of erodible tablets using gamma scintigraphy. Eur J Pharm Biopharm 2011;77:148–157.

48. Burke MD, Staton JS, Vickers AW, et al. A novel method to radiolabel gastric retentive formulations for gamma scintigraphy assessment. Pharm Res 2007;24:695–704.

49. Cora LA, Americo MF, Oliveira RB, et al. Disintegration of magnetic tablets in human stomach evaluated by alternate current biosusceptometry. Eur J Pharm Biopharm 2003;56:413–420.

50. Tuleu C, Khela MK, Evans DF, et al. A scintigraphic investigation of the disintegration behaviour of capsules in fasting subjects: a comparison of hypromellose capsules containing carrageenan as a gelling agent and standard gelatin capsules. Eur J Pharm Sci 2007;30:251–255.
51. Waterhouse ET, Washington C, Washington N. An investigation into the efficacy of the pectin based anti-reflux formulation-Aflurax. Int J Pharm 2000;209:79–85.
52. Washington N, Denton G. Effect of alginate and alginate-cimetidine combination therapy on stimulated postprandial gastro-oesophageal reflux. J Pharm Pharmacol 1992;47:879–882.
53. May HA, Wilson CG, Hardy JG. Monitoring radiolabeled antacid preparations in the stomach. Int J Pharm 1984;19:169–176.
54. Washington N, Washington C, Wilson CG. Gastric distribution and residence time of 2 antireflux formulations. Int J Pharm 1987;39:163–171.
55. Marciani L, Little SL, Snee J, et al. Echo-planar magnetic resonance imaging of Gaviscon alginate rafts in-vivo. J Pharm Pharmacol 2002;54:1351–1356.
56. Dettmar PW, Little SL, Baxter T. The effect of omeprazole pre-treatment on rafts formed by reflux suppressant tablets containing alginate. J Int Med Res 2005;33:301–308.
57. Washington N, Wilson CG, Williams DL, et al. An investigation into the effect of cimetidine pre-treatment on raft formation of an anti-reflux agent. Aliment Pharmacol Ther 1993;7:553–559.
58. Whitehead L, Fell JT, Collett JH, et al. Floating dosage forms: an in vivo study demonstrating prolonged gastric retention. J Control Release 1998;55:3–12.
59. Stops F, Fell JT, Collett JH, et al. The use of citric acid to prolong the in vivo gastro-retention of a floating dosage form in the fasted state. Int J Pharm 2006;308:8–13.
60. Steingoetter A, Weishaupt D, Kunz P, et al. Magnetic resonance imaging for the in vivo evaluation of gastric-retentive tablets. Pharm Res 2003;20:2001–2007.
61. Steingoetter A, Kunz P, Weishaupt D, et al. Analysis of the meal-dependent intragastric performance of a gastric-retentive tablet assessed by magnetic resonance imaging. Aliment Pharm Therap 2003;18:713–720.
62. Jackson SJ, Bush D, Washington N, et al. Effect of resin surface charge on gastric mucoadhesion and residence of cholestyramine. Int J Pharm 2000;205:173–181.
63. Jackson SJ, Bush D, Perkins AC. Comparative scintigraphic assessment of the intragastric distribution and residence of cholestyramine, Carbopol 934P and sucralfate. Int J Pharm 2001;212:55–62.
64. Thairs S, Ruck S, Jackson SJ, et al. Effect of dose size, food and surface coating on the gastric residence and distribution of an ion exchange resin. Int J Pharm 1998;176:47–53.
65. Burton S, Washington N, Steele RJC, et al. Intragastric distribution of ion-exchange resins: a drug delivery system for the topical treatment of the gastric mucosa. J Pharm Pharmacol 1995;47:901–906.
66. Hardy JG, Hooper G, Ravelli GP, et al. A comparison of the gastric retention of a sucralfate gel and a sucralfate suspension. Eur J Pharm Biopharm 1993;39:70–74.
67. Kedzierewicz F, Thouvenot P, Lemut J, et al. Evaluation of peroral silicone dosage forms in humans by gamma-scintigraphy. J Control Release 1999;58:195–205.
68. Liu F, Basit AW. A paradigm shift in enteric coating: achieving rapid release in the proximal small intestine of man. J Control Release 2010;147:242–245.
69. Davis SS, Hardy JG, Fara JW. Transit of pharmaceutical dosage forms through the small-intestine. Gut 1986;27:886–892.

70. Billa N, Yuen KH, Khader MAA, et al. Gamma-scintigraphic study of the gastrointestinal transit and in vivo dissolution of a controlled release diclofenac sodium formulation in xanthan gum matrices. Int J Pharm 2000;201:109–120.

71. Fadda H, McConnell E, Short M, et al. Meal-induced acceleration of tablet transit through the human small intestine. Pharm Res 2009;26:356–360.

72. Basit AW, Newton JM, Short MD, et al. The effect of polyethylene glycol 400 on gastrointestinal transit: implications for the formulation of poorly-water soluble drugs. Pharm Res 2001;18:1146–1150.

73. Basit AW, Podczeck F, Newton JM, et al. Influence of polyethylene glycol 400 on the gastrointestinal absorption of ranitidine. Pharm Res 2002;19:1368–1374.

74. Schulze JDR, Waddington WA, Ell PJ, et al. Concentration-dependent effects of polyethylene glycol 400 on gastrointestinal transit and drug absorption. Pharm Res 2003;20:1984–1988.

75. Ashiru D, Patel R, Basit A. Polyethylene glycol 400 enhances the bioavailability of a BCS class III drug (ranitidine) in male subjects but not females. Pharm Res 2008;25:2327–2333.

76. Sakkinen M, Marvola J, Kanerva H, et al. Are chitosan formulations mucoadhesive in the human small intestine? An evaluation based on gamma scintigraphy. Int J Pharm 2006;307:285–291.

77. Patel M, Shah T, Amin A. Therapeutic opportunities in colon-specific drug-delivery systems. Crit Rev Ther Drug Carr Syst 2007;24:147–202.

78. Hodges LA, Connolly SM, Band J, et al. Scintigraphic evaluation of colon targeting pectin-HPMC tablets in healthy volunteers. Int J Pharm 2009;370:144–150.

79. Sinha VR, Kumria R. Microbially triggered drug delivery to the colon. Eur J Pharm Sci 2003;18:3–18.

80. Yang LB, Chu JS, Fix JA. Colon-specific drug delivery: new approaches and in vitro/in vivo evaluation. Int J Pharm 2002;235:1–15.

81. Gupta VK, Beckert TE, Price JC. A novel pH- and time-based multi-unit potential colonic drug delivery system. I. Development. Int J Pharm 2001;213:83–91.

82. Khan MZI, Prebeg Z, Kurjakovic N. A pH-dependent colon targeted oral drug delivery system using methacrylic acid copolymers—I. Manipulation of drug release using Eudragit (R) L100-55 and Eudragit (R) S100 combinations. J Control Rel 1999; 58:215–222.

83. Xing L, Dawei C, Liping X, et al. Oral colon-specific drug delivery for bee venom peptide: development of a coated calcium alginate gel beads-entrapped liposome. J Control Rel 2003;93:293–300.

84. Sangalli ME, Maroni A, Zema L, et al. In vitro and in vivo evaluation of an oral system for time and/or site-specific drug delivery. J Control Release 2001;73:103–110.

85. Soutar S, O'Mahoney B, Perkins AC, et al. *Time-delayed release of paracetamol in man from the hydrophilic sandwich capsule*. International Symposium on Controlled Release of Bioactive Materials; 2001 23–28 June; San Diego: Controlled Release Society; 2001. p. 790–791.

86. Pecek D, Sime KA, Ronaldson VE, et al. Validation of a targeted release capsule using scintigraphic methods. AAPS J 2010;12:T3038.

87. Schiller C, Frohlich CP, Giessmann T, et al. Intestinal fluid volumes and transit of dosage forms as assessed by magnetic resonance imaging. Aliment Pharmcol Ther 2005;22:971–979.

88. Honkanen O, Marvola J, Kanerva H, et al. Gamma scintigraphic evaluation of the fate of hydroxypropyl methylcellulose capsules in the human gastrointestinal tract. Eur J Pharm Sci 2004;21:671–678.

89. McConville JT, Hodges LA, Jones T, et al. A pharmacoscintigraphic study of three time-delayed capsule formulations in healthy male volunteers. J Pharm Sci 2009; 98:4251–4263.

90. Clear NJ, Milton A, Humphrey M, et al. Evaluation of the Intelisite capsule to deliver theophylline and frusemide tablets to the small intestine and colon. Eur J Pharm Sci 2001;13:375–384.

91. Oo C, Snell P, Barrett J, et al. Pharmacokinetics and delivery of the anti-influenza prodrug oseltamivir to the small intestine and colon using site-specific delivery capsules. Int J Pharm 2003;257:297–299.

92. McInnes F, Clear N, James G, et al. Evaluation of the clearance of a sublingual buprenorphine spray in the beagle dog using gamma scintigraphy. Pharm Res 2008;25:869–874.

93. McInnes F, Clear N, Humphrey M, et al. In vivo performance of an oral MR matrix tablet formulation in the beagle dog in the fed and fasted state: assessment of mechanical weakness. Pharm Res 2008;25:1075–1084.

94. Ghimire M, McInnes FJ, Watson DG, et al. In-vitro/in-vivo correlation of pulsatile drug release from press-coated tablet formulations: a pharmacoscintigraphic study in the beagle dog. Eur J Pharm Biopharm 2007;67:515–523.

95. Schulze JDR, Peters EE, Vickers AW, et al. Excipient effects on gastrointestinal transit and drug absorption in beagle dogs. Int J Pharm 2005;300:67–75.

96. Kremser C, Albrecht K, Greindl M, et al. In vivo determination of the time and location of mucoadhesive drug delivery systems disintegration in the gastrointestinal tract. Magn Reson Imaging 2008;26:638–643.

CHAPTER 17

ROLE OF LARGE ANIMAL MODELS IN TRANSLATIONAL STUDIES OF IMAGING AND TARGETED DRUG DELIVERY

RAJESH K. UTHAMANTHIL AND MEI TIAN

17.1 INTRODUCTION

The tremendous advances made in understanding disease processes at the molecular level have not been translated effectively to clinical practice. A significant part of this "pipeline problem" is the lack of animal models in which the clinical, molecular, and genomic aspects/characteristics of humans/human conditions can be replicated [1, 2].

The majority of research animal models are rodents, specifically mice. The immense applicability of genetically engineered rodent models has helped to unravel the roles of genes and related molecular pathways in humans. However, owing to the significant differences between mice and humans, ranging from anatomical to the molecular level, translation of many of the basic science discoveries to the clinic has been limited [3]. The human body is ~2000–3000 times larger than that of a mouse, and humans live ~30–50 times longer than do mice. Given the lifelong turnover of cells in the body, humans experience about 100,000 more cell divisions in a lifetime than do mice, and mice have a basal metabolic ratio seven times that of humans [4].

Table 17.1 shows the weights of various organs in animals and humans. The difference between an organ in an animal and the same organ in a human affects the ability to translate drugs and therapies from the bench to the clinic. This is especially true in cancer, which has high rate of failure of drugs from first-in-human (phase 0) to registration trials [5]. Similarity of the animal model to humans in size, anatomy, and physiology is all the more important when studies involve imaging and targeted drug delivery, especially in clinical settings. Thus, effective translation of basic science discoveries requires the use of large animal models.

Drug Delivery Applications of Noninvasive Imaging: Validation from Biodistribution to Sites of Action, First Edition. Edited by Chun Li and Mei Tian.

TABLE 17.1 Organ Weights in Different Species

	Mouse		Rat		Rabbit		Dog		Rhesus monkey		Human	
Organ	kg	(%)	kg	(%)	kg	(%)	kg	(%)	kg	(%)	kg	(%)
Brain	4E-04	2	0.0018	0.7	0.014	0.6	0.08	1	0.09	1.8	1.4	2.0
Heart	1E-04	0.5	0.001	0.4	0.005	0.2	0.08	1	0.02	0.4	0.33	0.5
Liver	0.002	9	0.01	4.0	0.077	3.1	0.32	3	0.15	3.0	1.8	2.6
Lung	1E-04	0.5	0.0015	0.6	0.018	0.7	0.1	1	0.03	0.7	1	1.4
Spleen	1E-04	0.5	0.0008	0.3	0.001	0.0	0.025	0	0.01	0.2	0.18	0.3
Kidney	3E-04	1.5	0.0023	0.9	0.013	0.5	0.05	1	0.03	0.5	0.31	0.4
Remainder body	0.02	86.0	0.23	93.1	2.37	94.9	9.35	93	4.68	93.5	64.98	92.8
Total body	*0.02*	*100*	*0.25*	*100*	*2.5*	*100*	*10*	*100*	*5*	*100*	*70*	*100*

This chapter focuses on different aspects of responsible and efficient use of large animal models, as defined by the United States Department of Agriculture (USDA). We discuss the use of rabbits, dogs, pigs, and nonhuman primates (NHPs) in translational studies of imaging and targeted drug delivery. Our discussion includes selection of the proper animal model for specific studies, an overview of special techniques used to conduct research efficiently, and the ethical and regulatory aspects of using animals of higher order in research.

17.2 CHOOSING THE RIGHT ANIMAL MODEL

The choice of the right animal model depends on the specific objective of the study. There is not one "perfect model" that can be used for all research projects. For studies using imaging to assess drug delivery, the most suitable animal model may depend on the specific body systems targeted as well as on the ability of the model's body systems to metabolize the specific imaging agent. The major determinants for choosing the right model include the comparative anatomy, physiology, and size of the animal to those of humans; the genomic and molecular similarities of the animal species to those of humans; the similarity between the animal and humans in the etiology, development, and progression of the specific human condition targeted; the cost of conducting the study; ethical issues of using a large animal model in the specific type of study to be conducted; and the availability of specific molecular and other research tools for the specific animal species (Table 17.2).

One of the primary reasons for using a large animal model is its relative similarity in size, anatomy, and physiology to the human patient. This similarity allows the use of these models in clinical settings, with clinical standard-size scanners and drug delivery systems. The use of clinical standard-size scanners, which typically differ from rodent scanners in resolution and other attributes, is critical for the translation of imaging techniques to the clinic. The similarities of the comparative anatomy of the targeted organ as well as the relative topographic anatomy are important in choosing the right animal model. For example, pigs have been widely accepted as a suitable model for surgical and imaging studies of the cardiovascular and hepatobiliary systems [6, 7].

High similarity of the cardiovascular anatomy of pigs to that of humans makes pigs one of the most popular models for research in various cardiovascular conditions including myocardial infarction (MI) and aneurysms and in cardiovascular device development. Pigs are also a classical model for liver-related studies ranging from cirrhosis to transplantation. However, pigs differ significantly from humans in the structure and function of some organ systems, such as part of the urogenital system and the immune system. For example, male pigs do not have a large and easily accessible prostate gland, and the anatomy of the urogenital system differs from that of humans. In addition, unlike in a dog or NHP, in a male pig, it is extremely challenging to catheterize the urethra because of the pig's corkscrew-shaped tip of the penis. These two factors—the inaccessibility of the prostate gland and the difficulty of using a urethral catheter—rule out pig models for use in prostate studies. Development and function of the immune system of pigs also differs markedly from that of humans; differences include the diversity of antibodies, the anatomy of immune system-related organs, and the lack of passive immunity in piglets [8, 9].

TABLE 17.2 Use of Large Animal Models in Imaging and Targeted Drug Delivery—Advantages and Limitations

Species	Advantages	Limitations
Rabbit	Relatively smaller size, still usable in clinical sized machines Less expensive, easy to maintain Longer life span than rodents Various sizes/breeds available Ready availability of a transplantable tumor model Highly inbred—less animal to animal variability	Some organ systems significantly differ in anatomy and physiology from that in human—e.g., GI tract Molecular tools are limited since the species is widely used for antibody generation Difficult to handle for many procedures without anesthesia
Dog	Depth of knowledge exist in canine diseases, medicine, and comparative medicine Easy handling, sample collection, substance administration Many similarities with human in physiology and anatomy Powerful molecular tools available Most diseases well characterized Well adapted to group housing Canine pets share human living environment—great source for cancer research Experimental translantable cancer model available (cTVT) Whole genome sequenced. Microarray available	Expensive to purchase and maintain Special requirements for maintenance including daily exercise Ethical concerns Limited availability
Pig	High similarity in physiology and anatomy Easy to breed and propagate Various breeds (sizes) available Whole genome sequenced Microarray available	High maintenance expenses Difficult handling Molecular (e.g., antibodies) tools availability relatively limited
NHP	Closest animal species to human—high similarity anatomy, physiology, metabolism, and molecular biology Powerful microarray and other molecular tools available Different sizes and species available Most human molecular tools cross-reacts with macaque species. Only animal models available for some diseases Results are highly translational	Very expensive to purchase Zoonotic diseases Ethical concerns Need for higher levels of enrichment and standard of care Need skilled veterinary support for many procedures Less susceptibility to cancer (in comparison to human)

In addition to the limitations in anatomical and physiological similarities between animals and humans required to answer specific study or research questions, the choice of animal model is also limited by the scientific tools available in the specific model for a proper comparison. For example, the rabbit V×2 tumor model is widely used for tumor imaging studies and has a high potential for use in image-guided targeted drug delivery studies [10, 11]. However, from a molecular and cellular perspective, the rabbit V×2 tumor model has multiple deficiencies, including a high level of necrosis and a severe inflammatory response around the tumor. In addition, since rabbits are one of the most common models used for the preparation of polyclonal antibodies, the practical use of antibody-based molecular tools in this species might be limited.

NHP models are very important not only because of their close relationship to humans but also because of the high chance of cross-reactivity with human molecular tools and the extensive genomic and molecular data available. However, the use of NHPs for research is expensive, and thus NHP models should be used only for research projects that justify the use of this species.

Thus, choosing the most suitable animal model for an imaging or drug delivery study depends on a variety of factors, including the similarity of the model to humans with respect to the specific system(s) to be studied and the cost and practicality of the model in the specific institutional setup.

17.2.1 Rabbit Models

The laboratory rabbit (*Oryctolagus cuniculus*) is most commonly used in imaging and interventional radiology studies owing to the availability of transplantable tumor models. The life span of a rabbit is ~7–8 years. Rabbit models are used in image-guided interventional procedures and for targeted delivery of agents to different anatomical locations. Rabbits are a comparatively inexpensive model compared with other large animal models, easy to maintain, and large enough to be used in clinical settings.

17.2.2 Dog Models

Dogs played a major role in the early history of translational biomedical research. Dogs have a cooperative nature; are a good size for surgical, interventional, and imaging procedures; are readily available; and are easy to provide veterinary care; these factors led to the common use of dogs in biomedical research. However, the popularity of laboratory dogs has decreased lately, and the majority is now used in pharmacology and toxicology research. Compared with mice, dogs are much more homologous to humans. Because of the numerous breeds available, different sizes of dogs can be used for different types of studies.

Recently, the focus has shifted from laboratory dogs to the huge pool of companion dogs that share the environment with humans. Companion dogs also share many human diseases, making dogs an excellent model for translational research. This is especially true in cancer, where the incidence of cancer in pet dogs is very close to that in humans [12]. Many canine cancers, such as osteosarcoma, greatly resemble their counterparts in humans, thus providing a natural model of cancer suitable for translational research [13]. Currently, this potential for translational research is being investigated by the Comparative Oncology Program, which was established by the National Institutes of Health (NIH) and is run by a network of veterinary teaching hospitals around the United States [14].

17.2.3 Pig Models

Pig models are used extensively in imaging studies owing to these animals' similarity to humans in anatomy, physiology, and size. The similarity of the cardiovascular system makes the pig model useful for studies that involve blood pool imaging contrast agents. Extensive use of image-guided interventional techniques in this model makes it more amenable to novel imaging of drug delivery techniques [15–18].

Laboratory swine (*Sus scrofa domestica*) come in different breeds and sizes, ranging from body weights of 12 kg to more than 120 kg at sexual maturity. The normal life span of a pig is ~15 years. Because domestic pigs grow very rapidly, young animals are typically used in pig models so that the size relates more closely to a pediatric patient than to an adult. If the study requires adult physiology and/or chronic models, the use of mature miniature pigs (minipigs) should be considered. Female pigs are preferred for imaging studies since urinary catheterization is difficult in male pigs. There are a variety of breeds and sublines of laboratory minipigs, including Sinclair, Gottingen, Yucatan, and Ossabaw [6, 19], and the size of minipigs varies from body weights of 12 to 45 kg at sexual maturity.

17.2.4 NHP Models

Approximately 30 species of NHPs, ranging from the small New World monkeys to large apes like the chimpanzee, are used in research. NHPs' similarity to humans in their genomic, molecular, biological, and behavioral characteristics makes them the most relevant animal model for most research projects [2, 20]. The use of NHP models for pharmacokinetic/pharmacodynamic (PK/PD) studies of new radiotracers or other imaging agents may yield highly comparable data that can be used for early translation to the clinic. The most commonly used NHPs are Asian and African macaques [21].

The rhesus macaque (*Macaca mulatta*) is one of the most commonly used NHP models because of the accumulated knowledge of this species from its extensive use in research, easy breeding at various national primate centers, and better molecular and genomic characterization. The genome of the rhesus macaque has been mapped and has close similarity to that of humans [22]. The most commonly used large NHP species is the baboon (*Papio* spp.). Because of the baboon's large size and similarity to human anatomy, baboon models are very useful for device development [23]. Baboon models are also widely used in studies of xenotransplantation [24–26], osteoporosis [27], atherosclerosis [28, 29], and endometriosis [30].

Macaques are good models for image-guided targeted drug delivery studies as well as for imaging of drug delivery to cells [31–34]. Despite the high expense of NHP models, a simple PubMed search for "macaque" and "imaging" retrieved almost 2000 references, underscoring the relevance of this model in translational imaging research. Squirrel monkeys (*Saimiri* spp.) are one of the most commonly used small NHPs [35, 36]. Usually, they weigh less than 1 kg. Small-sized (500–750 g) squirrel monkeys can fit in the high-resolution scanners that are used for small animals, thus providing early research data that have greater translational potential than do data from small animal models.

Typically, specific pathogen-free (SPF) animals are preferred for sensitive and important translational studies involving NHPs. The definition of SPF varies within the scientific community with regard to the number of pathogens that animals must be free of to be considered SPF. The most common definition of an SPF primate is one that is free of *Mycobacterium tuberculosis* and of these four specific viruses: simian immunodeficiency virus, simian type D retrovirus, herpesvirus B (i.e., *Cercopithecine herpesvirus 1*), and simian T-cell lymphotropic/leukemia virus [37].

The limitations of NHP models include cost, management of risks such as zoonotic potential and aggressive behavior during the research study, and availability of specialized veterinary support when needed. For example, the purchase cost of a rhesus macaque ranges from $5000.00 to $8000.00, and the cost may be higher depending on the pathogen-free status of the animal. The cost of maintenance for a rhesus macaque ranges from $200.00 to $600.00 per month, depending on the location and the institution, whereas a larger NHP may have much higher maintenance costs. However, an NHP model may be cost-effective for some types of imaging-related studies as the same animals can be used for serial, repetitive imaging studies provided the imaging agents do not cause long-term physiological impairment.

Herpesvirus B is one of the most serious zoonotic infections that can be transmitted from an Asian macaque to humans [38]. While an infected animal can be completely asymptomatic, a human infection can be deadly. It is important for investigators using macaques to be trained in the safety aspects of primate research, including management of exposure to infectious pathogens. Irrespective of the pathogen-free status of the animal, if exposure occurs (such as through a bite, scratch, or mucus membrane exposure to another animal or cage material), the person and the animal should be screened for possible exposure to the virus or for the presence of the virus in the blood, respectively. If exposure occurs, immediate washing of the wound and reporting of the incident are required, and a timed screening of the exposed person's blood should be performed to check for the presence of antibodies.

Finally, a qualified veterinary support team is critical for carrying out complex experimental imaging and image-guided interventions in NHP models. The presence of such a team assures a high standard of care, including the safe handling of the animal, the proper induction and maintenance of anesthesia, and the recovery of the animal.

17.3 ANESTHESIA AND HANDLING OF LARGE ANIMAL MODELS

The use of appropriate restraint and handling techniques is important to reduce injuries to both personnel and animals and to reduce stress on the animals. Stress can significantly affect the outcome of research studies. Stress can affect the immune system, can lead to gastric ulcers in rabbits and pigs, and can have a marked effect on metabolism [39]. In rabbits, stress can cause severe effects, including cardiomyopathy, elevated blood glucose levels, gastrointestinal diseases, and decreased urinary parameters [40].

Proper restraint techniques differ among species. While dogs are one of the easiest species to handle, swine and NHPs require a special setup and skills. Small pigs, like dogs, can be handled without the use of a restraining apparatus. However, large pigs may require the use of a restraining apparatus such as the sling developed by Panepinto and colleagues [41]. Handling and restraining techniques for NHPs depend on the type of experiments performed and range from walking them with on leashes to restraining them in squeeze cages. Pigs and NHPs can be trained to help with research procedures, including allowing sample collection without anesthesia [42–44].

A wide variety of pharmaceutical agents are available for chemical restraint of laboratory animals. Typically, an injectable agent is used initially as a tranquilizer, sedative, or short-term anesthetic, which may be referred to as a preanesthetic medication. Atropine and glycopyrrolate are the typical parasympatholytic agents administered as premedications to counteract vagally mediated bradycardia and sympatholytic effects, including saliva secretion during the anesthetic period. It should be noted that atropine is typically not recommended for use as a premedication agent in rabbits because many animals produce atropine esterase, which rapidly breaks down atropine. Once anesthesia is induced, a tracheal tube is put in place, followed by long-term anesthesia maintenance using a gaseous anesthetic agent such as isoflurane or a constant infusion anesthetic agent such as propofol. The choice of the right agent depends on the species as well as the relevance of the physiological effect of the agent to the specific study. For a detailed list of agents that can be used in the restraint and anesthesia of large laboratory animals and the advantages and limitations of these agents, please refer to the laboratory animal medicine textbooks given in the reference list [45, 46].

Anesthesia significantly affects homeostasis. Constant monitoring and maintenance of normal body parameters are an important part of maintaining normal physiology during the experiment. The body temperature of the animal should be maintained using hot water blankets, warm air circulation (e.g., Bair Hugger), chemically activated body warmers, warm blankets, or a combination of these methods. A rectal thermometer probe can be used and secured with tape to the animal's tail so that movement of the table during imaging does not displace the position of the probe. If the temperature of the brain needs to be monitored, a tympanic probe placed in the external auditory canal can be used. Changes in blood O_2 and CO_2 levels can lead to significant metabolic impairment. Accurate measurement of blood gas levels can be monitored by blood gas analysis; however, the setup is very expensive. Thus, in most cases, the monitoring of blood gas levels can be estimated from expired (i.e., end tidal) CO_2 levels and pulse oximeter readings. Typically, the animal is placed on a ventilator during imaging studies to keep the blood gas levels within the normal range throughout the experiment.

In the authors' experience, placing two or more intravenous (IV) ports helps in data collection by providing separate ports for administering the compound and contrast agents and for collecting blood samples. In rabbits and pigs, ear veins are the easiest site in which to place IV ports. Placing IV ports in the jugular vein in rabbits might require a minor surgical procedure, and chronic blood sample collection and/

or substance administration will be aided by tunneling the port through the back of the neck and attaching it to the skin. In pigs, placing IV ports in the jugular vein can be done surgically using a cutdown technique or percutaneously using a modified Seldinger technique under ultrasound guidance.

Positioning of animals for imaging depends on the area to be imaged as well as the type of study to be performed. Various devices are available to maintain the proper position of the animals, including moldable bean bags, foam blocks, and flat stretchers with straps that can be used to position the body straight.

Placing a urinary catheter in large animal models during imaging studies, especially animals under long-term anesthesia, is highly recommended. A urinary catheter is advantageous for several reasons: (i) urine spillage on clinical equipment and in examination rooms can expose personnel to allergens, (ii) urine spillage can cause radioactivity if radioactive tracers were used on the animal, and (iii) the catheter allows easy, timed collection of urine for PK/PD or other types of studies.

Since clinical magnetic resonance imaging (MRI), computed tomography (CT), and positron emission tomography (PET) scanners are very expensive, most institutions allow use of clinical scanners for large animal studies during off or protected hours. The logistics of safely and effectively moving anesthetized large animals and the accompanying anesthetic equipment as well as proper attention to contamination and cleanup are important parts of this process. Animal dander, urine, and potential zoonotic agents are sources of safety concerns. Care should be taken to use blankets and/or plastic sheets on the scanners and to properly cover the entire animal when access to animal's body is not needed. The protocol for sanitization and disinfection should be clearly stated and approved by the health and safety/infection control and the veterinary departments of the institution. The animal and material transport should be well organized so that direct interaction does not occur between patients and patient support teams and the research animals and veterinary support teams.

Lack of compatibility of MRI animal anesthetic agents and monitoring equipment may be a problem in MRI-based drug delivery studies. MRI-compatible anesthesia machines are expensive and may not be easily affordable for many institutions. The solution for such a problem could be to place the anesthesia machine or vaporizer beyond the 5-G line and to use a long line for delivery of anesthetic gas. Alternately, IV anesthetic agents (e.g., propofol or fentanyl–propofol combination) could be another option. MRI-compatible pneumatic ventilators, which use IV anesthetic agents, are a cheaper option than purchasing MRI-compatible anesthesia machines for providing ventilator support. The pneumatic ventilators can be successfully used for ventilated support as well as breath-holds for thoracic and liver imaging in large animal models.

17.4 EXAMPLES OF TRANSLATIONAL IMAGING AND DRUG DELIVERY STUDIES IN LARGE ANIMALS

In the past decade, sufficient experience with specifically targeted therapeutic or diagnostic imaging agents has been gained to warrant moving the knowledge from bench to bedside. The development of quantitative *in vivo* imaging characterization of image-guided drug delivery in large animals is in high demand in the field of

TABLE 17.3 Special Concerns on Different Large Animal Species

Species	PPE	Overnight fast	Injection site	Bean bag	Warming method	Monitoring
Monkey	Tyvek, goggles, mask and double gloves, and shoe covers	Yes	Leg or arm	Yes	Heated circulating warmed air	Vital signs, RR, pulse SPO_2, CO_2, EKG, BP, BT
Pig	Basic PPE	Yes	Ear	Yes	Heated circulating warmed air	Vital signs, RR, pulse SPO_2, CO_2, EKG, BP, BT
Rabbit	Basic PPE	No	Ear	Yes/no	Heated circulating warmed air	Vital signs, RR, pulse SPO_2, CO_2, EKG, BP, BT

translational research. However, crucial prerequisites for the development of safe preclinical protocols in this research are needed since most previous experiments were carried out on rodents. In the past few years, our laboratory has developed several approaches for characterization of the target and delivery validation and therapy response in large animal models, as described below. Special concerns on different large animal species are summarized in Table 17.3.

17.4.1 PET/CT Imaging of ^{64}Cu-Labeled Gold Nanoparticles in Rabbits

Nanoparticles were designed as a novel platform for the delivery of anticancer agents [47]. However, a major challenge to successful biomedical application of nanoparticles is efficient delivery to targeted sites. An understanding of the fate and biological effects of nanoparticles in animals is critical for translation to the clinic. Pharmacokinetic and organ or tissue distribution properties of nanoparticles are of great interest from the clinical point of view because of the potential for cancer imaging and therapy. Our laboratory has developed a method to study the *in vivo* biodistribution of nanoparticles in large animals [48].

The initial study was to investigate the tumor uptake of hollow gold nanospheres (HAuNS) after intra-arterial (IA) and IV administration in a rabbit V×2 model with a tumor in the liver [48]. For the IA group, the agents were injected directly into the hepatic artery using fluoroscopic guidance. Approximately 1.0 mCi ^{64}Cu-labeled nanoparticles (1 ml) mixed with 0.4 ml contrast medium (Visipaque 320; GE Healthcare Inc., Princeton, NJ) was slowly injected through the catheter. For the rabbit receiving IV administration of ^{64}Cu-labeled nanoparticles, a 22-G winged IV catheter was placed in a marginal ear vein following induction of anesthesia. Subsequently, ~1.0 mCi ^{64}Cu-HAuNS suspended in 1.0 ml phosphate-buffered solution (PBS) was administered via the ear vein catheter, and the catheter was then flushed with 2.0 ml saline. A series of whole-body PET/CT scans were

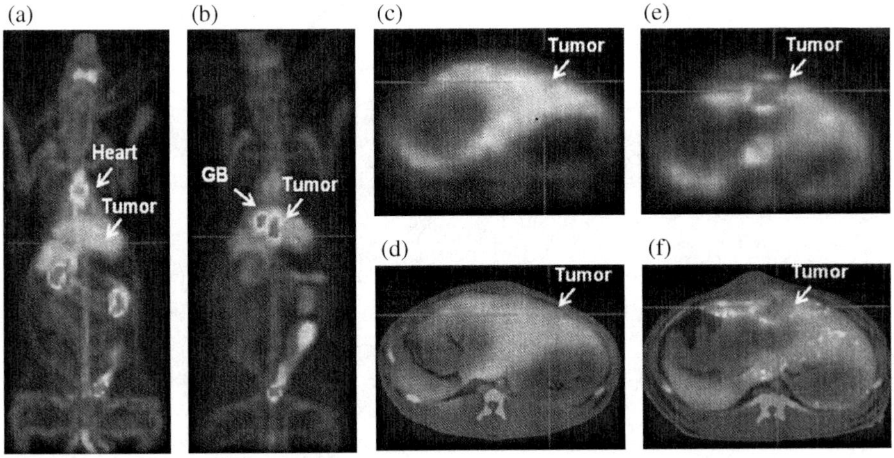

FIGURE 17.1 PET imaging of ^{64}Cu-labeled nanoparticles in liver tumor rabbits at 60 min after injection. Whole-body (a), transaxial (c), and PET/CT fused (d) images after IV injection of ^{64}Cu-HAuNS; whole-body (b), transaxial (e) and PET/CT fused (f) images after IA injection of ^{64}Cu-HAuNS-lipiodol. GB: gallbladder. For color details, please see color plate section.

obtained at 0–60 min after the injection of ^{64}Cu-labeled nanoparticles. Uptake of ^{64}Cu-labeled nanoparticles in tissues was calculated as the percentage of injected dose per gram of tissue (%ID/g).

At 1 h after injection, the PET/CT images demonstrated that the organs involved in the hepatobiliary and renal clearance pathways dominated the whole-body distribution of ^{64}Cu-labeled nanoparticles, with high radioactivity accumulation in the liver, kidney, and urinary bladder. The heart and aortic arch were still visible on the PET images at 1 h after injection of ^{64}Cu-labeled nanoparticles, which was due to slow washout of nanoparticles from the blood pool. In addition, heavily vascularized tissues, such as the retro-orbital plexus, showed higher accumulation than did the background tissues. Organs of the reticuloendothelial system (RES), including the spleen, demonstrated moderate radioactive accumulation especially in the rabbits after IA injection of ^{64}Cu-PEG-HAuNS, ^{64}Cu-RGD-HAuNS. The hepatic V×2 tumor had the highest radioactive accumulation and demonstrated significant intensive accumulation of ^{64}Cu at 1 h after IA injection of PEG-HAuNS + lipiodol (0.50 ± 0.16 and 0.22 ± 0.12%ID/g, respectively; $P < 0.0001$) compared with the IV group (Fig. 17.1).

The biodistribution of radioactivity 24 h following injection showed IA delivery of HAuNS-lipiodol resulted in the highest tumor uptake (0.3319%ID/g) among all the delivery methods ($P < 0.0001$). The tumor-to-normal liver ratio in the IA-HAuNS-ethiodol group was about 2.7-fold higher than that in other groups. IA administration of HAuNS along with lipiodol markedly increased tumor uptake and had a higher tumor-to-normal liver ratio compared with all other groups. PET/CT can be used to monitor the biodistribution of radiolabeled HAuNS after IV and IA administration in a noninvasive fashion.

17.4.2 Multimodality Imaging of Stem Cell Transplantation in a Myocardial Infarction Pig Model

Stem cell-based therapy is an exciting area of research that has shown promise in patients with acute MI or chronic ischemia. However, several fundamental questions are not well defined in the field of cardiac stem cell therapy, including the long-term fate of transplanted cells; the integration, proliferation, and differentiation of stem cells; and the optimal cell type, dose, delivery route, and timing of injection. Therefore, the successful introduction of potentially therapeutic stem cells into patients requires the development of sensitive, noninvasive, long-term repetitive imaging technologies to monitor the dynamics of stem cells in the myocardium and the persistence of stem cell-derived tissues.

One of the imaging strategies is reporter gene imaging, by which stem cells can be genetically engineered to express various reporter genes before transplantation; then the reporter gene can be detected by sensitive imaging devices such as PET, optical imaging system, and MRI. Currently, there are many potential stem cell sources for myocardial repair, including adult stem cells, embryonic stem cells (ESCs), and induced pluripotent stem cells (iPSCs). In particular, mesenchymal stem cells (MSCs) have been shown to reduce infarct size, improve myocardial blood flow, and increase cardiac function [49].

Our laboratory has been able to use herpesvirus type 1 thymidine kinase (*HSV1-tk*) for reporter gene imaging in a pig model of MI [7]. Autologous bone marrow was aspirated, and MSCs were isolated and characterized. The MSCs were then transduced with the MND-sr39HSV1tk-low-affinity nerve growth factor reporter (LNGFR) virus. Transduced MSCs expressing high levels of the intracellular-truncated LNGFR gene were designated as sr39HSV1tk-MSCs. An acute MI was induced in the pigs; 7 days later, autologous sr39HSV1-tk-MSCs were implanted in the left ventricular apical septum via nonfluoroscopic electromechanical mapping-mediated intramyocardial delivery. PET/CT studies using 2′-[^{18}F]fluoro-5-ethyl-1-beta-D-arabinofuranosyluracil ([^{18}F]-FEAU) were performed 4–6h and up to 148 days after IV administration of [^{18}F]FEAU. The HSV1-TK reporter enzyme phosphorylates the radiolabeled nucleoside analogs that are trapped inside the HSV1-tk transduced cell but not in nontransduced cells, and the cells can then be detected noninvasively by PET/CT imaging. In this study, PET/CT imaging was used for long-term *in vivo* monitoring (up to 5 months) of MSCs expressing the sr39HSV1-tk reporter gene after NOGA-guided transendocardial injection in a pig model of MI (Fig. 17.2). We have assessed the biodistribution, survival, and long-term engraftment of transplanted cells and found that transplanted cells exhibit a biphasic distribution pattern in the heart, peaking in the infarct region at 4–5 weeks after delivery. Furthermore, we detected transplanted MSCs at the interface of the scar and myocardial "penumbra," engrafted in the myocardial lymphatic vessels and lymph nodes, up to 150 days after injection.

17.4.3 PET/CT Imaging of T-Lymphocyte Trafficking in NHPs

Adoptive transfer of antigen-specific cytotoxic T lymphocytes (CTLs) has been successfully used to treat patients with various tumors. The analysis of molecular and biochemical events occurring *in vivo* during T-lymphocyte activation still requires

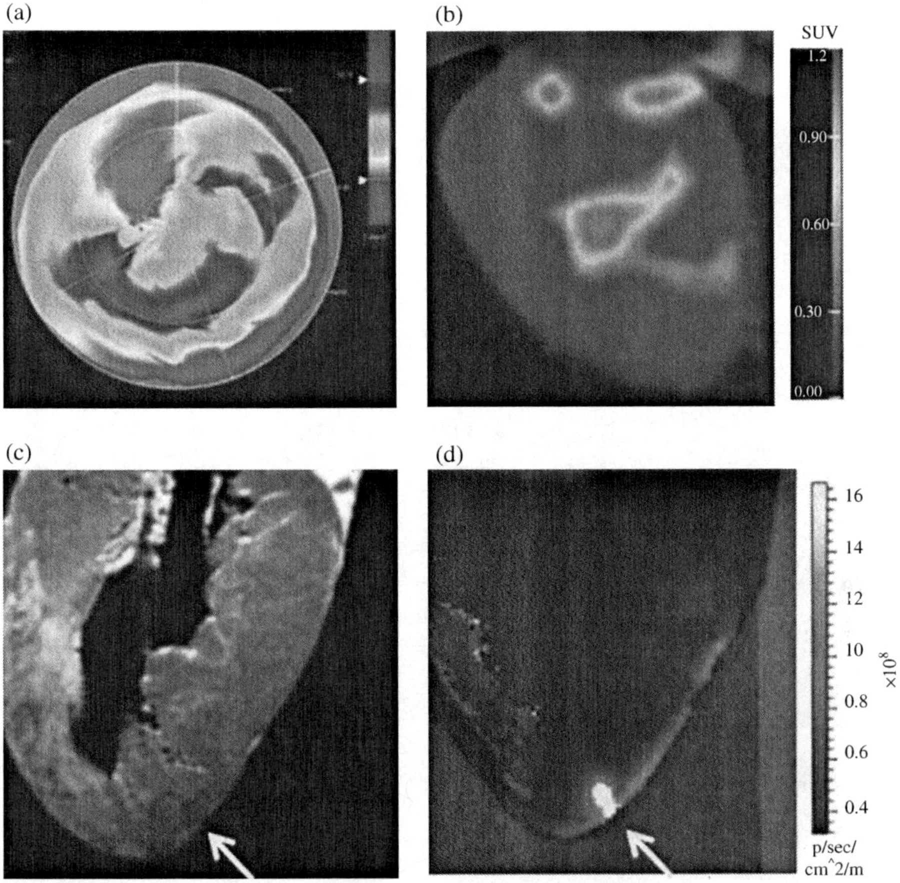

FIGURE 17.2 Representative images of intramyocardially injected sr39HSV1-tk-MSC cells. (a) NOGA bull's-eye map of the sr39HSV1-tk-MSC injection sites (brown dots) in the myocardium; (b) transaxial [^{18}F]FEAU PET/CT image of the injection sites (hot spots) after intramyocardial transplantation of sr39HSV1-tk-MSCs; MRI (c) and optical (d) images of the injection sites (arrow) after intramyocardial injection of a polymeric MRI/near-infrared fluorescent contrast agent PG-Gd-NIR813. For color details, please see color plate section.

invasive blood and tissue sampling. The limitation of repetitive invasive procedures often hinders and eventually precludes serial studies in the same subject and does not provide spatial–temporal distribution of activated T lymphocytes. This limitation can be overcome by application of noninvasive, accurate, sensitive whole-body imaging technologies that allow repetitive measurement *in vivo* of these cells. Recently, new imaging technologies have been developed to permit biochemical investigation *in vivo* and repeatedly over time if necessary. For example, repetitive PET/CT imaging with 2′-fluoro-2′-deoxy-1-β-D-arabinofuranosyl-5-iodouracil (FIAU), 9-(4-[^{18}F]fluoro-3-(hydroxymethyl)butyl)guanine [^{18}F]-FHBG, or [^{18}F]-FEAU of T cells has been genetically labeled to express the herpes simplex virus-1 thymidine kinase (HSV1-tk) reporter gene [50, 51]. Using this approach, HSV-tk^{+} CTLs have been efficiently detected in small animals after the infusion of [^{131}I]FIAU or [^{124}I]FIAU by serial images obtained by scintigraphy or PET, respectively.

Our group has developed the technique to monitor T lymphocytes after IV injection in NHPs [34]. Prior to T-lymphocyte infusions, autologous HSV1-tk^{+} T lymphocytes were generated, collected, washed twice with PBS, suspended in saline solution, and infused over 10–15 min using the left cubital vein. Repetitive PET/CT imaging studies were performed at baseline (1 week before T-cell infusion), 90 min, 1 day, and 1 week after the IV T-cell infusion. For each PET/CT imaging session, a whole-body PET scan (from head to mid-thigh) was started at 60 min after the IV injection of ^{18}F-FEAU with a dose of 10.9 MBq/kg (4 mCi/animal) using the right saphenous vein. PET images were acquired in two-dimensional mode for 3 min per bed position, with four bed positions in total. PET images were reconstructed using VUE Point reconstruction software, which uses an ordered subset expectation maximization (OSEM) algorithm. Emission data were corrected for attenuation, scatter (using CT), random events, and dead-time losses according to the manufacturer's instructions. The in-plane resolution was 5.47 × 5.47 mm with a Z axial slice thickness of 3.27 mm. Circular regions of interest (ROIs) were placed in the areas with the highest radioactivity as well as in the muscle tissue with low levels of activity. Maximal standardized uptake value (SUV) was calculated from each ROI using the formula SUV = measured activity concentration (Bq/g) × body weight (g)/injected activity (Bq), and the target-to-muscle SUV ratios (T:M ratios) were calculated. For the assessment of radioactivity accumulation in the lungs, the lung area in each transaxial plane on PET image was outlined manually based on corresponding CT images, and the mean SUVs per slice and for whole lungs were calculated. Lung-to-muscle ratios were calculated as well.

The repetitive ^{18}F-FEAU PET/CT imaging in NHPs demonstrated that activated T lymphocytes transduced with a retroviral vector encoded for HSV1-tk could be detected after IV infusion in discrete lymphoid organs and in a site of inflammation.

It is interesting that we did not find significant accumulation of T lymphocytes in the lungs of NHPs, as such accumulation is frequently observed in imaging experiments conducted in small animals [50]. However, we determined by quantitative analysis that in one monkey the whole lung radioactivity was 38.5% higher on the PET images obtained at 90-min post-T-cell infusion compared with images obtained 1 week later, suggesting only a transient diffuse retention of T cells in lungs. Our study differed from that of Yaghoubi et al. [52], which was direct injection of cells into the resected cranial tumor and followed by a single imaging time point. Our study was the first to demonstrate the feasibility of repetitive PET/CT imaging of spatial–temporal dynamics of distribution and trafficking of HSV1-tk reporter gene expressing T cells after IV administration in large animals, which is a proof of the principle that adoptively transferred T cells can be imaged after IV administration. The results of this preclinical large animal study support the application of this technology to noninvasive imaging spatial–temporal dynamics of biodistribution of intravenously administrated adoptively transferred CTLs in human patients. We conclude that PET imaging with ^{18}F-FEAU may be used in future clinical trials to evaluate the biodistribution of adoptively transferred CTLs modified to express the HSV1-tk reporter gene or its variants. This technology will ultimately allow for long-term monitoring of trafficking patterns and persistence of the T cells in patients and will facilitate critical improvements in T-cell therapy for cancer.

17.4.4 Biodistribution Kinetics, Metabolism, and Radiation Dosimetry Estimates of PET Imaging Agents in NHPs

For preparation of an investigational new drug (IND) application for clinical phase I imaging studies, the Food and Drug Administration (FDA) regulatory guidelines require the assessment of pharmacological and radiation safety of novel radiolabeled imaging agents. NHPs (i.e., rhesus macaques) are genetically close to humans and have very similar anatomy, physiology, and metabolism; therefore, rhesus macaques provide an ideal model for the evaluation of pharmacokinetics, metabolism, radiation dosimetry, and pharmacological safety of novel radiotracers [2].

Our laboratory has developed a method to study the *in vivo* pharmacokinetics, biodistribution, metabolism, and radiation dosimetry of new PET radiotracers using dynamic PET imaging and serial blood sampling in rhesus monkeys. Recently, we have successfully prepared an eIND for deterring the pharmacokinetics, biodistribution, metabolism, and radiation dosimetry of 4-[(3-iodophenyl)amino]-7-{2-[2-{2-(2-[2-{2-([^{18}F]fluoroethoxy)-ethoxy}-ethoxy]-ethoxy)-ethoxy}-ethoxy]-quinazoline-;acrylamide) (^{18}F-PEG$_6$-IPQA) in NHPs [20], which has been approved by the FDA for a phase I clinical trial. In this study, six rhesus macaques (three female and three male) were injected intravenously with 141 ± 59.2 MBq of ^{18}F-PEG$_6$-IPQA, and dynamic PET/CT images covering the thoracoabdominal area were acquired for 30 min followed by whole-body static images at 60, 90, 120, and 180 min. Multiple blood samples were obtained from the animals at different time points after the radiotracer was administered (Fig. 17.3). The whole-body static PET images were acquired in three-dimensional (3D) mode for 3 min per bed position. Also, a low-dose, non-contrast CT scan was performed immediately after the dynamic PET imaging and at the end of the last PET scan to assure accurate co-registration of PET and CT images. PET images were reconstructed using standard manufacturer-provided reconstruction algorithms, which incorporated OSEM and corrected for attenuation using data from the CT component of the examination; the emission data were corrected for scatter, random events, and dead-time losses using the manufacturer's software program. To ensure that the injected dose was accurate in terms of measured activity using PET, the dose calibrator was cross-calibrated with the PET/CT instrument. Radiolabeled metabolites in blood and urine were analyzed using high-performance liquid chromatography.

Repetitive physical examination, electrocardiograms, and hematological parameters were monitored during the imaging study and at different days after the imaging study. The ^{18}F-PEG$_6$-IPQA pharmacokinetic and radiation dosimetry estimates were determined using volume of interest analysis (VOI) of PET/CT image data sets and blood and urine time-activity data. Dynamic PET imaging revealed the pattern and kinetics of ^{18}F-PEG$_6$-IPQA distribution and clearance over 180-min post-IV administration. ^{18}F-PEG$_6$-IPQA exhibited rapid redistribution and was excreted via the hepatobiliary and urinary systems. ^{18}F-PEG$_6$ was the major radioactive metabolite. The critical organ was the gallbladder with an average radiation-absorbed dose of 0.389 mSv/MBq. The other key organs with high radiation doses were the kidneys, upper large intestine wall, small intestine, and liver. Lung tissue exhibited low uptake of ^{18}F-PEG$_6$-IPQA due to the low

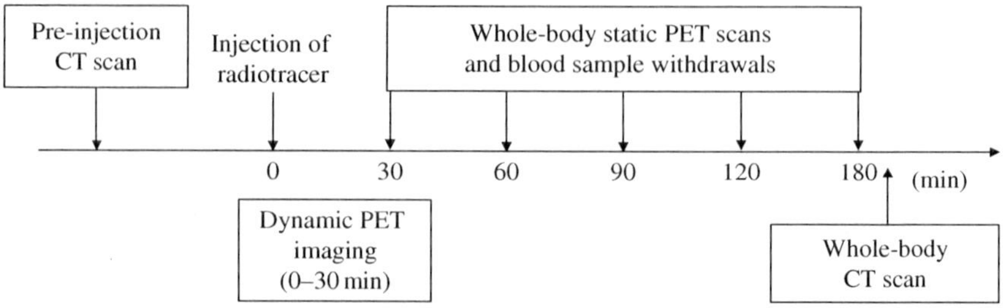

FIGURE 17.3 PET/CT imaging protocol for pharmacokinetic and pharmacodynamic study.

affinity of this radiotracer to wild-type epidermal growth factor receptor (EGFR) kinase. The whole-body radiation doses were 0.0121 mSv/MBq (or 0.0447 rem/mCi). No evidence of acute cardiotoxicity or acute or delayed systemic toxicity was observed. Based on our estimates, diagnostic doses of ^{18}F-PEG$_6$-IPQA up to 128 mBq (3.47 mCi) per injection should be safe for administration in the initial cohort of human patients in phase I clinical PET imaging studies. The whole-body and individual organ radiation dosimetry characteristics and pharmacological safety of diagnostic doses of ^{18}F-PEG$_6$-IPQA as determined in NHPs indicate that this radiotracer should be acceptable for PET/CT imaging studies in human patients.

17.4.5 Pharmacokinetics and MRI of PG-Gd in NHPs

Poly(L-glutamic acid)-benzyl-DTPA-Gd (PG-Gd) is designed as a gadolinium-based biodegradable macromolecular contrast media (MMCM) that is retained in the vascular system for an extended time relative to low molecular weight conventional gadolinium MR contrast agents. The relaxivity of the MMCM agent is significantly increased relative to conventional MRI contrast agents that, in concert with the increased blood concentrations, lead to considerably more blood signal enhancement on MRI scans at a lower injected dose [53]. In addition, PG-Gd localizes to necrotic tissues several days after administration, making it a promising agent for evaluating response to anticancer therapies [54]. Gadolinium-based contrast agents have recently been the subject of intense interest as their use has been putatively linked to a rare condition called nephrogenic systemic fibrosis (NSF) [55, 56]. The pathogenesis of NSF is unexplained, and the factor or factors triggering the onset of the disease are a matter of debate. It is thought that free Gd released from metal chelating agents via transmetallation may cause the toxic Gd effect on tissues of uremic patients. Therefore, for blood pool imaging agents, the use of lower dosages of a more stable Gd-chelate complex should be safer and preferred.

Colorimetric assay with arsenazo III was performed to compare the relative stability of PG-Gd and Magnevist, a DTPA-Gd solution used in the clinic. The amount of PG-BzDTPA required to replace 50% of the arsenazo III-Gd complex was only 75% of what was required for DTPA solution, implying that PG-BzDTPA forms a more stable complex with Gd than does DTPA. Previous investigators have shown

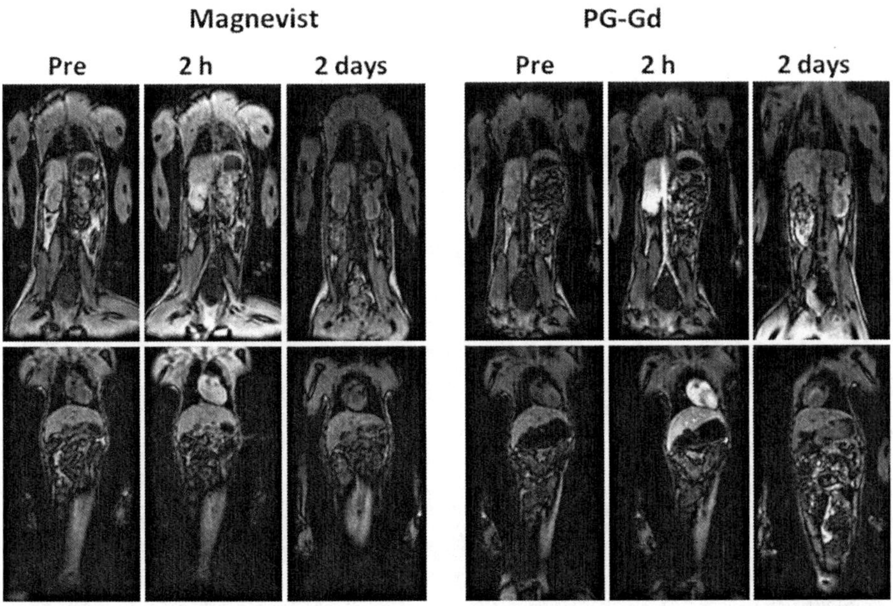

FIGURE 17.4 T1-weighted Fast Spoiled Gradient Echo images of the monkey before contrast injection, at 2 h, and 2 days after injection of Magnevist and PG-Gd at 0.02 mmol Gd/kg. Enhancement of the blood vessel, the heart, the kidney, and the liver were clearly visualized at 2-h post-contrast injection at doses from 0.02 mmol Gd/kg. By 2 days postinjection, most of PG-Gd was cleared from all major organs including kidney (Modified from Contrast Media Mol Imaging 2011;6:289–297 with permission).

that Bz-DTPA transmetallation decreases with the introduction of benzyl groups at the ethylene C-4 and C-5 positions owing to the steric hindrance of the benzyl group, resulting in more stable metal complexes than DTPA [57].

The pharmacokinetic characteristics of PG-Gd in the rhesus macaque model showed a small volume of distribution and vascular confinement of the contrast agent, which is superior than Magnevist where the agent is largely distributed to the extravascular extracellular space (outside the central nervous system) after IV administration. This is confirmed by the dynamic contrast enhanced (DCE) MRI data, which showed gradual reduction in signal intensity after Magnevist injection within 10 min after contrast injection, whereas stable enhancement was observed after injection of PG-Gd during the same period (Fig. 17.4). PG-Gd showed much higher initial blood concentrations within 2 h of administration than did Magnevist. Blood PG-Gd concentrations dropped to a significantly lower level by 2 days after drug administration, and mean residence times for PG-GD were within 15 h. PG-Gd had a longer elimination half-life than did Magnevist. Further studies are warranted to fully characterize the terminal disposition phase of PG-Gd in a larger number of animals. Nevertheless, our study did show greater systemic clearance of Magnevist as compared with PG-Gd.

17.5 ETHICAL AND REGULATORY ASPECTS OF USING LARGE ANIMAL MODELS

Over the past few decades, regulation of animal research has strengthened, resulting in better animal care, strong ethical considerations, and improved quality control of studies. However, for a researcher who is new to translational research and large animal models, the process could be stressful and confusing. Implementing animal welfare regulations and stringent oversights while not compromising the critical aspects of the study in animal models requires a close working relationship between different teams/groups within the institution. For a research team who plans to conduct translational studies in animal models, understanding this complex process is highly beneficial. Many institutions have training programs for researchers who work with laboratory animals, including training in working with the specific species and procedures. In addition, communication of the research to the public is another critical area that may need proper orientation and training for the researchers [58]. The institutional committee, typically named the Institutional Animal Use and Care Committee (IACUC), is the major institutional body involved in reviewing and approving research protocols involving animals. The major steps to prepare for research involving large animal models include approval of the IACUC and the safety (biosafety and/or radiation safety) committees/entities.

In complex large animal studies, it is highly advisable for a veterinarian experienced in laboratory animal medicine to be involved as early as possible in the research process, preferably at the planning stage of the study. In the long term, this assures better animal care, a good understanding of the potential and limitations of the facility and procedural support, and a reduction in unnecessary delays from multiple changes and rectifications to the protocols. In addition, a veterinarian experienced in large animal study support can provide valuable insight on choosing the right model and can offer input on the comparative anatomy and physiology that is relevant to imaging of drug delivery.

The two major federal regulations that cover animal research are (i) the Animal Welfare Act (AWA) and (ii) the Public Health Service (PHS) Animal Welfare Policy. AWA [59] covers all warm-blooded animals that are used for research except birds, rats (genus *Rattus*), and mice (genus *Mus*). AWA authorizes the USDA to develop regulations and standards as outlined in the act. Within the USDA, the Animal and Plant Health Inspection Service (APHIS) is responsible for the administration of AWA. The PHS policy is part of all federal research support through the U.S. PHS, which includes the NIH, Center for Disease Control (CDC), and FDA. The PHS policy implements the "US government principles for the utilization and care of vertebrate animals used in testing, research, and training" [60]. The implementation, interpretation, and evaluation of compliance with PHS policy by those receiving federal support through PHS is performed by the Office of Laboratory Animal Welfare (OLAW) at the U.S. Department of Health and Human Services.

The *Guide for the Care and Use of Laboratory Animals* (The Guide) [61] is a publication from the Institute for Laboratory Animal Research (ILAR), which is part of the National Research Council. The Guide provides nonbinding guidelines on the standards of laboratory animal care and is currently in its eighth edition [61]. The Guide is recognized by the PHS as well as the Association for Assessment and Accreditation of Laboratory Animal Care International (AAALAC International) as the standard reference on laboratory animal care and use programs. AAALAC International is a nonprofit organization that conducts voluntary peer review and accreditation of animal care and use programs.

Institutions that conduct animal research have two main responsibilities. First, they have a responsibility to the public to conduct research in a humane fashion and in accordance with applicable laws, regulations, and standards of care. Second, institutions have a responsibility to the laboratory animals to constantly seek alternatives, to refine the research methods, and to avoid or alleviate pain or distress whenever possible. The IACUC plays a major role in overseeing this responsibility. The minimum membership and types of members in the IACUC are defined by The Guide and AWA. The IACUC should meet at least once every 6 months. However, depending on the volume of research protocols, the IACUC may have more members and more frequent meetings. Irrespective of the frequency of the meetings, researchers must consider that proper review and approval of a protocol can take from a few weeks to months.

When reviewing a protocol, the IACUC should consider the following: the proper training of the research team to work with the specific species and techniques; details of the animal species, including breed, sex, and number of animals requested; the possibility of unnecessary duplication of previous research; clear and simple descriptions of all procedures; all agents to be used; veterinary involvement; attention to minimizing pain and distress; proper use of anesthesia and analgesia; and a relevant search for alternatives to the use of the animal model as well as American Veterinary Medicine Association-approved techniques for humane euthanasia specific for each species. While writing a protocol involving large animal models, it is important that an accurate description of methods is given. The increased commitment in time and effort required to create a detailed methods/procedures description will make the preparation of publications much easier at a later time. More important, including such details will allow meaningful comparisons with similar studies from different laboratories.

Other institutional committees that might get involved in the approval of animal research protocols are the institutional biosafety committee (IBC) and the radiation safety committee (RSC). The involvement of these committees depends on the use of biohazardous agents or radioactive agents in the animals. Since such agents are part of most targeted therapy studies, it is beneficial for the investigator to make sure approval is received as early as possible.

17.6 SUMMARY

Innovative translational research in the development of quantitative *in vivo* noninvasive characterization of image-guided drug delivery plays an important role in biomedical research and becomes an important opportunity to increase the efficiency, potentially saving millions of dollars in development cost. Translational research using experimental large animal fills the gap between preclinical research and clinical practice. A closer collaboration among basic science researchers, veterinarians, and clinical physicians is critical to design and conduct translational studies and to analyze and interpret the findings.

REFERENCES

1. Phillips KA, Van Bebber S, Issa AM. Diagnostics and biomarker development: priming the pipeline. Nat Rev Drug Discov 2006;5 (6):463–469.
2. Tian M, Welch MJ. Translational molecular imaging in drug development: current status and challenges. Curr Med Imaging Rev 2010;6 (1):51–55.
3. Rangarajan A, Weinberg RA. Opinion: comparative biology of mouse versus human cells: modelling human cancer in mice. Nat Rev Cancer 2003;3 (12):952–959.
4. Davies B, Morris T. Physiological parameters in laboratory animals and humans. Pharm Res 1993;10 (7):1093–1095.
5. Kola I, Landis J. Can the pharmaceutical industry reduce attrition rates? Nat Rev Drug Discov 2004;3 (8):711–715.
6. Swindle MM. *Swine In the Laboratory—Surgery, Anesthesia, Imaging and Experimental Techniques*. 2nd ed. Boca Raton, FL: CRC Press; 2007.
7. Perin EC, Tian M, Marini FC 3rd, et al. Imaging long-term fate of intramyocardially implanted mesenchymal stem cells in a porcine myocardial infarction model. Plos One 2011;6 (9):e22949.
8. Sinkora M, Butler JE, Holtmeier W, et al. Lymphocyte development in fetal piglets: facts and surprises. Vet Immunol Immunopathol 2005;108 (1–2):177–184.
9. McAleer J, Weber P, Sun J, et al. Antibody repertoire development in fetal and neonatal piglets. XI. The thymic B-cell repertoire develops independently from that in blood and mesenteric lymph nodes. Immunology 2005;114 (2):171–183.
10. Wang F, Fang W, Zhang MR, et al. Evaluation of chemotherapy response in V×2 rabbit lung cancer with 18F-labeled C2A domain of synaptotagmin I. J Nucl Med 2011;52 (4):592–599.
11. Lei Z, Ma H, Xu N, et al. The evaluation of anti-angiogenic treatment effects for implanted rabbit V×2 breast tumors using functional multi-slice spiral computed tomography (f-MSCT). Eur J Radiol 2011;78 (2):277–281.
12. Kelsey JL, Moore AS, Glickman LT. Epidemiologic studies of risk factors for cancer in pet dogs. Epidemiol Rev 1998;20 (2):204–217.
13. Rankin KS, Starkey M, Lunec J, et al. Of dogs and men: comparative biology as a tool for the discovery of novel biomarkers and drug development targets in osteosarcoma. Pediatr Blood Cancer 2012;58 (3):327–333.
14. Scientific and Humane Issues in the Use of Random Source Dogs and Cats. 2009. Available at http://dels.nas.edu/resources/static-assets/materials-based-on-reports/ reports-in-brief/ random_source_final.pdf.
15. Doyle B, Kemp BJ, Chareonthaitawee P, et al. Dynamic tracking during intracoronary injection of 18F-FDG-labeled progenitor cell therapy for acute myocardial infarction. J Nucl Med 2007;48 (10):1708–1714.
16. Granada JF, Kaluza GL, Wilensky RL, et al. Porcine models of coronary atherosclerosis and vulnerable plaque for imaging and interventional research. EuroIntervention 2009;5 (1):140–148.

17. Holz A, Lautamaki R, Sasano T, et al. Expanding the versatility of cardiac PET/CT: feasibility of delayed contrast enhancement CT for infarct detection in a porcine model. J Nucl Med 2009;50 (2):259–265.
18. Wood BJ, Zhang H, Durrani A, et al. Navigation with electromagnetic tracking for interventional radiology procedures: a feasibility study. J Vasc Interv Radiol 2005;16 (4): 493–505.
19. Panepinto LM, Phillips RW. The Yucatan miniature pig: characterization and utilization in biomedical research. Lab Anim Sci 1986;36 (4):344–347.
20. Tian M, Ogawa K, Wendt R, et al. Whole-body biodistribution kinetics, metabolism, and radiation dosimetry estimates of ^{18}F-PEG$_6$-IPQA in nonhuman primates. J Nucl Med 2011;52 (6):934–941.
21. Stewart CB, Disotell TR. Primate evolution—in and out of Africa. Curr Biol 1998;8 (16):R582–R588.
22. Gibbs RA, Rogers J, Katze MG, et al. Evolutionary and biomedical insights from the rhesus macaque genome. Science 2007;316 (5822):222–234.
23. Byrom MJ, Bannon PG, White GH, et al. Animal models for the assessment of novel vascular conduits. J Vasc Surg 2010;52 (1):176–195.
24. Cross N, Phillips CJ. Requirements for the welfare of baboons and pigs used in animal-to-animal xenotransplantation experiments. Aust Vet J 2009;87 (10):421–426.
25. Shimizu A, Yamada K. Pathology of renal xenograft rejection in pig to non-human primate transplantation. Clin Transplant 2006;20 Suppl 15:46–52.
26. Tseng YL, Sachs DH, Cooper DK. Porcine hematopoietic progenitor cell transplantation in nonhuman primates: a review of progress. Transplantation 2005;79 (1):1–9.
27. Havill LM, Mahaney MC, Czerwinski SA, et al. Bone mineral density reference standards in adult baboons (Papio hamadryas) by sex and age. Bone 2003;33 (6):877–888.
28. Vinson A, Mahaney MC, Cox LA, et al. A pleiotropic QTL on 2p influences serum Lp-PLA2 activity and LDL cholesterol concentration in a baboon model for the genetics of atherosclerosis risk factors. Atherosclerosis 2008;196 (2):667–673.
29. Shi Q, Hubbard GB, Kushwaha RS, et al. Endothelial senescence after high-cholesterol, high-fat diet challenge in baboons. Am J Physiol Heart Circ Physiol 2007;292 (6): H2913–H2920.
30. Braundmeier AG, Fazleabas AT. The non-human primate model of endometriosis: research and implications for fecundity. Mol Hum Reprod 2009;15 (10):577–586.
31. Heiss JD, Walbridge S, Asthagiri AR, et al. Image-guided convection-enhanced delivery of muscimol to the primate brain. J Neurosurg 2010;112 (4):790–795.
32. Krauze MT, McKnight TR, Yamashita Y, et al. Real-time visualization and characterization of liposomal delivery into the monkey brain by magnetic resonance imaging. Brain Res Brain Res Protoc 2005;16 (1–3):20–26.
33. Tian M, Wen X, Jackson EF, et al. Pharmacokinetics and magnetic resonance imaging of biodegradable macromolecular blood-pool contrast agent PG-Gd in non-human primates: a pilot study. Contrast Media Mol Imaging 2011;6 (4):289–297.
34. Dotti G, Tian M, Savoldo B, et al. Repetitive noninvasive monitoring of HSV1-tk-expressing T cells intravenously infused into nonhuman primates using positron emission tomography and computed tomography with 18F-FEAU. Mol Imaging 2009;8 (4):230–237.
35. Le Foll B, Chefer SI, Kimes AS, et al. Validation of an extracerebral reference region approach for the quantification of brain nicotinic acetylcholine receptors in squirrel monkeys with PET and 2–18F-fluoro-A-85380. J Nucl Med 2007;48 (9):1492–1500.

36. Zappe AC, Reichold J, Burger C, et al. Quantification of cerebral blood flow in nonhuman primates using arterial spin labeling and a two-compartment model. Magn Reson Imaging 2007;25 (6):775–783.
37. Morton WR, Agy MB, Capuano SV, et al. Specific pathogen-free macaques: definition, history, and current production. ILAR J 2008;49 (2):137–144.
38. Elmore D, Eberle R. Monkey B virus (*Cercopithecine herpesvirus* 1). Comp Med 2008;58 (1):11–21.
39. Anderson RA. Stress effects on chromium nutrition of humans and farm animals. In: Biotechnology in the Feed Industry. In: *Proceedings of Alltech's Tenth Annual Symposium*; Nottingham University Press; 1994. p 267.
40. Jenkins JR. Rabbit behavior. Vet Clin North Am Exot Anim Pract 2001;4 (3):669–679.
41. Panepinto LM, Phillips RW, Norden S, et al. A comfortable, minimum stress method of restraint for Yucatan miniature swine. Lab Anim Sci 1983;33 (1):95–97.
42. Blye R, Burke R, James C, et al. The use of operant conditioning of Göttingen minipigs for topical safety studies. *In 57th AALAS meeting, Poster presentation*, Salt Lake City. 2006 Oct 15–19.
43. Laule G, Whittaker M. Enhancing nonhuman primate care and welfare through the use of positive reinforcement training. J Appl Anim Welf Sci 2007;10 (1):31–38.
44. Wolf RF, White GL. Clinical Techniques Used for Nonhuman Primates. In *Non-Human Primates in Biomedical Research*. 2nd ed. Abee CR MK, Tardif SD, Morris T editors, Salt Lake City: Academic Press; 2012.
45. Fish RE, Brown MJ, Danneman PJ, et al. *Anesthesia and Analgesia in Laboratory Animals*. 2nd ed. Amsterdam:Academic Press; 2008.
46. Fox JG, Anderson LC, Loew FM, et al. *Laboratory Animal Medicine*. American College of Laboratory Animal Medicine Series. 2nd ed. San Diego: Academic Press; 2008.
47. Krauze MT, Vandenberg SR, Yamashita Y, et al. Safety of real-time convection-enhanced delivery of liposomes to primate brain: a long-term retrospective. Exp Neurol 2008;210 (2):638–644.
48. Tian M, Lu W, Zhang R, et al. Tumor uptake of hollow gold nanospheres after intravenous and intra-arterial injection: PET/CT study in a rabbit VX2 liver cancer model. Mol Imaging Biol 2013 Apr 23. [Epub ahead of print] PMID: 23608932.
49. Quevedo HC, Hatzistergos KE, Oskouei BN, et al. Allogeneic mesenchymal stem cells restore cardiac function in chronic ischemic cardiomyopathy via trilineage differentiating capacity. Proc Natl Acad Sci USA 2009;106 (33):14022–14027.
50. Koehne G, Doubrovin M, Doubrovina E, et al. Serial in vivo imaging of the targeted migration of human HSV-TK-transduced antigen-specific lymphocytes. Nat Biotechnol 2003;21 (4):405–413.
51. Dobrenkov K, Olszewska M, Likar Y, et al. Monitoring the efficacy of adoptively transferred prostate cancer-targeted human T lymphocytes with PET and bioluminescence imaging. J Nucl Med 2008;49 (7):1162–1170.
52. Yaghoubi SS, Jensen MC, Satyamurthy N, et al. Noninvasive detection of therapeutic cytolytic T cells with 18F-FHBG PET in a patient with glioma. Nat Clin Pract Oncol 2009;6 (1):53–58.
53. Wen X, Jackson EF, Price RE, et al. Synthesis and characterization of poly(L-glutamic acid) gadolinium chelate: a new biodegradable MRI contrast agent. Bioconjug Chem 2004;15 (6):1408–1415.

54. Jackson EF, Esparza-Coss E, Wen X, et al. Magnetic resonance imaging of therapy-induced necrosis using gadolinium-chelated polyglutamic acids. Int J Radiat Oncol Biol Phys 2007;68 (3):830–838.

55. Ledneva E, Karie S, Launay-Vacher V, et al. Renal safety of gadolinium-based contrast media in patients with chronic renal insufficiency. Radiology 2009;250 (3):618–628.

56. Perazella MA. Current status of gadolinium toxicity in patients with kidney disease. Clin J Am Soc Nephrol 2009;4 (2):461–469.

57. Laurent S, Botteman F, Vander Elst L, et al. Relaxivity and transmetallation stability of new benzyl-substituted derivatives of gadolinium-DTPA complexes. Helv Chim Acta 2004;87 (5):1077–1089.

58. Marris E. Animal research: grey matters. Nature 2006;444 (7121):808–810.

59. USDA. 1985. 9 CFR 1A (Title 9, Chapter 1, Subchapter A): Animal Welfare.

60. NIH. 1985. US government principles for the utilization and care of vertebrate animals used in testing, research, and training. Available at http://fmp-8.cit.nih.gov/oacu/guidepi/references/2govprinciple.pdf.Accessed 2013 Apr 22.

61. NRC. *Guide for the Care and Use of Laboratory Animals*. 8th ed. Washington, DC: National Academy Press; 2011.

CHAPTER 18

IMAGING TRANSCATHETER DELIVERY TO LIVER TUMORS

ADIL AL-NAHHAS, MEERAN NAJI, AND IMENE ZERIZER

18.1 INTRODUCTION

The management of primary liver cancer and secondary liver metastases spreading from other malignancies represents an extremely challenging issue in oncology and hepatobiliary surgery. When difficult to control, these afflictions cause significant morbidity and are responsible for a worldwide mortality of hundreds of thousands of patients each year.

Hepatocellular carcinoma (HCC) is the sixth most frequent cancer and the third highest cause of cancer deaths worldwide particularly among patients with liver cirrhosis [1]. HCC is characterized by a variable geographical distribution with high incidence in Southeast Asia and sub-Saharan Africa compared to North America and Europe. However, the incidence is increasing in Western countries with the age-adjusted incidence rates in the United States tripling between 1975 and 2005 [2]. The majority of cases of HCC develop as the final stage of the spectrum of cirrhosis caused by infection with the hepatitis virus B (Africa and Asia) and C (Japan, southern Europe, and United States) and as a consequence of alcoholism (Europe and North America) [3].

Secondary liver tumors have no distinct geographical distribution and are the most common liver tumors worldwide. A common route for metastases from the colon and rectum is through the portal venous system, but metastases from other organs and tissues such as the lung, prostate, breast, kidney, and genitourinary tract can also contribute to secondary liver tumors.

Colorectal cancer (CRC) is a leading cause of cancer-related death [4] with 412,900 new cases and 207,500 deaths in Europe in 2006 [5]. In 60% of cases, patients will develop liver metastases, and in 20% of these, the liver represents the only site of the disease at death [6]. This remains the main cause of morbidity and mortality among patients with CRC [7] and, if untreated, would reduce the median survival of patients to 6–8 months [8].

Drug Delivery Applications of Noninvasive Imaging: Validation from Biodistribution to Sites of Action, First Edition. Edited by Chun Li and Mei Tian.

18.2 TREATMENT OF LIVER TUMORS

The only curative treatment for liver tumors is surgical resection, but this modality is restricted to a limited number of patients, probably <10–15%, depending on the size, number, and site of liver tumor. Even with successful surgery, the majority of patients will experience a recurrence of their liver tumors within 3 years following resection.

External beam radiotherapy (EBRT) has a limited role in the treatment of unresectable liver tumors simply because the high-dose EBRT required to treat the tumors is bound to induce damage to the surrounding radiosensitive normal liver tissue. Modern chemotherapeutic regimens, including oxaliplatin or irinotecan, have improved response rates and can downgrade unresectable tumors to allow surgical resection. However, even after effective chemotherapy, the high rate of hepatic and extrahepatic recurrence is an eventual outcome in most patients [9]. There is therefore a high medical need for effective treatments of patients with unresectable liver tumors who have failed conventional chemotherapy regimens.

A number of palliative treatment modalities that can ameliorate symptoms and improve quality of life have been tried in this respect. Microwave coagulation therapy is a thermoablative therapy utilizing heat to induce tumor tissue necrosis that is effective in destroying tumor capsule and surrounding extracapsular invasion. Similar effect may be achieved with cryotherapy that induces rapid freezing leading to ice formation in the extracellular spaces leading to intracellular water shift and eventual dehydration and destruction of cellular structures. Laser and radiofrequency ablation are alternative methods of interstitial tumor therapy that cause tissue destruction by hyperthermic coagulative necrosis and high-frequency alternating currents, respectively. All these methods have been used with variable results but remain feasible only for patients with a limited number and certain size of lesions [10].

There is therefore a need to treat the large number of patients who have liver tumors that are difficult or impossible to remove surgically or modify by local ablative therapy and show progression under standard chemotherapeutic regimens. Successful treatment in such patients' group requires selective delivery of the therapeutic agent to individual tumors with the purpose of inducing effective ablation while preserving the normal liver parenchyma. This approach is currently feasible with selective internal radiation therapy (SIRT) with radiolabeled microparticles such as ^{90}Y-microspheres.

18.3 SELECTIVE INTERNAL RADIATION THERAPY (SIRT)

The selective delivery in SIRT is based on a well-established concept. It has been demonstrated that malignant liver tumors derive 95% of their blood supply from the hepatic arteries unlike normal liver parenchyma, which receives only 25–30% of its supply from the hepatic arteries, the rest being provided by the portal venous system [11]. This has been confirmed by CT and $C0_2$ microbubble ultrasound [12, 13]. The distribution of therapeutic agents injected directly into the hepatic arteries has also been shown to concentrate selectively in the malignant lesions [14].

Radiolabeled microspheres are available in two forms for commercial use. The microspheres are made of either glass beads (TheraSpheres) or resin beads (SIR-Spheres), with a mean diameter of 20–30 and 30–40 µrespectively. They are usually labeled with yttrium-90 (^{90}Y), which is a high-energy pure beta emitter. The emitted beta particles have a maximum energy of 2.27 MeV with a mean energy of 0.93 MeV. They travel through tissue with a maximum range of 11 mm (mean range of 2.5 mm) inducing damage to nearby tumor cells. The half-life of ^{90}Y is 64.2 h that provides effective treatment duration of 92.3 h. The interaction of beta particles with tissue can also produce gamma rays (Bremsstrahlung effect) and positrons that can be utilized to map the distribution of the delivered therapy using a standard gamma camera and a positron emission tomography (PET) scanner, respectively.

The delivery of ^{90}Y-microspheres is done through the hepatic arteries and requires two interventional angiography sessions, the first for initial mapping and assessment of hepatic-pulmonary shunt and the second for delivery of ^{90}Y-microspheres.

Depending on the administered ^{90}Y activity, the delivered target dose is usually in excess of 100 Gy over a period of 10–14 days, following which there is no further radioactivity. The microspheres are not metabolized or excreted and remain as permanent implants in the liver without exerting any adverse or mutagenic effect on residual liver tissue.

18.4 SELECTION OF PATIENTS FOR SIRT

The ideal setting for selecting patients to undergo SIRT is through a multidisciplinary group comprising clinicians, surgeons, radiologists, nuclear medicine specialists, and histopathologists. The patients must have thorough clinical, biochemical, and imaging assessment prior to consideration for therapy.

The main indications for SIRT are:

- The presence of multiple liver lesions such that total resection would leave insufficient liver tissue to sustain life.
- Tumors that have invaded, or in close proximity, to the hepatic venous confluence since surgery may compromise the drainage of the hepatic veins into the inferior vena cava.
- Tumor invasion of the porta hepatis with failure to preserve the portal veins.
- Recurrent and chemotherapy-resistant tumors in patients who are not suitable for surgery.

Contraindications to SIRT include:

- History of EBRT or sensitizing chemotherapy when additional radiation to the liver may induce radiation hepatitis.
- Clinical liver failure presenting ascites or deranged liver function tests.
- The presence of significant reflux of hepatic arterial flow to vital organs (stomach, pancreas, or bowel) or extensive hepatic-pulmonary shunt that may induce radiation pneumonitis.
- Widespread extrahepatic metastases, particularly to the lungs and bones.

The administered dose of ^{90}Y-microspheres ranges from 1.8 to 3 GBq. Several factors must be taken into consideration when calculating individual patients' doses. Since delivery through hepatic arteries is specifically enhanced to tumors compared to normal liver tissue, the dose is increased when a higher percentage of the liver is involved in order to achieve the maximum benefit. Conversely, a smaller dose is used when there is a significant hepatic-pulmonary shunt to avoid undesirable lung damage.

18.5 IMAGING IN PRETHERAPY ASSESSMENT

Routinely, a triple-phase contrast-enhanced CT (CECT) is performed prior to therapy, but other cross-sectional modalities such as MR and US are used if and when necessary. Imaging is an important step in determining the suitability of patients for treatment, assessing the number of lesions, patency of the portal vein, and calculating the volume of disease in relation to the whole liver. ^{18}FDG PET/CT has a vital role in staging patients prior to therapy and identifying sites of extrahepatic disease undetected by CECT that can potentially make patients unsuitable for treatment.

The optimal time of performing pretreatment imaging has not yet been determined, but from our experience a CECT and a PET/CT performed 4–6 weeks prior to therapy are accurate in the evaluation of patients before therapy and as a baseline for response assessment.

18.6 DIAGNOSTIC ANGIOGRAPHY

Diagnostic arterial angiography is performed in order to map the hepatic arterial supply and identify any arteries that may potentially result in microspheres being delivered to nontarget structures adjacent to the liver. Microspheres entering nontarget arterial territories can result in gastrointestinal complications such as gastric and duodenal ulceration and acute cholecystitis. Selective visceral angiography allows the identification of these extrahepatic vessels (right gastric, gastroduodenal, or falciform artery) and permits prophylactic coil embolization of these vessels to prevent complications. The technique of diagnostic angiography involves percutaneous access of the femoral artery followed by selective visceral angiography to assess the blood supply of the liver and the adjacent gastrointestinal tract. The superior mesenteric artery is usually assessed for the presence of accessory hepatic arteries, and the celiac trunk is selectively catheterized to identify the hepatic arterial supply of the tumor [15]. Angiography is vital not only to provide information on vascular anatomy but also to provide invaluable information on blood flow characteristics in various vessels, and this can also be repeated during the treatment delivery [16]. As ^{90}Y-microspheres have a strong embolic effect on vessels, blood flow may change or reverse during the treatment carrying a risk of possible delivery of the microspheres into nontarget areas. Therefore, blood flow must be continuously evaluated during the microspheres delivery [17].

Diagnostic angiography is also used to deliver ^{99m}Tc-macroaggregated albumin (^{99m}Tc-MAA) to assess hepatopulmonary shunting and detect extrahepatic passage of activity into the gastrointestinal tract [18]. It has been shown that liver tumors tend to demonstrate arteriovenous communications or shunts, through which microspheres may pass and possibly lodge within the pulmonary vasculature carrying a risk of radiation injury. It has been estimated that a hepatopulmonary shunt fraction >20% of the injected ^{90}Y dose may potentially expose the lungs to an absorbed radiation dose of more than 30 Gy with associated high incidence of radiation pneumonitis [19]. Maintaining the radiation exposure of the lungs below this dose is thought to significantly reduce the risk of this complication [20]. Because the size of the MAA particles (10–60 μm; average of 35 μm) is similar to that of the microspheres (29–35 μm), the ^{99m}Tc-MAA scan can provide a good indication of the expected distribution of the microspheres and allows to measure and quantify the hepatopulmonary shunting [10].

18.7 ^{99M}TC-MAA SCINTIGRAPHY

^{99m}Tc-MAA (150 MBq) is injected into the hepatic artery after coil embolization of the extrahepatic branches. The patient is then transferred to the nuclear medicine department within 30–60 min of the ^{99m}Tc-MAA injection for imaging. Planar images of the abdomen and thorax are usually obtained using a dual detector head camera. The hepatopulmonary shunt fraction can then be calculated as the ratio of counts in the lungs to total counts in the liver and lungs in regions of interest drawn on the planar images. Based on the manufacturer's recommendations, it has been suggested that a patient with a shunt of <10% is considered suitable for treatment without a need for dose reduction or modification [19]. The ^{99m}Tc-MAA scan may provide additional information and assist in detecting occult extrahepatic shunting into the gastrointestinal tract. In these cases, the culprit vessel could be retrospectively identified on the angiogram images and then embolized before treatment [19].

Although most centers use only planar or single photon emission computed tomography (SPECT) images when performing ^{99m}Tc-MAA scans, it has been suggested that integrating SPECT with low-dose CT may be more beneficial. In a recent study, Hamami et al. reported that ^{99m}Tc-MAA SPECT/CT scintigraphy increases the sensitivity and specificity of the scan in detecting extrahepatic shunting and therefore further reduces the risk of gastrointestinal complications [18].

18.8 ADMINISTRATION OF SIRT

^{90}Y-microspheres are shipped by the manufacture in a vial of 3 GBq dose [21]. The manufacturer recommends one of the two following methods for determining administered activity:

1. Body surface area (BSA) method using the DuBois formulae:

$$\text{BSA}\,(\text{m}^2) = 0.20247 \times \text{Height}\,(\text{m})^{0.725} \times \text{Weight}\,(\text{kg})^{0.425}$$

$$\text{Activity}\,(\text{GBq}) = (\text{BSA} - 0.2) + (\text{volume of tumor} / \text{volume of liver} \times 100)$$

TABLE 18.1 Adjustment of ^{90}Y-microspheres Dose According to Percentage of Lung Shunt

Percentage of lung shunt	Recommended SIR-Spheres dose
<10%	Full dose of ^{90}Y-microspheres
10–15%	Reduce dose of ^{90}Y-microspheres by 20%
15–20%	Reduce dose of ^{90}Y-microspheres by 40%
>20%	^{90}Y-microspheres are contraindicated (absolute)

Volume of the tumor is determined on the pretreatment CECT using three-dimensional (3D) measurement and is divided by the whole liver volume [22].

2. Empirical method.

   The activity is calculated for whole liver delivery based on involvement by tumor as demonstrated on CECT:

$$\text{Tumor volume} \leq 25\% = 2\,\text{GBq}, 25-50\% = 2.5\,\text{GBq}, 50\% = 3\,\text{GBq}$$

At our institution the BSA method is followed. The dose is further adjusted depending on the ^{99m}Tc-MAA shunt result as shown in Table 18.1.

Revascularization and changes to hepatic blood flow can occur rapidly, and therefore, the initial angiogram must be performed as closely as possible to the time of the SIRT. At our institution, the administration of SIRT is performed a week after the initial angiogram. A check arteriogram is also required just before administration of the SIR-Spheres to ensure that such revascularization has not occurred.

Traditionally, SIRT was performed as whole-liver infusion due to limitations in microcatheters and imaging. This approach is no longer recommended and recent developments in microcatheter technology allow lobar or sub-/segmental infusions that is the preferred method [23]. Following the check angiogram, the calculated dose of microspheres is injected via a three-way port into the hepatic artery segments under fluoroscopic guidance.

18.9 ASSESSMENT OF IMMEDIATE POST-THERAPY DISTRIBUTION

18.9.1 Post-Therapy Imaging: Bremsstrahlung

^{90}Y decays to stable zirconium-90 (^{90}Zr) with a half-life of 64 h. The interaction of the high-energy beta emissions with tissue results in the production of broad-spectrum Bremsstrahlung emissions, and this radiation can be detected following the delivery of ^{90}Y-microspheres by imaging with a standard gamma camera. The images can then be assessed to map the distribution of the injected microspheres in the body. It also helps in detecting extrahepatic activity that may predict impending gastrointestinal complications and serve as a quality assurance tool.

In a previous study, we evaluated the role of Bremsstrahlung imaging in 21 patients, who underwent SIRT for liver tumors and compared the results with angiography [24]. We found that Bremsstrahlung imaging is useful for confirming appropriate delivery of microspheres to the desired hepatic arterial territories and identifying nonuniformity of uptake resulting from faulty delivery technique. In another study of 37 patients, Ahmadzadehfar et al. evaluate the role of Bremsstrahlung imaging compared with ^{99m}Tc-MAA scans and found that their distribution was identical in all patients [25]. In addition, Bremsstrahlung imaging was successful in identifying and differentiating tumor lesions from surrounding liver parenchyma in 41% of the cases.

18.9.2 Post-Therapy Imaging: PET/CT

In addition to Bremsstrahlung, a minor branching fraction exists in the decay chain of ^{90}Y that results in the emission of a positron ($^{+}\beta$). This may be utilized for imaging the distribution of ^{90}Y-microspheres using PET/CT. However, there are two potential problems with PET/CT imaging of ^{90}Y: (i) low emission of positrons with relative abundance of 34 per million disintegrations may prevent the acquisition of adequate quality PET images, and (ii) the abundance of Bremsstrahlung has the potential to cause serious dead-time losses in the detectors with saturation effect on the PET scanner.

Despite the aforementioned problems associated with ^{90}Y PET/CT, there have been recent attempts to investigate the feasibility of imaging ^{90}Y using PET/CT in phantom studies and in patients [26, 27]. In a recent study, Lhommel et al. evaluated the feasibility of imaging the positron emissions of ^{90}Y with a PET/CT scanner in both experimental setting and in a patient following SIRT [27] using a high-end time-of-flight (TOF) PET/CT. The study showed that PET images correlated reasonably well with the baseline pretherapy ^{18}FDG PET images despite the differences in the mechanism of uptake and better reflected tumor heterogeneity compared with the traditional Bremsstrahlung SPECT. Initial results in clinical practice also demonstrated a high-resolution absorbed dose distribution and showed that ^{90}Y dosimetry correlated very well with the tumor response [27]. Interestingly, the authors used a 2.5-mm copper ring to overcome the problem of dead-time losses and prevent the saturation effect on the PET scanner. However, subsequent work conducted by Werner et al. suggested that good quality ^{90}Y PET/CT can be obtained without a need for additional copper ring or advanced TOF mode [28].

18.9.3 Adverse Effects

Following the administration of SIRT, the majority of patients may experience transient pain, fever, and nausea due to the embolic effect of the microspheres, all responding to simple conservative measures. Persistent pain requires further assessment and may result from inadvertent delivery of yttrium-90 microspheres to the gallbladder, stomach, or pancreas resulting in cholecystitis, gastric ulcers, and pancreatitis, respectively [29]. These adverse effects should be reduced to the minimum with prior embolization of corresponding vessels during assessment angiography. Radiation hepatitis may prove difficult to diagnose and may appear later in the course of the disease. The risk of developing this condition is higher with pretherapy hyperbilirubinemia, sensitizing chemotherapy, or high radiation dose of >150 Gy [30]. However, the majority of these adverse effects are self-limiting and resolve with appropriate measures. Radiation pneumonitis is a more serious consequence caused by high rate of hepatic-pulmonary shunt and should be avoided by abandoning the procedure if the shunt exceeds 20%. The implementation of routine measurements of shunts and subsequent dose adjustment has eliminated this adverse effect with no recent reported cases.

18.9.4 Imaging in Response Assessment

Response assessment to new anticancer treatments is highly important to justify their cost-effectiveness. Imaging studies play a major role in this evaluation, and several criteria have been introduced to objectively quantify the effect of anticancer therapy. Of these, the anatomical RECIST criteria have become the gold standard for evaluating response in trials and clinical practice. These criteria are based on measurements of tumors size pre and post therapy on CECT (Table 18.2) [31].

Although these criteria represent a robust tool for response assessment in several solid tumors treated with cytotoxic chemotherapy, their use in cytostatic treatments, biological agents, immunotherapy, and interventional therapy has been criticized [32, 33].

Several of these therapies particularly interventional therapies lead to intratumoral and peritumoral edema and hemorrhage leading to an apparent increase in the size of the lesions [34, 35]. Furthermore, necrosis of the treated lesions is not followed by immediate reduction in size, which only becomes apparent on prolonged follow-up. This can potentially lead to mislabeling conclusions of stable disease (SD). It has also been suggested that the RECIST criteria do not correlate with patient clinical outcome, and other methods of response assessment early in treatment may be used as surrogate of clinical end-point measures [36].

Additional functional information that complement the RECIST criteria have been suggested to assess response in several tumors such as evaluation of tumor enhancement using Hounsfield unit (HU) on CECT and development of mural nodules in gastrointestinal stromal tumors (GIST) treated with imatinib [37]. These combination criteria have been deemed to be more accurate measures of response assessment correlating with clinical outcome [37, 38]. Alternative criteria have also been proposed by the European Association for the Study of the Liver (EASL) in vascular interventional therapies and radiofrequency ablation of HCC [39]. These are

TABLE 18.2 RECIST Response Categories and Definitions

Group	RECIST criteria
CR	Disappearance of lesions, confirmed at 4 weeks
PR	30% decline in sum LD, confirmed at 4 weeks
SD	No PD or PR/CR
PD	Unequivocal new lesions, an increase of 20% in sum LD

Sum LD, sum of the longest diameter of target lesions; CR, complete response; PR, partial response; SD, stable disease; PD, progressive disease.

based on contrast enhancement characteristics of HCC pre and post treatment. The EASL panel of experts of HCC proposed assessment of response by spiral CT or MRI 4 weeks after treatment with measurements of the diameter of viable tumor as identified by areas of enhancement rather than longest diameter [39].

However, it has been well documented that tumor enhancement by contrast media is complex and is influenced by several parameters such as an intact blood supply to the tumor, background tissue circulation, permeability of the vessel wall, and the size of and osmotic pressure within interstitial spaces [40]. In addition, pretreatment embolization can affect the blood supply characteristics of the liver lesions influencing the degree of their enhancement.

A modern angiography suite that acquires rotational angiographic data to produce 3D images similar to CT has been described. This technique allows images to be analyzed to identify the distribution of injected tracers within target lesions and may be useful to identify areas of necrosis within the tumor [41]. This is a promising technique but may be labor intensive to and difficult to reproduce.

Perfusion imaging using dynamic contrast-enhanced MRI has also been suggested as a tool to measure response post radioembolization [42]. The advantages of using MRI over CT include more sensitivity to contrast enhancement due to a greater dynamic response of signal intensity to changing concentrations of contrast and a more simplified kinetic model that allows measurement of perfusion. However, it has been argued that quantification of signal change on MRI is difficult and the technique may not be reproducible being susceptible to patient factors [43]. MRI using diffusion and apparent diffusion coefficient (ADC) maps has been used in several solid tumors to quantify the diffusion of water molecules within tumor cells. This technique is sensitive to detect changes of necrosis that occur following treatment. However, its use in radioembolization has not been confirmed yet. This may be due to the presence of radioembolization coils that can lead to magnetic field irregularities leading to artifacts.

Contrast-enhanced ultrasonography (CEUS) is another technique that can evaluate the perfusion characteristics of liver lesions. CEUS has proven to be useful in response assessment post-radiofrequency ablation of liver tumors with improved detection and display of occult lesions, in the evaluation of residual and recurrent disease in the immediate follow-up period, and in the assessment of lesional geometry and orientation for a more accurate planning and guidance of repeat treatment [44].

^{18}FDG PET/CT is a powerful tool in response assessment in several solid tumors particularly lymphoma. It can detect functional changes in tumor following treatment, at a much earlier time than anatomical response criteria. It has therefore been proposed that ^{18}FDG PET/CT can depict metabolic changes post SIRT, which occur much earlier than reduction in tumor size. To date there are only few small studies that evaluated the efficacy of ^{18}FDG PET/CT in response assessment compared to other imaging modalities. An early prospective study of eight patients treated with ^{90}Y glass microspheres evaluated the role of ^{18}F-FDG PET compared to CT or MRI in response assessment [45]. There was a significant metabolic response assessed visually on ^{18}FDG PET post SIRT, which correlated with a decline in serum carcinoembryonic antigen (CEA). Anatomical assessment on CT and MRI did not produce a significant change in the size of the liver lesions and did not correlate with a decline in serum CEA. A later study by the same authors compared visual assessment to quantitative analysis and demonstrated a median decline in standard uptake values (SUV) from 10.455 at baseline to 9.075 after treatment ($P=0.011$). This translated into a 26% decline in patients who responded to treatment compared to 6% in nonresponders [34]. Popper et al. showed a similar decline in SUV that paralleled a decrease in tumor markers. An early study performed at our institution that compared ^{18}FDG PET and RECIST criteria for CT in 21 patients treated with SIRT demonstrated that early ^{18}FDG PET performed at 6 weeks showed a decline in SUV in 86% of patients while CT demonstrated a decline in size in only 13% of patients [46].

18.9.5 Clinical Experience with SIRT

Treatment with SIRT is palliative rather than curative aiming at improving survival and quality of life. A large study evaluated the use of SIRT across seven centers in 208 patients with unresectable CRC metastasis refractory to chemotherapy [47]. Response evaluated with CT, PET, and MRI showed partial response (PR) in 35% and SD in 55% of patients. A prospective multicenter phase II trial in 50 patients with unresectable CRC liver metastases showed median overall survival of 12.6 [48]. Similar results were noted in HCC [49–51]. Quality of life has been noticed to improve in patients treated with SIRT compared to those treated with conventional chemotherapy [52].

There is emerging evidence that the combination of SIRT with chemotherapy can lead to improved clinical outcome. A multicenter phase III study of 46 patients treated with 5-FU alone (group A) or combination therapy with 5-FU and SIRT (group B) demonstrated median time to liver progression (TTLP) of 2.1 in group A and 5.5 months in group B, respectively [53]. Similarly, a phase I study of SIRT with combined administration of chemotherapy was conducted in 20 patients with inoperable CRC liver metastases. The treatment was generally well tolerated and 18 patients had a PR while 2 had SD [36]. Currently, a multicenter randomized controlled phase II and III trial evaluating fluorouracil, oxaliplatin, and leucovorin (FOLFOX) with SIRT compared to FOLFOX alone in patients with liver metastases from primary CRC (SIRFLOX) is being undertaken with PFS and OS as study end points.

18.10 SUMMARY

Treatment of liver tumors with ^{90}Y-microspheres is a palliative procedure performed for patients who are unsuitable for surgery and have unsatisfactory response to chemotherapy. The procedure has minor inconvenience and minimal adverse effects and has been shown to improve quality of life and prolong survival. Phase II and III trials are being conducted to assess the value of combining SIRT with chemotherapy.

Imaging plays a major role in the selection and preparation of patients, delivery of therapy, immediate post-therapy assessment of distribution as well as long-term follow-up, and assessment of response.

REFERENCES

1. Parkin DM, Bray F, Ferlay J, et al. Global cancer statistics, 2002. CA Cancer J Clin 2005;55 (2):74–108.
2. Altekruse SF, McGlynn KA, Reichman ME. Hepatocellular carcinoma incidence, mortality, and survival trends in the United States from 1975 to 2005. J Clin Oncol 2009;27 (9):1485–1491.
3. Raoul JL, Boucher E, Rolland Y, et al. Treatment of hepatocellular carcinoma with intra-arterial injection of radionuclides. Nat Rev Gastroenterol Hepatol 2010;7 (1):41–49.
4. Bipat S, van Leeuwen MS, Ijzermans JN, et al. Evidence-base guideline on management of colorectal liver metastases in the Netherlands. Neth J Med 2007;65 (1):5–14.
5. Ferlay J, Autier P, Boniol M, et al. Estimates of the cancer incidence and mortality in Europe in 2006. Ann Oncol 2007;18 (3):581–592.
6. Weiss L, Grundmann E, Torhorst J, et al. Haematogenous metastatic patterns in colonic carcinoma: an analysis of 1541 necropsies. J Pathol 1986;150 (3):195–203.
7. McMillan DC, McArdle CS. Epidemiology of colorectal liver metastases. Surg Oncol 2007;16 (1):3–5.
8. Poston GJ. Surgical strategies for colorectal liver metastases. Surg Oncol 2004;13 (2–3):125–136.
9. Kuebler JP. Radioembolization of liver metastases in patients with colorectal cancer: a nonsurgical treatment with combined modality potential. J Clin Oncol 2009;27 (25):4041–4042.
10. Stubbs RS, O'Brien I, Correia MM. Selective internal radiation therapy with 90Y microspheres for colorectal liver metastases: single-centre experience with 100 patients. ANZ J Surg 2006;76 (8):696–703.
11. Taylor I, Bennett R, Sherriff S. The blood supply of colorectal liver metastases. Br J Cancer 1978;38 (6):749–756.
12. Kobayashi H, Hidaka H, Kajiya Y, et al. Treatment of hepatocellular carcinoma by transarterial injection of anticancer agents in iodized oil suspension or of radioactive iodized oil solution. Acta Radiol Diagn (Stockh) 1986;27 (2):139–147.
13. Koito K, Namieno T, Hirokawa N, et al. Enhanced sonography using carbon dioxide gas for small hepatocellular carcinoma: a comparison study between pure carbon dioxide gas and carbon dioxide microbubbles. Radiat Med 2005;23 (2):104–110.
14. Campbell AM, Bailey IH, Burton MA. Analysis of the distribution of intra-arterial microspheres in human liver following hepatic yttrium-90 microsphere therapy. Phys Med Biol 2000;45 (4):1023–1033.
15. Salem R, Lewandowski RJ, Sato KT, et al. Technical aspects of radioembolization with 90Y microspheres. Tech Vasc Interv Radiol 2007;10 (1):12–29.
16. Tehranipour N, AL Nahhas A, Canelo R, et al. Concordant F-18 FDG PET and Y-90 Bremsstrahlung scans depict selective delivery of Y-90-microspheres to liver tumors: confirmation with histopathology. Clin Nucl Med 2007;32(5):371–374.
17. Liu DM, Salem R, Bui JT, et al. Angiographic considerations in patients undergoing liver-directed therapy. J Vasc Interv Radiol 2005;16 (7):911–935.
18. Hamami ME, Poeppel TD, Muller S, et al. SPECT/CT with 99mTc-MAA in radioembolization with 90Y microspheres in patients with hepatocellular cancer. J Nucl Med 2009;50 (5):688–692.

19. Murthy R, Nunez R, Szklaruk J, et al. Yttrium-90 microsphere therapy for hepatic malignancy: devices, indications, technical considerations, and potential complications. Radiographics 2005;25 Suppl 1:S41–S55.
20. Leung TW, Lau WY, Ho SK, et al. Radiation pneumonitis after selective internal radiation treatment with intraarterial 90yttrium-microspheres for inoperable hepatic tumors. Int J Radiat Oncol Biol Phys 1995;33 (4):919–924.
21. SIRTeX. Available at http://www.sirtex.com/. Accessed 2013 Apr 22.
22. Van Hoe L, Van Cutsem E, Vergote I, et al. Size quantification of liver metastases in patients undergoing cancer treatment: reproducibility of one-, two-, and three-dimensional measurements determined with spiral CT. Radiology 1997;202 (3):671–675.
23. Lewandowski RJ, Sato KT, Atassi B, et al. Radioembolization with 90Y microspheres: angiographic and technical considerations. Cardiovasc Intervent Radiol 2007;30 (4): 571–592.
24. Khodjibekova M, Szyszko T, Singh A, et al. Treatment of primary and secondary liver tumours with selective internal radiation therapy. J Exp Clin Cancer Res 2007;26 (4): 561–570.
25. Ahmadzadehfar H, Sabet A, Brockmann H, et al. Significance of Bremsstrahlung scans in patients treated by selective internal radiation therapy (SIRT). J Nucl Med 2008;49 Suppl 1:344.
26. Gates VL, Esmail AA, Marshall K, et al. Internal pair production of 90Y permits hepatic localization of microspheres using routine PET: proof of concept. J Nucl Med 2011;52 (1):72–76.
27. Lhommel R, van Elmbt L, Goffette P, et al. Feasibility of 90Y TOF PET-based dosimetry in liver metastasis therapy using SIR-Spheres. Eur J Nucl Med Mol Imaging 2010;37 (9):1654–1662.
28. Werner MK, Brechtel K, Beyer T, et al. PET/CT for the assessment and quantification of (90)Y biodistribution after selective internal radiotherapy (SIRT) of liver metastases. Eur J Nucl Med Mol Imaging 2010;37 (2):407–408.
29. Szyszko T, AL Nahhas A, Tait P, et al. Management and prevention of adverse effects related to treatment of liver tumours with 90Y microspheres. Nucl Med Commun 2007;28(1):21–24.
30. Goin JE, Salem R, Carr BI, et al. Treatment of unresectable hepatocellular carcinoma with intrahepatic yttrium 90 microspheres: factors associated with liver toxicities. J Vasc Interv Radiol 2005;16 (2 Pt 1):205–213.
31. Therasse P, Arbuck SG, Eisenhauer EA, et al. New guidelines to evaluate the response to treatment in solid tumors. European Organization for Research and Treatment of Cancer, National Cancer Institute of the United States, National Cancer Institute of Canada. J Natl Cancer Inst 2000;92 (3):205–216.
32. Escudier B, Eisen T, Stadler WM, et al. Sorafenib in advanced clear-cell renal-cell carcinoma. N Engl J Med 2007;356 (2):125–134.
33. Curley SA, Izzo F, Delrio P, et al. Radiofrequency ablation of unresectable primary and metastatic hepatic malignancies: results in 123 patients. Ann Surg 1999;230 (1):1–8.
34. Wong CY, Salem R, Qing F, et al. Metabolic response after intraarterial 90Y-glass microsphere treatment for colorectal liver metastases: comparison of quantitative and visual analyses by 18F-FDG PET. J Nucl Med 2004;45 (11):1892–1897.
35. Marn CS, Andrews JC, Francis IR, et al. Hepatic parenchymal changes after intraarterial Y-90 therapy: CT findings. Radiology 1993;187 (1):125–128.

36. Sharma RA, Van Hazel GA, Morgan B, et al. Radioembolization of liver metastases from colorectal cancer using yttrium-90 microspheres with concomitant systemic oxaliplatin, fluorouracil, and leucovorin chemotherapy. J Clin Oncol 2007;25 (9): 1099–1106.
37. Choi H, Charnsangavej C, Faria SC, et al. Correlation of computed tomography and positron emission tomography in patients with metastatic gastrointestinal stromal tumor treated at a single institution with imatinib mesylate: proposal of new computed tomography response criteria. J Clin Oncol 2007;25 (13):1753–1759.
38. Benjamin RS, Choi H, Macapinlac HA, et al. We should desist using RECIST, at least in GIST. J Clin Oncol 2007;25 (13):1760–1764.
39. Bruix J, Sherman M, Llovet JM, et al. Clinical management of hepatocellular carcinoma. Conclusions of the Barcelona-2000 EASL conference. European Association for the Study of the Liver. J Hepatol 2001;35 (3):421–430.
40. Tofts PS, Brix G, Buckley DL, et al. Estimating kinetic parameters from dynamic contrast-enhanced T(1)-weighted MRI of a diffusable tracer: standardized quantities and symbols. J Magn Reson Imaging 1999;10 (3):223–232.
41. Wolchok JD, Hoos A, O'Day S, et al. Guidelines for the evaluation of immune therapy activity in solid tumors: immune-related response criteria. Clin Cancer Res 2009;15 (23): 7412–7420.
42. Goh V, Padhani AR. Imaging tumor angiogenesis: functional assessment using MDCT or MRI? Abdom Imaging 2006;31 (2):194–199.
43. Morgan B, Kennedy AS, Lewington V, et al. Intra-arterial brachytherapy of hepatic malignancies: watch the flow. Nat Rev Clin Oncol 2011;8 (2):115–120.
44. Leen E, Kumar S, Khan SA, et al. Contrast-enhanced 3D ultrasound in the radiofrequency ablation of liver tumors. World J Gastroenterol 2009;15 (3):289–299.
45. Wong CY, Salem R, Raman S, et al. Evaluating 90Y-glass microsphere treatment response of unresectable colorectal liver metastases by [18F]FDG PET: a comparison with CT or MRI. Eur J Nucl Med Mol Imaging 2002;29 (6):815–820.
46. Szyszko T, AL Nahhas A, Canelo R, et al. Assessment of response to treatment of unresectable liver tumours with 90Y microspheres: value of FDG PET versus computed tomography. Nucl Med Commun 2007;28(1):15–20.
47. Kennedy AS, Coldwell D, Nutting C, et al. Resin 90Y-microsphere brachytherapy for unresectable colorectal liver metastases: modern USA experience. Int J Radiat Oncol Biol Phys 2006;65 (2):412–425.
48. Cosimelli M, Golfieri R, Cagol PP, et al. Multi-centre phase II clinical trial of yttrium-90 resin microspheres alone in unresectable, chemotherapy refractory colorectal liver metastases. Br J Cancer 2010;103 (3):324–331.
49. Salem R, Lewandowski RJ, Kulik L, et al. Radioembolization results in longer time to progression and reduced toxicity compared with chemoembolization in patients with hepatocellular carcinoma. Gastroenterology 2011;140 (2):497–507.
50. Kulik LM, Carr BI, Mulcahy MF, et al. Safety and efficacy of 90Y radiotherapy for hepatocellular carcinoma with and without portal vein thrombosis. Hepatology 2008;47 (1):71–81.
51. Inarrairaegui M, Thurston KG, Bilbao JI, et al. Radioembolization with use of yttrium-90 resin microspheres in patients with hepatocellular carcinoma and portal vein thrombosis. J Vasc Interv Radiol 2010;21 (8):1205–1212.

52. Steel J, Baum A, Carr B. Quality of life in patients diagnosed with primary hepatocellular carcinoma: hepatic arterial infusion of Cisplatin versus 90-Yttrium microspheres (Therasphere). Psychooncology 2004;13 (2):73–79.

53. Hendlisz A, Van den EM, Peeters M, et al. Phase III trial comparing protracted intravenous fluorouracil infusion alone or with yttrium-90 resin microspheres radioembolization for liver-limited metastatic colorectal cancer refractory to standard chemotherapy. J Clin Oncol 2010;28 (23):3687–3694.

CHAPTER 19

DIRECT VISUALIZATION OF THERAPEUTIC RADIONUCLIDE DISTRIBUTION USING NUCLEAR MEDICINE IMAGING: I-131 IN METASTATIC THYROID CARCINOMA

HO YOUNG LEE, JUNE-KEY CHUNG, SO WON OH, JONG JIN LEE, KEON WOOK KANG, DO JOON PARK, BO YOUN CHO, AND MYUNG CHUL LEE

19.1 INTRODUCTION

Several radionuclides that emit beta ray or alpha particle have been used for radionuclide therapy. In these cases, gamma-emitting radionuclides can visualize the distribution and pharmacokinetics of the agents easily.

One typical example is I-131. In 1915, specific accumulation of iodide in the thyroid gland was found and immediately adapted to clinical medicine. After then, several radioiodines have been widely used in diagnosis and treatment of thyroid disorders including thyroid cancer. Among them, I-131 has been used as a treatment of choice in metastatic thyroid cancer during more than 60 years. I-131 emits beta ray, which breaks DNA double helix inducing apoptosis of cancer cells. In addition, I-131 emits gamma ray, which makes image of distribution of this radionuclide directly.

19.2 BASIC MECHANISM OF ACTIVE IODINE TRANSPORT

The mechanism of iodine specificity was determined recently. Iodide enters thyroid cells through a specific transporter with sodium, so-called sodium/iodide symporter (NIS). The driving force for iodide uptake is the transmembrane concentration gradient of sodium ion, which is generated and maintained by the sodium–potassium pump (ATPase). This transporter cotransports sodium ion and iodide ion, so named sodium/iodide symporter. NIS gene was identified in 1996 [1].

Drug Delivery Applications of Noninvasive Imaging: Validation from Biodistribution to Sites of Action, First Edition. Edited by Chun Li and Mei Tian.

Small amount of NIS is present in thyroid carcinoma cells, which causes radioiodine uptake in metastatic lesions from differentiated thyroid cancer. Due to NIS in the thyroid carcinoma cells, radioiodine could have a role in the diagnosis and treatment of differentiated thyroid cancer (DTC). However, iodine uptake is impaired in some thyroid carcinoma. Approximately one-third of DTCs and all anaplastic thyroid cancers do not concentrate radioiodine [2]. Furthermore, they have poor prognosis. Especially, the mean survival of patients with anaplastic carcinoma is only 2–6 months [3, 4].

The incidence of thyroid carcinoma is rapidly increasing in many countries including Korea, and thyroid carcinoma has become one of the most common cancers [5]. Most thyroid cancers are DTCs such as papillary thyroid carcinoma (PTC) and follicular thyroid carcinoma (FTC). DTC is characterized by slow growth and a good prognosis, and its 10-year survival rate is about 85–95% [5–8]. Improvements in diagnosis and treatment using radioiodine have reduced the death rates due to thyroid carcinoma [2, 8].

19.3 PREPARATION FOR RADIOACTIVE IODINE WHOLE BODY IMAGING AND THERAPY

I-131 scan and therapy were performed under a strict low-iodine diet and standard thyroid hormone withdrawal protocol. Each patient was required to hold levothyroxine supplementation 4 weeks before I-131 therapy and change thyroid hormone agent to triiodothyronine. After 2 weeks of triiodothyronine administration, all forms of thyroid hormone were prohibited 2 weeks before I-131 therapy [2].

However, this procedure induces transient hypothyroidism, and some patients are suffered from side effects. Sometimes it stimulates the growth of tumor mass. Intramuscular injection of rhTSH (0.9 mg two times) is alternative in these conditions.

At the same time, the patient was advised to have a low-iodine diet, which had lasted until I-131 therapy completed, for at least 2 weeks [2, 9].

19.4 IMAGE ACQUISITION

For the detection of recurrence and metastases, subsequent follow-up I-131 whole-body scans (WBS) are usually performed in regular interval with the measurement of serum thyroglobulin after total thyroidectomy. However, I-131 scan has two main weaknesses named stunning and low sensitivity. When I-131 more than 3 mCi is used for diagnostic scan, subsequent therapeutic dose of I-131 cannot sufficiently enter the cancer cells due to the stunning effect of previous beta ray from I-131. Although radioiodines are specific tracers to image local recurrence and distant metastases, sensitivity is rather low and depends on the radioactivities administered.

The advantages of I-123 are favorable gamma ray (159 keV) for imaging and the lack of beta energy, resulting in the lack of stunning. However, it is somewhat expensive and not available in some places.

19.5 NORMAL FINDINGS

Administered radioiodines accumulate in the lacrimal gland, salivary gland, stomach, and urinary system. Due to these physiologies, radioiodine scan shows normally positive uptake in nasolacrimal area, oral cavity, stomach and GI tracts, and urinary bladder.

False positive may occur. Breast, inflammatory lung lesions, tumors such as stomach and ovary have been reported as the causes of false accumulation of I-131.

19.6 EXPERIENCE AT SEOUL NATIONAL UNIVERSITY HOSPITAL

Historically, radioiodine treatment was started in Seoul National University Hospital. "Radioisotope Clinic" was opened in 1960 in this hospital and has played a central role in nuclear medicine in Korea. We tried to summarize the data of I-131 treatment on recurrence/metastasis of DTC in Seoul National University Hospital.

During the last 20 years, 2036 patients with thyroid carcinoma were treated in Seoul National University Hospital. 1873 patients had PTC (92%), and 163 had FTC (8%). In 2036 patients, 468 (22.9%) had lymph node metastasis and/or distant metastasis on I-131 post-therapy WBS. In detail, 313 patients (15.4%) had regional lymph node metastasis, 83 mediastinal lymph node metastasis (4.1%), 109 pulmonary metastasis (5.3%), and 25 had bone metastasis (1.2%). In PTC, 92 of 1873 patients (4.9%) and 17 of 163 patients (10.4%) in follicular cancer developed pulmonary metastasis, and 13 PTC patients (0.7%) and 12 FTC patients (7.4%) developed bone metastasis. Even though patient had metastasis, the choice of treatment should be I-131 therapy.

Patients with pulmonary metastasis underwent several courses of I-131 therapy (range: 3.7–7.4 GBq) over 40 months (range: 6–171 months). Patients underwent treatment on average 5.1 times (range: 114 times). During follow-up, pulmonary metastasis completely disappeared in 38 patients of 109 patients (34.9%). With pulmonary metastasis, 45 patients (41.3%) had a diffuse pattern of I-131 lung uptake on WBS. Combined nodular and diffuse uptake was observed in 35 patients (32.1%), and nodular uptake without diffuse uptake in 29 patients. When response to I-131 therapy was analyzed by pattern of I-131 uptake by pulmonary metastatic lesions, lesions with diffuse uptake showed better response compared to lesions with nodular uptake ($p<0.05$, Table 19.1). Figure 19.1 shows WBS of a patient after the first and third I-131 therapy who had complete response.

In 25 patients with bone metastasis from a DTC, 16 (64%) had multiple bone metastases in vertebrae and pelvic bones. A total of 105 bone lesions were detected on post-therapy scans; 40 in vertebra, 24 in pelvic bones, 12 in femurs, 10 in skulls, and others were detected in sternum, ribs, or clavicles. Of the 105 bone lesions, 75 were treated with I-131 alone. Three metastatic lesions in vertebrae and two lesions in skulls were treated with I-131 and external radiotherapy (Table 19.2). Twenty-five metastatic bone lesions were surgically removed. In six cases with a pathological fracture, metastatic lesions were removed before

TABLE 19.1 Response to I-131 Treatment According to I-131 Whole Body Scan Finding

	Patterns of I-131 WBS		
Response to RIT	Diffuse	Diffuse+nodular	Nodular
Complete remission	23	12	3
Partial remission	20	16	8
No change	1	6	9
Progression	1	1	9

(a) (b)

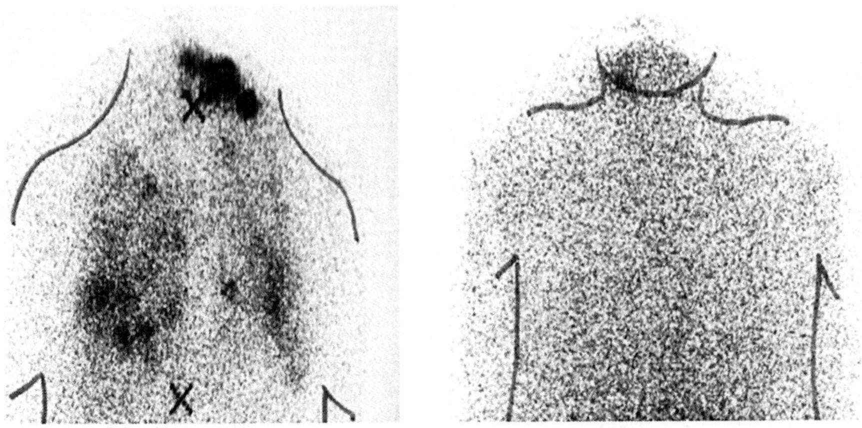

FIGURE 19.1 A 62-year-old male patient with postoperative papillary carcinoma. (a) The chest view of WBS after the first I-131 therapy showed multiple regional lymph nodes and lung metastases. (b) WBS after three times I-131 therapy (200 mCi) reached the complete response.

TABLE 19.2 Results of I-131 Treatment for Bone Metastasis

Treatment modality		Changes in I-131 WBS: Disappeared	Improved	No change or progressed	Total
I-131 only		0	34	41	*75*
I-131 with other treatment	Op+I-131	7	12	0	*19*
	Op+external RT+I-131	2	4	0	*6*
	External RT+I-131	0	5	0	*5*
Total		*9*	*55*	*41*	*105*

RT, radiation therapy.

I-131 therapy, and in eight cases with metastasis to the skull and sternum, resection was performed due to persistent bone pain, which was aggravated during I-131 therapy.

In 75 of the 105 bone lesions treated with I-131 alone, 34 lesions (45.3%) improved, and 41 lesions showed no response or were aggravated. In particular, 9 of 25 bone lesions treated by surgical resection and I-131 therapy completely disappeared. The other 16 lesions showed I-131 accumulation after combined resection and remained uncured after repeated I-131 treatment and radiotherapy. Five lesions treated with combined radiotherapy and I-131 achieved partial improvements.

Although metastatic lesion was detected in patients with DTC, I-131 therapy has a key role in the treatment of metastasis. Especially, the treatment to bone metastasis needs combination therapy with surgical resection or external radiation therapy [10–12].

19.7 DOSIMETRIC APPLICATION OF I-124

Several approaches for the selection of a therapeutic dose of radioiodine can be broadly classified into two groups as either (i) empiric fixed doses or (ii) dosimetrically determined doses [13, 14]. Given the heterogeneity of thyroid cancer patients, empiric dosing may not be associated with optimal results in all patients. One of the most frequently used sets of empiric fixed doses was proposed by Beierwaltes [15] in the range of 5.55–7.4 GBq (150–200 mCi). In case of dosimetric approach, traditional methods are somewhat complicated, resulting in few clinical applications. Some works have been developed to determine absorbed dose distribution of I-131 therapy using I-124 PET images [13, 16]. This capability of estimating I-131 absorbed dose can lead to patient-specific treatment planning for therapy.

REFERENCES

1. Smanik PA, Liu Q, Furminger TL, et al. Cloning of the human sodium iodide symporter. Biochem Biophys Res Commun 1996;226:339–345.
2. Choi CW, Moon DH, Lee MC, et al. Clinical study on thyroid cancer. Korean J Nucl Med 1986;20:59–65.
3. Haq M, Harmer C. Differentiated thyroid carcinoma with distant metastases at presentation: prognostic factors and outcome. Clin Endocrinol 2005;63:87–93.
4. El M, Jhiang SM. Long-term impact of initial surgical and medical therapy on papillary and follicular thyroid cancer. Am J Med 1994;97:418–428.
5. Ministry of Health and Welfare. 2002 Annual Report of the Korea Central Cancer Registry. 2003.
6. Machens A, Holzhausen HJ, Dralle H. The prognostic value of primary tumor size in papillary and follicular thyroid carcinoma. Cancer 2005;203:2269–2273.
7. Zeiger MA, Dackiw PB. Follicular Thyroid lesions, element that affect both diagnosis and prognosis. J Surg Oncol 2005;89:108–113.

8. Fernandes JK, Day TA, Richardson MS, et al. Overview of the management of differentiated thyroid cancer. Curr Treat Options Oncol 2005;6:47–57.
9. Kim YK, Chung J-K, Lee DS, et al. Ablation of remnant thyroid tissue with I-131 in well differentiated thyroid cancer after surgery. Korean J Nucl Med 1997;31:339–345.
10. Bernier MO, Leenhardt L, Hoang C, et al. Survival and therapeutic modalities in patients with bone metastases of differentiated thyroid carcinomas. J Clin Endocrinol Metab 2001;86:1568–1573.
11. Stojadinovic A, Shoup M, Ghossein RA, et al. The role of operations for distant metastatic well-differentiated thyroid carcinoma. Surgery 2002;131:636–643.
12. Shoup M, Stojadinovic A, Nissan A, et al. Prognostic indicators of outcomes in patients with distant metastases from differentiated thyroid carcinoma. J Am Coll Surg 2003;197: 191–197.
13. Van Nostrand D, Atkins F, Yeganeh F, et al. Dosimetrically determined doses of radioiodine for the treatment of metastatic thyroid carcinoma. Thyroid 2002;12:121–134.
14. Benua RS, Cicale NR, Sonenberg M, et al. The relation of radioiodine dosimetry to results and complications in the treatment of metastatic thyroid cancer. Am J Roentgenol Radium Ther Nucl Med 1962;87:171–182.
15. Beierwaltes WH. The treatment of thyroid carcinoma with radioactive iodine. Semin Nucl Med 1978;8:79.
16. Macaapinic H, Larson SM, Erdi AK, et al. Radiation dose assessment for I-131 therapy of thyroid cancer using I-124 PET imaging. Clin Positron Imaging 1999;2:41–46.

CHAPTER 20

IMAGING DRUG DELIVERY TO THE CNS USING TRANSLATIONAL POSITRON EMISSION TOMOGRAPHY STUDIES

CHI-MING LEE, KATARINA VARNÄS, AND LARS FARDE

20.1 INTRODUCTION

Drug delivery to the central nervous system (CNS) represents a common challenge in drug development. For drugs acting on targets in CNS, sufficient brain exposure is essential. Conversely, for drugs acting on peripheral targets, it is advisable to minimize brain entry to reduce the risk for CNS side effects. Thus, for efficient drug development, it is crucial to understand the regulation of drug passage across the blood–brain barrier (BBB) and to predict whether the drug candidate is likely to achieve relevant free CNS concentration to produce the desired pharmacological effect.

Positron emission tomography (PET) technology, a noninvasive molecular imaging technology, allows for direct assessment of radiolabeled drug molecules at organ level *in vivo* and has in recent years been increasingly used in early phases of CNS drug discovery for identification of candidate drugs with suitable brain exposure [1, 2].

In clinical pharmacokinetics, two kinetic parameters are commonly used to describe the access of a drug into the brain, namely, the rate and the extent of BBB penetration. These parameters are depicted in a schematic representation in Figure 20.1. The *rate* of brain access describes how fast a drug can be delivered from plasma to brain and is dependent on BBB permeability and brain tissue binding [3]. Higher rate of brain access generally translates into a faster onset of drug action. The *extent* of brain access is defined as the ratio of free brain to free plasma concentration at steady state and provides an estimate of how much of the drug being present in brain at steady-state conditions. For freely diffusible drugs, the *extent* is primarily a function of the unbound drug concentration in plasma and brain, whereas for some

Drug Delivery Applications of Noninvasive Imaging: Validation from Biodistribution to Sites of Action, First Edition. Edited by Chun Li and Mei Tian.

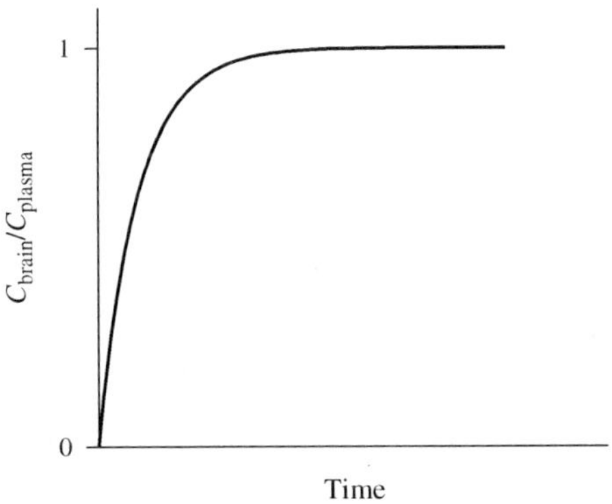

FIGURE 20.1 Schematic representation of the rate and extent of brain distribution of a drug as a function of time. The rate of brain access describes how fast a drug can be delivered from plasma to brain (represented by the slope) and is dependent on BBB permeability and brain tissue binding. The extent of brain access is defined as the ratio of brain to plasma concentration at steady state (i.e., when the influx rate equals the efflux rate).

drugs, active transport processes at the BBB and metabolism in brain also have to be taken into account.

In PET analysis of radiolabeled drugs, it is common to use compartment models to analyze the transfer of drugs across the BBB, where the compartments correspond to drug in brain and plasma, respectively. At time of steady state, the rate of influx equals the rate of efflux of drug from the brain compartment, as described by Equation 20.1:

$$K_1 C_{plasma} = k_2 C_{brain}, \tag{20.1}$$

where K_1 and k_2 correspond to the influx and efflux rate constants, respectively, for drug transfer through the BBB, and C_{plasma} and C_{brain} are concentrations of drug in plasma and brain compartments, respectively. In the early PET literature, which emerged from neurophysiology, the concepts *partition coefficient* or *distribution volume* correspond to K_1/k_2 (or C_{brain}/C_{plasma}) and have frequently been used instead of *extent*. It is thus important to be aware of the differences in nomenclature (see Summerfield et al. [4]; Table 1). Clearly, the field would benefit from common concepts in the future.

The assessment of the *rate* of passage across BBB and *extent* of CNS exposure of a drug candidate can be achieved directly by studying the drug candidate as labeled with a PET radionuclide. PET has been applied to study brain disposition and kinetics of a wide range of therapeutic drugs as well as of novel drugs under development for the treatment of CNS disorders [5–14]. This direct approach is often referred to as *microdosing* since injection of a tracer amount (<1 μg) of radiolabeled drug is generally sufficient to obtain a good signal. An indirect approach providing an estimate of brain exposure is to examine the competition of a drug candidate for binding to the target receptor (occupancy studies). In this chapter, we will review the impact of the BBB on CNS drug delivery and discuss the use of PET imaging to ascertain delivery and brain exposure.

20.2 BLOOD–BRAIN BARRIER (BBB) PHYSIOLOGY AND IMPACT ON DRUG DELIVERY TO THE BRAIN

The BBB is vital since it protects the CNS from many toxins. However, the same mechanism also limits the delivery of other exogenous molecules such as many potential therapeutic drugs. Poor permeability at the BBB is considered to be one of the main factors limiting success of CNS drugs [15, 16].

The absence of fenestrations and minimal pinocytosis at the endothelial cells of the brain's vasculature, together with the presence of tight junctions that seal the paracellular spaces between adjacent endothelial cells, results in a physical barrier that restricts the transfer of hydrophilic solutes from blood to brain [17]. In addition to being a passive barrier to diffusion, the BBB also regulates molecular transfer via active transporters. On the one hand, a number of nutrients and hormones (e.g., glucose, insulin, and most amino acids) are allowed to enter the brain through the assistance of transporters. On the other hand, there are ATP-driven efflux pumps (e.g., P-glycoprotein (P-gp/ABCB1/MDR1)), organic anion transporter (OAT), breast cancer resistance protein (BCRP/ABCG2,) and multidrug resistance-associated proteins (MRP/ABCC1) on the endothelial cells. Importantly, they can all pump out a range of hydrophobic molecules reaching the endothelial cells of the BBB by passive diffusion. This so-called ATP-binding cassette (ABC) of drug efflux transporters can handle a broad range of substrates from small drug molecules to polypeptides such as beta-amyloid (MW = 4200 Da).

For compounds that are not actively transported or that are effluxed by the capillary endothelium, the brain levels can generally be understood by models for passive diffusion. This dependency of diffusion is a reason for the common assumption that a CNS drug has to be sufficiently lipophilic in order to provide the required brain exposure. However, this is not always true! It is important to note that while high lipophilicity can assure passive diffusion across the BBB, it may also decrease brain delivery due to:

1. Increased plasma protein binding resulting in lower free drug concentration and potentially also increased nonspecific binding to proteins in brain.
2. Increased volume of distribution (lower plasma concentration due to peripheral disposition).
3. Decreased solubility (and thus decreased availability in the whole body due to incomplete absorption).
4. Increased metabolic clearance (lower AUC).

Thus, optimizing brain access will require a balancing trade-off among these contributing factors. For instance, there is a certain window of lipophilicity that allows adequate diffusion across biological membranes but not resulting in excessive nonspecific binding or suboptimal free concentration. Since efflux pumping may be overcome when passive diffusion is fast enough, a practical approach to limit the effects of P-gp on a CNS drug candidate is to find drug candidates with rapid passive diffusion [18]. In general, empirical data from medicinal chemistry have shown that passive diffusion can be increased by lowering groups that solvate in water, decreasing hydrogen bonding capacity, adding halogen groups, reducing molecular size, introducing constrained conformation, or promoting intramolecular hydrogen bonds (vide rev 18, 19).

20.3 PET ASSESSMENT OF THE FUNCTION AND REGULATION OF EFFLUX TRANSPORTERS AT THE BBB

The ABC drug efflux transporters can handle a broad range of substrates. Their constitutive presence at the BBB has been shown to restrict brain access of many CNS active drugs, for example, chemotherapeutics, opioids, antibiotics, and anti-HIV drugs. They also impose a formidable challenge to effective scaling to human subjects due to reported species [20] and pharmacogenetic [21] differences. The expression and functional activities of efflux pumps have also been reported to change in some brain diseases [22]. For instance, upregulation of P-gp in refractory epilepsy has been reported in animal models [23] and human subjects [24]. P-gp has been shown to be upregulated in cancer cells, making the tumors resistant to many small chemotherapeutic drugs. Thus, P-gp at the BBB has been postulated as a putative mechanism of drug resistance in brain cancer and refractory epilepsy [25]. However, despite the promising success of P-gp inhibitors in improving drug delivery to the brain in animal studies, no consistent clinically significant effect of P-gp inhibitors in multidrug-resistant cancer and medication-refractory epilepsy has yet been established [25, 26]. The potential reasons for the inconsistent observations include presence of multiple efflux transporters, dose-limiting side effects of existing P-gp inhibitors (such as cyclosporine A and verapamil), lack of suitable P-gp inhibitors with appropriate selectivity, and prevalence of genetic polymorphisms at the ABCB1 gene.

With the advent of radiolabeled probes for P-gp such as [^{11}C]-verapamil and [^{11}C]-N-desmethyl-loperamide, PET can now be used to study the function of P-gp *in vivo*. Work so far include examination of the expression and functional changes in P-gp *in vivo* and how they affect the permeation of P-gp substrates into animal and human brains in relation to aging, ABCB1 genetic polymorphisms, drug resistance, or disease progression [25, 27–31].

Notably, PET studies with [^{11}C]-verapamil indicate lower P-gp function with increasing age [32, 33]. The age-dependent reduction in BBB P-gp has been found to be associated with an increase in deposition of beta-amyloid in human brain [34]. A follow-up study further demonstrated that P-gp expression in elderly human brains was significantly and inversely correlated to the extent of cerebral amyloid angiopathy [35]. Interestingly, P-glycoprotein deficiency at the BBB had also been reported to increase amyloid-β deposition in an Alzheimer's disease (AD) transgenic mouse model [36].

Recent studies have shown that the expression or activity of efflux drug transporters can be regulated by several signaling pathways, e.g., the glutamate-NMDAR-COX2-EP1 signaling pathway in epilepsy, the TNF-α signaling pathway in brain inflammation, the E2-ER-PTEN-PI3K/akt/GSK3 signaling pathway in brain cancer, and the pregnane X receptor (PXR) signaling pathway in AD (rev 37, 38). Interestingly, since restoring P-gp through PXR activation substantially lowered brain level of beta-amyloid in an AD mouse model [39], it is conceivable that restoring P-gp in early AD patients may represent an alternative approach to reduce beta-amyloid.

It should be noted that the upregulation of P-gp via the glutamate-COX2 pathway appear to be a regional, rather than a global, effect [40] and may not be detected easily in studies that measure whole brain uptake of drugs. Thus, PET would be a good tool to study the regional regulation of P-gp. Indeed, a recent PET study has demonstrated a significant decrease of [^{11}C]-verapamil in frontal and temporal brain regions in major depressive patients compared to healthy controls [41]. Whether this regional increase in P-gp function may be related to treatment-resistant depression remains to be studied.

20.4 MEASUREMENT OF DRUG DISPOSITION/DISTRIBUTION BY PET MICRODOSING

PET allows for the detection of concentrations of radiolabeled molecules in tissue at the low picomolar range. The high sensitivity of PET thereby enables the study of drug pharmacokinetics after administration of subpharmacological doses (microdoses) without the requirement of the extensive safety evaluations associated with therapeutic dosing of the drug [42]. How the PET microdosing technique can be used for assessment of the brain exposure of novel drug candidates can be illustrated by two PET studies in which potential drug candidates were labeled with carbon-11 and administered intravenously in nonhuman primates (Fig. 20.2). The first example (left panel) shows an image of the averaged radioactivity and the curve for brain radioactivity versus time for a compound intended for the treatment of a CNS disorder. The second example (right panel) shows corresponding information for a drug candidate intended to target peripheral binding sites. The aim of the latter project was to achieve low brain exposure to limit undesirable CNS effects. The PET microdosing studies were undertaken to confirm low brain concentration of the drug candidate after intravenous administration.

In a PET microdosing study, it is not possible to distinguish parent radioligand from radioactive metabolites having sufficient lipophilicity to pass the BBB. Hence, the presence of radioactive metabolites in brain tissue will contribute to the total signal detected by the system. Characterization of the metabolism of the radiolabeled drug is therefore of crucial importance for the interpretation of results from a microdosing experiment [43].

An underlying assumption of the microdosing concept is that pharmacokinetic parameters derived after administration of microdoses of a drug can be extrapolated to the far higher pharmacologic doses. The limited number of studies thus far published support the utility of microdosing for predicting plasma pharmacokinetics at pharmacologic doses [44, 45]. Brain pharmacokinetic dose linearity has indeed been confirmed in a number of PET studies in human subjects examined after administration of microdoses of radioactive drug and after administration of microdoses in combination with a higher dose of nonradioactive drug [1, 2, 46]. It has moreover been shown for the radioligand [^{11}C]raclopride that the ratio of brain and blood radioactivity is similar at microdoses and pharmacologic doses [47]. A study comparing brain pharmacokinetics of [^{11}C]verapamil at microdose and

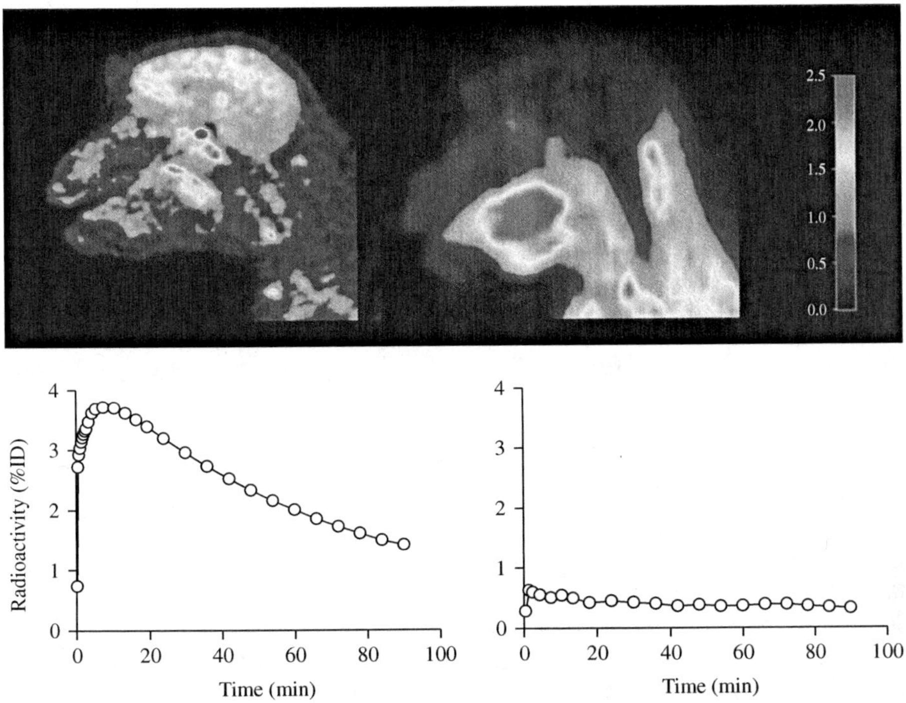

FIGURE 20.2 PET studies using two ^{11}C-labeled drug candidates injected as intravenous microdoses in cynomolgus monkeys. Left: Drug candidate targeting the central nervous system. Right: Drug candidate targeting peripheral tissues. Upper panel: Average PET images from 9 to 93 min after intravenous injection. For color details, please see color plate section. Lower panel: Curves for brain radioactivity, expressed as percentage of the total injected radioactivity (%ID), versus time.

pharmacologic dose showed no statistically significant differences in the influx and efflux rate constants between the two conditions [48]. Another PET study with [^{11}C]-labeled BMS-181101, a drug under development for the treatment of depression, showed similar brain and plasma kinetic properties after administration of [^{11}C]BMS-181101 at high and low specific radioactivity, respectively [49]. The relationship between brain and plasma exposure could, however, be expected to be nonlinear if plasma protein binding is saturable or if transfer across the BBB is regulated by transport mechanisms that can be saturated at higher doses. For instance, in a recent PET study with the P-gp inhibitor [^{11}C]tariquidar in rats, administration of a high dose of tariquidar increased brain radioactivity concentration without concomitant increases in blood radioactivity [50]. Further PET investigations of additional therapeutic drugs are therefore required to confirm dose linearity of brain pharmacokinetics.

The time curves for the brain radioactivity concentration are conventionally expressed relatively to the total amount of radioactivity injected, or as the standardized uptake value (SUV), defined as the local radioactivity per gram tissue divided by the administered radioactivity per gram body weight, where an SUV of 1 correspond to the concentration that would have resulted from an even distribution of radioactivity within the body. The majority of therapeutic CNS drugs examined with PET have shown SUVs or percentages of injected radioactivity in brain above 1 [8, 10, 12, 14]. A useful aspect of the microdosing technique is that it allows the derivation of brain pharmacokinetic parameters from the PET data. Brain pharmacokinetic measures may be used in relation to the plasma exposure and for comparisons among drug candidates. For instance, an early PET study comparing brain disposition and pharmacokinetics of ^{11}C-labeled opiates in brain of rhesus monkeys [10] showed differences in the maximum radioactivity and time to reach maximum radioactivity derived from ^{11}C-labeled morphine, codeine, heroin, and pethidine, respectively. Differences in rate and extent of brain access among these drugs were consistent with the rank order of lipophilicity, although other factors such as plasma protein and tissue binding and active efflux mechanisms [51–53] could have contributed to these differences.

A full quantitative analysis of drug delivery to the brain requires arterial blood sampling and detection of radioactive metabolites to generate the input function corresponding to the concentration of unchanged (parent) radioligand in plasma. The signal detected by the PET system consists of the summed radioactivity in brain tissue and in cerebral vasculature. If concentration of the radiolabeled drug molecule in brain is low relative to that in blood, radioactivity in cerebral vasculature will significantly contribute to the PET signal. Radioligand concentration in brain tissue ($C_T(t)$) can be estimated from radioactivity concentration in arterial blood using the following equation:

$$C_{\text{PET}}(t) = (1 - V_{\text{b}})C_{\text{T}}(t) + V_{\text{b}}\ C_{\text{a}}(t), \tag{20.2}$$

where $C_{\text{PET}}(t)$ corresponds to radioactivity detected by the PET system, V_{b} is the fractional volume of blood in tissue, and $C_{\text{a}}(t)$ is the radioactivity in arterial whole blood. Reference values of V_{b} varies from about 2% in white matter to 4–5% in the neocortex and the cerebellum and can be obtained from estimates of the cerebral blood volume reported in the literature [54–56].

The *rate* of entry of total (bound and unbound) radiolabeled drug probe to the CNS can be described in terms of the influx and efflux rate constants, K_1 and k_2, respectively, which can be derived by kinetic compartment analysis (57–59; see also Chapter 2 for PK modeling using PET/SPECT imaging). A PET compartment analysis uses the arterial plasma concentration as an input function to model tracer kinetics in a number of tissue compartments. The tissue compartments represent kinetically separate pools of tracer, commonly the free tracer in tissue, nonspecifically bound tracer, and tracer specifically bound to target sites (e.g., neuroreceptors, transporters, or enzymes). K_1 and k_2 describe the rate of transfer between plasma and the free brain tissue compartment. A common observation is that due to rapid equilibration between the free and nonspecifically bound tracer, kinetics in these compartments cannot be distinguished. These compartments are therefore commonly approximated by one single compartment [57]. A set of differential equations can be derived to describe the exchange of drug molecules between the different compartments. Thus, for the one-tissue compartment model, the rate of change in tracer concentration is defined by Equation 20.3:

$$\frac{dC_T(t)}{dt} = K_1 C_P(t) - k_2 C_T(t) \tag{20.3}$$

where C_T and C_P correspond to radioligand concentration in tissue and plasma, respectively. (Note that Eq. 20.1 given in the introduction corresponds to Eq. 20.3 at steady-state conditions.)

According to the model developed by Renkin [60] and Crone [61], K_1 can be expressed in terms of cerebral blood flow (F) and extraction fraction (E):

$$K_1 = FE \tag{20.4}$$

$$E = 1 - e^{(-PS/F)} \tag{20.5}$$

where *PS* is the permeability-surface area product. *PS* is of central interest for the imaging of drug delivery to the CNS as this parameter is a measure of BBB permeability. Thus, in PET studies using the P-gp radiotracers [^{11}C]verapamil and [^{11}C] N-desmethyl-loperamide, respectively, the influx constant K_1 [62–64] and the *PS* [63] have been shown to increase after pharmacologic inhibition of P-gp.

The *extent* of brain access is defined as the free brain-to-plasma partition coefficient at steady state. In PET studies, the total brain-to-plasma partition coefficient (termed the total volume of distribution) can be directly calculated by employing an infusion protocol to achieve steady-state tracer concentrations in brain and plasma [65]. In most PET studies, the tracer is, however, administered as a bolus injection. Under such conditions, the brain-to-plasma ratio may differ from that at steady state because of different clearance rates in brain tissue and plasma [10, 65]. Although steady state is not reached in a bolus injection study, the volume of distribution can be modeled using compartment analysis [57–59], or calculated using the Logan linear graphic method [66]. The nondisplaceable volume of distribution (V_{ND}; 67) corresponds to the ratio at steady state of the tracer concentration in the free and nonspecifically bound compartment (C_{ND}) to the plasma tracer concentration, and can be estimated for a region devoid of specific binding sites [9, 54, 68].

As the total brain-to-plasma partition coefficient does not take into account the binding of radioligand to brain tissue and plasma proteins, the ratio of unbound drug in brain to plasma is considered a more accurate predictor of CNS drug activity since only the free molecules can bind to the intended target (vide rev 69, 70).

The nondisplaceable volume of distribution can be expressed in terms of the concentration of free drug in brain (C_{FT}) and plasma (C_{FP}) as

$$V_{ND} = \frac{C_{ND}}{C_P} = \frac{f_P C_{FT}}{f_{ND} C_{FP}} \tag{20.6}$$

where f_P and f_{ND} correspond to the fractions of unbound drug in plasma and in tissue water, respectively [4, 67]. From Equation 20.6, it follows that the ratio of free brain and plasma concentrations can be estimated from V_{ND} and the brain and plasma free fractions [4, 9]. The parameter f_P can be estimated from measurements of the plasma protein binding of the radioactive drug by equilibrium dialysis *in vitro*. The parameter

f_{ND} cannot be measured with PET, but estimates of this parameter may be obtained from *in vitro* measurements, including equilibrium dialysis of the compound between buffer and brain homogenates [71] or brain slices [72]. Although free drug concentration in brain can be measured by microdialysis *in vivo*, this technology is too invasive for translational studies in human studies. Preclinical microdialysis studies have shown that free blood drug concentrations are sometimes different from the free drug concentration in brain [73] and thus cannot be used as a surrogate of free brain concentration.

Importantly, the C_{FT}/C_{FP} ratio provides information of the distributional impairment over the BBB since under freely diffusible steady-state conditions, this value should be close to unity. If the radioactive drug is a substrate of P-gp or other efflux proteins, C_{FT}/C_{FP} will be smaller than unity (i.e., lower free drug concentration in the brain), and if the drug is actively transported into the brain, this value will be larger than unity [69, 74].

20.5 IMAGING DRUG DELIVERY TO THE CNS BY PET TARGET ENGAGEMENT STUDIES

Target occupancy studies can provide an indirect estimate of the free concentration at target sites by comparison of drug target affinity values derived from *in vitro* and *in vivo* measurements. PET receptor occupancy or target engagement studies allow for the evaluation of whether the drug candidate enters into the CNS and also whether it reaches and binds to specific sites. This approach requires the use of an established radioligand to evaluate the degree of drug-induced occupancy after administration in therapeutic doses. In early drug discovery, this methodology can aid the screening of drug candidates suitable for further development. At the late stage of drug development, PET target engagement studies enable the selection of doses required for the aimed level of occupancy at the receptor in clinical trials.

Although PET receptor occupancy studies are frequently used in drug development, this approach is, however, restricted to drug candidates targeting well-established binding sites that can be studied using available PET radioligands. Of crucial importance in the field of neuropharmacology research is therefore the development of suitable PET radioligands for new CNS drug targets of interest. For instance, in a collaboration between Karolinska Institutet and AstraZeneca, the serotonin 5-HT_{1B} receptor selective radioligand [^{11}C]AZ10419369 has been recently developed [75]. This radioligand has been subsequently used in nonhuman primates and human subjects to study 5-HT_{1B} receptor occupancy with a selective novel 5-HT_{1B} receptor antagonist, AZD3783, under development for the treatment of depression. Using this experimental paradigm, it was demonstrated that AZD3783 enters into the brain and binds to central 5-HT_{1B} receptors with similar affinity in nonhuman primates and human subjects (Fig. 20.3; 76).

In a target occupancy study, multiple PET measurements are undertaken, at baseline and at post-drug administration conditions, respectively. Occupancy is calculated as the percentage reduction in radioligand binding after drug administration as compared to the baseline condition. To derive an estimate of the *in vivo* affinity of

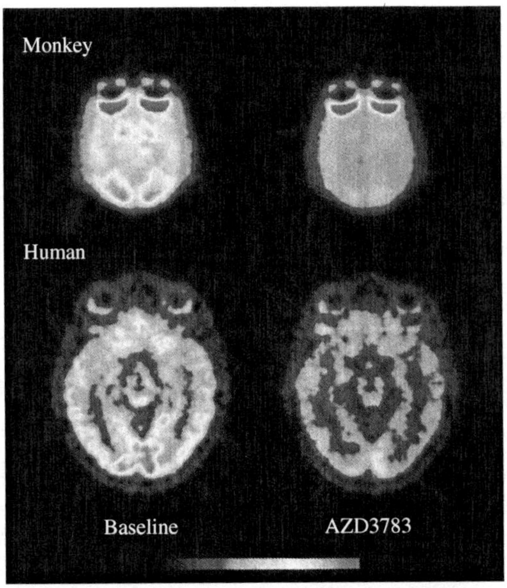

FIGURE 20.3 PET images of [^{11}C]AZ10419369 brain disposition in a cynomolgus monkey (upper panel) and a human subject (lower panel) at baseline (left) and after administration of the 5-HT$_{1B}$ receptor antagonist AZD3783 (right). For color details, please see color plate section.

the drug for the binding site, the relationship between receptor occupancy and drug plasma concentration, C_p, can be described by the following equation:

$$\text{Occupancy} = \frac{\text{Occ}_{\text{max}} \times C_{\text{p}}}{K_{\text{i,plasma}} + C_{\text{p}}}, \tag{20.7}$$

where Occ$_{max}$ is the maximal occupancy induced by the drug and $K_{i,plasma}$ the inhibition constant corresponding to plasma concentration required for half-maximum receptor occupancy [77]. $K_{i,plasma}$ is a useful measure for translation of findings from preclinical models to human condition. The $K_{i,plasma}$ value determined based on the free plasma concentration is of interest for comparison to *in vitro* affinity values since under freely diffusible steady-state conditions, this parameter corresponds to the affinity of the drug for its binding site (K_i) obtained *in vitro*. K_i can be measured from the shift in IC$_{50}$ (50% inhibition) that occurs during competitive inhibition using the Cheng–Prusoff equation [78].

PET has been extensively used to determine the occupancy range required for therapeutic efficacy of CNS drugs. Studies using this methodology have established the occupancy at dopamine D2 receptors required for therapeutic efficacy of antipsychotic drugs (79, 80; reviewed in 81). The availability of radioligands for a number of CNS drug binding sites, such as monoamine reuptake transporters [82–84], the GABA benzodiazepine binding site [85], and the mu-opioid receptor [86], enables the determination of the level of occupancy required for therapeutic efficacy [87–92]. These studies have established that in general antagonists require high

(60–90%) target occupancy, whereas the required occupancy level for agonist drugs may vary, depending on the target system and intrinsic efficacy of the drug [93]. Thus, in the evaluation of compounds interacting with established drug targets, occupancy–efficacy relationships derived from clinical PET studies may serve as guidelines for prediction of the dose range required for the intended pharmacologic effect.

In the study of novel drug targets, the therapeutic level of receptor occupancy is not known. For these targets, occupancy at efficacious doses in animal models can be useful for prediction of occupancy and exposure required for clinical efficacy. To validate the use of preclinical species for prediction of human *in vivo* pharmacology, it is necessary to compare plasma exposure–occupancy profiles across species. A number of translational imaging studies have found a good correspondence between $K_{i,plasma}$ values in preclinical species and humans. For example, it has been shown in rodent PET studies that the $K_{i,plasma}$ value of fluvoxamine at the serotonin transporter (6 ng/ml; Saijo et al. [94]) is similar to that determined in human PET studies (5 ng/ml; Takano et al. [95]). Also PET studies with the 5-HT_{1B} receptor antagonist AZD3783 shows a good correspondence between monkey ($K_{i, plasma}$ = 26 nmol/l) and human ($K_{i, plasma}$ = 21 nmol/l) *in vivo* affinity estimates [76]. Based on reference values for plasma protein binding of AZD3783, the unbound $K_{i, plasma}$ values were estimated to be 19 and 14 nmol/l for monkey and human, respectively, which is consistent with the unbound $K_{i, plasma}$ value determined in guinea pigs (11 nmol/l) and the K_i value obtained *in vitro* (13 nmol/l; Zhang et al. [96]).

Other examples have, however, shown between-species differences in apparent *in vivo* affinities. Thus, for instance, the $GABA_A$ receptor compound TPA023 has been shown to have similar *in vivo* affinity at central benzodiazepine receptor sites in preclinical species, including rats, rhesus monkeys, and baboons ($K_{i,plasma}$ = 19–30 ng/ml) but to be slightly more potent in humans ($K_{i,plasma}$ = 9 ng/ml; Atack et al. [97]). The 5-HT_{1A} receptor antagonist NAD-299 shows about 10-fold higher *in vivo* affinity in cynomolgus monkeys ($K_{i,plasma}$ = 5 nmol/l; Farde et al. [98]) than in human subjects ($K_{i,plasma}$ = 61 nmol/l; Andrée et al. [99]), a difference that could be accounted for by a 10-fold higher free fraction of NAD-299 in monkey than in human [99]. These examples illustrate that, although studies in preclinical species are useful for predicting initial clinical doses, subsequent studies in human subjects are important to confirm the exposure–occupancy relationships derived from preclinical species.

20.6 SUMMARY AND FUTURE PERSPECTIVES

A key challenge in CNS drug discovery and development is to define the effective dose for target engagement in the brain since free drug concentration in blood does not always correspond to free drug concentration in brain due to the unique characteristics of the BBB. This is particularly evident for drugs that are substrates of P-gp and other transporters at the BBB. PET is increasingly being used to measure the biodistribution and brain concentrations of drug candidates in view of its sensitivity, noninvasiveness, and applicability across species, which contrast with microdialysis and CSF measurements that are more limited in human studies due to their invasiveness. In addition to describing the basis for using microdosing and target occupancy studies to enable compound and dose selection in early clinical studies, we have also highlighted the importance of determining the free concentration in brain by supplementing PET microdosing (which measures the total drug concentration) with *in vitro* measurement of free and bound drugs by equilibrium dialysis, or, if a suitable PET ligand is available, to directly measure target engagement in brain using the target occupancy study approach. The current bottleneck is the lack of suitable PET ligands for the many interesting targets which have been identified for various CNS disorders. An ongoing effort to address this bottleneck is to promote the collaboration between academic and industrial scientists to develop and validate PET ligands for CNS targets as exemplified by some of the activities at the U.S. Biomarkers Consortium (www.biomarkersconsortium.org) and the European Innovative Medicines (www.imi.europa.eu). We are optimistic that more useful PET ligands will be available in the future to improve this translational tool for CNS drugs.

ACKNOWLEDGMENTS

We thank the members of the PET center of the Karolinska Institutet and the AstraZeneca PET working group for their stimulating discussions about the topics reviewed in this chapter. In particular, we wish to acknowledge Dr. Khanh Bui for his helpful discussions from a DMPK perspective.

REFERENCES

1. Bergström M, Grahnen A, Långström B. Positron emission tomography microdosing: a new concept with application in tracer and early clinical drug development. Eur J Clin Pharmacol 2003;59:357–366.
2. Lee CM, Farde L. Using positron emission tomography to facilitate CNS drug development. Trends Pharmacol Sci 2006;27:310–316.
3. Liu X, Chen C, Smith BJ. Progress in brain penetration evaluation in drug discovery and development. J Pharmacol Exp Ther 2008;325:349–356.
4. Summerfield SG, Lucas AJ, Porter RA, et al. Toward an improved prediction of human in vivo brain penetration. Xenobiotica 2008;38:1518–1535.
5. Baron JC, Roeda D, Munari C, et al. Brain regional pharmacokinetics of ^{11}C-labeled diphenylhydantoin: positron emission tomography in humans. Neurology 1983;33:580–585.
6. Bauer M, Langer O, Dal-Bianco P, et al. A positron emission tomography microdosing study with a potential antiamyloid drug in healthy volunteers and patients with Alzheimer's disease. Clin Pharmacol Ther 2006;80:216–227.
7. Bergström M, Yates R, Wall A, et al. Blood-brain barrier penetration of zolmitriptan—modelling of positron emission tomography data. J Pharmacokinet Pharmacodyn 2006;33: 75–91.
8. Gulyás B, Halldin C, Sandell J, et al. PET studies on the brain uptake and regional distribution of [^{11}C]vinpocetine in human subjects. Acta Neurol Scand 2002;106:325–332.
9. Gunn RN, Summerfield SG, Salinas CA, et al. Combining PET biodistribution and equilibrium dialysis assays to assess the free brain concentration and BBB transport of CNS drugs. J Cereb Blood Flow Metab 2012;32:874–883.

10. Hartvig P, Bergström K, Lindberg B, et al. Kinetics of ^{11}C-labeled opiates in the brain of rhesus monkeys. J Pharmacol Exp Ther 1984;230:250–255.
11. Ito H, Nyberg S, Halldin C, et al. PET imaging of central 5-HT_{2A} receptors with carbon-11-MDL 100,907. J Nucl Med 1998;39:208–214.
12. Lundberg T, Lindström LH, Hartvig P, et al. Striatal and frontal cortex binding of 11-C-labelled clozapine visualized by positron emission tomography (PET) in drug-free schizophrenics and healthy volunteers. Psychopharmacology (Berl) 1989;99:8–12.
13. Neu H, Hartvig P, Torstenson R, et al. Synthesis of [^{11}C-methyl]-(-)-OSU6162, its regional brain distribution and some pharmacological effects of (-)-OSU6162 on the dopaminergic system studied in the rhesus monkey by positron emission tomography. Nucl Med Biol 1997;24:507–511.
14. Smith DF, Jensen PN, Gee AD, et al. PET neuroimaging with [^{11}C]venlafaxine: serotonin uptake inhibition, biodistribution and binding in living pig brain. Eur Neuropsychopharmacol 1997;7:195–200.
15. Alavijeh MS, Chishty M, Qaiser MZ, et al. Drug metabolism and pharmacokinetics, the blood-brain barrier, and central nervous system drug discovery. NeuroRx 2005;2:554–571.
16. Pardridge WM. The blood-brain barrier: bottleneck in brain drug development. NeuroRx 2005;2:3–14.
17. de Boer AG, Gaillard PJ. Strategies to improve drug delivery across the blood-brain barrier. Clin Pharmacokinet 2007;46:553–576.
18. Raub TJ. P-glycoprotein recognition of substrates and circumvention through rational drug design. Mol Pharm 2006;3:3–25.
19. Reichel A. Addressing central nervous system (CNS) penetration in drug discovery: basics and implications of the evolving new concept. Chem Biodivers 2009;6: 2030–2049.
20. Ohtsuki S, Terasaki T. Contribution of carrier-mediated transport systems to the blood-brain barrier as a supporting and protecting interface for the brain; importance for CNS drug discovery and development. Pharm Res 2007;24:1745–1758.
21. Leschziner GD, Andrew T, Pirmohamed M, et al. ABCB1 genotype and PGP expression, function and therapeutic drug response: a critical review and recommendations for future research. Pharmacogenomics J 2007;7:154–179.
22. Taylor EM. The impact of efflux transporters in the brain on the development of drugs for CNS disorders. Clin Pharmacokinet 2002;41:81–92.
23. Rizzi M, Caccia S, Guiso G, et al. Limbic seizures induce P-glycoprotein in rodent brain: functional implications for pharmacoresistance. J Neurosci 2002;22:5833–5839.
24. Tishler DM, Weinberg KI, Hinton DR, et al. MDR1 gene expression in brain of patients with medically intractable epilepsy. Epilepsia 1995;36:1–6.
25. Loscher W, Potschka H. Drug resistance in brain diseases and the role of drug efflux transporters. Nat Rev Neurosci 2005;6:591–602.
26. Robey RW, Lazarowski A, Bates SE. P-glycoprotein—a clinical target in drug-refractory epilepsy? Mol Pharmacol 2008;73:1343–1346.
27. Elsinga PH, Hendrikse NH, Bart J, et al. PET Studies on P-glycoprotein function in the blood-brain barrier: how it affects uptake and binding of drugs within the CNS. Curr Pharm Des 2004;10:1493–1503.
28. Kortekaas R, Leenders KL, van Oostrom JC, et al. Blood-brain barrier dysfunction in parkinsonian midbrain in vivo. Ann Neurol 2005;57:176–179.
29. Syvänen S, Blomquist G, Sprycha M, et al. Duration and degree of cyclosporin induced P-glycoprotein inhibition in the rat blood-brain barrier can be studied with PET. Neuroimage 2006;32:1134–1141.

30. Takano A, Kusuhara H, Suhara T, et al. Evaluation of in vivo P-glycoprotein function at the blood-brain barrier among MDR1 gene polymorphisms by using ^{11}C-verapamil. J Nucl Med 2006;47:1427–1433.

31. Kannan P, John C, Zoghbi SS, et al. Imaging the function of P-glycoprotein with radiotracers: pharmacokinetics and in vivo applications. Clin Pharmacol Ther 2009;86: 368–377.

32. Toornvliet R, van Berckel BN, Luurtsema G, et al. Effect of age on functional P-glycoprotein in the blood-brain barrier measured by use of (R)-[^{11}C]verapamil and positron emission tomography. Clin Pharmacol Ther 2006;79:540–548.

33. Mangoni AA. The impact of advancing age on P-glycoprotein expression and activity: current knowledge and future directions. Expert Opin Drug Metab Toxicol 2007;3:315–320.

34. Vogelgesang S, Cascorbi I, Schroeder E, et al. Deposition of Alzheimer's beta-amyloid is inversely correlated with P-glycoprotein expression in the brains of elderly non-demented humans. Pharmacogenetics 2002;12:535–541.

35. Vogelgesang S, Warzok RW, Cascorbi I, et al. The role of P-glycoprotein in cerebral amyloid angiopathy; implications for the early pathogenesis of Alzheimer's disease. Curr Alzheimer Res 2004;1:121–125.

36. Cirrito JR, Deane R, Fagan AM, et al. P-glycoprotein deficiency at the blood-brain barrier increases amyloid-beta deposition in an Alzheimer disease mouse model. J Clin Invest 2005;115:3285–3290.

37. Bauer B. Targeting regulation of ABC transporters at the blood-brain barrier to improve pharmacotherapy of CNS disorders. International Drug Discovery 2010;August/September Issue:76–79.

38. Miller DS. Regulation of P-glycoprotein and other ABC drug transporters at the blood-brain barrier. Trends Pharmacol Sci 2010;31:246–254.

39. Hartz AM, Miller DS, Bauer B. Restoring blood-brain barrier P-glycoprotein reduces brain amyloid-beta in a mouse model of Alzheimer's disease. Mol Pharmacol 2010;77: 715–723.

40. Bauer B, Hartz AM, Pekcec A, et al. Seizure-induced up-regulation of P-glycoprotein at the blood-brain barrier through glutamate and cyclooxygenase-2 signaling. Mol Pharmacol 2008;73:1444–1453.

41. de Klerk OL, Willemsen AT, Roosink M, et al. Locally increased P-glycoprotein function in major depression: a PET study with [^{11}C]verapamil as a probe for P-glycoprotein function in the blood-brain barrier. Int J Neuropsychopharmacol 2009;12:895–904.

42. Farde L, Heminway S, Lee CM, et al. Using positron emission tomography (PET) microdosing to improve CNS drug development. Regulatory Rapporteur 2007;4:2–8.

43. Pike VW. PET radiotracers: crossing the blood-brain barrier and surviving metabolism. Trends Pharmacol Sci 2009;30:431–440.

44. Lappin G, Garner RC. The utility of microdosing over the past 5 years. Expert Opin Drug Metab Toxicol 2008;4:1499–1506.

45. Lappin G, Shishikura Y, Jochemsen R, et al. Pharmacokinetics of fexofenadine: evaluation of a microdose and assessment of absolute oral bioavailability. Eur J Pharm Sci 2010;40:125–131.

46. Wagner CC, Langer O. Approaches using molecular imaging technology—use of PET in clinical microdose studies. Adv Drug Deliv Rev 2011;63:539–546.

47. Farde L, Wiesel FA, Stone-Elander S, et al. D2 dopamine receptors in neuroleptic-naive schizophrenic patients. A positron emission tomography study with [^{11}C]raclopride. Arch Gen Psychiatry 1990;47:213–219.

48. Wagner CC, Simpson M, Zeitlinger M, et al. A combined accelerator mass spectrometry-positron emission tomography human microdose study with ^{14}C- and ^{11}C-labelled verapamil. Clin Pharmacokinet 2011;50:111–120.

49. Christian BT, Livni E, Babich JW, et al. Evaluation of cerebral pharmacokinetics of the novel antidepressant drug, BMS-181101, by positron emission tomography. J Pharmacol Exp Ther 1996;279:325–331.

50. Bauer F, Kuntner C, Bankstahl JP, et al. Synthesis and in vivo evaluation of [^{11}C]tariquidar, a positron emission tomography radiotracer based on a third-generation P-glycoprotein inhibitor. Bioorg Med Chem 2010;18:5489–5497.

51. Letrent SP, Pollack GM, Brouwer KR, et al. Effects of a potent and specific P-glycoprotein inhibitor on the blood-brain barrier distribution and antinociceptive effect of morphine in the rat. Drug Metab Dispos 1999;27:827–834.

52. Tunblad K, Jonsson EN, Hammarlund-Udenaes M. Morphine blood-brain barrier transport is influenced by probenecid co-administration. Pharm Res 2003;20:618–623.

53. Xie R, Hammarlund-Udenaes M, de Boer AG, et al. The role of P-glycoprotein in blood-brain barrier transport of morphine: transcortical microdialysis studies in mdr1a (–/–) and mdr1a (+/+) mice. Br J Pharmacol 1999;128:563–568.

54. Farde L, Eriksson L, Blomquist G, et al. Kinetic analysis of central [^{11}C]raclopride binding to D_2-dopamine receptors studied by PET—a comparison to the equilibrium analysis. J Cereb Blood Flow Metab 1989;9:696–708.

55. Leenders KL, Perani D, Lammertsma AA, et al. Cerebral blood flow, blood volume and oxygen utilization. Normal values and effect of age. Brain 1990;113:27–47.

56. Roland PE, Eriksson L, Stone-Elander S, et al. Does mental activity change the oxidative metabolism of the brain? J Neurosci 1987;7:2373–2389.

57. Huang SC, Barrio JR, Phelps ME. Neuroreceptor assay with positron emission tomography: equilibrium versus dynamic approaches. J Cereb Blood Flow Metab 1986;6: 515–521.

58. Mintun MA, Raichle ME, Kilbourn MR, et al. A quantitative model for the in vivo assessment of drug binding sites with positron emission tomography. Ann Neurol 1984;15: 217–227.

59. Wong DF, Gjedde A, Wagner HN, Jr. Quantification of neuroreceptors in the living human brain. I. Irreversible binding of ligands. J Cereb Blood Flow Metab 1986;6:137–146.

60. Renkin EM. Transport of potassium-42 from blood to tissue in isolated mammalian skeletal muscles. Am J Physiol 1959;197:1205–1210.

61. Crone C. The permeability of capillaries in various organs as determined by use of the "indicator diffusion" method. Acta Physiol Scand 1963;58:292–305.

62. Bankstahl JP, Kuntner C, Abrahim A, et al. Tariquidar-induced P-glycoprotein inhibition at the rat blood-brain barrier studied with (*R*)-^{11}C-verapamil and PET. J Nucl Med 2008;49:1328–1335.

63. Kreisl WC, Liow JS, Kimura N, et al. P-glycoprotein function at the blood-brain barrier in humans can be quantified with the substrate radiotracer ^{11}C-*N-desmethyl*-loperamide. J Nucl Med 2010;51:559–566.

64. Muzi M, Mankoff DA, Link JM, et al. Imaging of cyclosporine inhibition of P-glycoprotein activity using ^{11}C-verapamil in the brain: studies of healthy humans. J Nucl Med 2009;50:1267–1275.

65. Carson RE, Channing MA, Blasberg RG, et al. Comparison of bolus and infusion methods for receptor quantitation: application to [^{18}F]cyclofoxy and positron emission tomography. J Cereb Blood Flow Metab 1993;13:24–42.

66. Logan J, Fowler JS, Volkow ND, et al. Graphical analysis of reversible radioligand binding from time-activity measurements applied to [*N*-^{11}C-methyl]-(-)-cocaine PET studies in human subjects. J Cereb Blood Flow Metab 1990;10:740–747.

67. Innis RB, Cunningham VJ, Delforge J, et al. Consensus nomenclature for in vivo imaging of reversibly binding radioligands. J Cereb Blood Flow Metab 2007;27:1533–1539.

68. Laruelle M, Slifstein M, Huang Y. Relationships between radiotracer properties and image quality in molecular imaging of the brain with positron emission tomography. Mol Imaging Biol 2003;5:363–375.

69. Hammarlund-Udenaes M, Fridén M, Syvänen S, et al. On the rate and extent of drug delivery to the brain. Pharm Res 2008;25:1737–1750.

70. Liu X, Vilenski O, Kwan J, et al. Unbound brain concentration determines receptor occupancy: a correlation of drug concentration and brain serotonin and dopamine reuptake transporter occupancy for eighteen compounds in rats. Drug Metab Dispos 2009;37: 1548–1456.

71. Maurer TS, Debartolo DB, Tess DA, et al. Relationship between exposure and nonspecific binding of thirty-three central nervous system drugs in mice. Drug Metab Dispos 2005;33: 175–181.

72. Fridén M, Gupta A, Antonsson M, et al. In vitro methods for estimating unbound drug concentrations in the brain interstitial and intracellular fluids. Drug Metab Dispos 2007;35:1711–1719.

73. Hammarlund-Udenaes M. The use of microdialysis in CNS drug delivery studies. Pharmacokinetic perspectives and results with analgesics and antiepileptics. Adv Drug Deliv Rev 2000;45:283–294.

74. Smith DA, Di L, Kerns EH. The effect of plasma protein binding on in vivo efficacy: misconceptions in drug discovery. Nat Rev Drug Discov 2010;9:929–939.

75. Pierson ME, Andersson J, Nyberg S, et al. [^{11}C]AZ10419369: a selective 5-HT$_{1B}$ receptor radioligand suitable for positron emission tomography (PET). Characterization in the primate brain. Neuroimage 2008;41:1075–1085.

76. Varnäs K, Nyberg S, Karlsson P, et al. Dose-dependent binding of AZD3783 to brain 5-HT$_{1B}$ receptors in non-human primates and human subjects: a positron emission tomography study with [^{11}C]AZ10419369. Psychopharmacology (Berl) 2011;213:533–545.

77. Karlsson P, Farde L, Halldin C, et al. Oral administration of NNC 756—a placebo controlled PET study of D1-dopamine receptor occupancy and pharmacodynamics in man. Psychopharmacology (Berl) 1995;119:1–8.

78. Cheng Y, Prusoff WH. Relationship between the inhibition constant (Ki) and the concentration of inhibitor which causes 50 per cent inhibition (I50) of an enzymatic reaction. Biochem Pharmacol 1973;22:3099–3108.

79. Farde L, Wiesel FA, Halldin C, et al. Central D2-dopamine receptor occupancy in schizophrenic patients treated with antipsychotic drugs. Arch Gen Psychiatry 1988;45: 71–76.

80. Nordström AL, Farde L, Wiesel FA, et al. Central D2-dopamine receptor occupancy in relation to antipsychotic drug effects: a double-blind PET study of schizophrenic patients. Biol Psychiatry 1993;33:227–235.

81. Nord M, Farde L. Antipsychotic occupancy of dopamine receptors in schizophrenia. CNS Neurosci Ther 2011;17:97–103.

82. Huang Y, Zheng MQ, Gerdes JM. Development of effective PET and SPECT imaging agents for the serotonin transporter: has a twenty-year journey reached its destination? Curr Top Med Chem 2010;10:1499–1526.

83. Schou M, Pike VW, Halldin C. Development of radioligands for imaging of brain norepinephrine transporters in vivo with positron emission tomography. Curr Top Med Chem 2007;7:1806–1816.

84. Varrone A, Halldin C. Molecular imaging of the dopamine transporter. J Nucl Med 2010;51:1331–1334.

85. Pike VW, Halldin C, Crouzel C, et al. Radioligands for PET studies of central benzodiazepine receptors and PK (peripheral benzodiazepine) binding sites—current status. Nucl Med Biol 1993;20:503–525.

86. Henriksen G, Willoch F. Imaging of opioid receptors in the central nervous system. Brain 2008;131:1171–1196.

87. Fujita M, Woods SW, Verhoeff NP, et al. Changes of benzodiazepine receptors during chronic benzodiazepine administration in humans. Eur J Pharmacol 1999;368: 161–172.

88. Greenwald MK, Johanson CE, Moody DE, et al. Effects of buprenorphine maintenance dose on mu-opioid receptor availability, plasma concentrations, and antagonist blockade in heroin-dependent volunteers. Neuropsychopharmacology 2003;28:2000–2009.

89. Kent JM, Coplan JD, Lombardo I, et al. Occupancy of brain serotonin transporters during treatment with paroxetine in patients with social phobia: a positron emission tomography study with [^{11}C]McN 5652. Psychopharmacology (Berl) 2002;164:341–348.

90. Melichar JK, Hume SP, Williams TM, et al. Using [^{11}C]diprenorphine to image opioid receptor occupancy by methadone in opioid addiction: clinical and preclinical studies. J Pharmacol Exp Ther 2005;312:309–315.

91. Meyer JH, Wilson AA, Sagrati S, et al. Serotonin transporter occupancy of five selective serotonin reuptake inhibitors at different doses: an [^{11}C]DASB positron emission tomography study. Am J Psychiatry 2004;161:826–835.

92. Suhara T, Takano A, Sudo Y, et al. High levels of serotonin transporter occupancy with low-dose clomipramine in comparative occupancy study with fluvoxamine using positron emission tomography. Arch Gen Psychiatry 2003;60:386–391.

93. Grimwood S, Hartig PR. Target site occupancy: emerging generalizations from clinical and preclinical studies. Pharmacol Ther 2009;122:281–301.

94. Saijo T, Maeda J, Okauchi T, et al. Utility of small-animal positron emission tomographic imaging of rats for preclinical development of drugs acting on the serotonin transporter. Int J Neuropsychopharmacol 2009;12:1021–1032.

95. Takano A, Suhara T, Ichimiya T, et al. Time course of in vivo 5-HTT transporter occupancy by fluvoxamine. J Clin Psychopharmacol 2006;26:188–191.

96. Zhang M, Zhou D, Wang Y, et al. Preclinical pharmacology and pharmacokinetics of AZD3783, a selective 5-hydroxytryptamine 1B receptor antagonist. J Pharmacol Exp Ther 2011;339:567–578.

97. Atack JR, Wong DF, Fryer TD, et al. Benzodiazepine binding site occupancy by the novel GABAA receptor subtype-selective drug 7-(1,1-dimethylethyl)-6-(2-ethyl-2H-1,2, 4-triazol-3-ylmethoxy)-3-(2-fluorophenyl) -1,2,4-triazolo[4,3-b]pyridazine (TPA023) in rats, primates, and humans. J Pharmacol Exp Ther 2010;332:17–25.

98. Farde L, Andree B, Ginovart N, et al. PET-Determination of robalzotan (NAD-299) induced 5-HT$_{1A}$ receptor occupancy in the monkey brain. Neuropsychopharmacology 2000;22:422–429.

99. Andrée B, Hedman A, Thorberg SO, et al. Positron emission tomographic analysis of dose-dependent NAD-299 binding to 5-hydroxytryptamine-1A receptors in the human brain. Psychopharmacology (Berl) 2003;167:37–45.

INDEX

Note: Page numbers in *italics* refer to Figures; those in **bold** to Tables.

Drug Delivery Applications of Noninvasive Imaging: Validation from Biodistribution to Sites of Action, First Edition. Edited by Chun Li and Mei Tian.